Natural Area Tourism

ASPECTS OF TOURISM

***Series Editors*: Chris Cooper** *(Oxford Brookes University, UK),* **C. Michael Hall** *(University of Canterbury, New Zealand)* and **Dallen J. Timothy** *(Arizona State University, USA)*

Aspects of Tourism is an innovative, multifaceted series, which comprises authoritative reference handbooks on global tourism regions, research volumes, texts and monographs. It is designed to provide readers with the latest thinking on tourism worldwide and push back the frontiers of tourism knowledge. The volumes are authoritative, readable and user-friendly, providing accessible sources for further research. Books in the series are commissioned to probe the relationship between tourism and cognate subject areas such as strategy, development, retailing, sport and environmental studies.

Full details of all the books in this series and of all our other publications can be found on http://www.channelviewpublications.com, or by writing to Channel View Publications, St Nicholas House, 31–34 High Street, Bristol BS1 2AW, UK.

Natural Area Tourism

Ecology, Impacts and Management

2nd edition

David Newsome, Susan A. Moore and Ross K. Dowling

CHANNEL VIEW PUBLICATIONS
Bristol • Buffalo • Toronto

David Newsome is dedicating this book to two people who have been instrumental in fostering David's interest in travel and the natural world:

In memory of Kenneth Firth Newsome (1926–2010) and sunny days in Cornwall on the southern coastline of England, where, on family holidays and as a boy, I saw my first basking shark.

In memory of Jonathon Paul Welton (23/8/53–22/2/78) and bird watching in the Okavango Delta, Botswana, during March 1975.

Library of Congress Cataloging in Publication Data
Newsome, David, 1951–
Natural Area Tourism: Ecology, Impacts and Management/David Newsome, Susan A. Moore and Ross K. Dowling. — 2nd ed.
Aspects of Tourism: 58
Includes bibliographical references and index.
1. Tourism. 2. Tourism—Environmental aspects. 3. Natural areas. I. Moore, Susan A. II. Dowling, Ross Kingston. III. Title.
G155.A1N43 2013
338.4'791–dc23 2012036509

British Library Cataloguing in Publication Data
A catalogue entry for this book is available from the British Library.

ISBN-13: 978-1-84541-382-8 (hbk)
ISBN-13: 978-1-84541-381-1 (pbk)

Channel View Publications
UK: St Nicholas House, 31-34 High Street, Bristol BS1 2AW, UK.
USA: UTP, 2250 Military Road, Tonawanda, NY 14150, USA.
Canada: UTP, 5201 Dufferin Street, North York, Ontario M3H 5T8, Canada.

The policy of Multilingual Matters/Channel View Publications is to use papers that are natural, renewable and recyclable products, made from wood grown in sustainable forests. In the manufacturing process of our books, and to further support our policy, preference is given to printers that have FSC and PEFC Chain of Custody certification. The FSC and/or PEFC logos will appear on those books where full certification has been granted to the printer concerned.

Typeset by R. J. Footring Ltd, Derby
Printed and bound in Great Britain by the MPG Books Group

Contents

Figures

Tables

Boxes

Acknowledgements

The three authors would like to thank each other for a productive relationship spanning almost two decades. We have worked well together over the years, authoring and editing a number of books. As always, the experience of co-writing this book has been extremely productive and very enjoyable. We wish to acknowledge the support of our employers: Edith Cowan University, one of Australia's leading new-generation universities; and Murdoch University, one of Australia's leading research-intensive universities. Both our universities are committed to excellence in research, teaching and community engagement.

Comments and suggestions by a number of people significantly enhanced the calibre and comprehensiveness of both the original and this second edition. Valuable input to Chapter 2 was provided by Duncan McCollin, Karen Higginbottom, Richard Hobbs, Diane Lee, Darryl Moncrieff and Amanda Smith. Chapter 3 was refined using the ideas of Ralph Buckley, Michelle Davis, Darryl Moncrieff and Amanda Smith, Angela Arthington, Micha Liick, Paul Eagles, Yu-Fai Leung, Catherine Pickering, and Andrew Growcock. Chapter 4 benefited from inputs by Kelly Gillen, Simon McArthur and Steve McCool. Chapter 5 was assisted by contributions from Terry Bailey, Kerry Bridle, Kym Cheatham, David Cole, Tracy Churchill, Steve Csaba, Grant Dixon, Ian Grant, Michael Hall, Chris Haynes, Virginia Logan, Tony Press, Wayne Schmidt, Peter Wellings and Keith Williams. Chapter 6 incorporated insights from Rory Allardice, Pat Barblett, Kevin Keneally, Darryl Moncrieff and Amanda Smith. Chapter 7 received invaluable contributions from Don English, Richard Hammond, Yu-Fai Leung, Jeff Marion and Amanda Smith.

The scope and production of the book owe much to the support of the West Australian Department of Environment and Conservation, especially Darryl Moncrieff, Jim Sharp, Peter Sharp and Wayne Schmidt (retired). Chris Cooper, David Fennell, Michael Hall and Dallen Timothy have also played important roles in producing a high-quality publication.

Without the cartographic efforts of Alan Rossow, Colin Ferguson, Mike Roeger and Belinda Cale the diagrams and maps, which form the backbone of this book, would not have been possible. Noella Ross diligently entered and manipulated the references and formatted the text for the first edition and thanks also go to Amy Hodgson (tourism student, Edith Cowan University), who helped us order the references in the initial stages of proofing for the second edition. Ross Lantzke provided illustrative material and software assistance. Photographs provided by Stephen McCool and the Parks and Wildlife Service Tasmania, added to the visual quality of this publication. Support in the field was provided to the first author by the Director of Kruger National Park, Wilhem Gertenbach, Teresa Whitehead, Rory Allardice and Richard Baker.

David Newsome. I would like to thank my family – Jane, Ben and Rachel – for continuing to support my research efforts in many countries around the world. We all enjoy what the natural world has to offer and hope that natural landscapes and wildlife will be an integral part of human experience in 50 years' time and beyond.

Sue Moore. I would like to thank my family – Warren, Jess and Sam Tacey – for continuing to support my research efforts in Australia and elsewhere. These efforts include visitor loyalty to protected areas, community engagement in marine protected area establishment, and management and conserving biodiversity in agricultural landscapes under a changing climate. My family continues to share with me my passion for eclectic ideas and an ongoing desire to understand the world better and make it a better place.

Ross Dowling. I would like to thank the many Australian and international students from around the world who have participated in my undergraduate and postgraduate ecotourism classes. We have had a lot of fun and I have learned a lot about the industry from your research assignments and oral presentations. I wish to thank my wife, Wendy, for her unfailing love and support through this my tenth book in the last 10 years. I could not have achieved this without her. I also wish to thank my children and grandchildren for the contributions they have made, and continue to make, to my life. This book is part of my legacy for you all.

Preface

It is now over a decade since the first edition of this book was published as one of the first in Channel View Publications' initial foray into tourism publishing. Since that time we witnessed the book being used by researchers in many countries around the world and for teaching on both undergraduate and postgraduate courses in the UK, North America, Australia and New Zealand. Over recent years we were asked to update the book by researchers, policy makers, managers and teachers, so this new edition reflects our current thinking on the subject.

The first edition was born out of our individual and collective passion for the natural environment. All of us are environmental scientists who have spent a great part of our lives travelling in, researching about and teaching for a greater understanding of the global environment. Therefore it was only natural that at some stage we should wish to share our knowledge of, and enthusiasm for, natural areas and this has manifested itself through tourism. One of us has been a tour guide to national parks and wilderness areas for 35 years, another has been a natural area manager for more than a decade, while all of us have led field trips to natural areas as part of our tertiary teaching. We love the environment and believe that through tourism to natural areas people will be stirred within to act positively for their own environment upon their return to the predominantly urban areas from which they come.

This book focuses on tourism in natural areas, especially in relation to its ecology, impacts and management. There are many environmental books that briefly address tourism impacts as well as a large number of tourism texts that have chapters on environmental impacts. This book dwells on the nexus that exists between the environment and tourism and unashamedly fosters a positive link between the two. Our approach to the subject is embedded firmly in our shared belief that it is only through a greater understanding of the environment that tourism in natural areas will evolve to a place where it can be truly synergistic. Too often in the past the relationship has been one-sided, with tourism being the winner and the environment the loser. But it is our belief that with environmental understanding, informed management and an aware public, natural area tourism has the possibility of introducing people to the environment in an educative, ethical and exciting manner which will leave them with an indelible impression of the wonders of the natural world and the pivotal position that we humans have within it.

The book moves beyond the narrow niche of ecotourism and embraces the ideals of tourism in natural areas at a time when they are under pressure globally from competing resource uses. Obviously, any form of tourism development changes our relationship to the natural environment and moves our perception of it from viewing the environment as an attribute to regarding it as a resource. In that instant, our relationship with the environment has changed from an ecocentric basis to an anthropocentric one. With this firmly in mind, then, our goal is to foster tourism in environmentally sensitive areas in a manner which is conservation supporting, ecologically friendly and environmentally educative. It is our desire to lead the natural area tourism developers and managers, as well as the tourists themselves, towards these central elements of ecotourism. If this is fostered, then the promise of the symbiotic relationship between tourism and the environment will begin to be achieved.

The book has been written for a broad audience, including researchers in tourism and environmental management, tourism industry professionals, policy makers, planners and managers in natural area management, government agency employees and students pursuing university degrees and industry training programmes. As a research monograph and general text, it should be useful to researchers and students in a range of disciplines, including tourism, environmental science, geography, planning and regional studies. As a specific text it provides a practical guide for natural area managers, such as national and marine park managers, as well as tour operators. The applied approach to ecology and the understanding of environmental impacts makes this book also suitable for those from business, communications and marketing backgrounds, as well as those with more scientific leanings. The foundation on ecology and impacts is valuable, but of even greater value is the explanation of the practical aspects of managing natural area tourism. This includes planning frameworks for natural area tourism as well as a number of management strategies, including adaptive management. It also underlines the importance of interpretation in achieving positive outcomes as well as arguing for the need for ongoing monitoring of developments. Finally, at a time when global issues such as the loss of natural areas and the impacts of climate change are being highlighted on an almost daily basis, the book brings together the essential elements of ecology, impacts and management to comprehensively address the provision of sustainable natural area tourism.

Furthermore, it is important to note the parameters of the book, that is, the approach taken and hence the standpoints from which we have written it. Firstly, although we are familiar with and undertake research in the field of ecotourism, we have deliberately widened our view to include tourism in natural areas. This is because ecotourism is often viewed as a narrow niche or special-interest form of tourism, yet our research shows that ecotourism may be much wider and encompass many of the elements of what is commonly referred to as 'mass tourism'. In addition, other types of tourism also occur in natural areas, such as wildlife tourism, geotourism and aspects of adventure tourism, all of which involve an interest in the natural environment in some form or other. Another focus of the book is our approach to the topic of the impacts of tourism in natural areas, namely, from the perspective of their source. This is because it is the sources of impact that need to be managed in natural areas in order to reduce any adverse impacts. Therefore the sources of impact are identified, with suggestions for their amelioration. In a similar vein we do not tackle the broad area of environmental, tourism or environmental-tourism planning. Instead the book focuses on the realm of visitor planning, as this is a key task for managers of protected and natural areas. Such concepts are identified and a number of visitor planning frameworks are described. While a broad introduction is given to general management strategies and actions, the book narrows its focus to management actions for sites and visitors. The logic of this is to put a spotlight on the key strategies and specific tasks that are useful to manage tourism appropriately in natural areas. Having narrowed the approach to a number of topics, the book widens the scope of the importance of education to focus on interpretation. This is because we believe that interpretation holds a prominent position in natural area tourism, by fostering a link between tourists and the natural destinations they are visiting and experiencing. Our approach to monitoring is to discuss the monitoring of visitors to natural areas as well as the visitor impacts on natural areas. Once again, our desire is to use the science to inform management, so we have adopted a specific approach and targeted specific techniques. Finally, we champion both system-wide and integrated approaches to protected area management through assessments and integrated approaches to impact management.

1 Introduction

Tourism and the Natural Environment

Natural areas have always attracted people. A visitor writing in a guest book at a destination run by Conservation Corps Africa (CC Africa, now named &Beyond Africa) stated:

> My journey to Africa has been the odyssey of a lifetime. I have crossed great, beautiful landscapes and stood on, what seemed to me, the edge of the world as I knew it ... and seen my heart soar into its ancient sky, somehow humbler than I have ever felt. Somehow part of eternity. (CC Africa, nd)

This quote from a guest, evokes a feeling that many people share: the desire to see, touch, feel and connect with, and be inspired by, natural areas. The tourist visiting such areas is often passionate about the conservation of natural areas and the people and wildlife who inhabit these regions. A growing number of such tourists are seeking authentic, inspiring, transformational experiences in nature as they search for a greater sense of self and connection to planet Earth. This search for natural experiences is taking place around the planet, whether it is seeking solitude in the wilderness of the Antarctic, observing the migration of hundreds of thousands of herbivores cross the Serengeti in Tanzania into the Masai Mara in Kenya, or seeing the Iguazu Falls separating the Brazilian state of Paraná and the Argentine province of Misiones. All such experiences evoke powerful connections with nature and this is the essence of natural area tourism.

The United Nations World Tourism Organization (UNWTO) estimated that, in 2011, there were 990 million international tourist arrivals and tourism receipts generated US$1030 billion (UNWTO, 2012a). This represented a rise of 4.7% in tourist numbers over the previous year. According to the UNWTO, tourism has experienced continued expansion and diversification over the past six decades, becoming one of the largest and fastest-growing economic sectors in the world. The number of tourists grew from 25 million in 1950 to 277 million in 1980, 435 million in 1990, 675 million in 2000, to 990 million in 2011. A number of new destinations have emerged alongside the traditional ones of Europe and North America and in the world's emerging tourist regions the share in international tourist arrivals grew from 31% in 1990 to 47% in 2010.

Following the global financial crisis and economic recession of 2008–09, worldwide tourism recovered remarkably quickly. While the advanced economies had an average annual growth in international tourist arrivals of 1.8% for the period 2000–10, the world's emerging economies had a growth of 5.6%, with the Middle East (14%) and Asia and the Pacific (13%) UNWTO regions being the strongest growing (UNWTO, 2011). Overall, the fastest-growing region for international tourism was North East Asia, with growth in Japan and Taiwan being 27%. According to the UNWTO's Tourism 2020 Vision project, international arrivals are expected to reach 1.6 billion by the year 2020, with the fastest-growing regions predicted to be East Asia, the Pacific, South Asia, the Middle East and Africa (UNWTO, 2010).

The key tourist attractors possessed by many of these emerging economies are natural areas, and tourism to them is booming. It has been estimated that nature tourism has risen from approximately 2% of all tourism in the late 1980s (Ecotourism Society, 1998; Weaver & Oppermann, 2000) to approximately 20% today (Buckley, 2009). Thus, natural area tourism is undergoing explosive growth and, as such, it has the capability to change both natural areas as well as tourism itself. In this book we explore this phenomenon from the standpoint that natural area tourism can be beneficial to individuals, regions and countries – provided it is planned, developed and managed in a responsible manner.

The growing interest in conservation and the wellbeing of our environment over the last two decades has moved far beyond the realms of a concerned few and into the wider public arena. At the same time there has been a corresponding upsurge in tourism all over the world, leading to the phenomenon referred to as 'mass tourism'. With this unparalleled growth of the two it was inevitable that one day they would meet and interact. In natural areas, where tourism either already exists or is proposed, there is the potential for both beneficial and adverse environmental and socio-cultural impacts. Thus, there are two streams of thought regarding the environment–tourism relationship. The first is that the natural environment is harmed by tourism and hence the two are viewed as being in conflict. The second is that the two have the potential to work together in a symbiotic manner.

The environment–tourism relationship has been the subject of debate for the last three decades. The International Union for the Conservation of Nature and Natural Resources (IUCN; now known as the World Conservation Union) first raised the nature of the relationship when its director general posed the question in a paper entitled 'Tourism and environmental conservation: Conflict, coexistence, or symbiosis?' (Budowski, 1976). Thirteen years later the question appeared to remain unanswered when Romeril (1989a) posited the question 'Tourism and the environment – accord or discord?' Thus the environment–tourism relationship may be viewed from one of two standpoints – that it is one of either conflict or symbiosis. Either standpoint may be adopted and defended but it is argued here that, no matter which is espoused, the way to reduce conflict or increase compatibility is through understanding, planning and management, grounded in knowledge and understanding of environmental concepts. Such an approach will foster sustainable development.

The environment–tourism relationship is grounded in the sustainable use of natural resources, as fostered by the World Conservation Strategy (IUCN, 1980) and the sustainable development strategy of the World Commission on Environment and Development (WCED, 1987). This environment–development link often includes tourism as a bridge. The base of this partnership is resource sustainability and tourism must be fully integrated with the resource management process. This will require the adoption of resource conservation values, as well as the more traditional development goals. Central to the goals of environmental conservation and resource sustainability is the protection and maintenance of environmental quality. This primary goal in turn requires an awareness of environmental protection and enhancement while fostering the realisation of tourism potential. According to Shultis and Way (2006), tourism management in protected areas normally followed a reductionist, deterministic, linear view of nature and conservation research. However, they argue:

> Land managers need to adapt to a new paradigm that reflects and supports this philosophical change in conservation principles; this shift is also reflected in science itself, manifested by a move from normal to 'post-normal' science which embraces these new principles. This approach should link visitor expectations with dynamic, non-linear, self-organising natural processes in order to meet conservation objectives. (Shultis & Way, 2006: 223)

Natural Areas as a Focus for Tourism

At its core the word 'environment' simply means our surroundings. However, the environment is defined as including all aspects of the surroundings of humanity, affecting individuals and social groupings. At a broad scale, the environment may classified on a continuum between two major divisions, the natural and built environments. These two different aspects of the environment are not exclusive and can be viewed as being interrelated by human influence. Natural environments, on the whole, tend to retain their natural characteristics and are not modified to any large extent by human interference with the landscape or ecological processes. Such areas include patches of natural vegetation that either are found naturally in the landscape or are more likely to be conserved in protected areas. On the other hand, built environments are human altered areas where the natural environment has been modified to such an extent that it has lost its original characteristics. Such areas include urban landscapes.

Natural areas are regions that have not been significantly altered by humankind and this equates to intact natural landscapes that contain original vegetation, are unspoilt, are wild, are maintained by natural processes and the original biodiversity is present. Such areas contrast with areas that have a significant human imprint on the natural environment, through past and/or present use. A natural area, then, is one where the natural forms and processes have not been materially altered by human exploitation and occupation. Thus, the wildlife and ecological processes are largely in their natural state and the area comprises largely unmodified landscapes that preserve the integrity of natural vegetation, wildlife and landforms.

Nature and Naturalness

Many national parks were originally established with the dual mandate of fostering the protection of natural areas and the human enjoyment of them. However, modern approaches to their establishment and use through landscape ecology and conservation biology have 'demonstrated that parks are not the protected islands of virgin wilderness they were constructed to represent; rather than protecting these areas from disturbance, we now recognise that disturbance is a major component in ecological integrity' (Shultis & Way, 2006: 223). However, in an examination of the ecological integrity of Canada's national parks over the past decade, it was found that while there appeared to be some commitment to this approach, 'only time will tell whether management plans will focus on ecological integrity as the first priority in practice' (Wilkinson, 2011: 353).

Thus, the argument is now made to shift the focus of management on to a park's ecological integrity, in order to re-engage with landscape-level processes which have important outcomes in relation to both protected areas and sustainable tourism. Managing for naturalness is a complex concept for managers of such areas and the case can be made to move beyond this approach. Central to any approach is the need to investigate ways of managing such areas for conservation, and/or for human visitation, such as through tourism. A number of guiding concepts for park and wilderness stewardship in an era of global environmental change have been suggested by Hobbs *et al.* (2010). They argue that the major challenge to the stewardship of protected areas is to decide what interventions we should undertake to conserve their values. This is a value-laden concept which involves choices around preservation, conservation and sustainable development. It includes the maintenance and restoration of biodiversity, having regard to ecological integrity and resilience. In the quest for the management of ecological integrity, an understanding of environmental thresholds, monitoring and the measurement of impacts is essential. To achieve these ends Hobbs *et al.* (2010: 483) 'advocate a pluralistic approach that incorporates a suite of guiding principles, including historical fidelity, autonomy of nature, ecological integrity, and resilience, as well as managing with humility. The relative importance of these guiding principles will vary depending on management goals and ecological conditions.'

Protected natural areas are attractive for visitors because their protected status ensures their naturalness. They usually contain areas of exceptional natural qualities and their designation as protected national parks or World Heritage Areas confers a special status. For these reasons protected natural areas are now among the most sought after tourist attractions (Butler & Boyd, 2000). Today, a key focus of natural area tourism development is on enhancing the visitor's experience of nature. This has given rise to the increase in 'green' travellers, volunteer tourism or 'voluntourism', and a spectrum of ecotourist typologies along a continuum from casual or 'soft' ecotourists to hard-core or 'hard' ones (Weaver, 2008).

Human Approaches to Nature

People differ over their environmental views according to the different perspectives of the world they hold (Miller & Spoolman, 2008). Such views come in many forms but one basic distinction concerns whether or not we put humans at the centre of things. Two examples are the human-centred or anthropocentric view that underlies most industrial societies and the ecocentric or life-centred outlook. Key principles of the human-centred approach are that humans are the planet's most important species and we are apart from, and in charge of, the rest of nature. It assumes the Earth has an unlimited supply of resources, to which we gain access through the use of science and technology. Other people believe that any human-centred worldview, even stewardship, is unsustainable (Rowe, 1994). They suggest that our worldviews must be expanded to recognise inherent or intrinsic value to all forms of life, that is, value regardless of their potential or actual use to us. The life-centred or ecocentric view recognises the importance of biodiversity. The ecocentric perspective encompasses the belief that nature exists for all of Earth's species and that humans are not apart from, or in charge of, the rest of nature. In essence it posits that we need the Earth, but the Earth does not need us. It also suggests that some forms of economic growth are beneficial and some are harmful. In an ideal world our goals should be to design economic and political systems that encourage sustainable forms of growth and discourage or prohibit forms which cause degradation or pollution. A healthy economy depends on a healthy environment.

There are a number of major principles underlying the ecocentric or Earth-centred view (Miller & Spoolman, 2008). These are interconnectedness, intrinsic value, sustainability, conservation, intergenerational equity and individual responsibility. The first principle, of interconnectedness, focuses on the fact that humans are a valuable species. The second principle, of intrinsic value, is that every living thing has a right to live, or at least to struggle to live, simply because it exists; this right is not dependent on its actual or potential use to us. This principle includes the notion that it is wrong for humans to cause the premature extinction of any wild species or the elimination or degradation of their habitats. This focuses on the need for the third principle, conservation – the preservation of wildlife and the biodiversity principle. Conservation is one principle most understood by people in general. It recognises that resources are limited and must not be wasted. The fourth principle, sustainability, means that something is 'right' when it tends to maintain the Earth's life-support systems for us and other species, and 'wrong' when it tends to do otherwise (Miller & Spoolman, 2008). The fifth principle, intergenerational equity, suggests that we must leave the Earth in as good a shape as we found it, if not better. Inherent in the notion is that we must protect the Earth's remaining wild ecosystems from our activities, rehabilitate or restore ecosystems we have degraded, use ecosystems only on a sustainable basis, and allow many of the ecosystems we have occupied and abused to return to a wild state. The sixth and final principle is individual responsibility. We must ensure that we do not do anything that depletes the physical, chemical and biological capital which supports all life and human

Box 1.1 Naturalness, ecological integrity and natural experiences

Naturalness is a contested term and a 'natural experience' is not well defined in the literature (Newsome & Lacroix, 2011). Denoting what a natural area is and characterising both nature and naturalness are very complex and depend on a person's upbringing, education and worldview. A focus on the last factor alone provides a range of views, from preservation to use, with such views often being in conflict with one another. However, it is possible to define the main characteristics of a natural experience as involving three main perceptions – sights (such as of natural vegetation, wildlife and wilderness landscapes), sounds (such as bird song, insect and amphibian soundscapes, the calling of mammals) and smells (of wildflowers, seashores). However, the essence of a natural experience is a combination of all these things – the sights, sounds and smells – as well as the state of mind it induces (Newsome & Lacroix, 2011). The natural experience is sharpened if it also includes learning. Knowledge of nature is a vital component in visitor satisfaction, with environmental education and interpretation being recognised as essential elements of ecotourism (Fennell, 2008; Buckley, 2009).

Our interest in natural environments has been well documented (e.g. Wilson, 1984; Kellert, 1993; Bechtel, 1997). Visitor surveys conducted in Bako National Park, Sarawak, Malaysia, found that 'being close to nature' (78%), 'viewing wildlife' (72%), 'learning about nature' (70%) and 'viewing scenery' (71%) were all rated highly as part of the park experience (Chin *et al.*, 2000). Research in the south-west of Australia indicates that 90% of visitors to natural areas were there to 'be in and enjoy the natural environment' (Smith & Newsome, 2002). Further research in the region found that visitors 'wished to get away from the city' (87%) and enjoy 'outdoor activities' (89%) and they expressed a desire for 'solitude' (77%) (Smith, 2003). A later study found that 63% of visitors stated their preferred natural area experience was a very natural to totally natural landscape with limited or no facilities (Smith & Newsome, 2005).

People place considerable value on a natural experience and the restorative effects of experiencing nature. People often seek solitude, desire to see native flora and fauna, value learning about nature and wish to be substantially free from intrusive human noise and the visual impact of human-modified landscapes (Eagles, 1992; Bentrupperbäumer & Reser, 2003; Tao *et al.*, 2004; Ankre, 2009).

From Newsome and Lacroix (2011).

economic activities; the Earth deficit is the ultimate deficit. All people should be held responsible for their own pollution and environmental degradation.

Given this understanding and view posited by Miller and Spoolman (2008), then sustaining the Earth requires each one of us to make a personal commitment to live an environmentally ethical life. By extension, its application to natural area tourism is that governments, the tourism industry, operators, tourists and local communities should all play a part not only in conserving natural areas but also in their enhancement. In doing this, the very resource base which underpins the natural area tourism industry will be protected and able to be utilised in a sustainable manner which fosters environmental, social and economic wellbeing. The major challenge to stewardship of protected areas is to decide where, when and how to intervene in physical and biological processes, to conserve what we value in these places. To make such decisions, planners and managers must articulate more clearly the purposes of parks, what is valued and what needs to be sustained.

The maintenance and restoration of biodiversity are major goals for conservation. However, a broader range of values which are also important include ecological integrity, resilience, historical fidelity (i.e. the ecosystem appears and functions much as it did in the past) and autonomy of nature (Hobbs *et al.*, 2010). Hobbs *et al.* (2010) argue that, in the past, the concept of 'naturalness' was central to making conservation-related decisions in park and wilderness ecosystems management (Box 1.1). Their view is that to achieve the goal of nature conservation requires clear management objectives and so they advocate a multifaceted approach which incorporates this range of guiding principles.

Types of Natural Area

Despite population growth and widespread deforestation, infrastructure developments and the rising demand for energy, there are still many large tracts of natural area on the Earth. They include the polar regions, tundra, deserts, tropical rainforests, mountains, chaparral and temperate forests. In addition, they include marine areas such as parts of the Southern and Indian Oceans (see the World Database on Protected Areas at www.wdpa.org). Some of these areas represent wilderness, which are areas where the Earth and its community of life have not been seriously disturbed by humans and where humans are only temporary visitors. According to Holden (2008) two main perspectives on the meaning of 'wilderness' can be recognised. The first is a 'classical perspective', in which the view is taken that the creation of livable and usable spaces, such as urban areas, is a mark of civilisation and progress. The second approach is the 'romantic', in which untouched spaces have the greatest value, and wilderness assumes a deep spiritual significance.

In addition there are many smaller enclaves of natural areas surrounded by largely human altered environments. Such areas have generally been protected by humans. Protected areas are any region of land and/or sea that have legal measures limiting the use of the wildlife within it (IUCN, 2012). They include nature reserves, national parks, protected landscapes, multiple-use areas and biosphere reserves. The IUCN (2012) classifies protected areas according to their management objectives. The categories are recognised by international bodies such as the United Nations and by many national governments as the global standard for defining and recording protected areas and, as such, they are increasingly being incorporated into government legislation.

Protected areas which are recognised by this approach have certain traits (Saundry, 2009):

- They have clearly defined geographical space, with agreed and demarcated borders.

- They are 'recognised' in some fashion, such as in local, regional or national law; or by international classification such as a United Nations Educational, Scientific and Cultural Organization (UNESCO) World Heritage site, under the UNESCO Man and the Biosphere (MAB) Programme, or under the Convention on Wetlands of International Importance (the Ramsar Convention).
- They are 'dedicated' under covenant to some form of long-term conservation. For the IUCN, only those areas where the main objective is to conserve nature can be considered protected areas; these can include many areas for which there are other goals as well, at the same level, but where goals are in conflict nature conservation will be the priority.
- They are 'managed' in some fashion (even if that is a decision to leave the area untouched by humans).

According to the IUCN (2012) there are seven categories of protected area, including strict nature reserves, wilderness areas, national parks, habitat/species management areas and protected landscape/seascapes. The categories do not constitute a hierarchy but reflect the degree of human use acceptable in each case. The values of protected areas are to conserve nature and biological diversity as well as to offer humans opportunities for recreation, inspiration, education and understanding. It is the last goal which fosters the enjoyment of natural areas for people in the form of recreation and tourism.

Tourism and Tourists

Definitions of tourism share a range of common elements. The World Tourism Organization (WTO), forerunner of the United Nations World Tourism Organization (UNWTO), defined tourists (see Box 1.2) as people travelling to and staying in places outside their usual environment for not more than one consecutive year for leisure, business and other purposes (WTO, 1995). A more recent and comprehensive definition is that given by Weaver and Lawton (2010: 2; after Goeldner & Ritchie, 2006):

> Tourism may be defined as the sum of the processes, activities, and outcomes arising from the relationships and the interactions among tourists, tourism suppliers, host governments, host communities, and surrounding environments that are involved in attracting, transporting, hosting and the management of tourists and other visitors.

Thus, tourism is usually viewed as being multidimensional, possessing physical, social, cultural, economic and political characteristics. The tourism system has been described and modelled from several different perspectives. All include elements of demand and supply, linked by the interconnecting strand of travel. Gunn and Var (2002) have proposed a simple approach called the 'functioning tourism system', which consists of a number of interrelated components. Demand consists of the tourist market and incorporates people's interest in and ability to travel. Supply

Box 1.2 Definition and classification of 'tourists'

A tourist is someone who engages in tourism (Figure 1.1). Tourists are temporary visitors who visit a destination for leisure and who stay at least 24 hours (WTO, 1995). According to Holloway (2008), tourists can be categorised according to:

- motivation for the trip – holiday, business, other;
- characteristics of the trip – domestic (travel within a country) or international (overseas);
- modes of tour organisation – independent travel or packaged tours;
- composition of the tour – air, sea, rail or road;
- characteristics of the tourist – nationality, gender, age, lifestyle.

Tourists have been further classified in a number of typologies. The most well known classifications include one involving four tourist roles – the organised mass tourist, the individual mass tourist, the explorer and the drifter (Cohen, 1972). Another (Plog, 1973), classifies tourists as two types – 'psychocentric' (conservative visitors to safe destinations) and 'allocentric' (adventurous tourists to more remote destinations).

Central to tourism is the tourist 'experience' in the destination they are visiting and today tourism organisations are shifting their focus away from simply marketing a destination's attractions and amenities to promoting what it offers in the way of tourists' experiences.

Figure 1.1 Tourists in the remote Purnululu National Park, Western Australia, a World Heritage site (Photo: Ross Dowling)

components include transportation, attractions, services and information/ promotion. Transportation consists of the volume and quality of all modes of transport. Attractions are the quality resources which have been developed for satisfying visitors. Services include the variety and quality of food, lodging and other products. Information/promotion is essential to entice the tourist to visit the products offered.

A systems approach to tourism proposes the movement of tourists from their area of origin to their destination by way of a transit region (Leiper, 2004). This origin–destination approach emphasises the interdependence of the generating and receiving environments but it has been noted that many tourist flows are actually hierarchical in nature and may involve multiple, nested and/or overlapping destinations (Weaver & Lawton, 2010).

A tourism system model which embraces many of the elements of the existing models but which focuses on tourism's environmental aspects is outlined in Figure 1.2. It is based on the traditional view of a system incorporating inputs, processes, outputs and feedback. The inputs include elements of demand or markets, that is, the prospective tourist's motivation for and ability to travel, as well as supply, that is, the destination resource, with its attractions, services, information and hosts. Processes include economic, social and environmental interactions, which may have positive and/or negative outputs (impacts). Feedback allows for the planning of appropriate controls, capacities, policies and strategies for tourism growth while minimising adverse impacts.

According to the United Nations Environment Programme (UNEP), the tourism economy represents 5% of world gross domestic product (GDP), while it contributes to about 8% of total employment (UNEP, 2011: 418). International tourism ranks fourth (after fuels, chemicals and automotive products) in global exports, with an industry value of US$1 trillion a year, accounting for 30% of the world's exports of commercial services or 6% of total exports. At present, tourism is one of five top export earners in over 150 countries, while in 60 countries it is the number one export (UNEP, 2010). It is the main source of foreign exchange for one-third of developing countries and one-half of least-developed countries (LDCs).

Tourist arrivals globally have shown continuous yearly growth over the last six decades, with an average 4% annual increase during 2009 and 2010. This trend has held in spite of occasional short drops from international crises, such as pandemics, recessions and terrorism. There are around 4 billion domestic arrivals every year (UNWTO & UNEP, 2008) and international tourism arrivals reached 922 million in 2008; although they then dropped to 880 million in 2009, they recovered in 2010, with 940 million (UNWTO, 2011) and further increased to 990 million in 2011 (UNWTO, 2012a). Worldwide tourism is expected to continue its strong growth, at about 4–5% per year, and annual international tourist arrivals are estimated to reach 1.6 billion by 2020, with the emerging tourism regions – East Asia and the Pacific, South Asia, the Middle East and Africa – expected to grow at over 5% per year compared with the world average of 4.1%, while the more mature regions – Europe and the Americas – are anticipated to show lower growth rates (UNWTO, 2010).

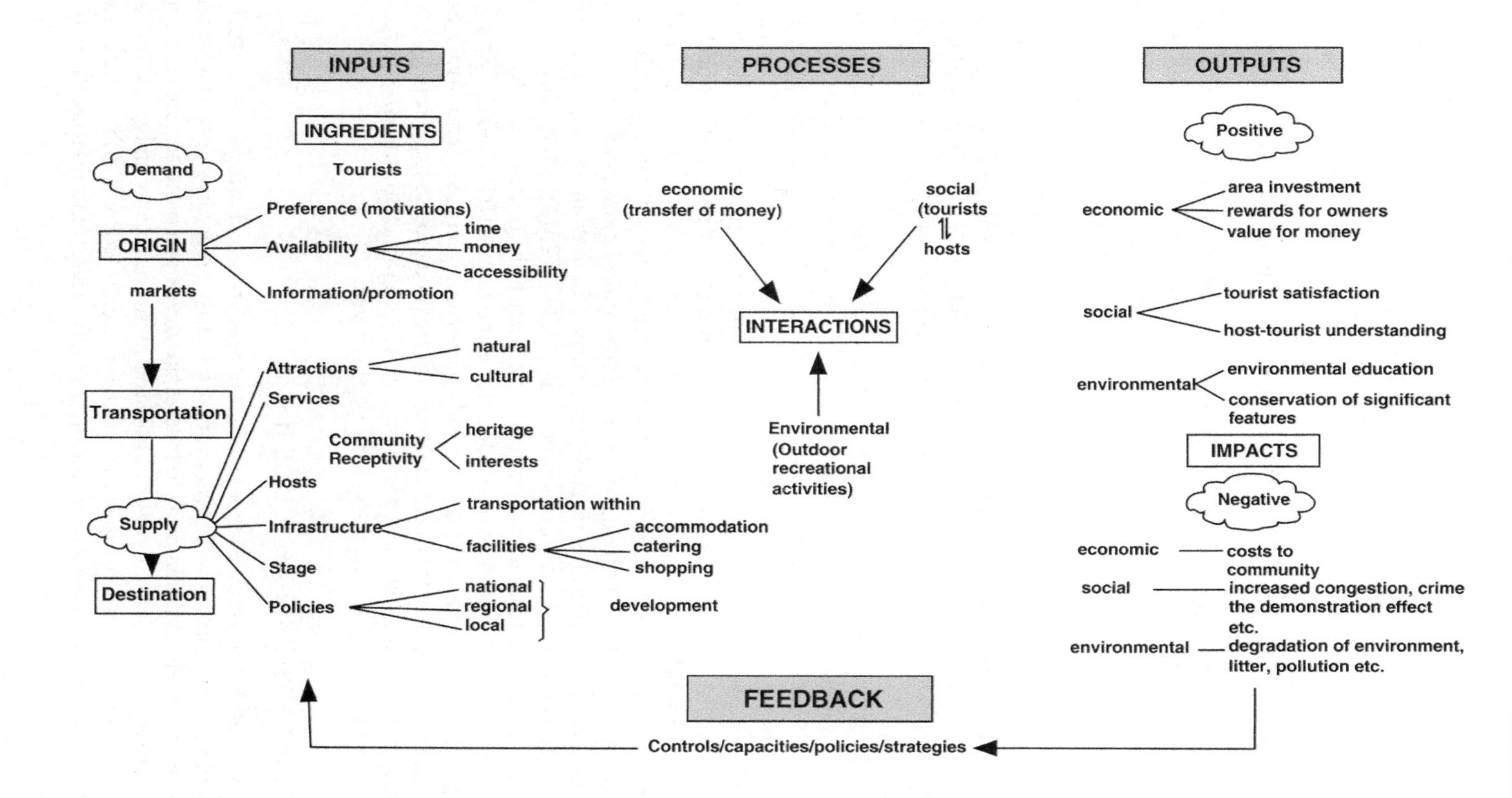

Figure 1.2 The tourism system

The UNWTO suggests that tourists of this century will be travelling further afield on their holidays (UNWTO, 2010). China will be the world's number one destination by the year 2020 and it will also become the fourth most important generating market. Other destinations predicted to make great strides in the tourism industry are Russia, Thailand, Singapore, Indonesia and South Africa. Product development and marketing will need to match each other more closely, based on the main travel motivators of the 21st century, namely the three 'e's: entertainment, excitement and education.

The development of tourism may have a wide variety of impacts on the destination, the transit route and the source of tourists. Tourism may also influence the character of the tourist, as tourism is primarily for leisure or recreation, although business is also important. It is generally understood that nature-based tourism is on the increase throughout the world and the WTO (1998b) estimated that it generated approximately 20% of all international travel expenditures. This increase has occurred in spite of increasing urbanisation and an associated increase in indoor recreational activities associated with television, the internet and video games (Pyle, 2003; Louv, 2005). A study of visitor trends in 280 protected areas in 20 countries found that visitation to protected areas is increasing in most countries (Balmford *et al.*, 2009).

A consideration of the values, attitudes and behaviour of people is fundamental when discussing genuine forms of natural area tourism or trying to identify potential natural area tourists. It has been suggested that people are not necessarily natural area tourists just because they visit a natural area (Acott *et al.*, 1998). It has also been revealed that natural area tourists who have a more ecocentric attitude to nature tend to prefer businesses that are environmentally friendly (Khan, 1997). Such tourists also expect knowledgeable personnel who are willing to instil a feeling of trust and confidence. This research found that natural area tourists showed a preference for services based on their attitude, behaviour, travel motivation and value.

Allied to the increasing interest by tourists in natural areas, they are also demanding the greening of tourism generally (UNEP, 2011). More than a third of travellers are found to favour environmentally friendly tourism and to be willing to pay between 2% and 40% more for this experience (UNEP, 2011). Traditional mass tourism has reached a stage of steady growth. In contrast, ecotourism, nature, heritage, cultural and 'soft adventure' tourism are predicted to grow rapidly over the next two decades. It is estimated that global spending on ecotourism is increasing at a higher rate than the industry-wide average growth.

Defining Sustainable Tourism

Sustainable tourism describes policies, practices and programmes that take into account not only the expectations of tourists regarding responsible natural resource management (demand), but also the needs of communities that support or are affected by tourism projects and the environment (supply) (UNEP, 2011). Thus, sustainable tourism aspires to be more energy efficient and climate friendly (e.g. by using renewable energy); consumes less water; minimises waste; conserves biodiversity, cultural heritage and traditional values; supports intercultural understanding and tolerance; generates local income; and integrates local communities with a view to improving livelihoods and reducing poverty. Tourism businesses which are more sustainable usually benefit local communities and raise awareness and support for the sustainable use of natural resources.

Tourist choices are increasingly influenced by sustainability considerations (UNEP, 2011). A survey by the Center on Ecotourism and Sustainable Development (CESD) and The International Ecotourism Society (TIES) (2005) found that a majority of international tourists are interested in the social, cultural and environmental issues relevant to the destinations they visit and are interested in patronising hotels that are committed to protecting the local environment. Increasingly, they view local environmental and social stewardship as a responsibility of the businesses they support. In 2007 TripAdvisor surveyed travellers worldwide ($n > 2500$) and 38% said that environmentally friendly tourism was a consideration when travelling, 38% had stayed at an environmentally friendly hotel and 9% specifically sought such hotels, while 34% were willing to pay more to stay in environmentally friendly hotels (Pollock, 2007). Three years later a survey of over 3000 travellers found that popular ecotourism holiday activities forecast for 2011 included visiting a national park (46%) and hiking (42%) (TripAdvisor, 2010).

In Uganda it has been found that biodiversity attributes increase willingness to visit tourism attractions, independently of other factors (Naidoo & Adamowickz, 2005). Research also indicates that consumers are concerned about the local environments of their travel destinations and are willing to spend more on their holidays if they are assured that workers in the sector are guaranteed ethical labour conditions in the places they are visiting (International Labour Organisation, 2010).

Sustainable tourism is a complex subject and it is evolving over time. However, it has been suggested that the concept of sustainable tourism is ambiguous, subject to multiple definitions and based upon fragile theoretical foundations (Sharpley, 2009).

Natural Area Tourism in Context

In relation to tourism in the outdoors, Buckley (2011) has divided outdoor tourism into three categories – consumptive, adventure and non-consumptive (nature-based). He states that consumptive outdoor tourism involves recreational hunting and fishing and adventure tourism uses natural environments for excitement-oriented (often thrill-seeking), highly active recreation. His view of non-consumptive nature-based tourism is that it takes place largely in protected areas, national parks, wilderness areas, and other public lands and oceans.

Tourism in the outdoors also includes many non-adventure activities focused on the abiotic, biotic and cultural (ABC) elements of the environment (Dowling, 2001). Abiotic activities include geotourism, for example, which focuses on geology and the landscape; biotic activities may, for instance, involve observing wildlife (fauna) and wildflowers (flora); finally, culturally based activities in natural areas are likely to focus on an area's past (historic or heritage) attributes and/or present (cultural) offerings. Natural area recreation and tourism are undertaken by local communities, independent travellers and commercial tour clients. However, it has been noted that 'nature, adventure, and ecotourism frequently overlap in practice, despite seemingly static definitions in academic literature' (Schoegel, 2007: 250).

The all-encompassing term used in this book for tourism in natural areas is 'natural area tourism' and it is necessary to explore the implications of the term, when the phenomenon could also be described as 'nature-based tourism'. According to Weaver (2008), nature-based tourism occurs in natural settings but has the added emphasis of fostering understanding and conservation of the natural environment. In addition it embraces viewing nature as the primary objective. The focus is usually upon the study and/or observation of the abiotic (non-living) part of the environment, such as rocks and landforms, as well as the biotic (living) component of it, that is, the fauna and flora. Essentially it is a form of tourism underpinned by the ecocentric philosophy, so that the natural environment provides the platform for environmental understanding and conservation. Nature-based tourism also embraces the sustainable approach and fosters 'responsible tourism'. Fennell (2008) describes nature-based tourism as any form of tourism which uses natural resources in a wild or undeveloped form. In adopting this stance, he noted that such tourism can encompass mass tourism, adventure tourism, low-impact tourism and ecotourism (ecotourism is discussed further on p. 16).

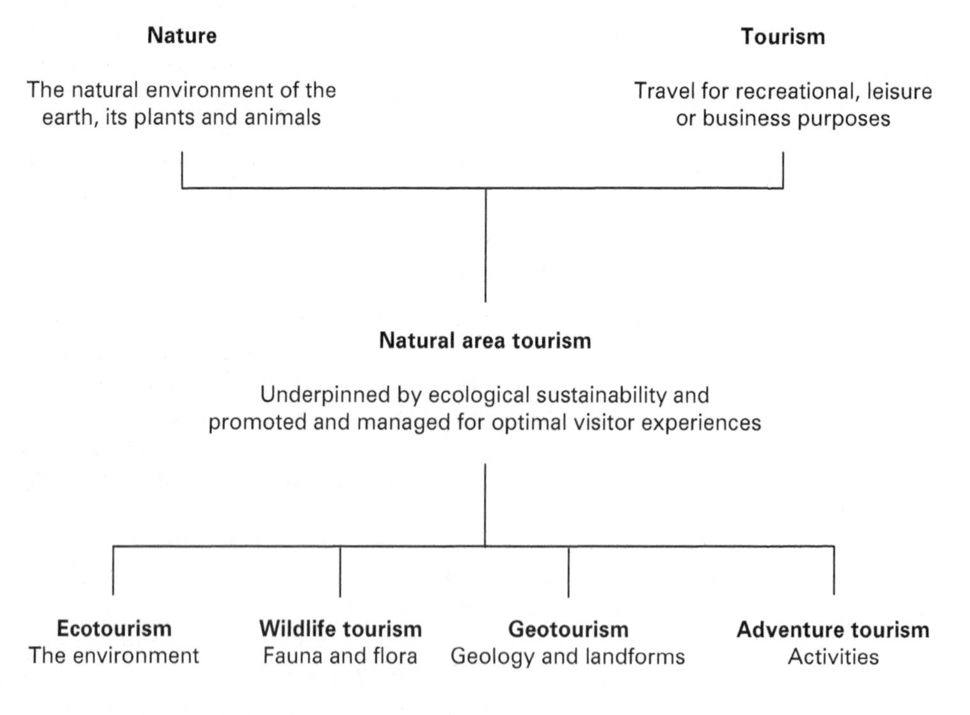

Figure 1.3 Characteristics of natural area tourism

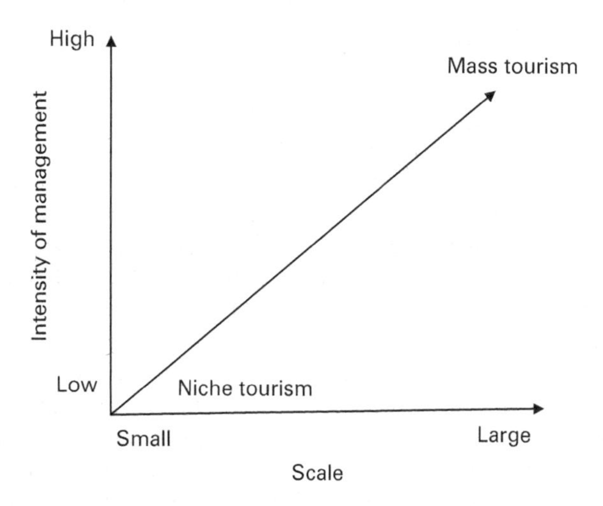

Figure 1.4 Management and scales of tourism: Natural area tourism can encompass the whole spectrum

Notwithstanding the views of Fennell (2008), natural area tourism, as defined in this book (Figure 1.3), is congruent with ecotourism, wildlife tourism, geotourism and adventure tourism (all discussed later in this chapter). Despite the framework indicated in Figure 1.3, we exercise caution in relation to the adventure tourism category because of its apparent dominant focus on activity-based experiences rather than time spent in nature in an appreciative and educative sense. Because of the vast range of opportunities that the natural environment provides and the complex nature of tourism and the tourism demographic, natural area tourism sits within a range of scales and management scenarios (Figure 1.4). At one end of the spectrum, there can be small numbers of people accessing a natural area with minimal facilities and this may be due to trip expense, logistical difficulties in locating features of target interest, niche interests and the nature of controls that might be in place. At the other end, a natural area may be subject to mass tourism, where millions of people travel to a site of interest, due to marketing, ease of access and the wide appeal of the site, as in the cases of Ha Long Bay World Heritage site in Vietnam and Yehliu Geopark in Taiwan. In the latter situation there is extensive management presence and infrastructure and up to 10,000 visitors on site during peak periods (Newsome *et al.*, 2012). Some sites, such as at Phillip Island in Australia (650,000 attendees at the penguin parade a year), span the entire spectrum at the same location, with guide-led small-group excursions at the niche end of the spectrum contrasting with the assembly of hundreds of people congregating at a viewing platform to watch little or 'fairy' penguins (*Eudyptula minor*) emerge at night under lighted conditions, at the mass tourism end of the spectrum (Newsome & Rodger, 2012a).

The Spectrum of Natural Area Tourism

What is ecotourism?

According to Weaver (2008) ecotourism existed well before the generally accepted introduction of the term in the 1980s. He suggests that during the latter part of the 20th century the term 'ecotourism' was used in the context of one of four phases or platforms that have been advanced for the field of tourism since the end of World War II (Jafari, 2001). With the advent of tourist air travel, and especially the introduction of the jet aircraft, which made long-haul travel more available, the modern mass tourism era began. Subsequently the study of tourism evolved through a number of phases or platforms, from *advocacy* (tourism is all 'good'), *cautionary* (tourism's 'good' is tempered by it having some negative elements), *adaptancy* (alternative types or forms of tourism were introduced under the banner of 'alternative [to mass] tourism') and finally the *knowledge-based* platform championed a more sustainable approach to tourism development generally (see Box 1.3). Weaver (2008: 6) noted that 'from an ecotourism perspective, a critical outcome has been the growing perception that this sector can legitimately occur as either alternative or mass tourism'. Weaver argued that this was a critical change and shifted ecotourism away from the traditional view of it being a 'form' or 'type' of tourism to it being an 'approach' to tourism.

Weaver (2005, 2008) suggested that arising from these approaches ecotourism can be further identified as either 'minimalist' or 'comprehensive'. The former is more likely to be focused on a particular site or species and involves only superficial learning, whereas the latter is wider in scope and encourages greater learning. Whichever way it is viewed, ecotourism has at least five key characteristics, as Weaver (2008: 17) indicated:

(1) It is a form of tourism.
(2) Attractions are primarily nature-based, but can include associated cultural resources and influences.
(3) Educational and learning outcomes are fostered.
(4) It is managed so that environmental and socio-cultural sustainability outcomes are more likely to be achieved.
(5) The importance of an operation's financial sustainability is recognised (it must at the least be financially viable).

The primary goals of ecotourism are to foster sustainable use through resource conservation, cultural revival and economic development and diversification. On an individual level it should add value to people's lives through their learning about the natural world. Ceballos-Lascurain (1998: 2) suggested that 'a lingering problem in any discussion on ecotourism is that the concept of ecotourism is not well understood, therefore, it is often confused with other types of tourism development'. Harrison (1997: 75) cautioned that 'in recent years ecotourism has become something of a buzzword in the tourism industry. To put the matter crudely, but not unfairly, promoters of tourism have tended to label any nature-oriented tourism product an example of "ecotourism", while academics have so busied themselves in

Box 1.3 Case study of ecotourism development: An industry association – Ecotourism Australia

Australia was an early adopter of ecotourism and the Ecotourism Association of Australia (EAA) was formed in Brisbane in 1991. It aimed to promote ecotourism, develop ethics and standards, promote understanding, appreciation and conservation of the natural and cultural environments visited, and facilitate interaction between tourist, host community, the tourism industry, government and conservation groups. A key goal was to ensure that ecotourism contributed to the conservation of places visited and to the preservation of biodiversity. The Association had a Code of Practice for Ecotourism Operators as well as a set of Guidelines for Ecotourists and in 1994 the Australian government established a National Ecotourism Strategy. Two years later the EAA launched the National Ecotourism Accreditation Program (NEAP), the first in the world. This programme was revised in 2000 and further amended and rebranded in 2003 as the ECO Certification Program. The scheme rates and certifies ecotourism products according to three levels – nature-based, ecotourism and advanced ecotourism. Around this time the Association launched a green EcoGuide Program.

In 2002 Australia featured heavily in the International Year of Ecotourism (IYE) and the Association convened its first international conference, in Cairns. In 2004 the Association was rebadged as Ecotourism Australia (www.ecotourism.org.au) and in 2008 it won the prestigious Tourism for Tomorrow *Conservation Award* of the World Travel and Tourism Council.

Today, Ecotourism Australia holds the Global Eco – Asia Pacific Tourism Conference annually and publishes *The Green Travel Guide*. However, its ECO Certification Program still remains its flagship and its logo is a globally recognised brand which assists travellers to choose and experience a genuine and authentic tour, attraction, cruise or accommodation that is environmentally, socially and economically sustainable. The ECO Certification Program assures travellers that certified products are backed by a strong, well managed commitment to sustainable practices and provides high-quality nature-based tourism experiences.

From Dowling (2013).

trying to define it that they have produced dozens of definitions and little else'. He went on to suggest that if sustainable tourism development is to occur, trade-offs are inevitable and often nature will be the loser. He also noted that ecotourism cannot solve all the problems of mass tourism and may in fact generate problems of its own. Harrison (1997) continued that it should not be considered to be a stepping stone to large-scale tourism, though it often proves to be so. He concluded that ecotourism is an ideal, but one worth working towards, because, at best, it fosters environmental conservation and cultural understanding.

It has also been stated that ecotourism is often nothing more than a marketing tool. In theory it should be an economically and socially sound means to conserve biodiversity, and also to provide revenue to improve the lives of people living in or near biologically important areas. Ecotourism is the fastest-growing segment of the tourism industry and constitutes a niche market for environmentally aware tourists who are interested in observing nature. It is especially popular among government and conservation organisations because it can provide simultaneous environmental and economic benefits. In theory it should be less likely than other forms of tourism to damage its own resource base but this is true only if such tourism is managed on a sustainable basis (Fennell, 2001).

Based on an analysis of 85 definitions of the term by Fennell (2001), he went on to define ecotourism himself as:

> [a] sustainable, non-invasive form of nature-based tourism that focuses primarily on learning about nature first-hand, and which is ethically managed to be low-impact, non-consumptive, and locally oriented (control, benefits and scale). It typically occurs in natural areas, and should contribute to the conservation of such areas. (Fennell, 2008: 24)

He argued that there are five criteria which should be used to define ecotourism – that it is a nature-based product, has minimal impact management, includes environmental education, contributes to conservation and contributes to community.

Fennell and Weaver (2005) argued that contemporary ecotourism is facing a crisis of credibility, with many 'ecotourism' products fostering minimal environmental attributes and often, in reality, merely being more traditional tourism products marketed with an ecotourism spin. They suggest that the core criteria of ecotourism should be that it is nature-based, includes learning opportunities and fosters a sustainability which is wider than just ecological. Minimalist approaches are more narrowly focused on specific (mega)fauna, superficial interpretation and minimal sustainability, while the comprehensive approach to ecotourism is more holistic, aims at interpretation and education leading to transformational participant attitudes and behaviours, and encompasses environmental, socio-economic and economic sustainable practices. The comprehensive model of ecotourism covers both the 'hard' and 'soft' ends of the ecotourism spectrum and embraces the four themes of research and education, ecological health, community participation and development, and partnerships.

In summary, ecotourism comprises a number of interrelated components, all of which should be present for authentic ecotourism to occur. The five key principles which we suggest are fundamental to ecotourism are that it is nature-based, ecologically sustainable, environmentally educative, locally beneficial and generates tourist satisfaction. The first three characteristics are essential for a product to be considered 'ecotourism', while the last two are viewed as being desirable for all forms of tourism (Dowling, 1996). This book adopts all of the features as being essential for the development of natural area tourism.

Nature-based

Ecotourism is based on the natural environment, with a focus on its biological, physical and cultural features. Ecotourism occurs in, and depends on, a natural setting and may include cultural elements where they occur in a natural setting (Figure 1.5). The conservation of the natural resource is essential to the planning, development and management of ecotourism.

Figure 1.5 Nine Bend River in the Mount Wuyi World Heritage site, Fujian Province, China. The area is famous for its red-coloured 'danxia' landscapes (Photo: Wendy Dowling)

Ecologically sustainable

All tourism should be sustainable – economically, socially and environmentally (Buckley, 2012). The sustainability of natural resources has been recognised by many countries as a key guiding principle in the management of human activity. Ecotourism is ecologically sustainable tourism undertaken in a natural setting. The challenge to ecotourism in any country or region is to develop its tourism capacity and the quality of its products without adversely affecting the environment that maintains and nurtures it. This involves ensuring that the type, location and level of ecotourism does not cause harm to natural areas.

The very incorporation of 'eco' in its title suggests that ecotourism should be an ecologically responsible form of tourism. Indeed, if this is not the case then the natural attributes upon which the tourism is based will suffer degradation to the point where tourists will no longer be attracted to it. The scale of such ecotourism activities implies that relatively few tourists will be allowed to visit the site and consequently supporting facilities can be kept to a minimum and will be less intrusive. Cater (1994) argued that ecotourism, with its connotations of sound environmental management and consequent maintenance of environmental capital, should, in theory, provide a viable economic alternative to exploitation of the environment.

This broadly defined travel oriented towards the natural environment is generally expected to respect and protect the environment and culture of the host country or region. However, according to Lawrence *et al.* (1997: 308) it is this larger goal of protecting or enhancing the environment that represents both its strength and weakness. Its strength is that ecotourism differentiates itself from the more traditional, consumptive forms of tourism, while its weakness is inherent in the tension that often prevails between achieving economic goals at the expense of ecological aims. They add that the ecotourism industry faces the paradoxical situation that the more popular the product becomes, the more difficult it becomes to provide.

Environmentally educative

The educative characteristic of ecotourism is a key element that distinguishes it from other forms of nature-based tourism. Environmental education and interpretation are important tools in creating an enjoyable and meaningful ecotourism experience. Interpretation is the art of helping people to learn and is a central tenet of ecotourism (Tilden, 1977; Weiler & Davis, 1993; Mancini, 2000). It is a complex activity that goes beyond making the communication of information enjoyable. Best-practice interpretation requires a thorough understanding and integration of audience, message and technique (Pastorelli, 2002; Mancini, 2008; Tilden & Craig, 2008).

Ecotourism attracts people who wish to interact with the environment in order to develop their knowledge, awareness and appreciation of it. By extension, ecotourism should ideally lead to positive action for the environment, by enhancing conservation awareness. Ecotourism education can influence tourist, community and industry behaviour and assist in the longer-term sustainability of tourist activity in natural areas. Education can also be useful as a management tool for natural areas. Interpretation helps tourists see the big picture regarding the environment (Crabtree, 2000). It acknowledges the natural and cultural values in the area visited as well as other issues, such as resource management.

Ecotourists expect high levels of ecological information. The quality of the environment and the visibility of its flora and fauna are essential features of their experience. Ecotourists also demand conservation (Chalker, 1994). Clear statements of the nature and aims of ecotourism need to be incorporated into literature and publicity material to educate and encourage active participation by stakeholders as well as the tourists themselves (Hall & Kinnaird, 1994). Lawrence *et al.* (1997) noted that a dominant part of ecotourism is for tourists to learn about and appreciate the natural environment in order to advance the cause of conservation.

Lee and Moscardo (2005) conducted a survey of tourists to see whether their resort experiences affected their environmental attitudes and behavioural intentions. They found that ecotourism accommodation could foster environmental understanding. The survey was conducted at Kingfisher Bay Resort and Village, on the World Heritage site Fraser Island, Queensland, Australia, in 2000. A sample of several hundred visitors found that the resort attracts guests who already have a high level of environmental awareness and concern, and that ecotourism products attract more environmentally aware customers. Particularly significant were the visitor awareness of, and involvement in, the resort's environmental management practices, as well as their participation in nature tour activities. However, the authors asserted that the resort can offer more environmentally effective information, interpretation and opportunities for guests in order to foster more responsible environmental attitudes and behaviours. Another survey compared ecotourists with more general tourists visiting the Annapurna Conservation Area in Nepal (Nyaupane, 2007). It found that the former were not necessarily motivated more by learning about nature and culture than general tourists and the author concluded that the tourists visiting general natural areas and ecotourism areas are not significantly different.

It has been noted that tour guides can contribute to the protection of natural areas by educating their tourists through interpretation and modelling appropriate behaviours (Randall & Rollins, 2009). This is carried out through strong leadership, motivation and interpretation.

Locally beneficial

The involvement of local communities not only benefits the community and the environment but also improves the quality of the tourist experience. These benefits should outweigh the cost of ecotourism to the host community and environment. Local communities can become involved in ecotourism operations, and in the provision of knowledge, services, facilities and products. Ecotourism can aid resource conservation in addition to having social and cultural benefits. Its contribution may be financial, with a part of the cost of an ecotour helping to subsidise a conservation project. Alternatively it could consist of practical help in the field, with the tourists being involved in environmental data collection and/or analysis, for example.

Drumm (1998) pointed out that local communities view ecotourism as an accessible development alternative which can enable them to improve their living standards without having to sell off their natural resources or compromise their culture. In the absence of other sustainable alternatives, their participation in ecotourism is often perceived as the best option for achieving sustainable development.

The term community-based tourism (CBT) emerged in the mid-1990s (Asker *et al.*, 2010). CBT is generally small-scale and involves interactions between visitor and host community; it is particularly suited to rural areas. CBT is commonly understood to be managed and owned by the community, for the community. It is a form of 'local' tourism, favouring local service providers and suppliers and focused on interpreting and communicating the local culture and environment. It has been pursued and supported by communities, local government agencies and non-government organisations (NGOs). Thus, the elements of CBT are integral to the development of ecotourism, especially in relation to its benefits to local communities.

According to Buckley (2011), the dynamics of tourism and conservation on community lands are both complex and contested. There are, though, many advocates of the need to integrate environmental conservation with tourism development in natural areas (e.g. Romeril, 1985; Wight, 1988; McNeely & Thorsell,

1989). According to Wight (1994), partnerships between tourism and conservation take many forms, including:

(1) donation of a portion of tour fees to local groups for resource conservation or local development initiatives;
(2) education about the value of the resource;
(3) opportunities to observe or participate in a scientific activity;
(4) involvement of locals in the provision of support services or products;
(5) involvement of locals in explanation of cultural activities or their relationship with natural resources;
(6) promotion of a tourist and/or operator code of ethics for responsible travel.

The implementation of ecotourism as an exemplar for sustainable development stems largely from its potential to generate economic benefits (Lindberg, 1998). These include generating revenue for the management of natural areas and the creation of employment opportunities for the local population.

Buckley (2011: 406) has reported on tourism's contributions to conservation through a combination of political, social and economic mechanisms. He noted that the ways in which tourism can contribute to conservation are linked to land tenure, which assigns different types of rights and responsibilities to different stakeholders. Examples include rights to use or sell wildlife, water or other natural resources, and to control or exclude access by various people for various purposes. In recent times commercial tourism has converted private or communal lands from primary production to conservation. Conservation tourism has limitations but also a number of successes and hence it seems likely that the role of tourism in supporting conservation will become increasingly critical (Buckley, 2011).

From a policy perspective, the role of tourism in supporting conservation both on and off reserves is becoming increasingly critical as human populations continue to expand, wilderness areas continue to shrink, and it becomes increasingly difficult for national governments to declare further public protected areas of any significant size. The effects of climate change, the need to increase the resilience of protected areas and the importance of landscape-scale connectivity conservation have rendered this area of research increasingly urgent (Buckley, 2011).

There is evidence that tourists seeking environmental and culturally differentiated destinations are willing to pay more for this experience (UNEP, 2011). This has been estimated to be between 25% and 40% (Inman *et al.*, 2002). The World Economic Forum (WEF) (2009) estimates that 6% of the total number of international tourists pay extra for sustainable tourism options and 34% would be willing to pay extra for them. One-third to one-half of international tourists (weighted towards the USA) surveyed in a CESD and TIES (2005) study said they were willing to pay more to companies that benefit local communities and conservation. Research by SNV (2009) records two studies where 52% of respondents in a UK survey would be more likely to book a holiday with a company that had a written code to guarantee good working conditions, protect the environment and support local charities, while some 58.5 million US travellers would pay more to use travel companies that strive to protect and preserve the environment.

One good example of community-based ecotourism occurs in the tropical rainforests, rivers and lakes of Malaysia and involves the participation of local communities in the tourism industry (Mamit, 2011). The rural communities provide tour packages for tourists, including ecotours and activities. The indigenous hosts also provide accommodation, meals, cultural shows, cooking and the sale of traditional arts and crafts. The involvement of indigenous communities in ecotourism is governed by their participation in the Home-Stay Programme, devised by the Ministry of Tourism, Malaysia, to enable rural indigenous communities to gain economic benefits from growing the tourism industry in the country. The home-stay destinations are the traditional villages of the indigenous communities, where rooms are provided as accommodation for tourists. In 2010 alone, the income derived from the Home-Stay Programme was US$4.1 million, from a total of 144 home-stay destinations.

Tourist satisfaction

Satisfaction of visitors with the ecotourism experience is essential to the long-term viability of the ecotourism industry. Included in this concept is the importance of visitor safety with regard to political stability. Information provided about ecotourism opportunities should accurately represent the product offered at particular ecotourism destinations. The ecotourism experience should match or exceed the realistic expectations of the visitor. However, client services and satisfaction should be second only to the conservation and protection of the resources on which the tourism is based.

Wildlife tourism

The desire for people to interact with wildlife in the natural environment continues to grow and visitation to sites with wildlife is on the increase (Newsome *et al.*, 2005; Rodger *et al.*, 2007; Newsome & Rodger 2012a). Mintel (2008) estimated that worldwide around 12 million trips were taken each year to view wildlife and that the annual growth rate was 10%. It was further estimated that this market was at the time worth approximately UK £30 billion, with up to 3 million people each year taking a holiday specifically to view wildlife.

While it is sometimes suggested that wildlife tourism includes the viewing of wildlife in both natural areas and captive settings such as zoos (Higginbottom, 2004), in this book wildlife tourism is bounded by the viewing of, and non-consumptive encounters with, wildlife solely in natural areas (Newsome & Rodger, 2012a). A useful illustration of the experiential spectrum that involves one species is the case of orang-utan viewing in Sabah, Malaysian Borneo. At one site orang-utan can be easily viewed at a feeding platform in a semi-natural state (Sepilok Orang-Utan Rehabilitation Centre). This contrasts with the much less reliable, but more adventurous, discovery and viewing of wild orang-utan along the banks of the Kinabatangan River in the same region (Newsome & Rodger, 2012b).

Often it is the quality of a natural area's living (or biotic) element – that is, the fauna and flora or wildlife – that plays a primary role in attracting tourists to specific destinations. Wildlife tourists often seek an experience that will enable them to explore, no matter for how short a time, a new ecosystem and its wildlife. Some tourists are lifelong wildlife enthusiasts and others merely take day trips to a wilderness area from a luxury hotel base. Many such visitors seek to be informed and educated, although others wish primarily to be entertained.

There are many different kinds of wildlife-watching holidays: tourists can, for example, choose between a luxury hotel-based safari in Kenya, wilderness backpacking in the Rockies or an Antarctic cruise to watch penguins and killer whales (Shackley, 1996). Accordingly, wildlife tourism can be divided into three main experiential categories. The first group are general nature-based tours that include a wildlife experience component. The second group comprises wildlife experience destinations, for instance breeding colonies and aggregations of wildlife such as coral reefs. The third group are specialised or dedicated wildlife tours that target specified groups, such as birds and marine wildlife, or individual species, such as monkeys or apes (Newsome & Rodger, 2012a) – examples include albatrosses in New Zealand (Higham, 1998), whale sharks in Australia (Figure 1.6; Mau, 2008) and polar bears in Canada (Lemelin, 2006).

A major issue for wildlife tourism is the diminishing number and size of undisturbed natural areas. This has spawned an emerging tourism trend in wildlife

Figure 1.6 Whale shark watching, Exmouth, Western Australia (Photo: Tourism Western Australia)

tourism known as 'last-chance wildlife tourism', where 'tourists explicitly seek vanishing landscapes or seascapes, and/or disappearing natural and/or social heritage' (Lemelin *et al.*, 2010: 478). This includes the desire to see unique, rare, special, enigmatic and difficult-to-view wildlife species, some of which are quickly disappearing. This form of wildlife tourism may increase visitation, development and revenue and has the potential to further profile the wildlife to a wider audience, raise awareness and provide money for conservation. However, the presence of large numbers of people and the need to manage tourists in environmentally sensitive situations needs forward planning, adaptive management, supervision and appropriate site interpretation, along with conservation-related education (Newsome & Rodger, 2012b).

Geotourism

Geotourism is emerging as a new global phenomenon. An early definition of geotourism as strictly 'geological tourism' has been refined as 'a form of tourism that specifically focuses on geology and landscape'. It promotes tourism to geosites and the conservation of geodiversity and an understanding of earth sciences through appreciation and learning. This is achieved through visits to geological features, the use of geotrails and viewpoints (Figure 1.7), guided tours, geo-activities and patronage of geosite visitor centres (Newsome & Dowling, 2010). Geotourists can comprise both independent travellers and group tourists; they may visit natural areas or urban/built areas wherever there is a geological attraction (Newsome *et al.*, 2012). This is the key distinction between geotourism and other forms of natural area tourism, as by definition natural area tourism takes place only in natural areas.

Geotourism has links with adventure tourism, cultural tourism and ecotourism, but is not synonymous with any of these. It is about creating a geotourism product that protects geoheritage, helps build communities, communicates and promotes geological heritage and works with a wide range of different people. The resources of geotourism include landscapes, landforms, rock outcrops, rock types, sediments, soils and crystals. The 'tourism' part encompasses visiting, learning about and appreciating geosites.

Geotourism may be further described as having a number of essential characteristics. These elements combine to shape geotourism in its present form. It comprises a number of interrelated components, all of which should be present for authentic geotourism to occur. In parallel with those applying to ecotourism itself (see above) there are five key principles which are fundamental to geotourism. They are that geotourism is geologically based (that is, based on the Earth's geoheritage), sustainable (i.e. economically viable, community-enhancing and fostering geoconservation), educative (achieved through geo-interpretation), locally beneficial, and generates tourist satisfaction. Again, the first three characteristics are considered to be essential for a product to be considered 'geotourism', while the last two characteristics are viewed as being desirable for all forms of tourism.

Geotourism attractions are now being developed around the world primarily as a tool for the development of local and regional communities. A major vehicle for such

Figure 1.7 Geotourism attraction, Granite Skywalk, Castle Rock, Porongurup National Park, Western Australia (Photo: Department of Environment and Conservation, Western Australia)

development is UNESCO's 'geoparks'. A geopark is a unified area with geological heritage of international significance and where that heritage is being used to promote the sustainable development of the local communities (UNESCO, 2011). Geoparks evolve through a series of levels, from 'aspiring', 'national', 'regional' (e.g. European or Asia-Pacific Regions) to 'global'. There are now 91 global geoparks in 29 countries (Figure 1.8 shows one example). The global geopark brand is a voluntary, quality label and while it is not a legislative designation, the key heritage sites within a geopark should be protected under local, regional or national legislation, as appropriate. UNESCO offers support to geoparks on an *ad hoc* basis via requests from

Figure 1.8 Tourist walkway, Langkawi Global Geopark, Malaysia (Photo: Ross Dowling)

member states. Global geopark status does not imply restrictions on any economic activity inside a geopark where that activity complies with local, regional or national legislation. The focus of geoparks is on geological heritage, geology and landscapes. These Earth heritage sites are part of an integrated concept of protection, education and sustainable development.

A geopark achieves its goals through conservation, education and tourism. It seeks to conserve significant geological features, and explore and demonstrate methods for excellence in conservation and geoscientific knowledge. This is accomplished through protected and interpreted geosites, museums, information centres, trails, guided tours, school class excursions, popular literature, maps, educational materials, displays and seminars. Geoparks stimulate economic activity and sustainable development through geotourism. By attracting increasing numbers of visitors, a geopark fosters local socio-economic development through the promotion of a quality label linked with the local natural heritage. It encourages the creation of local enterprises and cottage industries involved in geotourism and geoproducts.

In tourism terms, geoparks may be viewed as 'tourist destinations' similar in scope to World Heritage sites. Tourist attractions are generally thought of as sites

Box 1.4 Geotourism – Galapagos Islands, Ecuador

The Galapagos Islands are known around the world for their wildlife and place in the formation of the Darwin's theory of evolution. However, the islands have recently witnessed the birth of geotourism, with a number of tourist companies offering geological tours of the islands. The Galapagos are an archipelago of 19 volcanic islands distributed around the equator in the Pacific Ocean 972 km west of continental Ecuador. The islands were inscribed on UNESCO's World Heritage list in 1978 and as a Marine Reserve in 2001.

The Galapagos are one of the world's most active volcanic areas, with over 50 eruptions in the last 200 years. Volcanic processes formed the relatively young islands and most are volcanic summits, some rising over 3000 m from the sea floor (UNESCO, 2009). The most active volcanoes are the ones on Fernandina Island, which last erupted in April 2009, and Cerro Azul on Isabela (2008). All of the volcanoes are still active but most of the eruptions have been small.

A number of tour companies now offer geological expeditions to the islands, where tourists are offered the opportunity to examine its most significant geological sites and walk on the lava of the most recent eruptions. One such company is Metropolitan Touring (www.metropolitan-touring.com). Billed as 'Geology Expeditions in the Middle of the World', its tours are led by a field geologist and guests learn the story of the formation of the islands through an understanding of the various forms of volcanism. A typical expedition includes a visit by boat to Point Cormorant on Floreana (Charles) Island, where visitors witness a myriad of parasitic cones, evidence of a violent volcanic past. Spatter cones, cinder cones and tuff cones can be seen from all angles, especially when the vessel navigates along its coastline.

Santa Cruz (Indefatigable) Island, the third largest Galapagos island, has a varied geological history. Here tourists visit the twin pit craters geological formation, Los Gemelos. Genovesa (Tower) Island is the remaining edge of a large volcanic caldera that is now submerged. At anchor in the bay the walls of the caldera surround the tourist ship and many lava flows that make up this shield volcano can be seen. In the western part of Galapagos are the younger shield volcanoes of Isabela and Fernandina. Isabela is the largest island in the archipelago and was formed by the fusing together of six large volcanic domes. One of the volcanoes, Ecuador Volcano, rises from the water line with one of its halves completely collapsed and under water. On this island volcanoes are still forming, with eruptions occurring every three years or so, and tourists hike to the top of Darwin Volcano, which is a crater filled with salt water.

Fernandina (Narborough) Island is the youngest and westernmost island in the Galapagos and in 1969 its caldera dropped about 300 m to its new position. It is an active volcano and eruptions occur intermittently (Figure 1.9). Lava flows which entered the ocean have left dozens of lava tubes and flat beds where tidal pools form at low tide.

From Newsome and Dowling (2010).

Figure 1.9 Geotourists witnessing a volcanic eruption on Fernandina Island, Galapagos, Ecuador, in April 2009 (Photo: Ramiro Jacome, Galapagos Naturalist, Metropolitan Touring, Ecuador)

that appeal to people sufficiently to encourage them to travel there in order to visit (Holloway, 2008). Visitor attractions comprise natural environments, built structures or events. Therefore geotourism attractions comprise either geological environments or built structures or events based around such sites. Some examples (see Box 1.4) that have not been designated geoparks are the Grand Canyon, USA; the Valley of the Moon, Atacama Desert, Chile; the World of Fire, Vestmannaeyjar, Iceland; Mount Kinabalu, the Island of Borneo, Malaysia; the Seven-Coloured Earth of Chamarel, Mauritius; Al Hoota Cave, the Sultanate of Oman; and the Foz do Douro Geological Walk, Porto, Portugal (Dowling, 2011).

Adventure tourism

Adventure travel, like some other forms of tourism, is often divided into 'soft' and 'hard' dimensions. According to Christiansen (1990), soft adventure activities, for example, bike riding, photographic safaris, walking tours and certain wildlife watching activities, such as gorilla watching, are pursued by those interested in an adventure with a perceived risk but in fact little actual risk. Hard adventure activities, by contrast, are known by both the participant and the service provider to have a high level of risk, for example, mountain trekking, rock climbing, downhill mountain biking and rafting. Having said this, the spectrum is complex, as adventure tourism may encompass bicycle tours, expeditions, nautical tourism, volunteer tourism (voluntourism) as well as consumptive forms of tourism such as sport fishing, deer hunting and big game hunting. Adventure tourism accordingly can mean different things to different people and this makes its fit into natural area tourism problematic. Adventure tourism and nature-based tourism thus share similarities but are different aspects of tourism, with the adventure aspect generally encompassing high levels of exertion, an element of risk, and the use of specialised skills to participate successfully and safely in the activity (Buckley, 2006). With adventure tourism the natural environment is often the backdrop or place in which the activity takes place and not the focus of tourism activity, as in appreciating wildlife, geology or natural landscapes from an experiential and learning perspective.

Of particular concern is the recent trend to conduct commercially sponsored sporting and competitive events in natural areas. Adventure racing is a new component of adventure tourism which involves teams of people competing in challenges in natural environments, generally involving running, walking, bike riding and similar activities. It often attracts large numbers of visitors, including tourists as spectators to protected areas. Newsome and Lacroix (2011) state that the recent increase in interest in the use of protected areas for more action-based activities is a cause for concern, as such activities have the potential for significant negative impacts. In particular, they have the potential to impact on the experience of other users who are interested in enjoying the natural values of the area, and they cause additional and significant problems for management, especially in a context where management capacity is already limited. They suggest that the increasing popularity and pressures of adventure-style activities, and their impacts, require a discussion about how these activities should be managed and whether they should be encouraged in protected areas.

The use of natural areas by adventure tourists adds a new level of complexity to the way natural areas are used from a recreational perspective and raises the question as to whether such users are environmentally responsible. One survey indicated that income levels and moral obligation are the best predictors of tourists being environmentally friendly (Dolnicar, 2010). In that study of environmentally friendly individuals, to see if they are also environmentally friendly tourists, Dolnicar (2010) found that the correlation was high. The only factor to reduce this relationship was the lack of identification with the destination. It was concluded that 'by attracting tourists who behave in an environmentally friendly manner, the environmental footprint of tourism at the destination can be reduced' (Dolnicar, 2010: 730). This work supports the findings of a study of individuals with a strong nature orientation who were found to have more positive views of environmentally responsible practices by tourism businesses than tourists who were not so nature oriented (Andereck, 2009). Furthermore, environmental actions by businesses which were viewed as being important included recycling programmes, renewable energy systems, composting toilets, items made of recycled materials, grey water systems, water use reduction programmes, local architecture compatible with the environment and landscaping with native plants. The study confirmed the importance of the natural world as a motivator for tourism experiences as well as the high level of importance and value tourists place on environmentally friendly practices at tourism sites and businesses. A corollary was that such tourists also place importance on seeing environmentally responsible practices being implemented by tourism businesses. How such observations might fit with action-based interests and adventure racing events remains to be researched and is lacking in data.

Key Issues for Natural Area Tourism in the 21st Century

Natural area tourism faces a number of challenges in the 21st century. Some examples include adaptive management, sustainable development and climate change.

Adaptive management

Adaptive management has its genesis as a means to accept and embrace uncertainty in understanding the environmental impacts of new projects (Holling, 1978). Thus, it aided in the prediction of how part of an ecosystem would respond as a result of the implementation of a management decision (plan or policy). When applied to other situations, the characteristics of the approach include collaboration of interests, identification of values and continuous learning. Organisations which are adaptive have well defined mandates, an innovative membership and a range of participatory systems; they integrate and coordinate related processes. Underpinning these principles is the ability of stakeholders to implement change while recognising institutional limitations; that is, once the new knowledge has been identified, they have the will and capacity to act on the information (Newsome *et al.*, 2005).

The key characteristics of adaptive management are: the inclusion of both the natural and social sciences; recognising uncertainty, complexity and long time scales; regarding policy and management interventions as objective-driven and experimental, with monitoring an integral part; the inclusion of stakeholders; and the use of feedback (Dovers, 2003). When applied to the planning of natural areas, the focus is on setting goals, monitoring and feedback. Monitoring is discussed as an integral part of management as it is only with monitoring that the success or otherwise of a management strategy can be judged. The feedback provided by monitoring is part of adaptive management. Feedback makes learning, adaptation and change possible. Stankey (2003) suggests that adaptive management can help close the gap between managers and scientists, making them partners in formulating problems, developing management strategies and associated monitoring programmes, and evaluating outcomes.

Thus, adaptive management is:

> a deliberative and purposive process through which questions are framed, hypotheses proposed, implementation is designed to enhance learning opportunities, results are critically evaluated, and, if appropriate, subsequent actions and policies are revised and applied ... in a manner ... to enhance the continuing process of learning. (Stankey, 2003: 175)

Most importantly, adaptive management also involves taking risks, but protected area agencies have become risk-averse, making it difficult for managers to experiment without facing censure (Beckwith & Moore, 2001). In relation to wildlife tourism, the most promising adaptive management opportunity derives from its multiplicity of stakeholders, which enables this approach to succeed when it has failed in many other settings (McLain & Lee, 1996). Adaptive management also relies on and embraces wide participation and indigenous knowledge (Berkes & Folke, 1998), both features of natural area tourism in many parts of the world. These features, which often impede resource management elsewhere, may actually facilitate, through the use of adaptive management, the achievement of sustainable natural area tourism (Newsome *et al.*, 2005). Those involved should extend beyond scientists and managers to include any stakeholders whose values are involved or who may be affected by decision making (Kruger *et al.*, 1997).

Sustainable development

The future of sustainable natural area tourism lies in its planning, development and management. Planning for natural area tourism enables developers and managers to foster tourism in these areas in such a way as not only to protect the natural environment but also to bring about a greater understanding of it. The key lies in the activity of planning *for* natural areas rather than solely planning *in* them. This is best carried out in an inclusive manner which embraces the interests of, and input by, key stakeholders. In addition, such planning should be iterative and flexible so as to allow objectives and strategies to be achieved while still providing a means for consistent management.

There are a number of ways of managing tourism in natural areas, some of which focus on the available site management whereas others focus on visitor management techniques. A key management strategy is zoning, in which activities are separated by space and/or time. Management is the strategies and actions taken to protect or enhance natural areas in the face of impacts from tourism activities. Strategies are defined as general approaches to management, usually guided by an objective, for example reserving and/or zoning a natural area as a national park. A strategy can also be a group of actions, for instance site management and its associated actions.

Planning and management need to reflect a balanced approach to how natural resources are used and to include local communities in the development process. A more 'sustainable' approach to tourism development is required and it is through natural area tourism that this may be achieved (Holden, 2012). While it is imperative that the Earth's environmental elements are not perceived solely as attributes, the reality is that if natural areas are to survive they must be 'valued' more through developments such as tourism. Often it is only tourism that will foster conservation of such areas. So while there is undoubted concern at the increasing demand for tourism to natural areas, this may just be the one activity that ensures their continued survival.

According to Asker *et al.* (2010) sustainable tourism development comprises a number of key elements. First, it makes optimal use of environmental resources that constitute a key element in tourism development, maintaining essential ecological processes and helping to conserve natural heritage and biodiversity. It also respects the socio-cultural authenticity of host communities, conserves their built and living cultural heritage and traditional values, and contributes to intercultural understanding and tolerance. Third, it ensures viable, long-term economic operations, providing socio-economic benefits to all stakeholders that are fairly distributed, including stable employment and income earning opportunities and social services to host communities, and contributing to poverty alleviation. It should also maintain a high level of tourist satisfaction and ensure a meaningful experience for the tourists, raising their awareness about sustainability issues and promoting sustainable tourism practices amongst them (UNWTO, 2005). However, Balmford *et al.* (2009) argue that nature-based tourism and recreation will become sustainable only where there is effective planning, management and local participation (see Box 1.5). Then it has the potential to make a growing contribution to both conservation and sustainable development.

The UNWTO has forecast that by 2015 China will be the major tourist generating and destination receiving country in the world (Ma *et al.*, 2009). China's national parks have been developed largely on the premise of their approach to nature being rooted in their cultural connectedness with the land. National parks are viewed as resources to secure economic returns through tourism. Thus, there is a high degree of 'urbanisation' permitted within national parks for both accessibility

Box 1.5 Case study: Sustainable tourism – Kumul Lodge, Papua New Guinea

Kumul Lodge is a well known ecotourism and bird-watching lodge located in Enga Province in the Highlands of Papua New Guinea (http://kumul-lodge.com). Made out of local materials, entirely owned and managed by local people, and employing local guides, it was established by a husband-and-wife team who are owners and managers of the lodge. The lodge is the second most visited bird-watching destination in Papua New Guinea and is home to many rare and iconic species, such as the birds of paradise. The lodge also offers other nature-based activities, such as trekking and guided bird-watching tours.

The lodge conducts conservation of the surrounding natural area in order to maintain the birds' habitat, as they are the lodge's primary attraction. Adjoining landowners are paid US$4 for every guest who stays at the lodge and they are educated on the importance of protecting the birds. The lodge maintains tracks and bird-watching facilities in the area. Its long-term aim is to have the government establish a national park in the region. The peak tourist season runs from June to September but the Lodge provides its staff with financial security by employing them year round. The rest of the year is used for developing the facilities of the lodge. The owners of the lodge realise that in order to protect the environment of the local area, and with it their own core business assets, they need to ensure they have the support of the local community. So they have spread some of the economic benefits of the tourism enterprise through the local community and thus been able to ensure that the community as a whole has a stake in protecting the environment.

From Asker *et al.* (2010).

and accommodation, as, the more commercially successful a park is, the more it can support environmental conservation. Thus parks become central to tourism development (Ma *et al.*, 2009). However, the obvious contradiction here is the risk of overdevelopment.

Included in the many benefits of sustainable tourism development are those to local communities and conservation. Income from tourism has contributed to community wellbeing and sometimes also to nature conservation. There are well known examples in sub-Saharan Africa, especially in Namibia, Botswana and South Africa, where companies, such as &Beyond and Wilderness Safaris, have successfully funded community conservation through commercial tourism (Buckley, 2011).

In Queensland, Australia, Tourism Queensland and the Queensland Parks and Wildlife Service have introduced a Tourism in Protected Areas (TIPA) programme (Thomas & Morgans, 2011). The key objective is to provide for sustainable nature-based tourism in the state's national parks. A three-year roll-out will see it established in parks, including Fraser Island World Heritage site, the Whitsunday Islands on the Great Barrier Reef and Daintree National Park in the Wet Tropics World Heritage site. Key objectives of the programme are to achieve best-practice standards in nature-based tourism, to identify sustainable visitor limits, to manage capacities to maximise commercial opportunities, to increase certainty for industry to invest for the long term, and to target compliance and monitoring. The idea is to position nature-based and ecotourism in the state's national parks as a world-class experience of nature.

A similar scheme has been introduced in Western Australia, called 'Naturebank' (Tourism WA, 2011). The initiative prepares sites for development of quality, environmentally sensitive tourist accommodation experiences in the state's national parks. So far eight sites have been identified for ecotourism development across the state (www.tourism.wa.gov.au/naturebank). Successful proponents are offered a performance-based lease with social and environmental performance conditions that reflect the values of the areas. The programme places considerable emphasis on developers providing best-practice ecotourism operations (Quartermain & Telford, 2011).

Weaver (2011) positions sustainable mass tourism as an emergent and desired outcome for most destinations, based upon the amalgamation of ongoing support for growth and sustainability. He suggests that there are a number of paths leading to sustainable mass tourism and that these are a conventional demand or market-driven 'organic' path, a supply-side or regulation-driven 'incremental' path, and a hybrid, government-driven, 'induced' path. His conclusion is that all three paths are converging towards a goal of sustainable mass tourism based on a claim that five external factors will guide the sector into an 'emergent norm of sustainability', combined with the 'entrenched norm of support for growth'. These factors are: conventional/renewable resource price convergence; climate change awareness; the global financial crisis; institutionalised environmentalism; and the internet. However, a rejoinder to Weaver's paper critiques his operationalisation of sustainability and his assumption that sustainable mass tourism will be the 'emergent norm' due to external factors (Peeters, 2012). Instead, Peeters (2012: 1040) suggested that:

it is now time to systematically analyse and present the problems of tourism, and to clarify to stakeholders and politicians that we do face a serious problem but that the solutions are obvious and technically and economically achievable. What fails is political will.

Climate change

Climate change is already impacting on nature-based tourism. Global warming appears to be particularly affecting the Arctic region (ACIA, 2005) and the reduced cover of sea ice may now allow greater and longer tourist access to the polar regions (IPCC, 2007a; UNEP, 2007). This nexus also raises the question of sustainable approaches to transport in tourism, as the long-haul nature of polar destinations is likely to result in significant carbon emissions (UNEP, 2007).

Global warming is being manifested as a reduction in the extent and persistence of sea ice and the melting of glaciers, plus increasing precipitation and shorter and warmer winters in the polar regions (IPCC, 2007a). The loss of sea ice is likely to impact on ice-dependent species such as the polar bear, which is already the focus of increasing tourism activity. The polar fringe is subject to contraction and flooding of the area of tundra, with a subsequent redistribution of tundra plants and animals. Changes in seasonality and the nature and abundance of food resources are likely to affect the birds and mammals that seasonally migrate to and within the tundra ecosystem. Ocean temperature changes and alterations in the abundance of krill, which form the basis for the marine food web, especially in the Antarctic region, are likely to have significant impacts on dependent populations of birds, pinnipeds and cetaceans. There has been a widespread reduction in the extent and persistence of ice and snow on land, as well as large-scale melting of glaciers in mountainous regions (IPCC, 2001).

Park and wilderness areas are not large enough to sustain our natural heritage by themselves. Conservation planning must extend beyond the boundaries of park and wilderness areas, and climate change makes this scale of planning even more imperative (Hobbs *et al.*, 2010). Global sea surface temperatures are predicted to rise between an average of 1.10°C and 4.60°C over the next 90 years or so (European Environment Agency, 2004). Other ocean–climate interactions include changes in climate patterns and ocean currents and acidification of the ocean (IPCC, 2007b). Polar regions have higher rates of predicted temperature change (IPCC, 2007b), with slower effects in temperate and tropical regions (MacLeod, 2009). Thus, the sustainability of whale watching could be significantly affected by alterations in sea surface temperatures impacting on the occurrence, abundance, range, migratory patterns and fecundity of cetaceans (Lambert *et al.*, 2010).

According to Buckley (2011) skiing seasons are already shorter, and snow quality poorer, in many heavily frequented ski resorts in a number of countries. He noted that beach tourism destinations may be affected by increasing storminess in some coastal areas, and dive tourism destinations are being affected by damage to coral reefs associated with increasing ocean temperatures and acidity. Finally, national parks and wilderness zones in forest and woodland areas may suffer higher risks of fire and consequent closure, preventing recreational access. All of these climate change impacts have affected the tourism industry.

All forms of motorised transport consume energy and contribute to climate change. The principal contribution is from air transport of tourists and service items. The climate change impacts of tourism thus depend largely on patterns in air travel, which are influenced by a range of factors, such as fuel prices and carbon taxes (Buckley, 2011). Many travel agents, airlines and car rental companies now sell carbon offsets, but only about 1% of travellers buy them, even though prices are very low. Offset programmes claim to fund physical measures to reduce atmospheric concentrations of greenhouse gases, but there is little evidence that they actually do so, and it seems that travellers do not trust them (Buckley, 2011).

According to UNEP (2011) the sustainability drivers of climate change are: the costs of greenhouse gas emissions (driven by post-Kyoto rules), concern by customers about their carbon footprint, host government policies and priorities (climate change mitigation and energy), the uptake of corporate social responsibility (CSR) and the impact of climate change on tourism sites. UNEP suggests that the likely implications for climate change will be the increased substitution of fuels towards electricity, particularly increased investment in passive solar collectors, photovoltaics and alternative fuels for vehicles. Added to this will be an increased number of project developers orienting business strategies towards smaller carbon footprints, the increased expectations of a broader stakeholder base and the demand for carbon offsets and other mechanisms to compensate for residual emissions.

Climate change is viewed as a major threat to the Great Barrier Reef and the marine tourism industry it supports (GBRMPA, 2009). Tourism contributes approximately AU$5 billion to the national economy and provides over 50,000 jobs; the reef is visited by approximately 2 million people per year (GBRMPA, 2009). Climate change has already had a significant impact on the reef, including higher water temperatures, which has caused mass coral bleaching and an increase in the occurrence of coral diseases (Prideaux, 2009).

A survey of tourists visiting the Great Barrier Reef found that they had a range of views concerning the impact of climate change and their role in mitigating these impacts (Paris *et al.*, 2011). Some of the 81 respondents felt that climate change will result in an increase in tourism to the region, while others thought there would be a decrease. Respondents who felt that visitation would increase due to climate change suggested that if the region's warm weather was warmer for longer, then high seasonal arrivals would increase overall. Conversely, the majority of respondents stated that there would be a significant decrease in tourism in the region if climate change were to adversely impact its natural resource assets. They argued that the destruction of the reef would result in decreased numbers of tourists in the long term.

Tourists were also asked for their opinions on what they could do to mitigate the impacts of climate change on the region's natural resources and its tourism industry. The general consensus was that tourists could reduce pollution and littering, promote education, minimise transportation use and follow rules established by the local heritage sites (Paris *et al.*, 2011).

Outline of the Book

This book provides a comprehensive description of tourism in natural areas as an applied science. The aim is for the reader to understand the scope of, complexities arising from and possibilities of undertaking successful tourism developments in natural areas. One objective of the book is to overcome the existing perception that tourism developments in environmentally sensitive areas are inherently adverse; rather, it offers the view that, with adequate foresight, planning and management, such developments represent vehicles to bring about increased awareness and conservation. One underlying theme is that natural area tourism is an appropriate vehicle to bring about greater environmental understanding. For this reason such tourism separates itself out from other forms of tourism. If people in the developed world seek to ascribe some form of legacy to the Earth, then one of the best ways of doing this is by gaining an understanding of our environment, leading to its appreciation, which inspires action for the environment.

This book reviews the environment–tourism relationship through the context of natural area tourism, a rapidly growing form of tourism. It involves the clarification and definition of terms and concepts such as sustainable tourism, natural area tourism, ecotourism, wildlife tourism, geotourism and adventure tourism. The book focuses on the principles and characteristics of natural area tourism and illustrates these through the context of ecology, impacts and management.

Chapter 2 provides a synthesis of the ecological knowledge that is required to define the nature-based tourism resource, prevent its degradation and maintain its sustainability. This knowledge is based on the concept of 'ecological connectivity', in which all living things are viewed as being interconnected. It is an ecological axiom that all living and non-living parts of the environment are interdependent. Therefore the nature and characteristics of ecosystems are examined in relation to the structure of ecosystems, ecosystem function, ecological communities, ecological disturbance and succession and landscape ecology. This is followed by an overview of tourism in a range of valued ecosystems, including islands, coral reefs, tropical rain forests, savanna and modified ecosystems. Wildlife tourism then comes under our focus, especially with regard to the disturbance of wildlife caused by tourism. The chapter concludes with a plea for a greater understanding of key ecological processes in order to effectively manage tourism in natural areas.

The third chapter provides an account of the environmental impacts of tourism and recreation in natural areas based on an understanding of the area's ecosystems. The sources of impacts are reviewed, including those caused by trampling, access roads and trails, the use of off-road vehicles, the use of built facilities and camping areas, and the use of water edges. It outlines the impacts of recreation and tourism on abiotic environments (mountains and caves) as well as the biotic (wildlife and wildflowers) and also introduces the social impacts of natural area tourism. The chapter concludes with an overview of the impacts of natural area tourism in the context of wider environmental issues such as cumulative impacts, the landscape matrix, river systems, agriculture, urbanisation and protected-area security.

Chapter 4 describes planning for visitor management in natural areas as a means of achieving efficient and cost-effective sustainable tourism. The subjective nature of planning and the need to engage stakeholders throughout the planning process are emphasised. Reasons for planning are also given. A number of concepts central to visitor planning in natural areas, including carrying capacity, 'acceptable' change and the spectrum of recreation opportunities, are described before the planning frameworks central to the chapter are explored. Most important of the frameworks are the recreation opportunity spectrum, limits of acceptable change, visitor impact management and the tourism optimisation management model. Each framework is described using diagrams and examples. The chapter concludes with suggestions for how to choose the 'best' planning framework. The chapter is central to the book, in that it acts as a bridge between the impacts generated by recreation and tourism and the management strategies employed to reduce the adverse impacts and advance any beneficial impacts created by tourism in natural areas.

The fifth chapter describes a number of management strategies and actions available to assist in achieving sustainable natural area tourism. The management strategy of creating and designing protected areas is outlined, and an overview is given of the extent and types of protected areas. Governance and joint management are recognised as other key management strategies as governments are now beginning to share their management responsibilities with local communities and NGOs. This is followed by an introduction to zoning, a key strategy for managing protected areas. Two broad groups of strategies and associated actions, site and visitor management, then dominate much of the remainder of the chapter. Site management relies on designing and then managing linear features such as roads, tracks and trails, whereas visitor management is explored through either direct regulation or communication and education. Factors influencing the choice of actions by managers, such as the cost and extent of impacts, are explored. Several means of managing/working with the tourism industry, including voluntary and regulatory approaches, are described. Voluntary approaches are particularly interesting and include codes of conduct as well as certification. Combined approaches to management are often employed; these include reservation, zoning, co-management, visitor regulation and education. Such an approach is outlined in a case study of Kakadu National Park in northern Australia, with its application of a broad range of management strategies and associated actions to managing tourism in a large natural area. The chapter concludes by exploring ways to manage the tourism industry's use of natural areas through both voluntary and regulatory strategies.

Interpretation as a means of educating and communicating ideas to the natural area tourist is explored in Chapter 6, where it is championed as an integral part of best-practice ecotourism. The principles and application of interpretation are considered in relation to providing minimal-impact messages and thus fostering sustainable tourism through a number of stages of the interpretive experience. Examples of interpretive techniques are illustrated through publications and websites, electronic educational resources, visitor centres, self-guided trails and guided touring. The role and effectiveness of interpretation are highlighted in visitor impact management and in improving environmental awareness among tourists. The chapter concludes by evaluating the roles of the tour guide and tour operator.

Chapter 7 describes monitoring and argues it has been a long-neglected element of natural area management. It is defined and then principles to guide the development of monitoring programmes are suggested. A number of approaches to monitoring visitor impacts on natural areas are described for built facilities, campgrounds and campsites, roads and trails, and water bodies. A number of methods of monitoring visitors to natural areas are outlined, including visit counts, questionnaires and personal interviews, focus groups and other interactive techniques, such as task-forces and the internet. Finally, system-wide and integrated approaches are outlined, including protected-area management effectiveness assessments, benchmarking, and integrated approaches to impact management. Crucial to all these descriptions are accompanying details on sampling strategies, and the selection of indicators and standards and assessment procedures, which enable monitoring of protected area systems. A case study of Warren National Park in south-western Australia is presented in Box 7.10 to illustrate an integrated approach to monitoring, with data from both visitor impacts and from the visitors themselves collected and used.

The final chapter, a conclusion, explores the future of natural area tourism from a number of perspectives and the important issues of sustainability and climate change are again emphasised. Emphasis is also placed on natural area tourism's ecological underpinnings, the nature of its impacts, the need to adopt appropriate planning and management strategies, the essential place of monitoring in natural area management, and the role of interpretation as an essential bridge between tourist visitation and connection to natural environments.

2 The Ecological Perspective

Introduction

Patches of natural vegetation, peri-urban reserves, nature reserves, national parks, islands, coastal areas, mountains, caves, concentrations of animal life and larger wilderness areas continue to be important tourism resources (e.g. De Lacy & Whitmore, 2006). Some of these areas are substantially natural, as in the case of some deserts and the arctic tundra, or where large tracts of undisturbed forest remain, as in Russia and parts of Brazil. Other areas, such as many types of forest around the world, are more fragmented, while some landscapes are almost entirely modified, as in the case of the patchwork of regenerated woodlands and agricultural land in the British Isles. Even in already altered and semi-natural landscapes, the continuing impacts of people have the potential to bring about unwanted change to any remaining natural components (see Chapter 3, p. 190).

Given sustained tourism interest in, and continued high levels of visitation to, natural and protected areas over the past 20 years, the call of Tyler and Dangerfield (1999) remains pertinent. They pointed out that the wider application of ecological knowledge had been largely neglected in predicting and managing tourism-related disturbance in natural environments. Their paper emphasised the importance of a holistic approach, where the ecological system that supports a particular valued species is taken into consideration in the maintenance and management of sustainable tourism. A diversity of reef fish, for example, requires an intact and healthy coral reef. Similarly, the endemic bird life of many islands is dependent on specific plant communities. Tyler and Dangerfield (1999) also recognised that the nature of tourism itself complicates the ecological assessment of tourism impacts. These include the type, intensity, duration and spatial extent of various activities (examples and the specific nature of such impacts are the subject of Chapter 3). More than a decade later Monz *et al.* (2010) reiterated the complex relationship between recreational use and ecological conditions. They noted the vital importance of understanding the nature of abiotic (soil and water) and biotic (plant and animal communities) components in establishing the ecological effects of the amount, type and distribution of recreational/tourism use in any given ecosystem.

The principal aim of this chapter, therefore, is to outline the important elements of ecology that can be applied to understand, predict and mitigate any change that might be brought about by tourism. For example, certain plant and animal species are especially prone to disturbance, such as the endemic bird life of oceanic islands and species with colonial nesting habits. Roads, in particular, have the potential to act as barriers to movement and disrupt migration routes. In these cases an ecological definition of the physical requirements of living organisms, and the environment in which they live, can be used to predict the possible disrupting effects of tourism development and tourist activity.

Managing visitors also requires an ecological perspective. Tourists may impact directly on wildlife and vegetation and may also change aspects of the physical environment, for example by straying off designated trails and trampling vegetation. Educating visitors about ecological sensitivity can help them appreciate ecological conditions and enhance their understanding of the specifics of managing natural area tourism. This is achieved, for instance, by providing information on the biology of an organism, a story of how it survives and interesting facts on how it interacts with other organisms and the physical environment. The how and why of tourism management can also be explained to visitors, with the aim of gaining support in managing a site sustainably. Interpretation is therefore used as a means of protecting the natural environment, especially when visitors are made aware of the fragility of some natural ecosystems.

For the purpose of exploring the ecological context of natural area tourism, it is important to appreciate how ecosystems fit into the landscapes that form tourism resources. The ecological view in this chapter is therefore presented largely in terms of the composition and functioning of ecosystems. A justification for this view may be exemplified by an analogy with a car or clock. In both cases the component parts are very important but we nevertheless need to appreciate how the whole fits together as a total functioning system.

Ecology is now a well established discipline and its content is large and complex. It can be explored at the organism, population, community, ecosystem or landscape level (Figure 2.1). The treatment here is necessarily brief and our objective is first to provide an ecological framework that shows what ecosystems are made of and then briefly to explain how ecosystems work. This is followed by a consideration of how various landscapes change as a result of natural and human disturbance.

Because tourism takes place in many different environments it is impossible to cover the full range of tourism situations from the ecological standpoint. We have therefore selected some of the most important destinations for natural area tourism. Many tropical environments are important and some of their particular ecological characteristics are therefore briefly described. Other specific environments are included in the remaining chapters of this book.

The purpose of including selected tourism destinations, as presented later in this chapter, is to show a range of important natural area tourism situations and their differing ecological contexts. From these examples, general principles, similarities and differences in ecosystem structure and function can be appreciated. Furthermore, in

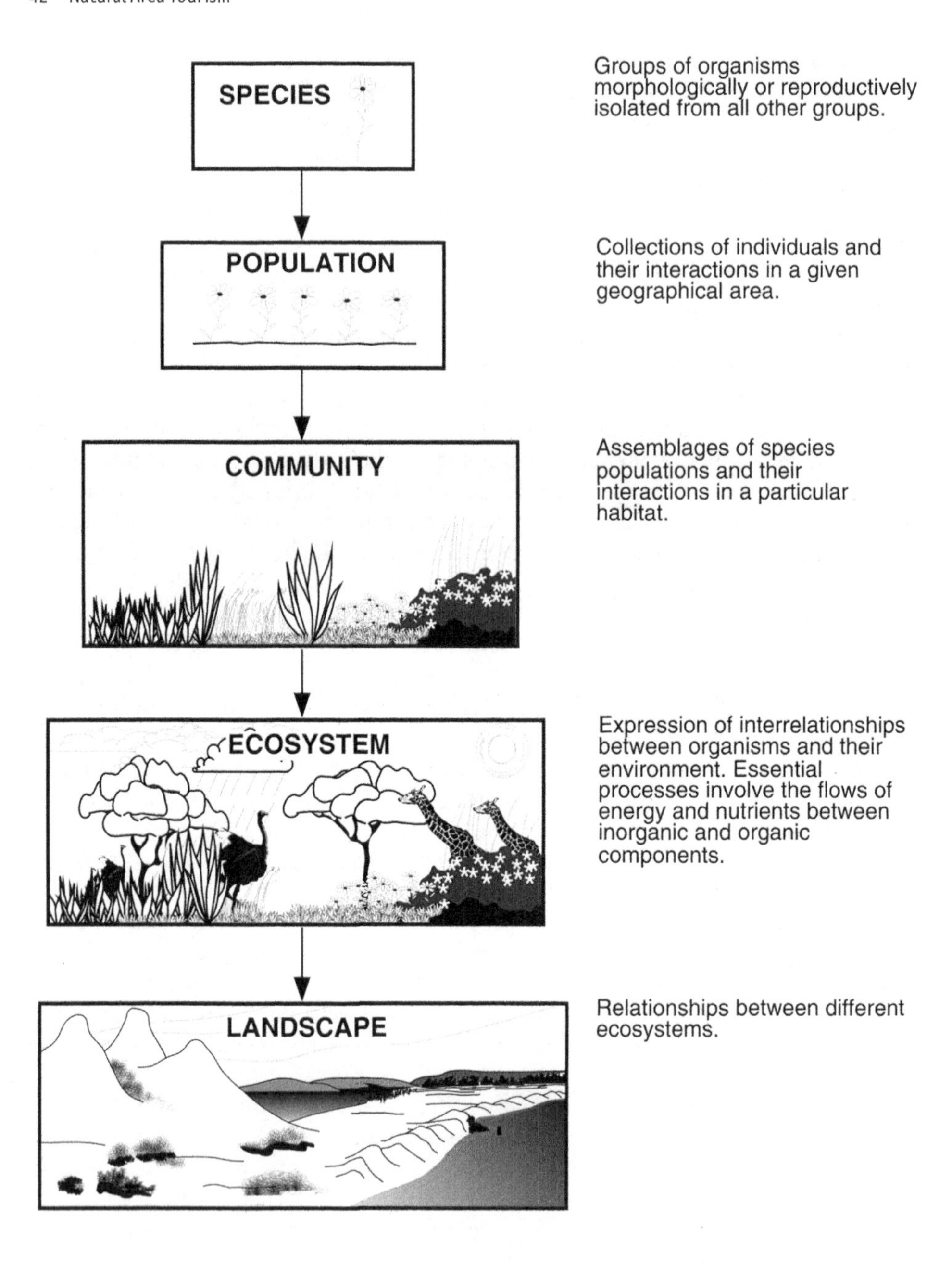

Figure 2.1 Levels of organisation within the science of ecology

order to understand the ecological effects of tourism, this chapter also necessarily flags some of the discussion on impacts that are more fully covered in Chapter 3 as well as some management issues that are more fully explored in Chapter 5.

An Introduction to Ecosystems and Landscapes

Ecology is concerned with the structure and functioning of ecosystems and thus how plants and animals interact together and with the physical environment. Although most ecosystems are very complex and difficult to study as entire systems, their essential structure and function are well established.

Ecosystems comprise structural components such as living organisms (biotic components), soil and landforms and other non-living features (abiotic components), such as wind, rain and water flow. Energy and materials, such as water and nutrients, flow through this combined system, resulting in ecosystem function (Figure 2.2).

The structure of ecosystems

Ecology can be studied from the perspective of populations or communities of plants and animals or whole ecosystems (Figure 2.1). A population refers to the number of organisms of the same species which inhabit a defined area, while a community consists of a group of populations of different species interacting with one another in a defined area. An ecosystem, on the other hand, represents a community of organisms interacting with the environment.

The scale at which an ecosystem can be identified ranges from the microbial community inside the gut of a cow through to a pond, ocean, forest or the entire planet. Despite this issue of scale, we can conveniently separate ecosystems into the two major divisions of terrestrial and aquatic (Table 2.1). These can then be sub-divided further according to the occurrence of specific plant and animal communities and the differing physical factors that determine their existence. On land, for example, differences in moisture and temperature give rise to different ecosystems in the form of forest, woodlands, grasslands and deserts. These comprise the major terrestrial 'biomes', which can be further divided according to specific conditions of precipitation, temperature, the occurrence of fire and differences in soil conditions. An example of this can be seen in the global occurrence of forests, which can be divided into tropical rainforests, tropical seasonal forests, temperate rainforests, temperate forests and northern coniferous forest, or taiga. Despite such differences in detail it is possible to define the structure of any ecosystem according to its major biotic (living) and abiotic (non-living) components (Figure 2.2).

Biotic components

The biotic part of an ecosystem is its component plant and animal populations; these make up the biotic community. In terrestrial ecosystems this comprises plants, such as grasses, herbs and trees, and associated populations of animals. The 'plant' component in aquatic ecosystems consists of rooted aquatic plants and phyto-plankton.

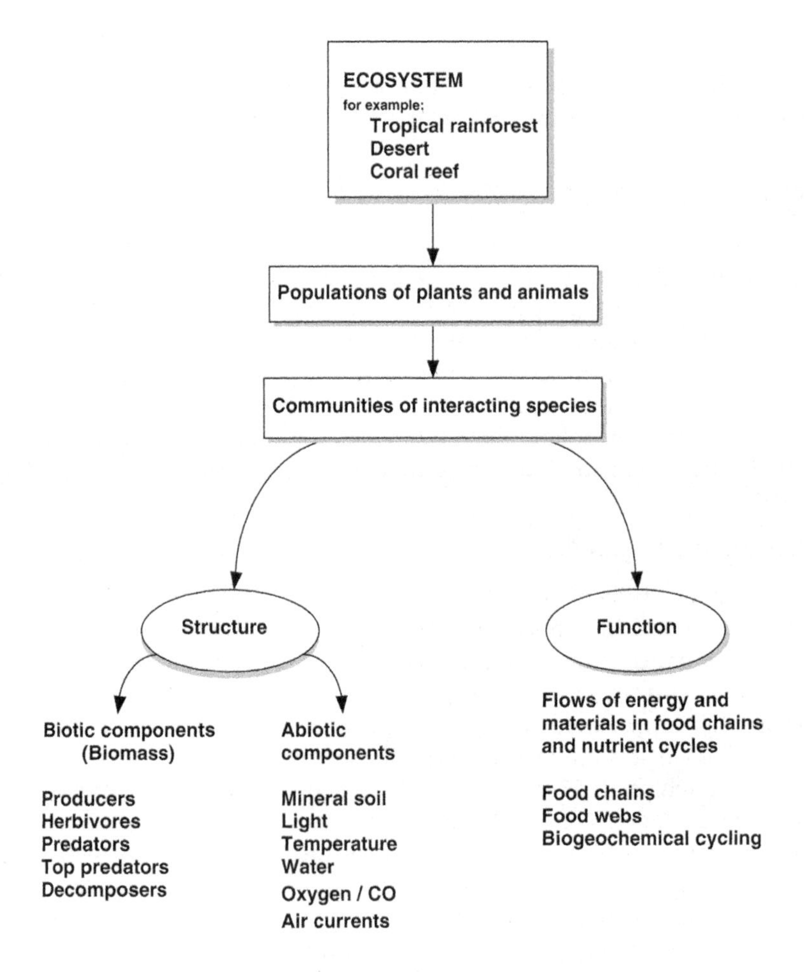

Figure 2.2 Components of an ecosystem

The communities of plants and animals that make up different ecosystems are often centred around one or several dominant species. In terrestrial ecosystems these are usually the larger plants, such as trees, and they largely determine what other plants and animals comprise the biotic community. The community structure of terrestrial ecosystems can thus be characterised by the predominant form of plants, as in the comparison between forests and grasslands. In aquatic ecosystems the massive growth forms of kelp beds and coral reefs help to define community structure. Phytoplankton, however, which form the basis of most aquatic ecosystem food chains, do not accumulate obvious massive structures as the vascular plants do on land (see Figure 2.3).

Table 2.1 The major types of ecosystem

Ecosystem	*Example of nature-based tourism destination*
Aquatic	
Ocean	Whale shark tourism, Indian Ocean, Western Australia
Seashores	West Coast National Park, South Africa
Coral reefs	Great Barrier Reef, Australia
Mangroves	Bako National Park, Sarawak, Malaysia
Estuaries	Mawddach Estuary, Wales, UK
Streams and rivers	Iguazu Falls National Park, Argentina
Lakes	Lake Naivasha, Kenya
Freshwater marshes	Parc naturel régional de Carmague, France
Terrestrial	
Chaparral/Mediterranean	Eucalypt forests and heathlands of south-west Western Australia
Grassland/savanna	Serengeti National Park, Tanzania
Deserts	Mojave Desert, California, USA
Arctic/alpine tundra	Nordvest-Spitsbergen Nasjonal Park, Norway
Coniferous forest	Rothiemurchus Forest, Scotland, UK
Deciduous forest	Shenandoah National Park, Virginia, USA
Tropical forest	Reserva de la Biosfera del Manu, Peru

The biotic composition of an ecosystem also exists in a state of constant interaction. Organisms are dependent on one another and also competing for abiotic and other biotic resources. This relationship can be defined as the autotrophic–heterotrophic system (Figure 2.3). The autotrophic system consists of the photosynthetic function of plants. This is where light energy is used in the conversion of carbon dioxide and water to carbohydrate and oxygen. Plant biomass is built in this way and converted into animal biomass as plant material is eaten, in the heterotrophic system, through a sequence of herbivory and carnivory. Different ecosystems have different capacities to do this, according to prevailing abiotic conditions. For example, water, light and temperature are usually the major limiting factors; where they are not limiting, as in tropical rainforests, the production of plant biomass is at its greatest relative to other terrestrial ecosystems on Earth.

Abiotic components

The physical and chemical factors that affect the distribution and activity of plants and animals comprise the abiotic structure of ecosystems. At a fundamental level, climate, geology and soils determine the different ecosystems around the world. These major physical environmental conditions significantly determine the kinds of

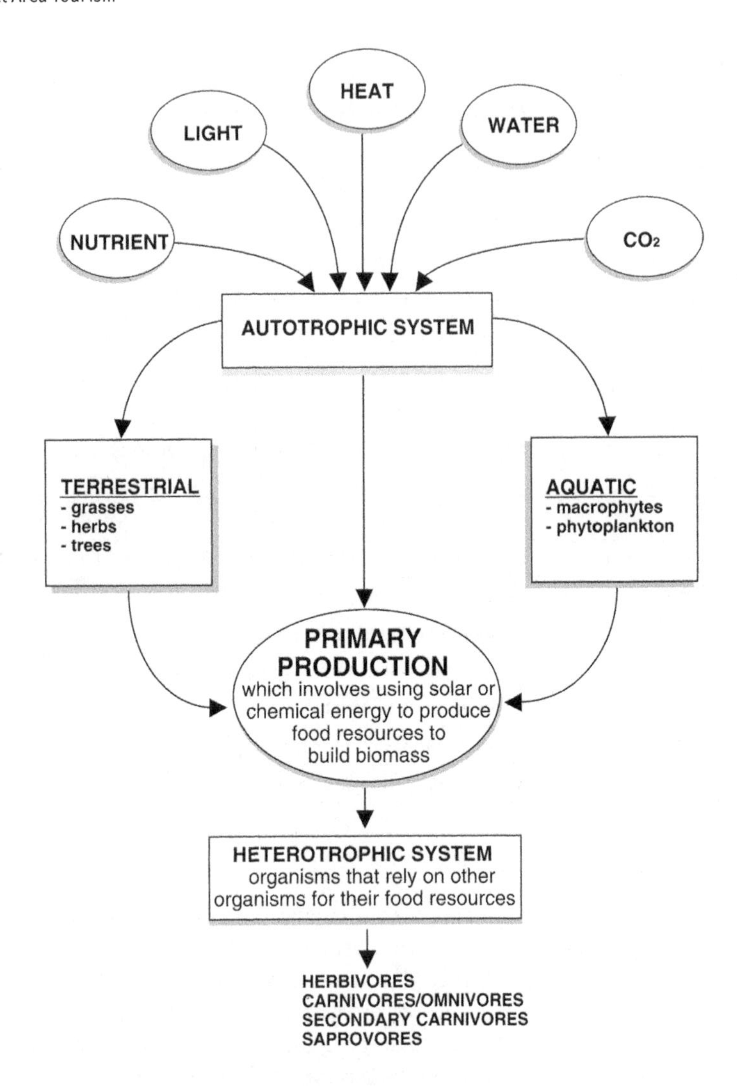

Figure 2.3 Interaction of the biotic components in ecosystems

organisms that are present and the degree to which they are organised into communities. Factors of latitude and altitude have resulted in a series of gradients between ecosystems around the world. For example, temperature gradients typify ecosystem change from the arctic to the tropics. Similarly, moisture gradients determine the structure of vegetation within latitudinally determined climatic zones. The tropical climatic zone, for example, contains a range of vegetation formations that are determined by the amount of reliable rainfall (Figure 2.4). Mountain ranges also define temperature and moisture gradients, through either altitude or aspect. These

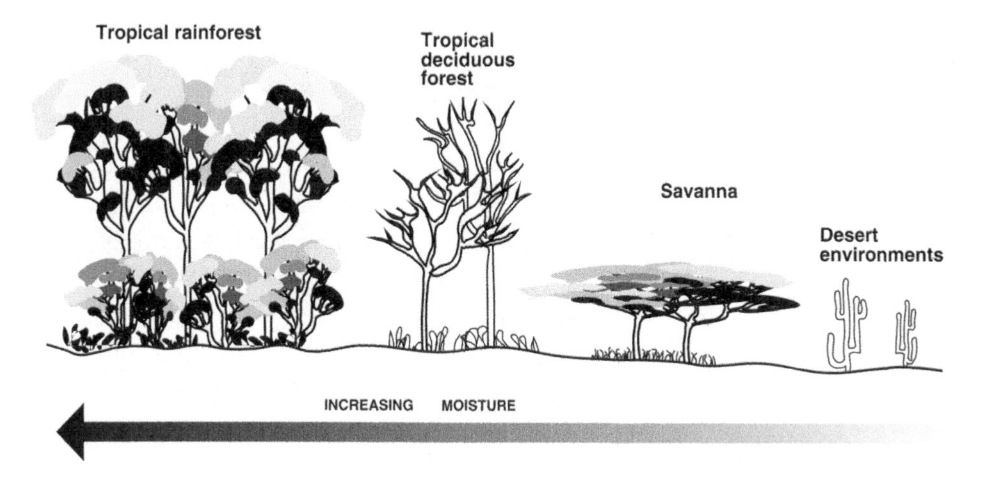

Figure 2.4 Moisture largely determines the development of terrestrial vegetation in the tropical climate zone

changes are often gradual. Sudden changes are, however, evident where aquatic and terrestrial ecosystems meet at an interacting edge, known as an ecotone.

Ecosystem structure is thus made up of the biological community and the varying quantities and distribution of abiotic materials. It was noted above that some of these materials, for example water, act as limiting factors. Accordingly, the capacity of an ecosystem to produce biomass is primarily dependent on temperature, light conditions, water, supplies of nutrients and the efficiency with which energy and materials are circulated (Figures 2.3 and 2.5). For example, in cold environments mean annual temperature, duration of snow cover, waterlogging in summer, wind speeds and soil rooting depth all act as a control over plant growth.

In aquatic ecosystems temperature and the availability of light and nutrients play a pivotal role in plankton biomass production. To this end, the amount of suspended solid material, water currents and the level of dissolved oxygen and salts can also influence aquatic ecosystem biota. The interaction of living organisms with these and other abiotic components constitutes in large the second major ecosystem property, that of ecosystem function.

Ecosystem function

The necessary components for ecosystem function are abiotic factors, such as light, temperature, water, various minerals, oxygen and carbon dioxide, and biotic factors, which comprise the living components. The sun, being the ultimate source of heat and light, is the driving force for all ecosystem function on Earth. While it is possible to see many of the structural elements of an ecosystem, the processes of material and energy transfer cannot be directly observed. Ecosystem function

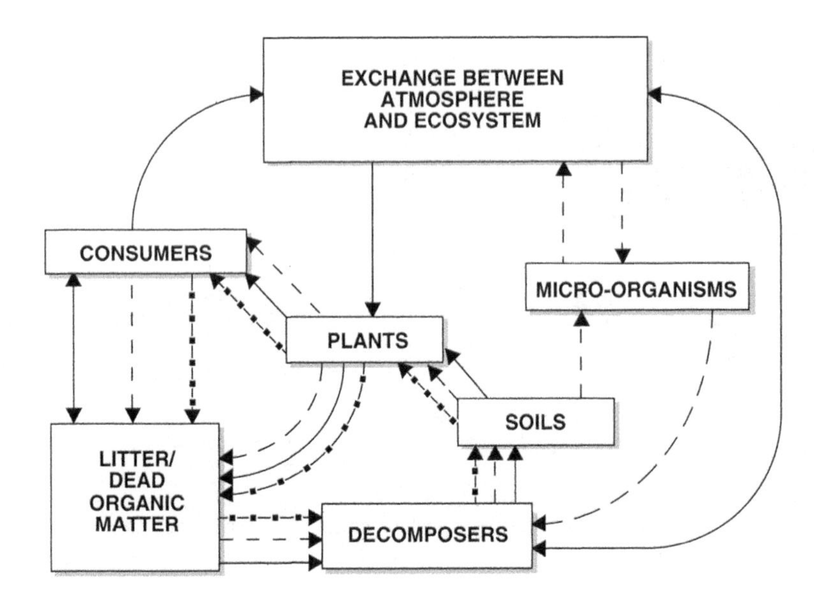

Figure 2.5 Generalised models of nutrient cycling (modified from Etherington, 1975; and Krebs, 1985). *Key:* —▸ Global cycle of C, O and H. ---▸ Global nitrogen cycle. -·-▸ Local-scale nutrient cycle (P, K, Ca, Mg and micronutrients) occurring at the ecosystem level, for example within a forest

operates through processes of nutrient and energy transfer. The transfer of organic matter, for example, occurs through biotic intake, growth, excretion, reproduction and death. Nutrients, any material that an organism needs (e.g. carbon, hydrogen, oxygen, potassium, nitrogen, phosphorus and sulphur), move through the biota in this way (Figure 2.5).

Flow of energy through ecosystems

Energy flows through all ecosystems but the amounts 'captured' by plants and other autotrophic organisms varies greatly between different ecosystems. Energy is then transferred from one organism to the next (e.g. plant–caterpillar–bird–fox) through food chains. There is a loss of heat at each stage in the process as energy is shifted from one level (trophic level) to the next (Figure 2.6). Because of physiological respiration and maintenance, a large quantity of energy at any trophic level is not passed to the next level. The end result is that less energy becomes incorporated into biota at the next level and an ecological pyramid is formed (Figure 2.6).

Looking at this another way, ecological pyramids are also formed because the primary producers outweigh the organisms that consume them. This is an essential relationship, in that populations of organisms that are dependent on the primary

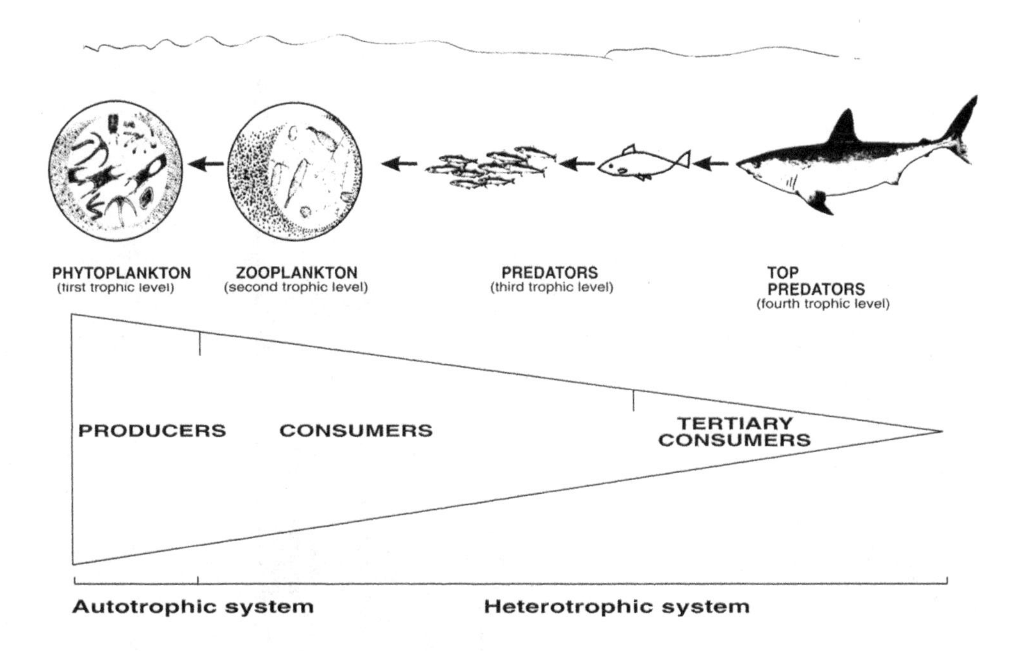

Figure 2.6 Simplified marine ecosystem food chain and ecological pyramid

producers do not generally exhaust the food supply that supports them. Thus, there is always more energy available at the trophic level that supports the next (upper) trophic level. Herbivores that rely on primary producers, such as algae and plants, therefore must in turn outweigh the carnivorous species that are supported by them. This relationship explains why ecosystems are structured with a lesser mass of consumers (herbivores and carnivores) than producers (vegetation/algae). At the top of the food chain carnivores are naturally uncommon, with secondary carnivores occupying the apex of the pyramid (Figure 2.6). The ecological pyramid is made up of a number of food chains, which are frequently interconnected into a food web (Figure 2.7).

The universal food chain model consists of autotrophs (first trophic level), herbivores (second trophic level) and carnivores (higher trophic levels). It is thus through the process of hebivory and predation that energy moves along a food chain. Energy and materials are then eventually passed into the decomposer part of the ecosystem via death and excretion.

Ecosystems are thus essentially maintained by the shifting of energy and nutrients through them. Figure 2.8, a conceptual diagram of ecosystem function, encapsulates how ecosystems work. However, while energy is continuously being lost as heat, nutrients can cycle indefinitely through an ecosystem.

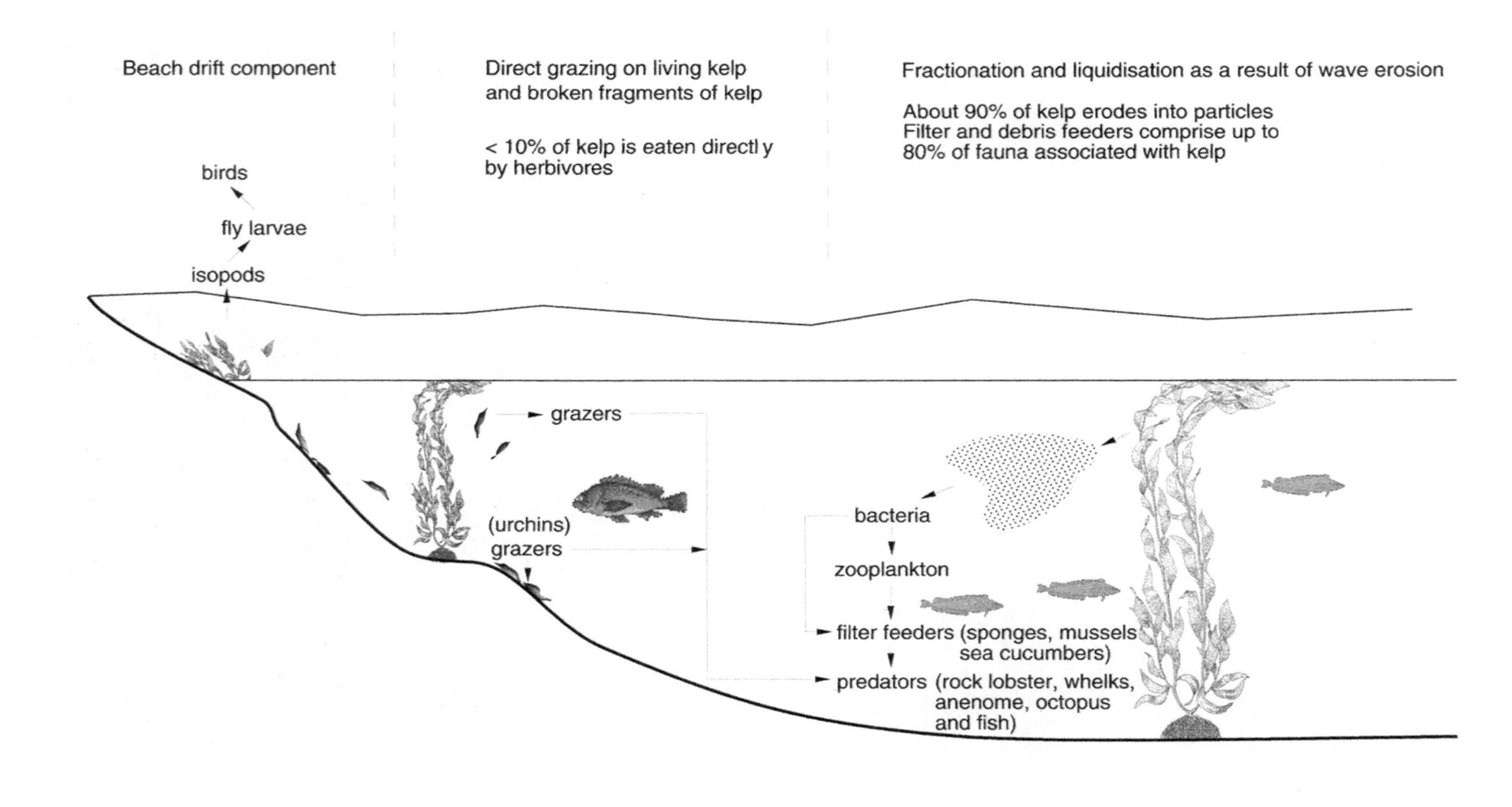

Figure 2.7 Example of a food web: Kelp in marine ecosystems (Source of data: Two Oceans Aquarium, Cape Town, South Africa)

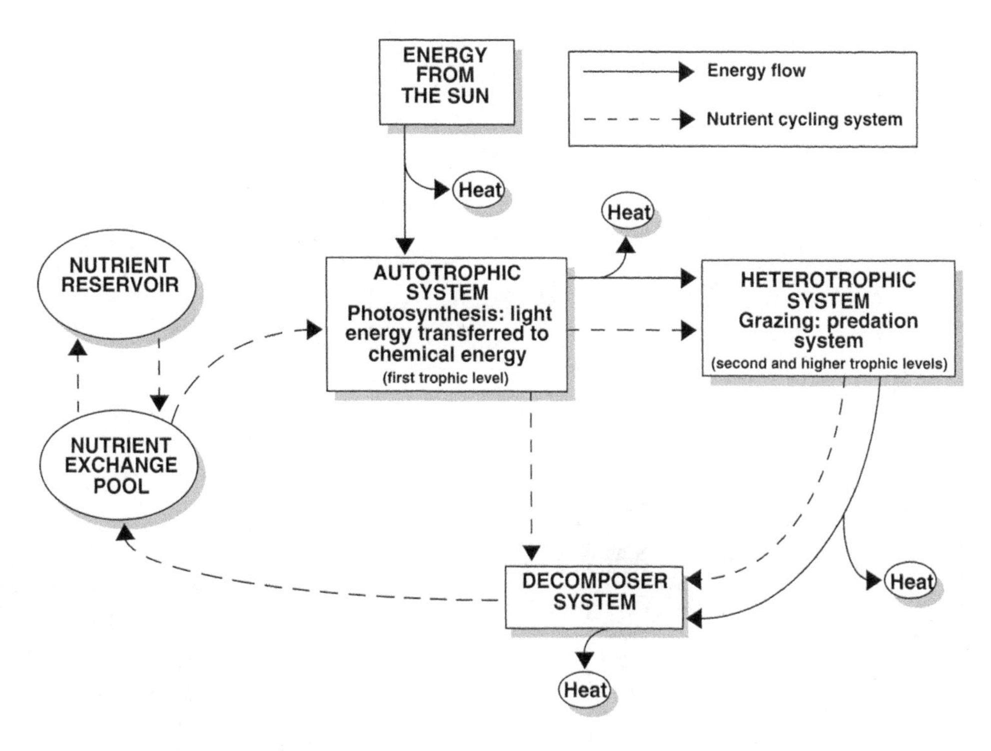

Figure 2.8 Conceptual model of ecosystem processes

Biogeochemical cycles

Nutrients are used to construct the bodies of living organisms and are the chemical elements and compounds essential to life. They occur in three categories – major, macro- and micronutrients – depending on the extent to which they are incorporated into the tissues of plants and animals.

A number of important nutrient elements cycle between the abiotic and biotic components of an ecosystem, in what are termed biogeochemical cycles. In contrast, nutrients such as carbon, hydrogen, nitrogen, phosphorus and calcium tend to be retained or recycled within the biota (Figure 2.5). Biogeochemical cycles operate at a global scale and at smaller scales within the various biomes. The forest biome, which includes a number of types of ecosystem, can even contain nutrient cycles which operate at the local level of a river drainage basin.

Nitrogen is a particularly important nutrient as it is required by all organisms to manufacture proteins and the functioning of many ecosystems is limited by its availability. The reservoir for nitrogen is the atmosphere and in biota, but the main exchange pool is the soil. Soil bacteria determine the form that nitrogen is in and its concentration in the soil. Nitrogen enters land plants via the soil and decomposer part of terrestrial ecosystems (Figure 2.5). Animals obtain nitrogen by eating plants

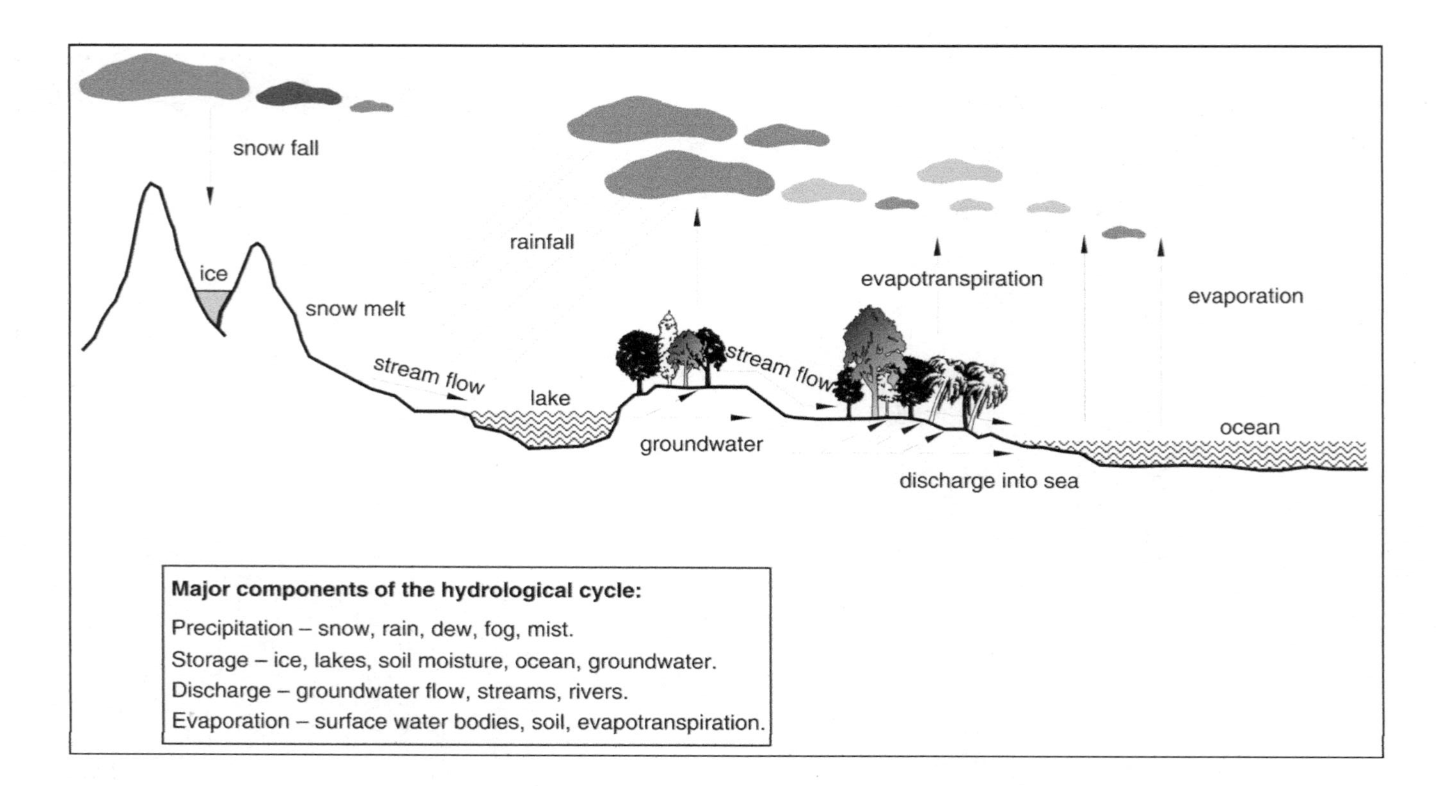

Figure 2.9 Conceptual diagram of the hydrological cycle (local and global scale)

and a subsequent transfer along the food chain. Nitrogen is then returned to the reservoir and exchange pool through excretion and the death of biota and is thus used again and again.

Water in ecosystems

Of all the biogeochemical cycles, the water cycle, or hydrological cycle, is the most important (Figure 2.9). In terms of the global reservoir some 97% of all water is in the oceans, but in terms of nutrient supply most of it is unavailable to terrestrial organisms because of its salt content. The biologically important freshwater component for land organisms is only a very small percentage of the total and its distribution across the Earth is not even.

Water movement through an ecosystem is driven by solar energy, through the processes of precipitation, evaporation and transpiration, the extents of which vary according to climate. Freshwater is stored in the atmosphere, soils and rocks, lakes and rivers and as snow, ice and permafrost. Freshwater in the form of ice or permafrost is 'locked' and not available for circulation. This means that less than 1% of all water on Earth is currently being cycled through the world's terrestrial ecosystems at any one time.

All terrestrial organisms have a minimum water requirement, below which stress, dehydration and death will occur. Living organisms that occupy arid environments have evolved mechanisms to avoid water loss and/or to conserve the water in their bodies. Plants and animals have also become adapted to high-water environments such as lakes and other aquatic ecosystems.

Where there is an excess of water in otherwise relatively dry environments, this can also present problems for organisms, particularly plants. This is because waterlogging leads to oxygen stress in soils and reduces the activity of other biogeochemical cycles. It also relates to the water cycle being closely connected with other cycles. Soil water, for example, contains dissolved nutrients such as nitrogen, phosphorus and magnesium that circulate through living organisms. Plants therefore rely on soil water in order to acquire many nutrients by way of their root systems. Plants also shift water from the soil, which is then moved into the atmosphere in gaseous form through the mechanism of transpiration. This is achieved through the pumping of water from the soil–root system to the atmosphere via the leaves. Figure 2.9 shows the main stores and flows that comprise the global hydrological cycle.

Ecological communities

Community ecology is concerned with how plant and animal species interact with one another. The features of ecological communities that can be affected by tourism development and activity include competition, predator–prey relationships, symbiosis, niche occupancy and keystone species.

Competition

Competition is basically concerned with how the availability of resources, such as the food and space utilised by various organisms, is reduced by other organisms.

Tourism and recreation can result in the transfer of plants and animals to locations where they do not normally occur. In these situations the 'alien' species are often at an advantage, because the new environment is usually devoid of any natural controls that the 'invader' would have evolved with in its original environment. Alien plants compete with indigenous species for space, light, nutrients and water. The introduction of alien plants can result in the disruption and impoverishment of natural plant communities. This has occurred in South Africa, for example, where introduced Australian shrubs have and are degrading species-rich fynbos plant communities in the Southern Cape region.

Predation in ecological communities

The predator–prey relationships of ecological communities need to be understood. A predator is an animal that kills and utilises another animal for food. Predators exert a significant control over herbivores. They also compete with one another and this has resulted in various behavioural and anatomical specialisations among the predators themselves. Predators will usually attack individuals that have been weakened as a result of disease and/or ageing. This pressure has 'strengthened' prey species over time and has also resulted in the evolution of an entire spectrum of anti-predator detection and defence mechanisms, such as toxins, camouflage, regimes of activity, spines and teeth, and flocking and herding for combined protection, as well as various escape responses.

Predators which are deliberately or accidentally introduced outside their normal environments can have enormously detrimental effects. The problem is particularly acute in island settings, where a wide variety of naturally occurring predators and complex food webs are typically absent. Island species are thus particularly susceptible to disturbance from the sudden arrival of a new predatory species. When 'alien' predators, such as dogs and cats, are introduced into different ecosystems, such as islands, their impact can result in a severe reduction in or even extinction of native species. This is because, in the absence of a co-evolved predation pressure, isolated and endemic island faunas have not evolved evasion and escape mechanisms. Island faunas then become easy prey for such introduced predators. This has happened on the Australian continent, where introduced cats and foxes pose a severe threat to existing populations of native marsupials in national parks and nature reserves.

Symbiotic relationships

Symbiosis is another essential feature of ecological communities and is defined as a close association between two different organisms. Symbiotic relationships are widespread. Parasitism is one type of symbiotic relationship where one organism obtains its nutrition at the expense of a host. There are many types of parasites that live on or in another species. Perhaps the most significant from a tourism point of view is the potential for humans to transmit viruses and bacteria to animals that they visit in natural settings. An example of this is the sensitivity of gorillas to human diseases. Accordingly this issue is considered in more detail in Chapter 6.

Commensalism, on the other hand, is where one organism benefits from its association with another, such as where sea anemones provide protection for clown fish or where epiphytic orchids are attached to rainforest trees. In these relationships, however, neither the sea anemone nor the trees appear to receive any particular benefit from the relationship.

Mutualism is where both species benefit, as in the case of the relationship between coral polyps and symbiotic algae. If this relationship is broken, the coral will die, as evidenced when coral is stressed due to coastal development, pollution or increases in ocean temperature beyond optimal conditions. In terrestrial ecosystems an intimate association between plant roots and bacteria or fungi allows many species of plant to obtain additional nutrients under conditions of plant root competition, nutrient poverty and hostile substrates. Any disruption of this relationship will reduce a particular plant's capacity to obtain essential nutrients such as nitrogen and phosphorus.

Mutualism is also widespread in terms of plant–insect relationships. There are many examples of this, for example the insect-specific pollination of orchids and the widespread close plant–insect pollination systems in tropical rainforests. These features mean that a number of specialist, as opposed to generalist, species can be present in any particular ecosystem.

Specialist and generalist species

Specialist species are particularly efficient in utilising their resources but in doing so have become specialised and adapted to a particular way of life. Such species are said to occupy a narrow niche (Box 2.1). An example of this, as mentioned above, would be the specialisation of fauna to specific feeding strategies in rainforest environments. Examples include narrow-spectrum frugivorous (fruit-eating) birds, specific flower-adapted hummingbirds and insects which are plant-specific herbivores. Such species are vulnerable to disturbance because of the tight relationship these animals have with their food supply. If the food supply situation changes then the animal, because of its specialisation, has no alternative source of food. It may then die out because its food supply has been lost, or it may perish in the face of competition for food from other species. Generalist species, on the other hand, are more adaptable to such changes because they are not restricted to a particular food source.

Box 2.1 Relation of organisms to their environment: Habitat and niche

Habitat is the geographically located physical environment in which a species occurs. Occurring at various scales, habitats can be restricted areas such as a pool in a cave or within a bromeliad perched high in the tree canopy of a Costa Rican forest. Conversely, habitats can occur at much larger scales, as in soil, and be distributed at the continental level. The world's coral reefs, for example, are habitat to a wide array of marine organisms.

Niche refers to the role that a species has in the community (the food it eats, its position in the food chain and the ways it competes for resources), in combination with how it interacts with its environment. In this sense, each species has its own specific set of functions – its niche. Although niches might appear to overlap, no two species will occupy the same ecological niche.

Niche and habitat

Distinct abiotic and biotic needs and specific roles in the ecological community define the niche of an organism. What these conditions amount to is how an organism lives and interacts with other species (Box 2.1). For example, up to six different species of monkey can coexist in the West African rainforest because they all have a different niche (Mader, 1991). Mader points out that such diversity exists because the monkeys occupy different levels in the forest and have different diets. The red colobus (*Colobus badius*), for example, lives in the forest canopy and feeds on leaves, flower buds and some insects. By contrast L'Hoest's monkey (*Cercopithecus lhoesti*) lives mostly towards the forest floor and feeds on fruits, herbs, mushrooms and insects. In this way, by occupying slightly different habitats, these species avoid direct competition for food resources and space. While an organism's habitat is where a species can be found and involves some physical aspect of the ecosystem (Box 2.1), the concept of niche also embraces feeding and breeding activity and the way the organism itself influences abiotic and other biotic components of the ecosystem.

Dominant and keystone species

A dominant species helps to determine the nature of the ecological community and its removal would result in a change in the community. In terrestrial ecosystems the dominants are trees and shrubs but animals, for example some herbivores and carnivores, can also exert a dominant influence in communities. Trees are dominant in forest ecosystems, in that they create microclimates and provide food and habitats for a range of other organisms. If a dominant species declines or is lost, it is possible for another to take its place. An example of this is in the situation of co-dominant trees or grasses. If the dominant one is, however, a keystone species, then the entire ecological community changes if this single species is lost. Keystone species are plants or animals that exert an important controlling influence in the ecosystem. This is achieved by their presence determining the structure and/or composition of the community of species. The ecological importance of keystone species is thus much greater than would be expected simply from the organism's overall abundance in the ecosystem.

Figure 2.10 illustrates this relationship in response to the reintroduction of the grey wolf (*Canis lupus*), a keystone predator, to Yellowstone National Park in the USA. The wolf was formerly lost from Yellowstone primarily due to hunting pressure. The loss of this top carnivore resulted in a shift in the composition of the animal community. Smith *et al.* (2003) reported that 31 grey wolves were introduced to Yellowstone in 1995–96 and in 2002 the population was estimated to be around 216. Elk (*Cervus canadensis*) is the wolf's primary prey and the predation and displacement of elk from stands of densely wooded areas has favoured the recovery of aspen (*Populus tremuloides*) and willow (*Salix spp.*) communities (Ripple & Beschta, 2007). The decline of coyote (*Canis latrans*) associated with wolf predation has a potential positive effect in the numbers of scavengers (especially raven, *Corvus corax*) that use wolf kills. A reduction in herbivore pressure has led to the recovery and growth of willow communities, especially in relation to increases in stand stature, a situation

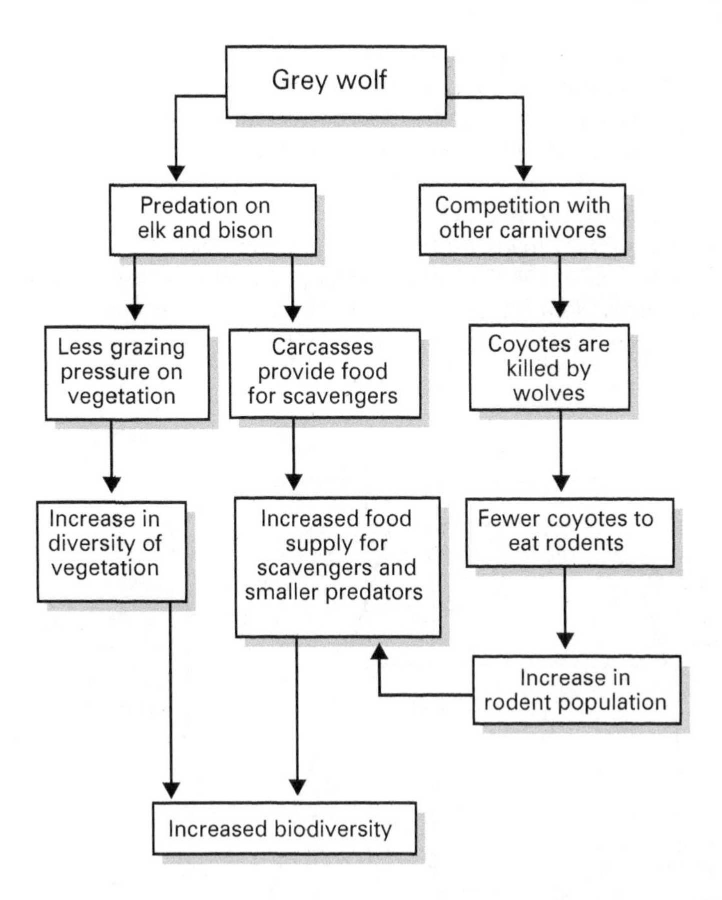

Figure 2.10 The potential role of the grey wolf as a keystone predator in Yellowstone National Park, USA

that favours beavers (*Castor canadensis*) and associated bird communities. Ongoing studies aim at predicting the impact of wolves on the biodiversity of Yellowstone National Park. Research is directed at the role that wolves are playing with regard to modification of biotic communities (predation of herbivores and competition with other predators) in a complex ecosystem subject to natural perturbations (severe winters, fire, disease) and within the broader influences of climate change (Smith *et al.*, 2003). Moreover, Kiernan (2011) reports that wolves are now one of the biggest attractions in Yellowstone, with tourists seeking sightings and listening to their calling, as well as learning about their role in the ecosystem.

An animal can also have a keystone role as a result of it physically modifying its environment. Naiman (1988) reported on how the disturbance of woody plants by increased numbers of African elephant (*Loxodonta africana*) resulted in a large reduction in trees and shrubs, which in turn affected the food supply and populations of other animals, leading to an alteration in local biogeochemical cycling. Where a population of elephants is reduced, trees and shrubs are able to re-establish in grasslands that, in the absence of elephants, would contain more woody species. Major structural components of the entire ecosystem are therefore dependent on the density of elephants (Figure 2.11).

Similarly, the holes made by the red-naped sapsucker (*Sphyrapicus nuchalis* – a type of woodpecker) in the USA provide nesting habitat for two species of swallow and food resources (sap) for a suite of other species. These species are in turn food resources for various predators (Daily *et al.*, 1993).

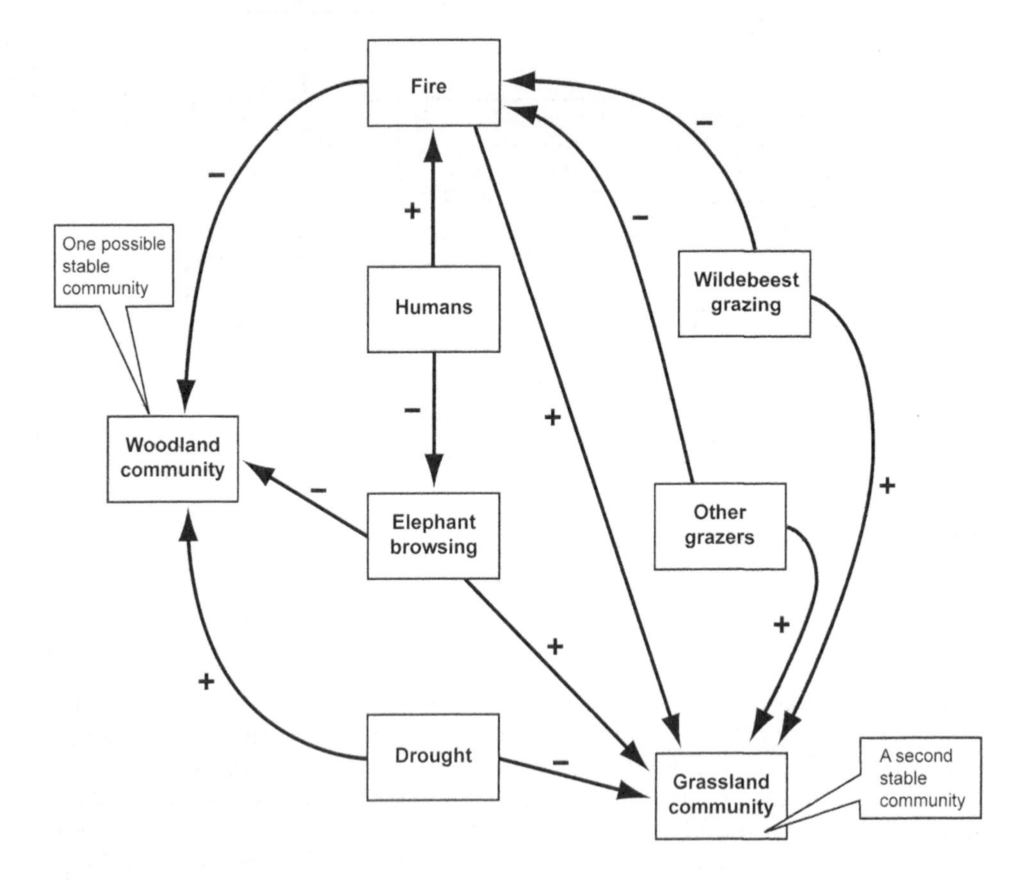

Figure 2.11 Factors affecting the establishment of woodland or grassland communities in the Serengeti–Mara region of East Africa. This community exists in two alternative stable community states, woodland or grassland. Trees are killed by elephants, humans and fire, which promotes the development of grasslands. Wildebeest and other grazers remove biomass, reducing the risk of fire. Shallow, impermeable soils become waterlogged and do not favour the growth of trees and deep-rooted trees are able to survive drought. (Derived from Krebs, 2008)

Disturbance and succession in ecosystems

Disturbance is the alteration of ecosystem structure and/or function arising from the loss of biotic components as might be caused by human activities such as clearing land, removing vegetation and accelerated erosion. Severe natural disturbances – for example wild fires, volcanism and severe storms, usually infrequent but a normal part of ecological processes in landscapes – can have a similar effect. Succession is the process of ecosystem recovery following such disturbances. Natural area tourism, along with recreational activities and tourism development, has the capacity to change ecosystem structure and function through disturbance (also see Chapter 3).

According to Krebs (2008), communities of plants and animals can occur on a gradient with stability at one end and instability at the other. Those communities that are continually subject to disturbances, where disturbance intervals are shorter than recovery time, are said to represent non-equilibrium conditions. Communities are often heterogeneous and can represent a spectrum of states, from equilibrium to non-equilibrium. Disturbance can lead to patchiness. Patchiness is an inherent property of natural communities. A stable condition may be prevented by physical factors such as temperature (climate change), salinity (land clearing) or fire (climate change, fire management practices and/or recreational activity). Some communities may exist in multiple stable states and these may be confused with non-equilibrium assemblages. If a community is disturbed sufficiently, it may change to a new configuration, at which it will remain even when the disturbance has stopped (see Figure 2.11).

Change in ecosystems

Plant cover is fundamental to the structure of most terrestrial ecosystems and when it is damaged or removed the assembly of the community of plants and dependent animals is changed to varying degrees. Loss of plant cover can also result in associated abiotic changes, such as the loss of surface soils or an alteration in the physical and chemical properties of soils. Changes in ecosystem function as a result of disturbance can manifest as increased or reduced water availability and disruption to the normal pattern of nutrient cycling. The energy flux and microclimate of a particular site are also altered as a result of removing vegetation.

Human disturbance plays a pivotal role in the creation of patches and corridors and thus landscape heterogeneity (see Figure 2.12, p. 62). The edges of patches are more prone to wind acceleration, erosion, trampling and grazing, and an increased frequency of disturbance results in the presence of generalist plants that can tolerate these conditions. An interesting aspect of edge habitats is that they can act as a sink for wildlife. An example of this is where herbivores preferentially graze and browse edge vegetation and this in turn attracts predators. The 'open' nature of edges therefore means that they often provide good locations for people who wish to view wildlife. Edges can also function as barriers where dense and spiny vegetation restricts access and thus reduces any source/conduit function arising from disturbance corridors.

Roads, tracks and hiking trails are corridors that are usually maintained in an altered state by continuous or repeated disturbance. Roads in particular show a central area of major disturbance with an outer edge composed of generalist species and usually an assemblage of exotic weeds. These road verge floras are subject to the influence of the conduit function of the road and tend to be more resilient to the additional wind, heat, particulate and pollutant imports from adjacent ecosystems. Where tracks are subject to less frequent disturbance, successional processes may take place.

Recovery from disturbance: Succession

Ecological succession is where plant and animal communities change through time; these changes comprise a recovery response following disturbance. Succession involves the replacement of species and changes in the availability of abiotic resources. The colonisation of disturbed ground, for example by pioneer species, contributes organic matter and also allows soil micro-biota to re-establish. As the vegetation community develops, colonisation by larger plants changes the microclimate by providing shade and protection of the soil surface. Any disruption to nutrient cycling that has occurred can be restored in this way. Most succession involves the recovery of sites that previously supported vegetation and this process is termed secondary succession. In contrast, primary successions are where vegetation gradually develops on a new land surface, as in the case of recent glacial or volcanic deposits.

The way the succession develops (the successional pathway) is dependent on the intensity, timing and frequency of disturbance, the area disturbed and the original state of the environment. The intensity of disturbance relates to the amount of biomass of a particular species or population that can be damaged or eliminated from an ecological community. Disturbance intensity therefore sets the baseline condition for succession. For example, in Australian fire-prone ecosystems, vegetation can burn at very low intensities and only the ground cover and shrub layer are affected. A particularly intense fire, however, is more damaging to the vegetation and can reach into the forest canopy and incinerate all of the foliage. These two fire scenarios therefore have different implications for forest recovery.

The capacity for total recovery

The frequency at which disturbance can occur is also variable and has a bearing on how much time an ecosystem has to recover before the next disturbance occurs. An ecosystem that is subject to a high frequency of disturbance is likely to contain species that do not need a long time to mature and that are able to reproduce at a young age. Species that require a long time to develop, mature and reproduce (e.g. many trees, large birds and a number of large mammals) will not be able to maintain occupancy under high-frequency disturbance regimes.

The actual timing of a disturbance can be critical in influencing which species are impacted or susceptible to it, as in the case of disturbances occurring during the breeding season of seabirds which become concentrated at certain locations.

The area of disturbance is also significant in terms of recolonisation of the site. When a natural ecosystem is disturbed, recovery can take place from the re-sprouting of underground parts of plants, dormant soil microbes re-establishing themselves and the arrival of propagules (seeds and spores) from adjacent areas. The mobility of propagules and proximity to recolonisation sources exert control over the nature and speed of succession. Seeds and spores are more efficient in reaching a disturbed area if they are sourced in the immediate vicinity of the disturbance. Where natural areas are isolated, as in a matrix of agricultural land, then new sources of native seed may be limited. This situation is especially critical when an entire species is impacted by constant disturbance and where exotic weed species are widespread in the matrix.

The successional pathway can result in: a return to the original ecological community; a structure that is only part of the original; or a completely new community. The endpoint is strongly dependent on the previously considered factors and the original state of the environment. At the same time, the process of disturbance and succession can result in species replacement, changes in the population structure of component organisms and an alteration in abiotic factors.

As already noted, the invasion of exotic species can result in competition for available resources and help to eliminate those species that are sensitive to disturbance. Changes in abiotic factors, for example when soil is eroded, can affect plant productivity and nutrient cycling. A significant change in the ability of the abiotic environment to support vegetation can lead to loss of biodiversity and a reduced capacity for ecosystem function.

Disturbance and succession, however, are features of all ecosystems and each system, whether it is a forest, savanna, swamp, rocky shore or coral reef, will have its own response and sequences of successional development. In addition to this, some ecological communities exhibit more resilience (capacity to recover) to different disturbance regimes than others. Furthermore, Rapport *et al.* (1998) assert that resilience is the capacity of an ecosystem to maintain structure and function in the presence of stresses with which it has evolved and is therefore viewed as an important indicator of ecosystem health.

Landscape ecology

Landscapes are mosaics of different ecosystems and land uses. The landscape mosaic can be conveniently divided further into patches, corridors and matrix (Figure 2.12). The major issue in appreciating such a structural arrangement in landscapes is that there are boundaries to these various elements and these elements may be linked by flows of water, particulates, pollutants and organisms.

Patches in the landscape

Patches consist of uniform, non-linear areas of vegetation that differ from the surrounding landscape. Forman (1995) recognises five different types of vegetation patch, all of which clearly can vary in size and general shape. He sees vegetation patches comprising the following:

(1) disturbance patches, which usually occupy small areas within a large area of natural vegetation;
(2) remnant patches, which occur within a disturbed (e.g. agricultural) matrix;
(3) environmental patches, where rock type or soils differentiate a patch;

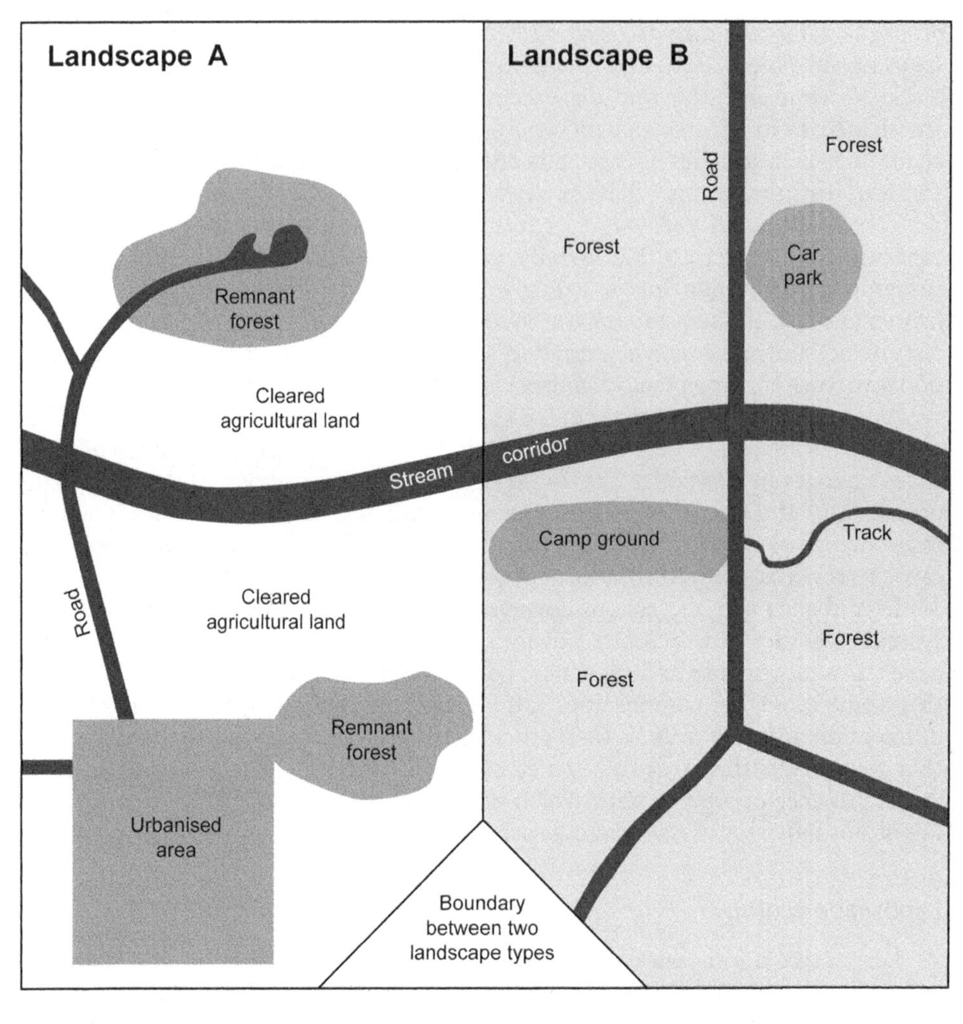

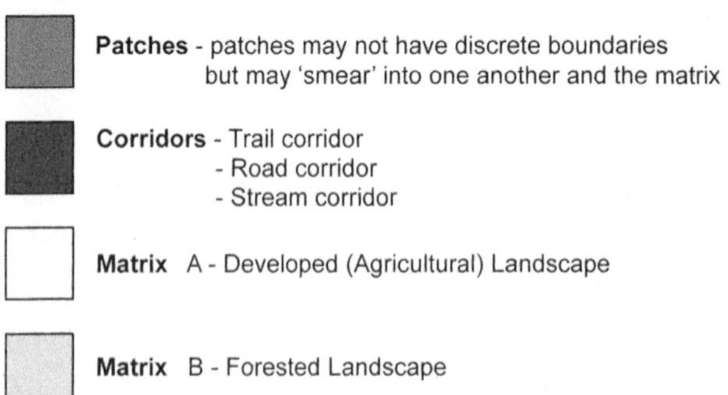

Figure 2.12 A comparison of patches and corridors in the landscape matrix. (A) Agricultural matrix; (B) natural forest cover matrix

(4) regenerated patches, which reflect regrowth following disturbance;
(5) introduced patches, which are generally urban, cultivated areas or plantations.

These vegetation patches change according to successional processes.

Large patches of natural vegetation are often the core habitat for important species of tourist interest. Patches of vegetation that function in this way include remnant forest in Indonesia, such as the Pangrango-Gede and Ujung Kulong National Parks in Java (see Figure 2.13). Such large patches thus provide important landscape functions for tourism. This function is also achieved indirectly, for example by protecting soils with stream networks from unsightly erosion and sedimentation, as well as reducing the risk of flooding. In this way large patches act as a buffer against wider ecological degradation problems and help to promote wider sustainable land uses. At the same time, small patches of natural vegetation, like Bukit Timah Nature Reserve in Singapore (Figure 2.13), additionally play a role in the provision of habitats for plants and animals. This is especially so when they are numerous and connected to one another by corridors of remnant vegetation.

Corridors in the landscape

Corridors can be viewed as differing from and permeating the landscapes in which they occur; that is, they are strips that differ from the surrounding landscape (Figure 2.12). They vary in their width, length and in the nature and degree of activity that occurs along them. From a tourism perspective they are often key habitats, as in river systems, and sites where recreation can take place. Corridors can both separate and connect patches in the landscape. A disturbance corridor such as a road can act as a filter preventing organisms moving from one patch to another. Conversely, the same road could act as a conduit for and source of organisms, by allowing movement along the corridor and even dispersal into the matrix. The efficiency with which organisms can move from patch to patch will depend on the width of the corridor, as in the case of remnant vegetation, and its connectivity in the landscape. The more connected corridors are in the landscape, the greater the conduit and source function is likely to be.

Corridors can be strips of vegetation, roads, walking trails and rivers; they are of different size and width and show variable degrees of connectivity. Forman (1995) also notes that corridors can change diurnally, as in the day/night movements of animals, according to seasonal cycles and in response to successional processes.

Ecological significance of disturbance corridors in landscapes

Forman (1995) provides a comprehensive overview of the role of corridors in the landscape. Five types of corridor are recognised: environmental (e.g. river with riparian vegetation); disturbance (e.g. trails and roads); remnant (e.g. strips of woodland/natural road reserves); regenerated (e.g. European hedgerows) and introduced (e.g. windbreaks). Of particular significance in natural area tourism are disturbance corridors such as roads, paths and trails. From a tourism perspective these corridors serve as a means of access to interesting sites and wildlife, as well as being a focus of recreational activity. Some natural corridors, such as streams and rivers, are

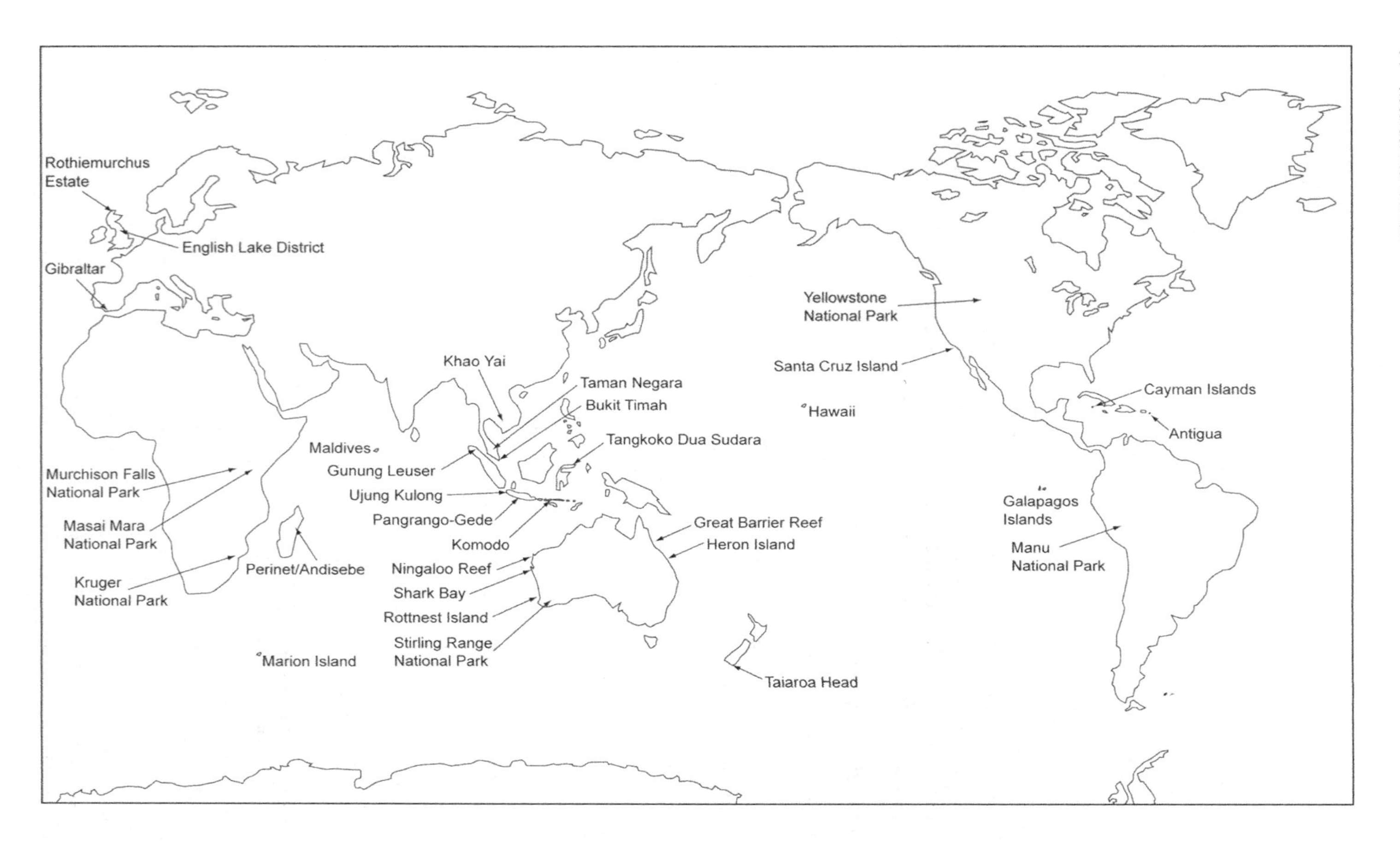

Figure 2.13 Location map of important nature-based tourism destinations referred to in this chapter

tourism resources in their own right and strips of natural vegetation along roadsides in Western Australia have become a significant wildflower tourism resource.

Because a large proportion of tourism centres on corridors, it is important to understand the ecological functions of corridors. Five such functions are recognised: habitats, conduits, filters, sources and sinks (Forman, 1995). Disturbance corridors (roads, tracks and pathways) will provide a useful focus here, as it is where many tourism and recreational impacts occur. Disturbance corridor habitats mostly comprise disturbance-tolerant, edge and invasive species such as weeds. These species of course perform an ecological function in their own right, in being able to tolerate the environmental conditions brought about by a change in the original habitat. Exposed soils can be stabilised in this way and these species can pave the way for the invasion of other species. If, however, the disturbance is continuous, as it normally is in these corridors, then a permanent community of disturbance specialists often persists.

Road corridors

In the case of roads and trails, the open areas provide a conduit for the transfer of energy and wind. Plants can enter in the form of seeds and animals also use these corridors as a means of moving through the landscape. Disturbance corridors thus act as a source by which plants, animals and tourists can spread into a reserve or wilderness area (Box 2.2, Figure 2.14). At the same time, a network of roads and/ or trails can also act as a filter, in that organisms can become separated into patches within a particular natural area. Propagules, sediments and animal life can accumulate in environmental corridors; in the case of roads (a major disturbance corridor), the corridor can become a sink if predators and scavengers are attracted to road kill.

The role of roads as disturbance corridors is in fact highly significant. Figure 2.15 illustrates this according to the five ecological functions of corridors mentioned above. By comparison, tracks and pathways, because of a general lack of vehicles and softer surface, tend to be less noisy, less polluted and less well defined. They can, however, like roads, be a significant source of ecological impact in the form of soil erosion, vegetation damage, weed invasion and disturbance to wildlife. The degree and nature of these impacts are strongly connected with intensity of usage, extent and location of the trails and their management.

River and stream corridors

River and stream corridors are particularly important because of their tourism associations and hydrological flow function. Water is a 'magnet' for tourists and recreational and other human activities have the potential to alter water quality. Polluted water can move easily along stream and river corridors from one ecosystem to the next (see Figure 2.12). Additionally, particulates, such as eroded soil sourced in the matrix, can enter, accumulate and be transported in stream corridors. River corridors also function as habitats for a wide variety of specialised and generalist species. Generalists often occupy the edges of streams and rivers where natural disturbance events take place due to fluctuations in water levels. Stream and river corridors also act as conduits for a wide spectrum of plants and animals.

Box 2.2 Case study: The spread of an introduced pathogen through tourism corridors in the Stirling Range National Park, Western Australia

This case study highlights the significance of ecological linkages and demonstrates the potential ecological damage that can be brought about by apparently benign tourism activities. The Stirling Range National Park (see Figure 2.13) is a large area (115,600 ha) of natural vegetation and a zone of exceptional biodiversity. It contains 1500 plant species, of which 87 are endemic to the park. As with many national parks around the world, the Stirling Range provides for tourism, recreation as well as the conservation of biodiversity. The mountainous scenery and rich array of wildflowers, visible from August to November, make the Stirling Range an increasingly popular tourism destination. Almost all visitors hike in the park, with a significant percentage undertaking self-guided touring by car. Roads for tourist access were initiated around 1920 and since that time further road improvements and the development of walking tracks has taken place (CALM, 1996). A number of mountain peaks provide for hiking, wildflower viewing and the appreciation of scenery.

Active management is necessary to maintain tourism facilities and prevent degradation such as path erosion. The long summer dry season gives rise to a bush-fire hazard and this is managed according to a programme of prescribed burning and firebreak corridors (McCaw & Gillen, 1993). There is a risk of ecological degradation in the form of edge effects from surrounding agricultural land. This includes increased fire hazard, weed invasion and the immigration of feral animals from adjacent farmland. Roads and tracks in the park are now acting as habitats for weeds and conduits and sinks for introduced predators such as the European fox (*Vulpes vulpes*). The single most serious threat to the integrity of the Stirling Range ecosystem, however, is the introduced pathogen *Phytophthora cinnamomi* (CALM, 1992a; Wills, 1992; Newsome, 2003). The pathogen is considered to have been introduced around 1960, when a number of roads, footpaths and fire management tracks were constructed (Gillen & Watson, 1993; CALM, 1996).

Phytophthora cinnamomi is a soil-borne water mould that enters the roots of susceptible plants. Vegetative reproduction produces sporangia, which then release mobile zoospores, which infect the roots of host plants. It can then spread by root-to-root contact, by soil being moved from one place to another and in water (Shearer, 1994; Shearer *et al.*, 2004). Widely dispersed infection can result from the movement of soil on footwear and the wheels of vehicles. Soil-disturbing activities such as road building and the maintenance and construction of firebreaks would have been the initial mechanism through which the disease spread through the park (Gillen & Watson, 1993; CALM, 1996). The pathogen has also spread along walking trails as a result of infected soil being carried in the boot tread of hikers (Wills & Kinnear, 1993). There is a significant correlation between the distribution of the fungal infection and the more accessible tourist peaks (CALM, 1996). The presence of the disease in upland areas is of particular significance because of the potential for downhill spread of infectious zoospores in water through surface and subsurface run-off (Gillen & Watson, 1993; CALM, 1996).

Infected host plants are killed as a result of impaired root function followed by the death of photosynthetic tissues. A wide range of species, mostly in the plant families Proteaceae, Mytaceae, Papilionaceae and Epacridaceae, are susceptible (Wills, 1992; Shearer, 1994; Shearer *et al.*, 2004). These plant families comprise a very high proportion of the plants found in the Stirling Range. Moreover, the health of the wider Stirling Range ecosystem is under threat from the total effect of *P. cinnamomi* on plant communities. The disease has a major impact on the species-rich under-storey that is an essential component of mallee heath, and an important plant community in the park. Wilson *et al.* (1994) reported that as many as 60% of the component species present can be destroyed following infection by *P. cinnamomi*.

Because the plant family Proteaceae contributes much of the floristic structure to many plant communities in the park, their loss due to *P. cinnamomi* infection will cause a decline in species richness, changes in community structure and a reduction in plant biomass. Such observed and predicted changes to vegetation are illustrated in Figure 2.14. The resultant change in community structure and function degrades the vegetation and its capacity to provide suitable habitats and resources for fauna. Wilson *et al.* (1994) provided an overview of the likely impacts on wildlife. They point out the potential consequences for arboreal mammals and birds that rely on the canopy of dominant species, such as *Banksia spp.*, for food and shelter (Figure 2.14). Widespread infection by *P. cinnamomi* therefore has the capacity to degrade the conservation and tourism value of the Stirling Range National Park.

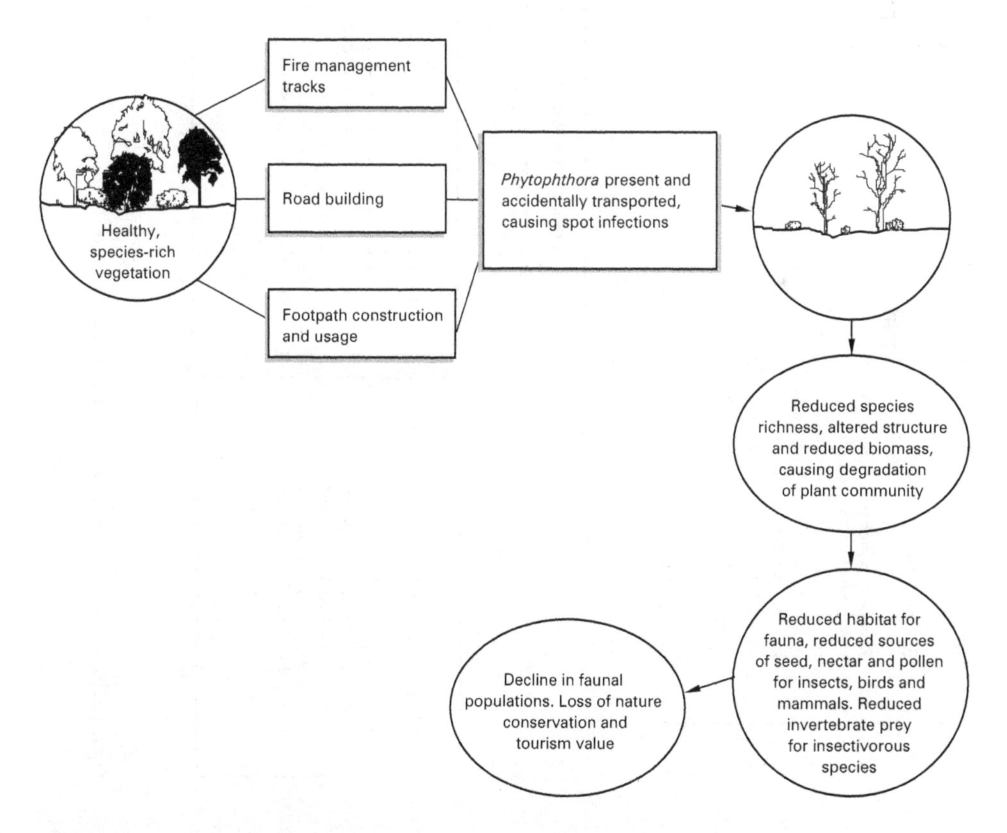

Figure 2.14 Predicted ecological impact caused by *Phytophthora cinnamomi* infection in the Stirling Range National Park, Western Australia. (Adapted from Newsome, 2001)

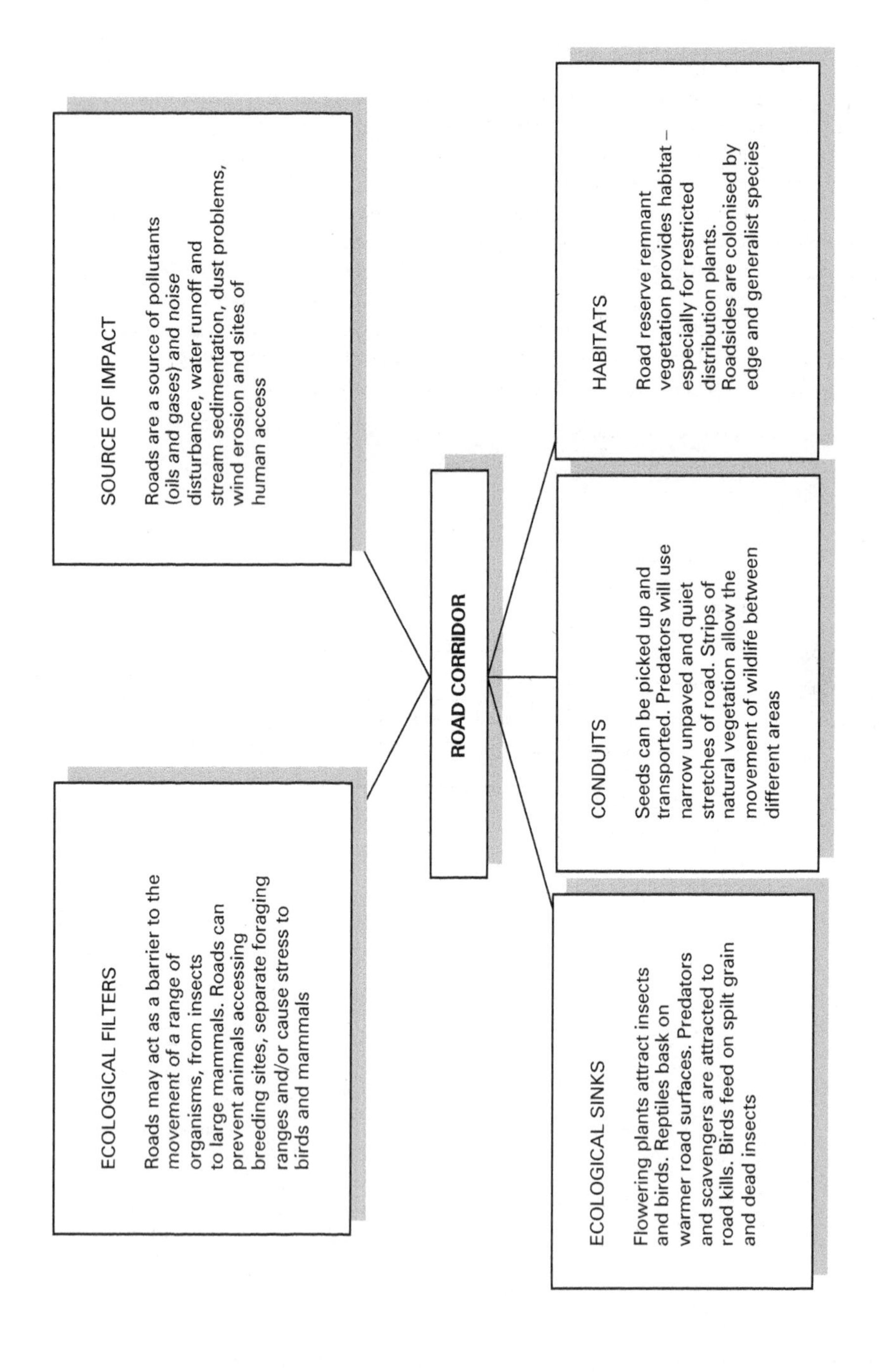

Figure 2.15 Ecological function of roads (After Foreman, 1995)

The landscape matrix

In contrast to discrete patches and corridors, the matrix is the background ecosystem or land use in a landscape mosaic. The important features of the landscape matrix are that it is usually an extensive area containing patches and corridors. The landscape matrix may be a cultural landscape comprising urban and agricultural land in which lie various corridors and patches of vegetation. Conversely, it may consist of forest or some other vegetation type in which patches and corridors occur as a result of changes in soil type or manifest as disturbed areas (Figure 2.12, p. 62).

Energy, materials and species flow between the various components of such a landscape mosaic. In this way different ecosystems can be interlinked by the movement of water, particulates and organisms between them. Exotic species that invade native ecosystems are a particularly significant issue because of their capacity to degrade tourism resources and natural areas (Box 2.2). Such invaders commonly include non-native weeds and feral animals, which profit from human disturbance. Opportunistic species such as non-native pioneer plants (weeds) that are adapted to and respond to disturbance are able to proliferate in the absence of their usual diseases and predators.

Ecological Characteristics and Tourism Activity in Different Types of Ecosystem

This section explores some of the specific ecological characteristics of selected ecosystems from around the world (Table 2.2). As noted previously, because of the large range of environments utilised for recreation and tourism it is impossible to survey all of them in this chapter. Instead, some of the most highly valued ecosystems for tourism are considered here, along with some specific impacts. Polar regions are only briefly considered here, but because islands occur in many contexts (e.g. sub-antarctic through to tropical environments) there is an emphasis on them, as important wildlife tourism destinations. There is a specific focus on tropical rainforests because they have become a very important nature-based tourism resource, both domestically and for international visitors seeking specific wildlife tourism experiences. Details and ecological conditions, impacts and management problems at tourism destinations such as other forest ecosystems, ice- and snow-dominated landscapes, deserts, mountains, lakes and rivers and caves are not covered in this chapter but are, however, periodically explored in the remaining chapters of this book.

Recreation and tourism in the polar regions

Polar regions offer unique tourism experiences in the form of extensive areas of ice and snow and large expanses of tundra in the Arctic, in association with being a critical habitat for cetaceans, pinnipeds and migratory and breeding birds. Large mammal species occurring in the Arctic include polar bear (*Ursus maritimus*), musk ox (*Ovibas moschatus*) and walrus (*Odobenus rosmarus*). Bird populations comprise alcids such as puffins (*Fratercula arctica*), razorbills (*Alco torda*) and other auks.

Table 2.2 Comparison of some important ecological characteristics of selected ecosystems

Selected ecosystem property	*Ecosystem type*			
	Polar environments	*Coral reef*	*Tropical rainforest*	*African savanna*
Major structural components	Extensive areas of snow and ice. Also includes tundra and some mountain environments. Arctic vegetation comprises dwarf trees, low-growing shrubs, grasses, lichens and bryophytes Antarctica characterised by bryophytes, lichens and algae on land	Growth of soft corals Sediments stabilised by algae Coral polyps build calcareous skeletons which accumulate as reef structures	Four major levels of vegetation dominated by large trees	Grassland with scattered trees and shrubs
Aspects of ecosystem function	Cold ocean waters are rich in mineral nutrients supporting phytoplankton and zooplankton as basis of polar food chains Insect biomass supports many species of resident and migratory land-based birds and mammals in the northern tundra ecosystem	Warm ocean waters are low in mineral nutrients Photosynthetic activity driven by zooanthellae Blue-green algae fix nitrogen Nutrients conserved by algae	Nutrient stored mostly in vegetation and recycled via the soil Insects are major grazers, consume large quantities of plant material and support complex food webs Termites important in the recycling of woody materials	Food chains based largely on grasses Large nutrient store in herbivore biomass Termites are key organisms in the recycling of plant litter
Components creating habitat and niche	Ice-free islands during summer provide breeding sites for birds and mammals Sea ice and ice flows provide resting and refuge areas for birds and mammals	Architectural complexity and diversity of corals	Complex vertical structure of woody vegetation Epiphytes Forest-floor microclimate Diversity of trees	Diversity of grasses River channels and surface water Scattered trees and shrub patches Topographical complexity
Keystone organisms	Copepods in the marine environment and Diptera (flies, midges and mosquitoes) on tundra	Coral polyps Coral-feeding fish	Large fruit-bearing trees such as fig trees Important fruit dispersers such as bats	Elephants

Characteristic wildlife of the Antarctic zone includes species of penguins and albatross and pinnipeds such as elephant seals (*Mirounga leonina*) and leopard seals (*Hydrurga leptonyx*). The International Association of Antarctic Tour Operators (IAATO) noted that polar tourism had expanded rapidly in the two decades 1987–2007, with an increase of 430% over 14 years for ship-based tourists and a 10-year increase of 757% for land-based visitors (IAATO, 2007). In recent times cruise liners with a capacity of up to 3500 passengers have entered the polar tourism industry and the number of sightseeing over-flights has increased. In Alaska there may be as many as 10,000 passengers a day disembarking at Skagway during the peak tourist season in July, while the Canadian arctic also indicates a growing trend in cruise tourism, with 22 cruise ships visiting the area in 2006 (Lück *et al.*, 2010). At the same time, travel to Antarctica increased from 6704 passengers in 1993 to 32,637 in 2007/08 (Crosbie & Splettstoesser, 2009).

Concerns have been raised over a number of negative environmental impacts brought about by the increased volume of ship and aircraft traffic and land-based traffic concentrated at seabird breeding colonies, pinniped haul-out sites and sensitive migration routes (Crosbie & Splettstoesser, 2009). Potential impacts include disturbance at bird colonies, passenger landings at breeding sites and trampling of low-resilience vegetation, noise disturbance by aircraft such as helicopters, which may cause panic flights and increased egg/chick mortality, and pollution from oil spills or rubbish (UNEP, 2007). For information on management approaches such as codes of conduct, interpretation and licensed tour guiding see Chapter 5 and Box 5.9.

Island ecosystems

Islands are highly desired as recreation and tourism destinations for a number of reasons. In the first instance, special social values can be associated with the journey and a sense of remoteness. A boat trip can conjure up a sense of adventure and this is particularly so if the island is uninhabited or if there is an element of danger, as with island volcanoes. In contrast to this, the palm-clad beaches of the humid tropics attract visitors especially from north-western Europe and the USA. People who visit these sites are often in search of the 'recreational beach life' for a few weeks. This latter type of tourism can, however, occur in sensitive natural environments. Examples include coral reefs, turtle nesting beaches and seabird breeding areas and management based on ecological knowledge is therefore crucial.

Many tourists wish to see such wildlife and islands all around the world have become the focus of wildlife-centred tourism. It is possible, for example, to swim and snorkel with sea lions in Australia and the Galapagos Islands (Figure 2.13). Seabird breeding colonies and other birds on many islands are also a focal point of tourism in Australia (e.g. North Stradbroke and Moreton Islands), the USA (e.g. Santa Cruz and Pribilof Islands), the UK (e.g. the Shetland and Farne Islands), Norway (Lofoten archipelago), Peru (Ballestas and Chincha Islands) and Indonesia (e.g. Komodo and the Banda Islands) and at many other island locations around the world.

Island tourism often embraces the desire to see unusual endemic species. Classic examples of these are the Galapagos Island marine iguanas (*Amblyrhynchus cristatus*) and giant tortoises (*Testudo elephantopus*), the Komodo Island dragons (*Varanus komodoensis*) and rare and endangered bird life on Hawaii. Penguins and seals are attracting an increasing number of visitors to South Georgia, the South Orkneys and the South Shetland Islands in the Antarctic region.

The vulnerability of island biota and the problem of invasive species

Many island ecosystems are ecologically fragile due to their relative small size and unique evolutionary development. Some of this fragility is also related to the isolated nature of these plant and animal populations, as any renewal of species that are lost is dependent on sources from beyond the island itself. Uniqueness and especially small endemic populations are thus major contributing factors to the sensitivity of island biotas. It should be noted that the accidental or deliberate introduction of non-native plants and animals to islands could occur as a result of tourism visitation and development.

Exotic plants can be transported to islands attached to visitors' clothing and as seed in soil stuck in the tread of footwear (Pickering & Mount, 2010). Small mammals can be accidentally transported on boats and then reach land either by swimming a short distance or by being carried in luggage or boxes of food. The house mouse (*Mus musculus*) can be readily transported in this way. The significance of such an introduction depends on how the 'invader' interacts in the local island ecosystem. For example, on Marion Island in the sub-Antarctic (Figure 2.13) the house mouse competes with lesser sheathbills (*Chionis minor*) for terrestrial invertebrates as a source of winter food. The reduced availability of food resources, due to this competition, is considered to be the cause of population decline in Marion Island lesser sheathbills (Huyser *et al.*, 2000).

Larger, especially herbivorous, species of animal have usually been introduced as part of an island history of colonisation and development. Species deliberately transported in this way include food species such as rabbits, goats and sheep. The ecological significance of introduced herbivores is in their impact on plant community structure and on any fauna that particularly rely on plants that are being preferentially consumed. For example, on Santa Cruz Island, California (Figure 2.13) the alteration of plant community structure has changed habitat structure, resulting in the loss of several endemic subspecies of ground-nesting birds (Van Vuren & Coblentz, 1987). Similarly, the loss of Mamane Forest on Hawaii (Figure 2.13), due to sheep grazing has impacted on endemic Hawaiian birds. Mamane Forest is the main food supply of the palila (*Loxioides bailleui*), for example, and its loss threatens the bird's survival (Snowcroft & Griffin, 1983).

Many island plants are thought to be vulnerable because their populations have evolved in the absence of mammalian herbivores. Bowen and Van Vuren (1997) demonstrated this in an investigation of the morphological and chemical defences of the same species of plants which occurred on Santa Cruz Island and mainland California. Some of the island species were found to have reduced defences against herbivory. The significance of these findings were tested in a feeding trial using sheep. The sheep were observed to consume a greater biomass of the island populations of four out of six species of plant tested. It was concluded that the island populations are more vulnerable to herbivory because they lacked the better-developed defences of their mainland counterparts. Bowen and Van Vuren (1997) emphasised that, because of this, island populations of mainland species and endemic plants are particularly sensitive to the introduction of exotic herbivores.

The evolutionary effects of isolation on animal life include flightlessness in insects and birds, gigantism and fearlessness. These three features combined give rise to a high sensitivity to disturbance. Many island endemic birds have evolved flightlessness in the absence of mammalian predators. Examples of this include the Galapagos cormorant (*Phalacorcorax harrisi*) and the kiwi (*Apteryx australis*) and kakapo (*Strigops habroptilus*) of New Zealand. Because there has been little or no evolution to reduce their vulnerability to predation, such species can be easily killed by introduced predators such as domestic dogs and cats. Fearlessness, especially in birds, is a behavioural trait that has also evolved in the absence of mammalian predators. This feature in itself, however, has also become a tourist asset, as it allows visitors to approach and photograph island wildlife.

Human visitation to islands can cause significant disturbance as a result of walking and trampling, for example through the collapse of burrows, as in the case of breeding petrels and shearwaters (GBRMPA, 1997). Seabirds use islands because they are safe from predators and the impact of feral animals can be severe on ground-nesting birds. Introduced herbivores can compete for scarce resources and alter ecosystem structure by preventing the regeneration of larger, woody plant species. Introduced rats (*Rattus rattus*) are well known as egg predators and their introduction to island ecosystems causes significant mortality on burrow- and ground-nesting birds (Marchant & Higgins, 1990; Birdlife International, 2000).

Nature-based tourism in island settings is dependent on ecologically intact and well managed ecosystems. For example, the introduced black rat has been eradicated from 12 of Antigua's offshore islands (Figure 2.13). Since eradication in 1998 the populations of six species of birds have increased on the islands. Caribbean brown pelicans (*Pelecanus occidentalis*) had increased from a breeding population of two pairs in 1998 to 60 by 2007. The breeding population of laughing gull (*Leucophanaeus atricilla*) had increased from less than 20 pairs to 500 pairs by 2007. Similarly, the numbers of the Antiguan racer (*Alsophis antiguae*), one of the world's rarest snakes, grew from a total population of 50 to 500. Coincidental with the increase in fauna, tourism to Great Bird Island increased from 17,000 visitors in 1995 to 50,000 in 2007 (*Fauna and Flora International Update*, 2011).

Seabird breeding islands

Natural ecological factors that determine the success of seabird breeding fall into the following categories: variation in the abundance of prey; capacity of the birds to obtain food and adequately feed their chicks; nest disruption by other animals; predation by gulls and raptors; disease and parasitic infections; and severe weather conditions such as storms (Nelson, 1980). Additionally, a number of tourism-related and non-tourism-related disturbances have the potential to determine the success of seabird breeding in island settings. Fishing and mainland coastal development can result in the indirect alteration of food supply for seabirds; long-line fishing in particular and marine pollution have emerged as significant threats to seabirds (especially albatrosses) in recent years (Burchart *et al.*, 2010). The alteration of islands by commercial development and human habitation often results in siltation and pollution of waterways or directly impacts on seabird breeding habitat. With permanent tourism development, residents and business operators often bring in pets, while exotic plants and rats can be introduced accidentally.

From a tourism standpoint, studies have shown that seabird response to human disturbance is variable and is highly dependent on a particular species' behavioural ecology and the specifics of the tourism activity (Hockin *et al.*, 1992; GBRMPA, 1997; Sekercioglu, 2002; Buckley, 2004c; Barter *et al.*, 2008; Steven *et al.*, 2011). Where seabirds are nesting, the disturbance caused by tourism activity can result in an increased risk of egg and chick predation, increased nest desertion, reduced hatching success and increased stress for the birds (Newsome *et al.*, 2005). Any combination of these responses to disturbance can lead to a decline in the population of seabirds and even a change in bird community species composition. It has also been observed that some birds will shift the location of the breeding colony in response to disturbance (Barter *et al.*, 2008). This has been reported for albatross colonies on the Galapagos Islands and at Taiaroa Head, New Zealand (Roe *et al.*, 1997; GBRMPA, 1997). Reproductive success may consequently decline as a result of breeding colonies being 'pushed' into less favourable sites for the birds.

Despite the fact that it is known that some species are reasonably tolerant of human activity, with some species even benefiting, it is important to manage and study tourism activity on seabird breeding islands to prevent loss in numbers and/or community structure (Newsome *et al.*, 2005). The importance of understanding this is exemplified by an important study carried out by Burger and Gothfeld (1993), who concluded that visitors to seabird islands could not be properly managed without adequate information available to managers about the potential impacts tourism has on birds. Burger and Gochfeld (1993) studied the impacts of tourism on boobies in the Galapagos Islands. The study set out to determine the potential for disturbance to breeding habitat and breeding behaviour. The investigation posed whether tourists on walk trails changed the behaviour of nesting boobies, whether boobies avoided trails when nesting and whether boobies abandoned their stations when visitors passed. They showed that masked boobies (*Sula dactylatra*) that were 2 m from the trail flew off when tourists passed by. Blue-footed boobies (*Sula nebouxii*) and red-footed boobies (*Sula leucogaster*) in contrast simply moved away. However, with repeated passages of tourists these birds were seen to be in an agitated condition for quite a long time. All species had fewer nests adjacent to walking trails, reflecting an alteration in territory boundaries and available nesting sites. The greater sensitivity exhibited by the masked booby is thought to relate to the early stage that this bird was at in its breeding cycle, when it is more prone to disturbance. These findings clearly demonstrate that the birds are reacting to the presence of tourists. Furthermore, many groups of birds are more susceptible to disturbance if they are at the crucial initial stages of breeding such as display and courtship. The reaction exhibited by these birds means that disturbance could be reducing the time that boobies need for display and courtship and also for incubation and care of the young. There are implications here for tourism planning and management (Chapters 4 and 5) in terms of the pressure to open up further walk trails if birds move away from existing trails and the need for the creation and maintenance of no-go conservation zones.

Plans to develop infrastructure and/or mitigate tourism impacts on islands requires relevant baseline studies and monitoring of ecological conditions. For example, on Rottnest Island, a major tourism destination in Western Australia, the peak tourism season coincides with peak numbers of migratory shorebirds. In relation to the nexus between tourism and protecting bird habitats Saunders (2010) reported that a proactive monitoring system could identify areas of tourism–conservation conflict, such as fencing off areas where fairy terns are breeding as the birds use different locations from year to year. Saunders (2010) also noted that monitoring could have mitigated the magnitude of past tourism-related activity, such as the construction of a road around a salt lake that impacted on the feeding areas for waders and the construction of a powerline within the flight path of birds moving between salt lakes which resulted in significant mortality arising from collisions. Saunders (2010) went on to report that it took 15 years before the powerlines were fitted with warning devices, a situation which could have been avoided if a tourism impact monitoring system, linked to adaptive management, had been in place (see Chapters 5 and 7).

Island wildlife as tourism icons

Tourists are attracted to islands that host examples of gigantism, as in the case of tortoises (*Testudo elephantopus*) on the Galapagos Islands and the Komodo dragons (*Varanus komodoensis*) on the island of Komodo in Indonesia (Figure 2.13). Highly valued species such as these are then subject to much attention and potential attendant tourism development/activity. When such animals become the subject of constant attention by humans they may become stressed or habituated. In some instances hand-outs of food result in changes in the animal's natural behaviour which can manifest as aggression and/or disease for the animal concerned.

On Komodo Island, concerns over the ecological impacts of artificial feeding of Komodo dragons led to the practice of attracting dragons with bait being terminated (Walpole, 2001). Although leading to reliable sightings of dragons, artificial feeding was resulting in an abnormal concentration of dragons at the designated feeding site. Walpole (2001) also noted that the provision of food to concentrate species can lead to degraded habitats, increased dependency on provisioned food and higher levels of aggression between members of the same species; it also poses risks of people being bitten where animals are dangerous and not wary of tourists. The cessation of food provisioning resulted in less reliable sightings of dragons but the trade-off was that the dragons could then be seen under natural conditions and able to properly occupy their niche as a significant predator on the island.

It would seem, however, that the dragon-feeding debate remains unresolved. Local community income declined following cessation of the feeding programme because the supply of goat meat for dragon feeding was a profitable exercise. Debate has since centred around providing water points to attract dragon prey, the reintroduction of a sporadic feeding programme and the provision of viewing hides located where dragons nest. Information available online suggests that feeding continues to occur. Following the cessation of tethered goat feeding the idea was that visitors would be encouraged and assisted in observing the animals under more natural conditions. Ranger-led excursions have been developed where tourists join a trek to a fenced enclosure where they can take photos of dragons at an egg-laying site and/or while feeding (Indonesia Attractions, 2002). In another case rangers entice dragons to a viewing area with meat wrapped in a cloth (Schonhardt, 2011). It would seem that food provisioning has not ceased despite concerns regarding ecological impacts on the dragons.

This account of an island tourism icon illustrates the point that wildlife tourism is a complex mix of policy decisions surrounding the maintenance of a tourism profile, servicing visitor satisfaction, the need for tourist education and informed management and the welfare of the species of interest.

Coral reef ecosystems

Coral reefs have a wide distribution in clear, warm, shallow seas and extensive coral formations can be found along the coastlines of the Caribbean, east Africa, tropical south-east Asia, and western and eastern Australia. Coral reefs also occur in shallow waters that surround isolated islands and island archipelagos in the tropical zone. Coral needs light and a water temperature between 20°C and 28°C in order to survive. Because of the sensitivity of coral to temperature conditions, coral systems tend to favour slightly warmer and more sheltered eastern coastlines. Their presence on western coastlines, for example the Ningaloo Reef in Western Australia (Figure 2.13), is due to a warm current – in that instance derived from warmer tropical seas that lie further north.

Coral formation can be divided into three main types. Coral reefs that occur close to the shoreline are called *fringing reefs* and become well developed where there is little or no sedimentation from local river systems. They are believed to evolve into *barrier reefs* through a combination of land subsidence with the coral growing upwards towards the light as the landward side of the reef sinks down. This process is accompanied by more rapid coral growth on the seaward side, where the water is better oxygenated and ocean currents bring in nutrients. Over time the reef evolves into a barrier reef and occurs some distance offshore, separated from the coastline by a zone of deeper water. The Great Barrier Reef, which lies 65–190 km offshore from north and central eastern Australia (Figure 2.13), is a striking example of barrier reef development. Many islands are surrounded by fringing reefs and barrier reefs can form extensive offshore coral formations, as with the Great Barrier Reef. The third type of coral reef development is the *coral atoll*, which is a ring of coral enclosing an inner lagoon. Coral atolls can occur far out to sea in deep water and they form first as a fringing reef around an island but over time the island gradually sinks. The coral has, however, continued to grow upwards by the same process as described above.

The various species of coral colonies grow by forming a variety of structures; these include branches, rounded features, shelf and plate-like formations. This makes the reef architecturally complex and provides shelter, breeding sites and sources of food for a diverse array of marine organisms (Table 2.2). The coral reef itself is composed of the calcium-rich skeletons of coral polyps. When the coral polyp dies these skeletons become the base for new, living coral polyps which grow upwards towards the light. The dead coral material, which lies beneath, gradually becomes compacted and forms coral rock. Coral reefs are eroded by wave action and are subject to grazing by various species of fish, molluscs and echinoderms. It is through these activities that parts of the reef are converted to coral sand and rubble. This loose material is often moved around by currents and deposited to form the basis for a new reef or coral island.

Diverse animal communities

Coral reefs are biologically rich and contain a significant proportion of the world's marine biodiversity (Allen & Steene, 1999). For example, at the Ningaloo Reef in Western Australia 250 species of coral, 500 species of fish and 600 species of mollusc have been identified (Storrie & Morrison, 1998). The most diverse reef systems in the world occur in the Indo-West Pacific zone, which ranges from the Australian Great Barrier Reef to the Philippines (Veron, 1986).

One fundamental characteristic of all coral reef systems is the symbiotic relationship between the coral polyp and single-celled organisms called zooanthellae, which occur inside the coral tissue (Figure 2.16). Because zooanthellae are photosynthetic,

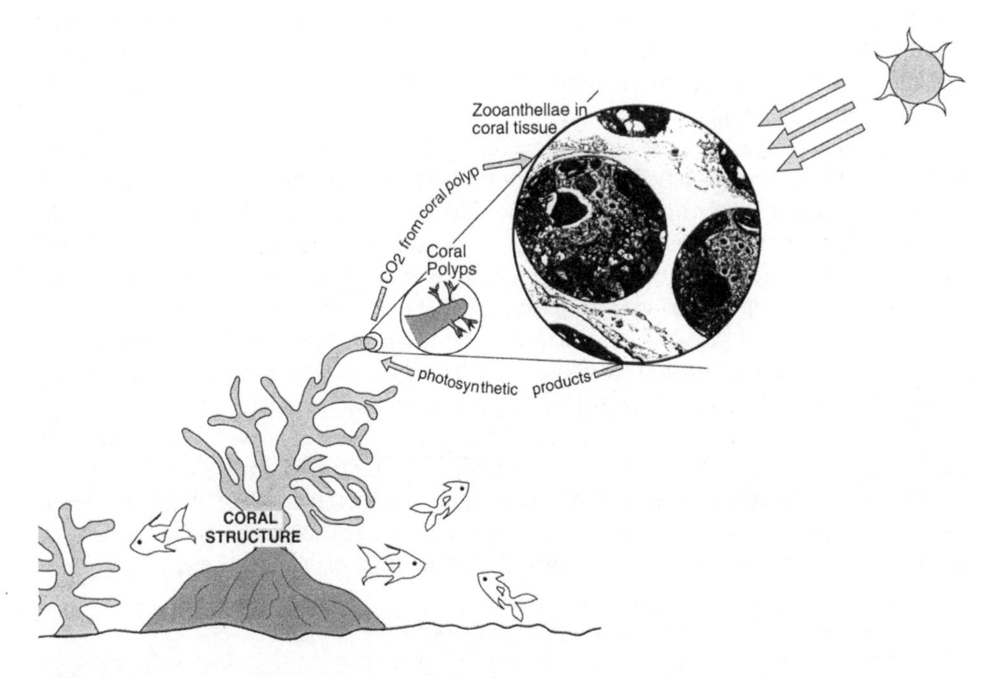

Figure 2.16 The coral polyp ingests a dinoflagellate, which then becomes incorporated into the gastric endodermis and forms a symbiotic relationship with the polyp

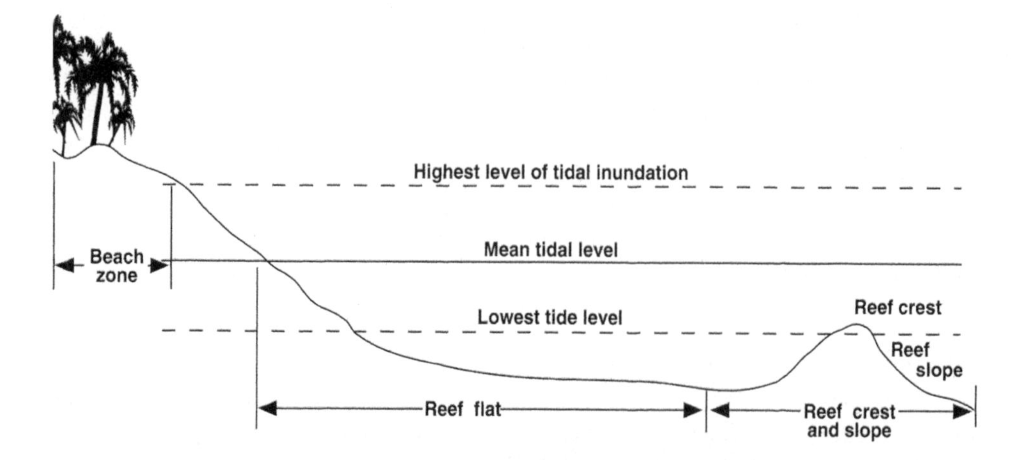

Figure 2.17 Ecological zones of a fringing coral reef

light is essential to them. Zooanthellae utilise carbon dioxide which is given off by the coral while at the same time the coral polyp is able to obtain some of the nutrient products of photosynthesis. The relationship between the two organisms is so profound that if the partnership is disrupted the coral will die. The photosynthetic activity of coral reefs allows them to proliferate and grow under relatively low nutrient conditions.

Fringing reefs occur in two distinct ecological zones (Figure 2.17). The zone which lies under the influence of the highest and lowest tide levels is the reef flat and is, in part, subject to drying every time the tide recedes. The reef flat is a zone of relatively shallow water and is also influenced by wave action and inputs of freshwater as a result of rainfall. As a consequence of these environmental conditions only the more resilient coral species can be found here. These include rounded and flexible soft corals. The sand flats also provide a different habitat for animals, and burrowing echinoderms, crustaceans and molluscs can be found in this zone. Some species of fish specialise in utilising the reef flats by moving to and fro with the tide.

Further out to sea, where the edge of the reef meets deeper water, the reef crest and slope zone can be distinguished (Figure 2.17). The reef crest is subject to wave action, which limits the presence of many species of coral; the reef slope, however, is always covered with water. Here occur corals, such as the staghorn coral, that are intolerant of wave action and desiccation. The habitats of many species of invertebrates and fish can also be found here.

Coral reef tourism

The diversity of animal life, brightly coloured organisms and structural complexity of the coral attracts many people to coral reef environments. Visitor activity includes swimming, boating, snorkelling and scuba diving. Certain species are of particular interest and this had led, for example, to swimming with manta rays (*Manta birostris*) at Coral Bay and whale shark (*Rhiniodon typus*) tourism based at Exmouth on the Ningaloo Reef in Western Australia. Indeed, the shark diving industry has expanded in many coral reef environments around the world.

Ong and Musa (2011) describe scuba diving as one of the world's fastest-growing recreational activities. For example, PADI (2011) reported that the global number of certified divers was 2.5 million in 1988 but had increased to over 20 million in 2011. Salm (1986) reported that resorts in the Maldives catering for divers increased from 3 to 37 in the period 1972 to 1981. Garrod and Gössling (2008) note that there are 28 million divers worldwide and in Thailand alone divers contribute US$130 million annually (Dearden *et al.*, 2006).

A major coral destination is the Great Barrier Reef Marine Park in Australia, which recorded 1.7 million tourist visitor days in 2000 (Harriott, 2002). The Great Barrier Reef is accessed by boat-based operations, with tourist pontoons used as a base for day trippers. Tourism is concentrated on two accessible reefs, such that 85% of tourists visit only around 7% of the entire reef area. Pontoons, sometimes moored up to 60 km offshore, cater for up to 400 visitors at a time (50% of total tourist load on the reef system).

Specific coral reef systems elsewhere in the world also have high levels of usage, for example 350,000 visits a year to the Cayman Island reefs (Tratalos & Austin, 2001) and 250,000 visitors a year at Eilat in Israel, with diving activity occurring along a 4 km length of fringing reef (Zakai & Chadwick-Furman, 2002). Such intensity of use has the potential for significant tourism-related impacts.

Ecological consequences of coral reef tourism

Tourism in coral reef environments generates pressure to develop tourist facilities and accommodation. Associated problems can include fishing, pollution and the collecting of marine life as souvenirs. Coral reefs are susceptible to damage and become stressed when there is too much sedimentation, high nutrient levels, high water temperatures and massive inputs of fresh water, which alters optimum salinity conditions (Johannes, 1977). Whereas the first two of these problems can be caused by tourism development, the latter two are more associated with natural fluctuations in climate. Siltation and eutrophication are common problems associated with tourism development. Sediments are derived from runoff when adjacent land is cleared of vegetation. Eutrophic conditions are brought about by the addition of nutrients into the ecosystem from sewage outfalls and/or fertiliser runoff from coastal golf course maintenance, for example.

Also of widespread significance are the problems brought about by visitor activity on and around coral reefs. Research has demonstrated this in various parts of the world. Visitor impact problems on Australian coral reefs were recognised nearly 40 years ago by Woodland and Hooper (1977). They investigated damage caused by walking on exposed coral at Heron Island, Australia (Figure 2.13). They performed an experiment in which people walked on a designated transect and then the experimenters collected the broken coral. They found that there was an 8% reduction in coral cover after 18 traverses. A total of 607 kg of living coral was seen to be destroyed and 37.5% of this was broken off in the last eight traverses. Experiments such as this clearly show the potential for damage if there is intensive use of an area. More recent work detailing the impacts of recreation and tourism on coral reef systems is further considered in Chapter 3.

The reason why coral is particularly susceptible to damage is that the polyps live in the outer layers of the coral structure. This living veneer will not tolerate constant trampling and breakage. The deposition of mud and silt on coral surfaces blocks out the light needed by the zooanthellae. Similarly, the addition of extra nitrogen and phosphorus into the ecosystem leads to an excessive growth of epiphytic algae. The growth of these algae is normally limited by low concentrations of nutrients. The transport and input of additional nutrients by currents is therefore a significant ecosystem process and any changes can lead to disruption of the normal coral reef nutrient cycle When nutrient levels are increased, algae grow in greater profusion on the surfaces of the coral and block out light. This results in death of coral polyps.

The loss of coral means a loss of habitat structure and food supply for dependent biota and therefore a reduced capacity to support associated animal communities. These changes have the potential to alter the ecological structure of the reef. That is, if large areas of the structural elements afforded by the coral are lost, with a subsequent loss of animal diversity, disruption of symbiotic relationships and ecological function through the simplification of food chains will occur. Coral reefs that become heavily damaged will therefore support less diversity and be less appealing as tourism resources (see Chapter 3).

Tropical rainforests

The tropical forest biome contains both wet and dry forests while an annual rainfall of over 2500 mm defines a tropical moist forest. Within the moist forest classification there are tropical rainforests and tropical deciduous forests. The latter consist of the Asian monsoon forests, the seasonally dry forests of South America, West Africa and north-east Australia and various other subtropical forest formations. Tropical rainforests, however, are mostly distributed in the equatorial climatic zone (Figure 2.18) and receive between 4000 and 10,000 mm of rainfall per annum. Average temperatures range from a minimum of 23°C to a maximum of 32°C along with constant high relative humidity. Tropical rainforests that occur in mountainous environments above an altitude of 1000–1500 m are referred to as montane rainforests and they are usually wetter and somewhat cooler that their lowland counterparts.

The world's tropical rainforests all exhibit the same basic structure. Because of the non-limiting inputs of solar energy and moisture there is abundant growth of vegetation and rainforest trees are able to attain heights of 50 m or more. This is despite the fact that the soil nutrient content of many rainforest ecosystems is low, due to the leaching effects of heavy rainfall. Symbiotic relationships between rainforest trees and fungi or bacteria, however, enhance the uptake of some nutrients. In many situations the decomposition of fallen leaves, flowers and fruit is rapid and the released nutrients are taken up from surface root mats and rootlets which grow into

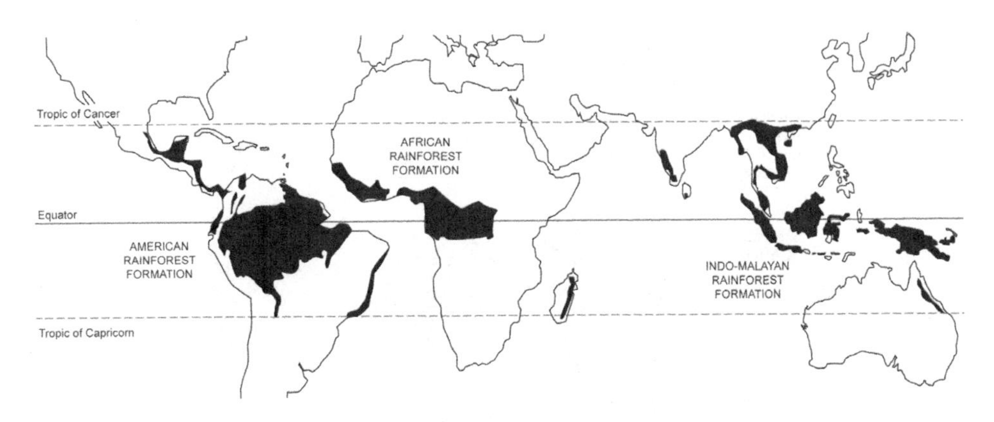

Figure 2.18 Potential global distribution of tropical rainforests (large areas have now been cleared and fragmented for agriculture and plantation forestry)

the leaf litter. Many rainforest species also have thick, unpalatable leaves to resist loss of nutrients through herbivory.

An essential feature of tropical rainforest is the closed nature of the canopy, which results in reduced light penetration to the lower layers of the forest. Vertical differences in light intensity give rise to different zones of vegetation. Five zones can be recognised: the forest floor, lower layer, middle layer, sub-canopy and the canopy with emergent trees (Figure 2.19). The main ecological feature of these various zones is a stratification of various habitats and the adaptations of plants and animals to live in these various zones. The forest floor receives the least light and maximum temperatures are a few degrees cooler (27–29°C). There is little air movement and the humidity is maintained at around 90%. Plants that occur here are much more shade tolerant than species which occur in the higher strata. In contrast to this, the canopy can be up to 30% less humid and a few degrees warmer, and is the site of intense photosynthetic activity.

Climbers and epiphytes reflect the struggle for light and typify tropical rainforest ecosystems (Figure 2.19). Epiphytes grow at all levels and comprise some 25% of all plant species in lowland rainforest ecosystems, with some 15,500 species in tropical America alone (Collins, 1990; Primack & Corlett, 2006). They provide additional habitats for insects, spiders and small vertebrates in the middle and upper strata by trapping, retaining and localising water and organic matter. Climbing plants germinate on the forest floor and then climb towards the canopy; depending on the species, they attach themselves to trees via spikes, thorns, tendrils, clasping roots or adhesive hairs. Many climbers produce woody stems and when these larger species mature and reach the canopy they are often supported by several tree crowns. Buttresses or flanges typify the basal structure of many trees and these provide additional support for their shallow roots (Figure 2.19). Such structures most likely also give rise to more root area to optimise uptake of nutrients and assist in oxygen uptake in those regions where seasonal flooding of the forest floor takes place.

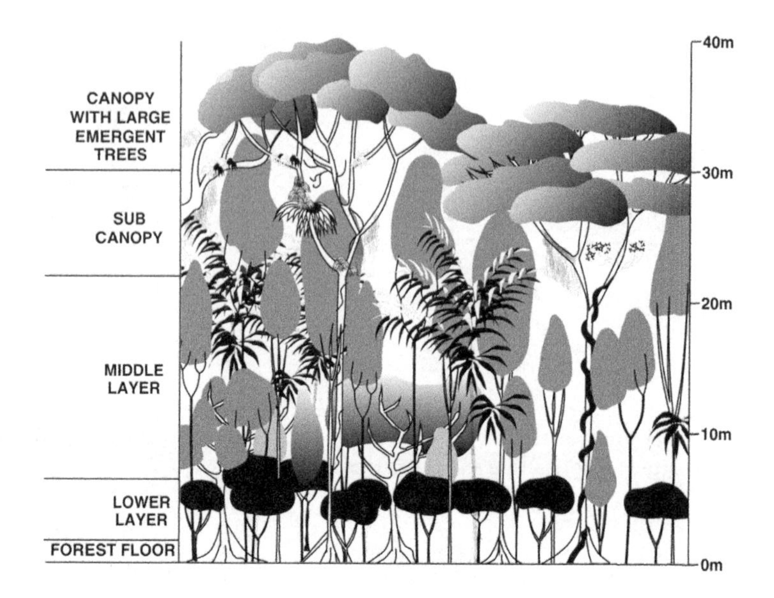

Figure 2.19 Different layers of vegetation which form the structure of a tropical rainforest

The complex structure of tropical rainforests therefore provides many niches. Niche differentiation is achieved through the separation of species according to diet, space or activity periods (Box 2.1). The forest floor provides a suitable habitat for organisms that favour more humid conditions. Other species are confined to and attracted to the concentration of flowering and fruit production in the canopy zone. Fig trees provide a source of food for many insects, mammals and birds. The fruits of strangler figs are mostly dispersed by frugivorous birds and germinate in notches and depressions in other rainforest trees. As they grow, aerial roots descend down towards to forest floor and establish growing points from which further branches ascend into the host tree. The many leafy branches that thicken and interlace act like a 'blanket' over the host tree and it eventually dies, leaving the strangler fig standing in its place.

Tropical rainforests are typified by a diversity of tree species; for example, one hectare of rainforest in Malaysia contains up to 180 tree species (Collins, 1990; Primack & Corlett, 2006). This, coupled with numerous specific insect–plant associations and the large range of animal niches, provides for the most structurally complex and biodiverse terrestrial ecosystem on Earth. Furthermore, although, across the Earth's rainforests, plants and animals demonstrate a fundamental similarity in community structure, there is a wide variation in species composition between different geographical regions. For instance, the tropical rainforests of South America are characterised by the tree family Lecythidaceae, species of *Cecropia* and bromeliads, while Asian tropical rainforests, by contrast, are typified by the tree family Dipterocarpaceae, species of *Macaranga* and pitcher plants (*Nepenthes* spp.).

Rainforest tourism

Tropical rainforests are a diminishing but important tourism resource. They are one of the most diverse biomes on Earth and therefore of great biological interest. Widespread logging and the conversion of primary or virgin forest to secondary (re-growth forest) and plantations are major problems; because of this, forests of all kinds remain the subject of more environmental concern than any other type of vegetation (see Table 3.18, Chapter 3). An increasing number of people are interested in the specific 'atmosphere' and structure of tropical rainforest ecosystems. Such features include distinctive botanical structures such as buttresses, epiphytes, lianes, strangler figs, cauliflory (flowers and fruits that grow directly on the trunks of trees) and palms. However, despite these unique features it is niche wildlife tourism that currently dominates interest in tropical rainforests in many parts of the world.

Tropical rainforests have grown in importance as both a domestic and an international nature-based tourism resource during the last 10 years. There is now a range of well established tourism products globally which focus on a combination of visual features, soundscape and wildlife. Rainforests are now presented to the visitor via boardwalks, observation towers, canopy walkways, self-guided trails and day and night interpretive tours, all of which are supported by well designed panels, pamphlets, books and visitor centres (Chapter 6).

Tropical rainforest tourism includes high-value package tours to view spectacular, enigmatic and highly sought after species, for example lemurs and birds in Madagascar, gorillas in eastern and central Africa, orang-utans in Indonesia and Sabah, and birds in South and Central America. Ecolodges, tours and facilities to provide for the 'rainforest experience' have been developed and are in operation in South America, Central Africa and Asia (see Table 2.3).

Ways of experiencing the rainforest include self-guided access along boardwalks with interpretive signage (Mulu National Park, Sarawak, Malaysia), guided walks along designated trails during the day and at night (Danum Valley, Sabah, Malaysia), boat-based tour guiding (Kinabatangan River, Sabah, Malaysia), viewing the forest from elevated viewing structures (Lamington National Park, Australia), use of hides (Khao Yai National Park, Thailand) and forest canopy walkways (Taman Negara, Malaysia; Figure 2.20). In recent years there has been a proliferation in the construction and marketing of zip lines in forested environments, coupled with the capacity to give the visitor a canopy experience by spending the night in a tree house (e.g. the Gibbon Experience, Laos).

Ecotourism providers located within or on the edge of protected areas offer a range of services. For example, the Cristalino Jungle Lodge in Brazil provides for hiking, river trips/canoeing, bird watching, butterfly observation and special-interest guided tours (birds and insects) for groups of eight or fewer people. Tourism support facilities include an observation tower, trail network and a tree house for observing mammals visiting a salt lick. Staff servicing the lodge are trained guides and the lodge utilises green technologies and waste disposal systems. The Borneo Rainforest

Table 2.3 Selected nature based tourism destinations in tropical East Asia

Country	*Protected area/ ecotourism destination*	*Key environmental characteristics for ecotourism*	*Tourism products and activities*
Malaysia	Taman Negara	Primary forest, birds, mammals, rivers	Hiking, bird watching, river trips, canopy walkway, wildlife hides, camping, extended treks
Brunei	Ulu Temburong National Park	Primary forest, birds	Hiking on elevated wooded walkways, bird watching, river trips, canopy walkway
Indonesia (Sumatra)	Gunung Leuser National Park	Primary and secondary forest, birds	Hiking, bird watching, visiting orang-utan rehabilitation centre and feeding platform
Vietnam	Cuc Phuong National Park	Primary and secondary forest, 320 species of bird, Delacour's languar	Hiking, bird watching, visiting caves, visiting endangered-primate centre
Laos	Bokeo Nam Kan National Park	Primary and secondary forest, waterfalls, black gibbon (*Nomascus concolor*), birds	Hiking, bird watching, zip-line access to tree canopy, accommodation
Thailand	Khao Yai National Park	Primary forest, birds, waterfalls, Asian elephant sightings, bat caves	Hiking, bird watching, night drives
Singapore	Sungei Buloh Reserve	Remnant tropical forest, birds, mangrove vegetation, mudskippers, butterflies	Bird watching, ecotours, mangrove boardwalk with interpretation

Lodge in Sabah also practises green initiatives and employs trained guides. Rainforest experiences include bird watching, hiking along a trail network, a tree canopy walk, night walks and a night safari drive.

Ecological consequences of rainforest tourism

A study by Turton (2005) of the recreation and tourism impacts in the tropical rainforests of Australia provides an insight into some of the consequences of visitor use. Recreational and tourism activities in the Australian wet tropics include hiking, visits to day use areas, camping and travel via off-road vehicles. The environmental impacts associated with rainforest tourism in north-east Australia were found to be trail degradation, creation of non-approved trails, trampling of vegetation, soil compaction and erosion, contamination of water and the spread of weeds, pathogens and feral animals along trail and road corridors (see Chapter 3). Such impacts are also evident in Asian national parks. For example, erosion, creation of bare ground,

Figure 2.20 Canopy walkway, Taman Negara, Malaysia (Photo: David Newsome)

the development of non-approved trail networks and disturbance to wildlife as a result of food provisioning have been reported from Thailand (Nuampukdee, 2002; Sangjun *et al.*, 2006). The listed impacts are common to many different environments; however, some impacts might be regarded as forest specific. Such impacts comprise the loss of forest canopy leading to increased erosion and the barrier effects of roads on arboreal animals (Turton, 2005). Fragmentation by roads and tourism infrastructure (see Chapter 3) may also impact on the dispersal and colonisation of sites by plants and animals, and where forest is lost due to disturbance (e.g. at large campsites) a change in microclimate at the forest edge may increase the mortality of some trees (Primack & Corlett, 2006).

Because wildlife is a particular focus of tourism in many tropical rainforest environments, case studies from Madagascar, Sumatra and Sulawesi are considered here in order to illustrate some aspects of rainforest visitation. Trail impacts, water contamination and damage to soils and vegetation are considered in more detail in Chapter 3.

(1) Indri watching in Madagascar

The endemic wildlife and diminishing forests of Madagascar are promoted as a unique wildlife and nature-based tourism opportunity. Some 20 years ago Stephenson (1993) examined the impacts of tourism in the increasingly popular Perinet Reserve (Figure 2.13). He observed that, despite the presence of established walkways, visitors were straying from paths in order to maximise opportunities to photograph wildlife, lemurs in particular. Besides path erosion and the presence of litter it was found that additional trails were being created and the trampling of native vegetation favoured the invasion of herbs, weeds and rats. Stephenson posited that the altered structure of natural vegetation by trampling caused a reduction in the number of small endemic mammals. Furthermore, Stephenson found that there was direct interference with reptiles and the endangered lemur, the indri (*Indri indri*).

Given that it is often difficult to observe wildlife in forested environments, it is often necessary to actively search for animals. It was argued that the shy and apparently easily disturbed indri was the subject of daily forest searches. Stephenson (1993) also reported that several species of reptile and small mammals were regularly caught by guides for tourists to observe and photograph at close range. Stephenson (1993) questioned what impacts these activities might have on wildlife.

Since 1993, protected areas in Madagascar have become high profile and highly desired wildlife tourism destinations (e.g. Naturetrek, 2012). Major features of current nature-based tourism in Madagascar and at Perinet (Andasibe National Park) include bird watching, lemur watching, viewing chameleons, nocturnal excursions and appreciation of endemic flora. The real focus of attention, however, at Perinet/Andasibe are the reliable sightings of indri, the largest lemur in Madagascar. Guide-led excursions to view indri and other species of lemur take place on a daily basis. Tour guides listen for calling indri with the aim of locating family groups. At the sighting, tourists congregate under trees where indri are moving or resting in order to take photographs. A family group of indri engaging in territorial calling, which is a unique aspect of lemur watching in Madagascar, is much anticipated by visitors. In 2011 Newsome observed tourists leaving designated trails and this was even encouraged by tour guides. Furthermore, over time, off-trail observation of indri has resulted in a proliferation of informal trails and consequent trampling of vegetation. These informal trails, however, also serve as access points that reduce trampling of vegetation across a wider area. Some tourists were observed to call out at the indri in the hope that the indri would look down towards them so that an appealing photograph could be taken. Some people tried to encourage territorial calling and some guides participated in this via the use of taped calls. There can be as many as 50 people at an indri sighting, with people coming and going all the time during the morning viewing sessions, which last from around 7am until midday. The indri groups exposed to tourism are habituated to humans and show no alarm or disturbance behaviour in their presence. Family groups were observed by Newsome to be resting in the forest canopy and were visible to tourists at distances of only 10–15 m.

What has changed since 1993 is that there is a very active guides association (the Natural Tour Guide Association of Madagascar), whose aim is to conduct training, give skills in the interpretation of wildlife and share knowledge, resulting in official qualifications and employment. In terms of impacts, some of the problems highlighted by Stephenson (1993), such as trampling, still exist. The ecological implications of informal trail development and trampling in this case, however, remain unclear. If guaranteed sightings and successful tourism mean that the forest will be protected in Madagascar, then the trampling could be viewed as an acceptable impact and a trade-off for successful conservation of the wider ecosystem. Littering and direct interference with fauna such as chameleons and snakes were not observed in 2011, suggesting improvements have been made over the past 20 years. In the future, management of pushy visitor behaviour (shouting at indri and encouraging the use of taped calls to elicit calling), client interpretation and ecological studies of forest under-storey impacts and tree regeneration will require research and attention.

(2) Watching wildlife in a Sumatran rainforest, Indonesia

Griffiths and van Schaik (1993) investigated the possible impacts of wildlife tourism in Gunung Leuser National Park, north Sumatra (Figure 2.13). Gunung Leuser is one of the largest complexes of lowland and montane rainforest in the world. The park contains a rich array of wildlife, including tiger (*Panthera tigris*), clouded leopard (*Neofelis nebulosa*), Sumatran rhinoceros (*Didermocerus sumatrensis*) and orang-utan (*Pongo pygmaeus*). Griffiths and Van Schaik (1993) set out to study the impact of human traffic on the activity and abundance of wildlife by comparing a pristine site with a site that was heavily travelled by people. They found that animals such as the tiger, clouded leopard and Sumatran rhinoceros avoided heavily travelled areas. The tiger also changed its activity period from diurnal to nocturnal. Other species, however, remained unaffected; wild pigs (*Sus scrofa*), for instance, would flee only a short distance from humans, and macaques (*Macaca nemestrina*) became habituated to people (see Chapter 3).

Griffiths and van Schaik (1993) considered some of the wider ecological implications of these changes. It was suggested that habituated species, such as macaques, could increase in numbers but at the same time reduce their foraging range by staying in one place. A reduced foraging range could also lead to an altered pattern of seed predation and dispersal, which in turn could impact on the local dispersal of tree species. As many plant species rely on mammals for their seeds to be transported some distance from mature parent trees, this type of impact, if sustained, could be significant in changing the species composition of trees in a particular area of rainforest.

(3) Watching primates in Sulawesi, Indonesia

Wild primates are a major tourist attraction and this is reflected by tourism interest in the crested black macaque (*Macaca nigra*) of Sulawesi, Indonesia.

Kinnaird and O'Brien (1996) examined the tourism impacts on these animals as part of a wider study in the Tangkoko Dua Sudara Nature Reserve (Figure 2.13). The interaction of tourist groups with three different groups of human-habituated crested black macaques was monitored and recorded. The results of their work show that the macaques reacted to the visitors by showing various responses, including running, screaming and retreating into trees. The degree of response, however, varied according to different groups of visitors and between different groups of macaques. Macaque group 1 (the smallest of the groups) demonstrated a lack of tolerance to tourist groups of seven or more. The second group of macaques was even less tolerant than group 1 and always displayed disturbance behaviour and retreated when approached by groups of more than five tourists. The third group of macaques, which was the largest and received 60% of tourist visits, was the most tolerant of humans, as indicated by a high level of 'no responses' recorded with groups of up to seven people. When tourist groups comprised more than seven people, up to 80% of macaques displayed disturbance reactions by splitting into subgroups, with the females moving furthest away. It was also observed that some tourist groups consisted of up to 17 people and some tourists would talk loudly and even pursue macaques away from the sighting. Given the importance of reducing flee responses and stress in the monkeys, Kinnaird and O'Brien (1996) recommended that visitors need to be quiet and considerate (slow and stooping) in their approach and that tourist group size should be small (four or five people).

Continuous disturbance as described by Kinnaird and O'Brien (1996) is likely to disrupt the daily activities of Sulawesi macaques and reduce foraging and feeding activity. Moreover, it was considered that if unsupervised tourists were allowed to feed the macaques, significant habituation could take place resulting in bold and aggressive behaviour towards tourists.

The implications of tourist visits and behavioural changes in primates have been reported from Belize (Marron, 1999). Howler monkeys (*Alouatta* spp.), considered to be an important component of the Belize nature-based tourism industry, have been the subject of long-term monitoring studies. The data show that howler monkey groups that are visited by a large number of tourists have fewer females with young than monkey groups with much lower tourist visitation rates. Although primates differ in their responses, there is always the risk of negative interactions when tourists and primates mix (Newsome & Rodger, 2008). For example, the long-tailed macaque is easily habituated to humans and this creates problems (with macaques stealing food, damaging property and behaving aggressively) for tourists visiting Bako and Tanjung Pai National Parks in Malaysia and Gunung Leuser National Park in Indonesia. The experience demonstrates that any tourism experience focusing on primates needs to prohibit feeding by the visitor.

African savanna

The viewing of wildlife in savanna settings is a major aspect of the nature-based tourism product in Africa. The 'safari corridor' extends from Kenya in east Africa, and includes Tanzania, Zambia, Malawi, Zimbabwe and Botswana, and extends to the southern coastline of South Africa. There is generally good road access, and lodges and resorts are widely available. Large concentrations of wildlife can be seen which are the subject of strong profiling in magazines, films and documentaries. Safari tours are common and some locations, for example the Kruger National Park, are subject to high visitor numbers and management geared towards tourism.

Ecologically, savanna represents an ecosystem that lies somewhere between forest and grassland. It can vary from treeless plains through to densely wooded vegetation. Generally, however, it consists of open country characterised by drought-resistant trees with a grass under-storey. Savanna is a widespread and typical vegetation formation in east and southern Africa.

There are five major physical or abiotic determinants of savanna. First is the amount and seasonality of rainfall: a prolonged dry season (lasting four to seven months) allows a fire risk to develop. Secondly, fire (which mostly occurs at the end of the dry season), coupled with seasonal aridity, has a strong controlling influence over the development of woody vegetation, in that it prevents the growth of young plants and colonisation by woody invaders. The occurrence of fire is dependent on the build-up of fuel loads that are derived from dead and dry grasses (see Figure 2.11). Where savanna is less dominated by grasses, fires are less frequent and more limited in extent. Thirdly, adverse soil characteristics, such as shallowness, salt levels or hardpans, prevent the growth of woodland and forest. Fourthly, and allied to the third factor, geomorphological factors such as well drained and dry slopes support grassland, while wetter or low-lying areas prone to waterlogging enable the development of woody vegetation. Finally, frost is also a physical determinant where savannas occur in upland areas because of its role in eliminating frost-sensitive woody species.

The major biotic determinant in savanna ecosystems is grazing pressure. Millions of years of prolonged grazing have resulted in the co-evolution of vegetation and herbivores, which means that grazing pressure is now essential for the maintenance of savanna ecosystems. The action of grazing stimulates storage of plant root assimilates, which in turn stimulates continuous plant growth under grazing pressure.

In Africa there is a wide variety of herbivores and their evolution is connected with the wide spectrum of plant species that are available. The fact that different mammal species specialise in different plants or parts of plants also results in dispersal of the grazing pressure. An example of such niche differentiation is the case of reduced competition for food between impala (*Aepyceros melampus*), which eat acacia leaves, and wildebeest *(Connochaetes taurinus)*, which eat short-growing species of grass. In order to maximise seasonal feeding opportunities, many of these herbivores migrate in large herds in response to rainfall events that stimulate the growth of grasses in eastern and southern Africa. Seasonal migrations in response to rainfall and growing seasons mean that the vegetation can recover between visits of herbivores.

National parks, which are often surrounded by fencing or agricultural land, are now the only safe refuge for large concentrations of wildlife in Africa. Here, because of the lack of natural migration systems, animal densities can be unnaturally high. The resultant constant grazing pressure means that there is no capacity for the savanna to recover. There is though a potential for an increase in the abundance of unpalatable species; such plants would normally be controlled by fire but the overgrazing of grasses reduces fuel loads and thus capacity for fires to develop (Figure 2.11). Overgrazing also reduces plant cover, exposing soils to erosion. Woody shrubs are also more likely to invade because of a reduction of water usage by grasses. The surplus water becomes available to deep-rooted shrubs because more of it penetrates into the deeper sub-soils.

Tourism in the Masai Mara National Park, Kenya

Ceballos-Lascurain (1996b) reported on the potential impacts of viewing and photography on big cats in the Masai Mara (Figure 2.13). Tourist activities consist of vehicle congregation, encirclement and close positioning in order to get good photographs. It was found that leopard (*Panthera pardus*) would walk or run away when approached at a distance of less than 5 m. Moreover, encirclement by vehicles would prevent cats from getting away altogether. This type of disturbance can reduce the feeding activity of diurnal hunting cats and has been observed to be a particular issue for the cheetah (*Acinonyx jubatus*). In heavy tourism periods/zones, cheetahs may be forced to feed in the middle of the day, when the tourists have left, and not at their preferred feeding time of early morning. Cheetah subject to constant attention of this sort can become stressed and less able to raise young and compete with other predators. However, the situation can be complex, as Burney (1982) observed cheetah responses to vehicles to be variable, with some individuals being more tolerant of tourist vehicles than others. The observed tolerance was found to be best when vehicles were slow moving and approached cheetah indirectly. The cats were most disrupted when cars moved quickly and directly towards them, when large numbers of vehicles created encirclement and when people actually got out of the vehicle. In the latter case, cheetah would move away to a distance of 300 m or more. Burney (1982) was of the view that cheetah that were more tolerant of vehicles would have a higher success rate when hunting because habituated cats would not be driven away from a kill. A succession of vehicles, however, might be more disruptive of feeding behaviour. These observations show that the simple activity of viewing has the capacity to cause stress and interfere with prey searching, seeking cover and even mating. The lion (*Panthera leo*) is less affected because it largely hunts and feeds at night and is mostly at rest during the day.

Related to cheetah viewing in east Africa are recent wildlife tourism developments in South Africa that include packages that provide tourists with the opportunity to go on a tracking drive and be involved in the darting of animals for research and translocation purposes. Nsele Reserve in South Africa has a cheetah and leopard habituation programme where the cats are progressively habituated to wildlife-viewing vehicles (Nsele Reserve, 2011). Once the habituation process is complete the cats are captured and translocated to other reserves in South Africa. Burney (1982) raised an important point about habituated cheetah on the Masai Mara: where there only few tourists, such cats will be susceptible to poachers. Distributing tourist facilities and ranger posts across the protected area network could discourage poachers and aid in protecting the cats.

Because seasonally dry grasslands, such as the Masai Mara, are at risk of desertification, Onyeanusi (1986) set out to examine the erosion risk caused by off-road driving. In the Masai Mara tourists are allowed to go off-road and on tracks, where the terrain is dry and grassy, to view and photograph wildlife. In order to quantify any ecological damage, Onyeanusi investigated the impacts caused by an off-road vehicle on experimental strips. There was loss of vegetation cover and damage to underground plant structures, especially from the action of turning vehicles. It was concluded, however, that the total ecological impact on the park was small, due to the rapid regrowth and recovery of grasses. Although the Masai Mara grassland appears resilient to off-road vehicle damage there remains the problem of informal tracks being converted to permanent tracks and impacting on the aesthetics of the landscape. Moreover, the development of further tracks and such off-road driving also constitute an extended disturbance to wildlife. Nevertheless, as pointed out by Burney (1982), careful management of visitor behaviour, employment of viewing codes of conduct, ranger presence and strategic location of tourism facilities can be employed to facilitate conservation and sustainable tourism.

Tourism in the Kruger National Park, South Africa

The Kruger National Park (KNP) (Figures 2.13 and 2.21) contrasts with the Masai Mara–Serengeti complex in terms of road access, the presence of extensive boundary fencing separating the park from highly modified areas and the nature of ongoing tourism and wildlife management. These differences also highlight a number of ecological issues that are concerned with tourism. For example, much of the extensive 7926 km road network in the KNP is sealed, with the southern portion of the park having the greatest road density and potential fragmentation (Freitag-Ronaldson & Foxcroft, 2003). Although potentially detracting from a wilderness or 'natural' experience, such roads reduce the impact of vehicles on soils and vegetation, help to confine visitors, minimise disturbance to wildlife (where road densities are low) and are cheaper to maintain in the long run. Despite these advantages Freitag-Ronaldson and Foxcroft (2003) raise concerns about road kill (especially invertebrates), road avoidance by sensitive species, barrier effects and weed invasion along road corridors. These authors also raise a relevant concern about total road-affected area and point out that 12,500,000 m^3 of gravel was excavated over the road-building period, leaving 1000 unrehabilitated gravel pits of differing sizes which detract from natural values and hold water, creating unnatural water sources.

The recreation, tourism and conservation function of the KNP was recognised some 80 years ago and over time boundary fences have been erected to prevent human–animal conflict and protect wildlife in the park from diseases carried by domestic animals, as well as poaching. The KNP was completely fenced by 1976 but since that time segments of fence line have been removed to allow for the expansion and connection of private game reserves with the park. Much of the western boundary of the KNP is juxtaposed with human modified landscapes, principally rural and agricultural land uses (Box 2.3). Boundary fences, however, bring their own problems by acting as barriers to natural migration systems and also preventing any in-migration of wildlife. In the case of the KNP, east–west migratory patterns

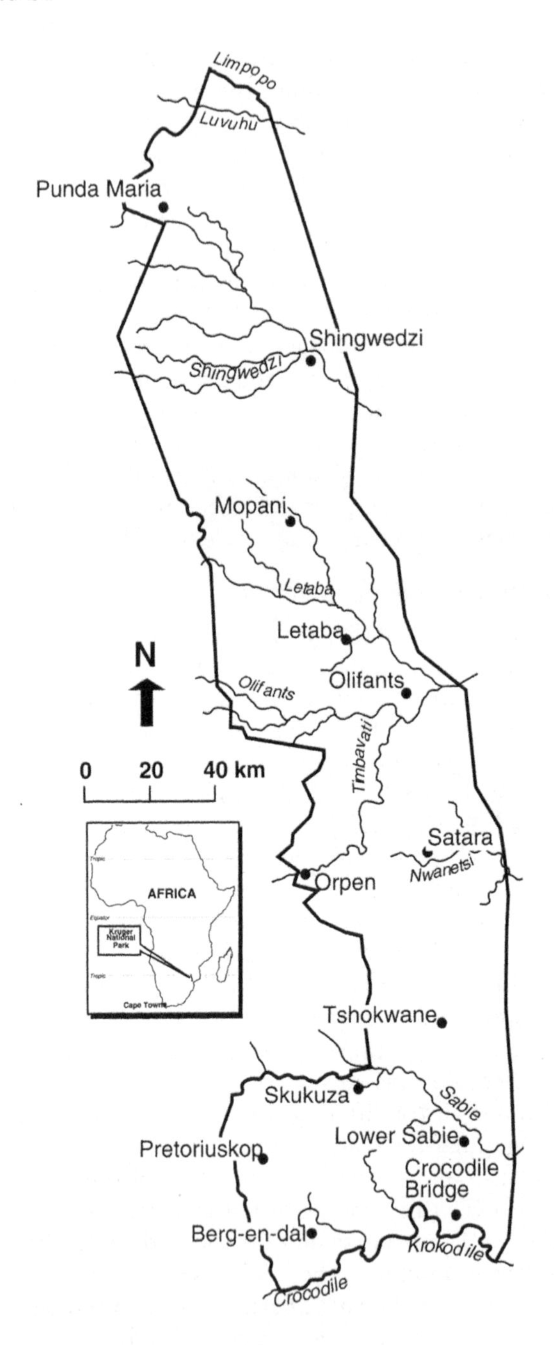

Figure 2.21 Kruger National Park, South Africa

Box 2.3 The Kruger National Park in the landscape matrix

The Kruger National Park illustrates some important ecological impacts associated with river corridors. The park is juxtaposed against a number of activities, such as irrigated crop systems, forestry and industrial development. This is especially so along the western border and the park is influenced by fires, erosion and pollution that are sourced beyond its boundaries. The ecological function of environmental corridors becomes significant here, in that a number of rivers flow west to east through the Kruger National Park (Figure 2.21).

Much agricultural activity takes place in the upper reaches of these rivers (Deacon, 1992). The dependence of plants and animals in the Kruger National Park on these natural corridors is significant, as reflected in the problem of reduced water volume due to water abstraction upstream, especially in the Crocodile River catchment. The loss of water volume can cause hippopotamus (*Hippopotamous amphibius*) to concentrate in areas of deeper water, which in turn leads to localised overgrazing of riverine vegetation. Poor flushing of the river systems also results in an increase in stagnant water, which can provide a habitat for malarial mosquitoes which can then impact on the human population.

The conduit function of natural stream and river corridors is exemplified by the downstream transport of weeds, which have colonised the banks of the Crocodile River in the Kruger National Park (Deacon, 1992). Land clearing along the western boundary has also resulted in soil erosion and downstream sedimentation in the Olifants River (Figure 2.21).

Fish kills have been directly associated with sedimentation events, which in turn reduce the prey available for fish-eating birds. The death of fish in the Kruger National Park river system has also been linked to eutrophication caused by nutrient inputs sourced from sewage derived from rural areas adjacent to the park. Pesticides, originating from intensive farming activities, are known to impact on the breeding success of bird life and have been reported to have entered the park through these natural corridors (Bannister & Ryan, 1993).

have been disrupted by boundary fencing (Freitag-Ronaldson & Foxcroft, 2003). Because natural migration patterns have been disrupted, wildlife cannot leave the park in times of drought or when populations of herbivores increase in good years, hence the policy of providing artificial watering points in the park. Because of the ecological impact of artificial water holes, disruption of natural migration patterns and the controversy surrounding culling, boundary fences along the border with Mozambique have been removed to create the Greater Kruger National Park. The north-eastern section of the KNP is now connected to the Limpopo National Park (Mozambique) and the Gonarezhou National Park (Zimbabwe), creating a much larger ecological unit, totalling 35,000 km^2.

It is important, however, not to lose sight of the role that fencing a protected area can play and the various positive and negative impacts of a decision either to have or not to have fencing. For example, in the Australian conservation biology context, fencing plays a crucial role in protecting rare and endangered species from feral predators. In the human dimension context, a recent account of human–wildlife conflict in Mozambique by Dunham *et al.* (2010) again raises the need for fencing in terms of the high incidence of human–wildlife conflict near the Limpopo National Park and the KNP. Wildlife-caused human mortality is deemed to have been created by facilitating unrestricted movement of wildlife between the two protected areas, with suggestions that the eastern boundary of the Limpopo National Park be fenced to reduce conflict (Anderson & Pariela, 2005).

The KNP has primarily been managed according to principles concerning its wildlife-carrying capacity, which is assessed according to the condition of the savanna grassland ecosystem. Culling has taken place in the past because migration has not been an option during drought years when the population of elephant, buffalo and hippopotamus has been high. The philosophy behind culling is that it prevents sudden population crashes and prevents the mass death of wildlife due to natural causes when populations have increased due to restricted migration.

Population increases in elephants are particularly significant in this regard because of their keystone function (see Figure 2.11). Whyte *et al.* (2003) note that elephants occurred at low densities and were hunted to extinction prior to the area becoming a national park. Natural colonisation from Mozambique occurred over time, with significant recovery of the population recorded in the 1960s. Concerns over the rising population and elephant damage to trees, such as marula (*Sclerocarya birrea*), resulted in an active culling programme taking place from 1967 to 1994. The problem is that tourists find the shooting of elephants very distasteful. Because of this the control of elephant populations in the KNP became controversial and various other control strategies were researched, such as the use of contraceptives (Bannister & Ryan, 1993; Whyte *et al.*, 1998). With the cessation of culling and only a small number of animals being translocated in the period 1993–2001, elephant management remains an ongoing problem as the population continues to rise. The large, dispersed nature of the population also means that contraception is not a viable solution and Whyte *et al.* (2003) advised adaptive management in maintaining biodiversity in accordance with a manageable population of elephants in the KNP.

Smit *et al.* (2007) provide a concise account of the history of the development of artificial water holes in the KNP. They note that rivers were dammed, earth dams were constructed and at least 300 bore holes were sunk in order to provide for strategic water points. The idea was to provide permanent water sources for species that were unable to migrate due to the barrier effects of boundary fencing. The provision of additional water holes ostensibly was to prevent deaths under drought conditions and to also concentrate and facilitate tourist viewing of wildlife. However, trampling and heavy grazing pressure during the dry season eliminate herbaceous plants, resulting in loss of vegetation and accelerated soil loss around water holes for a radius of as much as 250 m (Thrash, 1998). Walker *et al.* (1987) also reported that starvation due to overgrazing in a private reserve bordering the KNP was related to the presence of a high density of permanent water holes. Because of a subsequent loss in fuel loads due to overgrazing, these areas are not likely to burn and woody perennial species will then be able to invade the less trampled, overgrazed areas over time (Figure 2.11). The subsequent lack of dense grass constitutes a reduction in habitat diversity and is likely to impact on those herbivores that require dense grass as cover for the protection of their young. Thrash (1998) suggested that a uniform spread of water holes would extend this type of impact over a wider area. His recommendation was to cluster the watering points in specific areas of the park to reduce trampling and overgrazing.

A decline of the roan antelope (*Hippotragus equinus*), which occurs in the northern part of the KNP, was traced to the provision of artificial water holes. Harrington *et al.* (1999) reported that these water holes attract wildebeest (*Connochaetes* spp.) and zebra (*Equus burchelli*) during drought conditions. These animals in turn had attracted predators such as the lion (*Panthera leo*), which also prey on the roan antelope. Numbers of this locally endangered antelope appeared to recover when 50% of artificial water holes were closed in the northern part of the park. However, Grant *et al.* (2002) observed that rare antelope had declined further since the closure of water holes. Work by Smit *et al.* (2007) has confirmed that grazers, and particularly wildebeest and zebra, are associated with water holes. Furthermore, heavily grazed areas are especially favoured by wildebeest because of their preference for short grass. Given that the provision of artificial water holes can alter the distribution of mammals in the savanna landscape, policy and planning surrounding water hole provision in the KNP and elsewhere in Africa needs to take into account the ecological implications of such decision making.

In 2008/09 a total of 4,374,739 people visited South African national parks (Strickland-Munro *et al.*, 2010) and today the KNP is a highly valued tourism destination, with an annual visitation of around a million. Issues have been raised over infrastructure development, pollution and overcrowding but the KNP encompasses such a large area that the development footprint amounts to less that 3% of its land surface (Freitag-Ronaldson & Foxcroft, 2003). Concerns about the impacts of tourism include water use, the generation of waste, electricity consumption, vehicle emissions and the impact of roads (see Chapter 3). In terms of natural resource depletion to support tourism activity, such as campfires, fuel wood consumption (see Chapter 3) reached such proportions that the practice had to be replaced with commercially available gas, charcoal and wood (Freitag-Ronaldson & Foxcroft, 2003).

Tourism in modified and semi-natural ecosystems: The British countryside

This example differs from the previous ones in that the natural environment of Britain has been substantially modified by the presence of humans for thousands of years. The end result is that hardly any of the present landscape can be considered pristine in the ecological sense. There is a long history of agricultural land use, grazing in particular. Moreover, intensive agriculture and the grazing of marginal lands are largely responsible for the vegetated landscapes that are evident today (Figure 2.22).

Once virtually entirely covered by forest and woodland, plant communities remain only as small semi-natural remnants or, in the case of the English New Forest, one of the few larger areas of forest cover. Many vegetation types represent various successional stages following disturbance by clearing or grazing. Interestingly, these plant communities have become valued for their specific characteristics and biodiversity and are now managed to prevent them reverting to woodland as a result of natural succession.

Figure 2.22 Rural landscape in Leicestershire, England. The landscape consists of many patches and corridors comprising crops, grazing land, hedgerows, woodland remnants, roads and tracks (Photo: David Newsome)

The landscape mosaic described above is typical of many areas in Britain that are highly valued tourism resources. One such is the English Lake District (Figure 2.13), a national park that comprises mountains, moorlands, lakes, streams, woodlands, agricultural land, grasslands and human settlements in the form of farms, small villages and tourism facilities. In contrast to wildlife or the other specific ecological attributes described in previous sections, places like the Lake District attract visitors because of a 'sense of wildness' that is derived from the landscape and because of their geological interest. Some of these values are born out of the fact that mountainous areas confer a more natural element to the landscape, coupled with the need for people to escape urban environments that are substantially devoid of 'natural' experience. The Lake District landscape mosaic thus provides for hiking, mountain-top viewing, bird watching and general nature study, rock climbing, boating, pleasure driving and other recreational activities.

The Rothiemurchus estate in Scotland caters for both domestic and international tourism in a mountainous setting, among lakes and semi-natural pine forest (Figure 2.13). Tourism facilities consist of roads, car parks, walkways, nature trails, picnic sites, camping and caravan sites and tourism information centres. Part of the tourism experience also includes fishing, off-road driving, orienteering, cycling, mountain-bike riding and boating. This is a different type of tourism when compared to more remote and less intensively developed locations with fewer activities, such as rainforest treks in Malaysia or wildlife viewing in Africa.

Accordingly, the Rothiemurchus estate provides for a range of tourist experiences in a less natural setting.

Despite the fact that tourism in Britain takes place in largely modified environments there is still scope for detrimental ecological effects. Indeed, it could be argued that any potential ecological effect is more significant because many ecological qualities have already been lost from the landscape and what remains is in special need of protection. Loss of biodiversity and a reduction in bird populations, for example, are ongoing problems in Britain and tourism can put additional pressure on what remains. Important remnant populations of plants and animals occur in areas that are visited by tourists. Moreover, some natural communities and wild animals are present and protected largely because of tourist interest in them and/or their habitats.

The intensity of visitation is high in many parts of 'natural' Britain and as a result important remnant vegetation and wildlife remain under threat from development and pollution. Sensitive mountainous areas are particularly at risk of further damage and the sheer congestion that arises during peak holiday periods constitutes a social impact of overcrowding and possible visitor dissatisfaction. This can lead to negative attitudes about the area concerned or even result in irresponsible activities such as camping in undesignated zones and walking in untracked or restricted areas. As already noted, sites can become eroded, vegetation is at risk of damage and wildlife may be disturbed. An appreciation of ecology, environmental sensitivity and informed management is therefore important in all situations, whether it be a national park in Madagascar or a bird reserve in Britain. In doing so, sustainable tourism can be achieved and different natural environments around the world can be maintained as important tourism resources.

Wildlife as a Specific Component of Ecosystems

Wildlife tourism is a large and diverse component of natural area tourism and needs to be considered as a subject in its own right (e.g. Reynolds & Braithwaite, 2001; Higginbottom, 2004; Newsome *et al.*, 2005; Rodger *et al.*, 2007; Higham & Liick, 2008; Newsome & Rodger, 2008a, 2008b; Steven *et al.*, 2011; Newsome & Rodger, 2012a, 2012b). Nonetheless, it is essential to consider important aspects of wildlife in the tourism context. Wildlife is part of the ecological perspective in terms of its dependence on abiotic and biotic conditions and in relation to the special interest that tourists have in wildlife, for example as considered previously in the context of tropical rainforests and African savanna. It is therefore important to focus on how the daily activities of wildlife can be disturbed and have some insight into any associated ecological effects. Wild animals can respond to human attention according to the widely recognised behavioural responses of avoidance, attraction and habituation and such responses are considered further in Chapter 3. Here, the major categories of wildlife disturbance and related ecological responses that are briefly considered include some physiological aspects of stress, disturbance of normal feeding activity and potential tourist disruption of the maternal care of wildlife.

A physiological response of ecological significance: Stress

In many species the avoidance of humans starts with alarm behaviour and alertness, followed by agitation and then escape to a safer distance. This is the obvious behavioural response to stress caused by the presence of humans. These reactions are frequently associated with physiological responses in the form of hormonal changes that result in increases in heart and respiration rates, elevated blood sugar levels and a rise in body temperature. Gabrielsen and Smith (1995) stated that these responses are more intense when animals are nesting or caring for their young, when humans are visible while walking and when people move off existing trails. Some species, as in the case of many birds, have the opportunity to take immediate flight but others will not move until absolutely necessary, for example in the case of camouflaged birds sitting on eggs or during incubation at penguin rookeries and pelican breeding sites (e.g. Barter *et al.*, 2008). Fowler (1999) noted that although some birds, such as penguins, appear unperturbed by the presence of humans, physiological reactions such as increased heart rate have been detected.

The stress responses that are invoked can have detrimental health effects for the species concerned and could be particularly significant for those species existing under

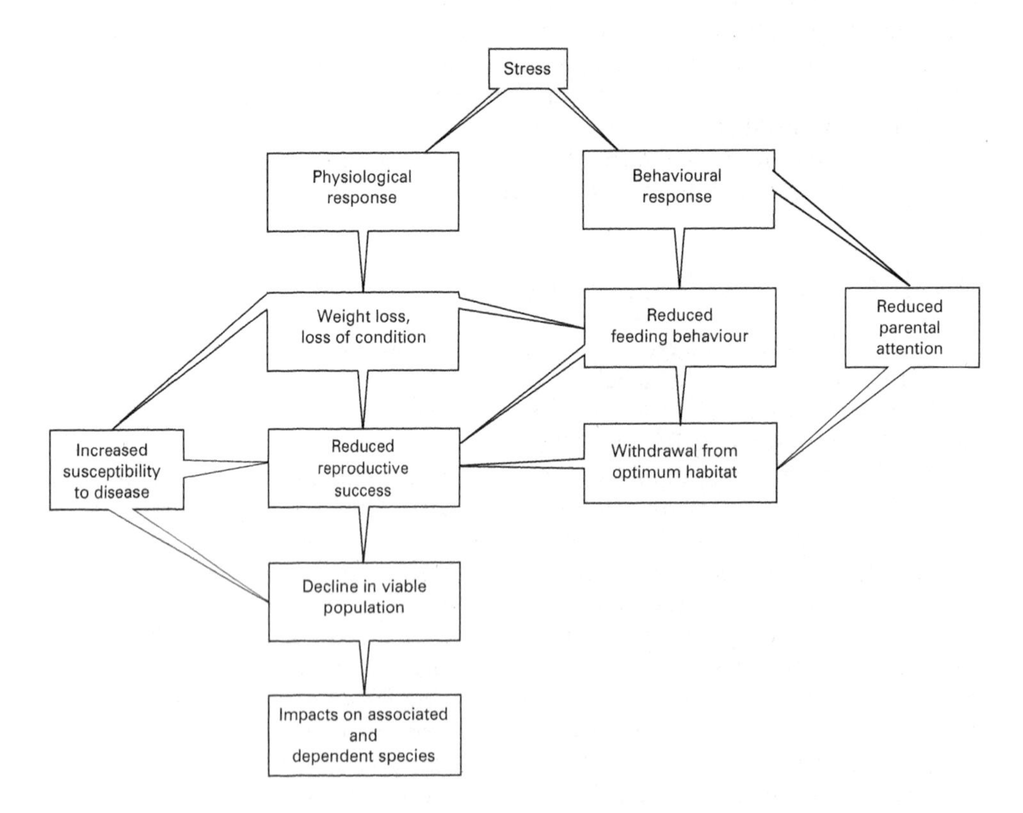

Figure 2.23 Potential ecological implications of stress caused by disturbance to wild animals

naturally stressful environmental conditions. Antarctica, for example, continues to be an in-demand and increasingly important tourist destination. Of particular note is that only 2% of Antarctica is utilised as breeding areas for seabirds and these are also the sites of tourist interest. The fact that penguins are reluctant to move in the presence of tourists means that visitors are able to get very close to birds that are incubating eggs. Regel and Putz (1997) have shown that penguins exposed to human disturbance exhibit stress, manifested by increased stomach temperatures that result in additional energy expenditure by the birds. The conservation of energy is critical in cold environments and any unnecessary, and especially continuous, losses will result in reduced vigour and breeding success. Such disturbance may result in changes that have implications at the population level which could ultimately translate into a loss in biodiversity (Figure 2.23).

Disturbance of normal feeding patterns as a result of food provisioning

There is the risk of wildlife becoming accustomed to and dependent on humans for food in tourism areas. Depending on the nature of the food provisioning this could have serious health effects (obesity, loss of condition, tooth decay), ecological impacts (changes in species composition of an area, increased predator activity) and behavioural implications (disrupted foraging habits, increased aggression), particularly for a rare species or species with a restricted population (Newsome & Rodger, 2008b, 2012a). For instance, disturbance in the form of food provisioning can result in lower levels of foraging behaviour, as demonstrated by barbary macaques (*Macaca sylvanus*) which occur in Gibraltar (Figure 2.13). Macaques normally spend at least 50% of their time foraging, but those receiving food from humans have been recorded as only spending 7% of their time foraging (Fa, 1988). Predictable sources of food might reduce the home range of an animal and result in reduced learning behaviour in searching for food and thus a reduced capacity to find food in the wild (Newsome & Rodger, 2008b).

Besides the potential risk of visitors being bitten, food provisioning may lead to aggression between members of the same species and social grouping. Brennan *et al.* (1985) reported that vervet monkeys (*Cercopithecus aethiops*) that had access to human food had more evidence of physical injury (scars, torn ears, hair loss) than their wholly wild counterparts. The reason for this is that food provisioning leads to abnormal concentrations of animals competing for the same resource, resulting in aggression.

An increase in the population of opportunistic, scavenging and predatory species due to food provisioning can also have undesirable effects. Increases in various species of gull and crow, in tourist car parking areas and accommodation sites, can impact on less robust species of bird through the predation of eggs and their young. This can reduce diversity at a local level and threaten the survival and recovery of any endangered species. The ways in which natural feeding behaviour can be disrupted is summarised in Figure 2.24 and is considered further below in relation to opportunistic predation.

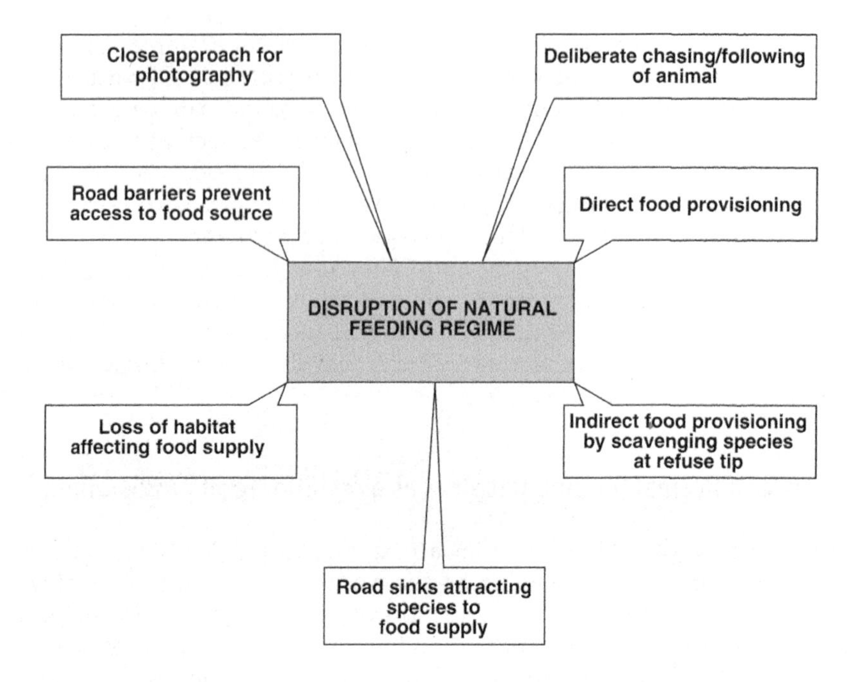

Figure 2.24 Ways in which tourists can impact on the natural feeding activities of wildlife

Tourist activity resulting in the avoidance of optimal resting and feeding areas

Animals need to rest and secure protection from predators or conflict within the same species. Particular problems can arise when the peak tourist season corresponds with a period of restricted or preferred food availability for a particular animal. Roe *et al.* (1997) provide the example of the giant otter in Manu National Park in Peru (Figure 2.13). Here, tourists have been observed chasing otters around lakes in order to get good views. Such lakes are optimal feeding areas for the otters and this, coupled with disturbance from trails that are located close to the lake shoreline, reduces the chance of otters maximising opportunities to obtain food. Continuous disturbance of this nature, through the effects of stress and disrupted feeding activity, has the capacity to reduce the density and diversity of species in a particular location.

Disturbance to feeding and the problem of opportunistic predation

Tourism can displace predatory birds and mammals from hunting and capturing their prey. Additionally, night driving and vehicle noise may interfere with the hunting strategy of an animal that relies on hearing in order to detect its prey. A study carried out by Wilson (1999) provided some insight into the potential effects of spotlighting on nocturnal mammals. She found that spotlighting disrupted the social and foraging behaviour of possums. Avoidance behaviour had resulted in a decline in the numbers of possums sighted at night. It was suggested that the animals may need a up to 30 minutes to recover their night vision following a spotlighting session.

Increased vulnerability to predators is recognised by many as an issue where human disturbance occurs on seabird breeding islands. Predatory birds such as gulls and skuas attend breeding colonies of puffins (*Fratercula arctica*), razorbills (*Alca torda*) and guillemots (*Uria aalge*) in the northern hemisphere and penguin rookeries are attended by skuas, gulls and sheathbills in the sub-antarctic. Displacement of adult birds from nests increases the opportunity for predatory birds to attack unattended nests and exposed chicks.

The eggs of reptiles are also subject to predation. Edington and Edington (1986) report on the increased mortality of crocodile eggs due to tourism in the Murchison Falls National Park, Uganda (Figure 2.13). The close approach of tourist boats to river banks, which are breeding sites for the Nile crocodile (*Crocodylus niloticus*), causes the crocodiles to enter the safety of water but in doing so their unprotected nests are at risk of predation. It was found that undisturbed sites suffered a predation rate of 0–47% while the disturbed sites were predated at a rate of 54–100%.

Disturbance of reproduction and maternal care

As discussed above, the reproductive success of an animal is likely to be affected by stress and poor body condition (e.g. Regel & Putz, 1997; Giese, 2000). Fa (1988) reported that in the case of the barbary macaque (*Macaca sylvanus*) in Gibraltar the feeding of high-calorie foods can distract monkeys from their normal breeding activity and this may lead to a decline in numbers. In addition to the issues surrounding the feeding of barbary macaques, research undertaken by Marechal *et al.* (2011) in Morocco revealed that the monkeys demonstrate stress responses in the presence of tourists. Using visual stress signals, such as self-scratching, and faecal glucocorticoid levels, they found that macaques were more anxious in the presence of tourists, especially when the interactions between tourist and monkey were of an aggressive nature.

Problems that can arise in cases of disturbance at the courtship stage have been discussed by Burger and Gochfeld (1993) and in the case of nesting birds and during stages when young are vulnerable to predation by Anderson and Keith (1980) and Yalden and Yalden (1990). The disruption to parent–offspring caring and learning stages can also lead to infant mortality. This has been recorded at the Monkey Mia dolphin feeding site, Shark Bay, Western Australia. In this case there was the loss of a young dolphin to a shark in the presence of its non-attentive food-provisioned mother (Mann & Barnett, 1999). Further issues surrounding the response of wildlife to tourists and tourism are discussed in Chapter 3.

Conclusion

The focus of this chapter has been on the importance of understanding ecology in terms of maintaining nature-based tourism resources. Various potential problems and impacts have been documented so that we are in a better position to anticipate tourist pressures in natural areas. At the same time, a knowledge of ecology deepens our appreciation of nature by helping us to understand how plants and animals 'solve' their own problems of survival in various ecosystems. Accordingly, from a human standpoint, this helps in the conservation of nature and enhances our quality of life due to the psychological benefits derived from natural experiences (Newsome & Lacroix, 2011).

Recreation and tourism can alter the composition of biotic communities through habitat loss, the introduction of exotic species and pollution. Disturbance caused by trampling and erosion can reduce plant productivity and biomass at a particular site. Humans can also alter the structure of a community by directly or indirectly affecting dominant and/or keystone species. Indeed, the structure of many ecological communities is a direct product of human disturbance, which also includes recreation and tourism.

Many natural areas around the world are, however, subject to intense tourism pressure and there is a need for studies of tourism's ecological impact and especially thresholds of change so the appropriate management can be instigated. Graham and Hopkins (1993) highlighted the need for knowledge of ecology in understanding recreational activity in tropical rainforests, especially in terms of a forest's capacity to absorb various impacts. They emphasise that ecological information is fundamental to the accurate prediction of disturbance effects but at present the ecological determinants of many rainforest organisms are poorly understood.

Furthermore, Tyler and Dangerfield (1999) have emphasised that, in understanding how a natural area might respond to tourism and recreational activities, the emergent properties of ecosystems need to be taken into account. These include concepts of stability and resistance. Stability is the capacity of an ecosystem to remain unchanged, whereas resistance is the ability of an ecosystem to 'absorb' impacts. Understanding these concepts requires knowledge of ecology. For example, as already noted, the sensitivity of coral reefs means that they are susceptible to damage and therefore have much less resistance to tourism impacts than other aquatic ecosystems, such as estuaries. Tyler and Dangerfield (1999) gave savanna as an example of a resistant terrestrial ecosystem because significant disturbance is required to cause degradation. This can be contrasted with the sensitivity of the Stirling Range ecosystem in Western Australia, which has now been severely impacted by the accidental introduction of *Phytophthora cinnamomi* (see Box 2.2). Much of the *Banksia coccinea* population has been lost from the park due to dieback infection and once the pathogen is established it is very difficult to control. The Stirling Range comprises a biodiversity hotspot (Myers *et al.*, 2000) and the intractable nature of *Phytophthora cinnamomi* infection constitutes a serious impact on biodiversity and the tourism dependent on this in Western Australia (Newsome, 2003; Buckley *et al.*, 2004).

Effective ecological management of tourism therefore depends on an understanding of key ecological processes along with some assessment of ecosystem resistance to change (Tyler & Dangerfield, 1999). Monz *et al.* (2010a) have reiterated that ecosystem structure and function can be impacted by recreation and tourism due to the relationships between the components of ecosystems. They go on to state that an expansion of research that focuses on the ecology of recreation and tourism, to include a broader geographical scope and greater spatial and temporal scales, will strengthen the role of recreation ecology in providing data for the management of a vast array of natural areas that serve as resources for tourism.

Further reading

We direct the reader to key textbooks, monographs and the major research paper databases in order to review recent research and the state of knowledge on a particular subject area. For additional information on the fundamental nature of ecology as described in this book, Furley and Newey (1983), Colinvaux (1993) and Dickinson and Murphy (1998) are still useful, with the recent text by Krebs (2008) bringing the reader up to date. Connell and Gillanders (2007), Hutchings *et al.* (2008) and Short and Woodroffe (2009) provide further reading on corals reefs, islands and the marine environment. Stattersfield *et al.* (1998) remains a good source of information on the problems faced by isolated and endemic populations of birds on islands. Updated information on the conservation and status of many species of bird can be found in del Hoyo *et al.* (2010) and related volumes.

The work of Forman (1995) remains a comprehensive overview of landscape ecology. Given the increasing importance of rivers as tourism destinations and because of their function in the landscape matrix, the text by Prideaux and Cooper (2009) is recommended. The book is useful in containing an account of the physical geography and ecology of rivers, along with an international perspective on the human use of rivers, juxtaposed with a complex array of tourism interests, ranging from religious tourism, heritage values, fishing tourism and adventure activities such as rafting through to the problems of pollution and human interest in rivers worldwide. The text illustrates how rivers are the subject of multiple interests, of which nature-based tourism can be one, at the same time as a range of other activities.

Primack and Corlett (2006) cover the ecology and biogeography of tropical rainforest environments while Corlett (2009) provides an important perspective on the ecology of tropical east Asia.

Solbrig *et al.* (1996) remains a useful text for savanna ecosystems while du Toit *et al.* (2003) provide a comprehensive account of savanna ecology and management from the perspective of the Kruger National Park. Although not dealt with in any great detail in Chapter 2, readers can find more information on polar tourism in UNEP (2007), Maher *et al.* (2009), Stonehouse and Snyder (2010) and Lück *et al.* (2010).

3 Environmental Impacts

Introduction

Natural and outdoor areas tend to be visited by people keen to get away from busy urban life and to be in and experience nature. Recreation and tourism focused on the outdoors can involve many different activities within natural ecosystems. The marked increase in natural area tourism that has taken place over the last 20 years continues to be promoted by the wide availability of travel guides, natural history periodicals, websites and a plethora of travel magazines that emphasise outdoor activities and nature experiences. Moreover, events, meetings, onsite attractions and experiences can be communicated widely to others via social media such as YouTube, Facebook, Twitter, TripAdvisor, online forums, blogs and mobile phone applications that promote sites of interest and experiences.

Tourists also see and 'use' nature differently, the consequences of which are that recreation and tourism occupy a complex spectrum, with some forms blurring into one another (see Chapter 1, p. 15). The type of experience gained will depend on the site and its ecological condition, combined with personal expectations, modes of access and travel, the number of people present in a group, the activities undertaken and the number and behaviour of other visitors at the destination. Some people will simply undertake a 'country' drive and picnic in a semi-natural landscape, while others require the challenge and solitude of more inaccessible wilderness areas (Figure 3.1). Some visitors to natural areas are generally interested in seeing wildlife as part of a walk among natural vegetation, along riverbanks or in mountainous areas. In contrast to this, bird watchers will target specific species and make every effort to see that species, whether it be in a remote area or at a city sewage farm. Others wish to see dangerous, rare or spectacular assemblages of wildlife. There may, though, be walkers, mountain bikers and off-road vehicle enthusiasts all wishing to use the same natural area at the same time. Additionally, over the last 10 years or so there has been an increasing trend to conduct organised activities, such as mountain bike challenges, fun runs and other sporting events (also promoted using social media and in magazine articles) in various natural areas. These new activities, based around the promotion of health and sporting interests, now take place alongside the 'softer', passive, eco-based recreational interests. The implications of this complex spectrum

Figure 3.1 Valley leading to the Jostedalen ice cap and Tunsbergdalsbreen glacier, Norway. Few people visit this remote area that is accessed across snowfields which are present throughout the northern summer. Beneath the snow are fragile and slow-growing lichen communities that are easily damaged by trampling (Photo: David Newsome)

of recreation and tourism are that an immense range of activities can be undertaken in a wide range of natural areas and their associated ecosystems.

With increasing numbers of people visiting spatially diminishing, potentially degraded and modified natural areas, it is important they do not add to any problems or have recreation-specific negative impacts. In this context, however, it is important to realise that not all tourism has the potential to cause problems. Indeed, there are many examples of sustainable tourism operations and positive impacts. Degraded and disturbed areas can be restored and repaired, nature reserves created and national parks expanded as a result of actual and anticipated interests in natural area tourism. However, negative impacts do occur and the nature and degree of these impacts can be complex and vary in significance depending on the activity and situation. Impact significance, for example, can depend on the type and source of the problem, environmental sensitivity, other cumulative pressures and the effectiveness of any management that is in place. Moreover, what is a well recognised and significant impact in one country or environment may not be a problem elsewhere. Accordingly, this chapter attempts to provide an overview and a global perspective on these issues and expands upon some of the impacts already described in Chapter 2.

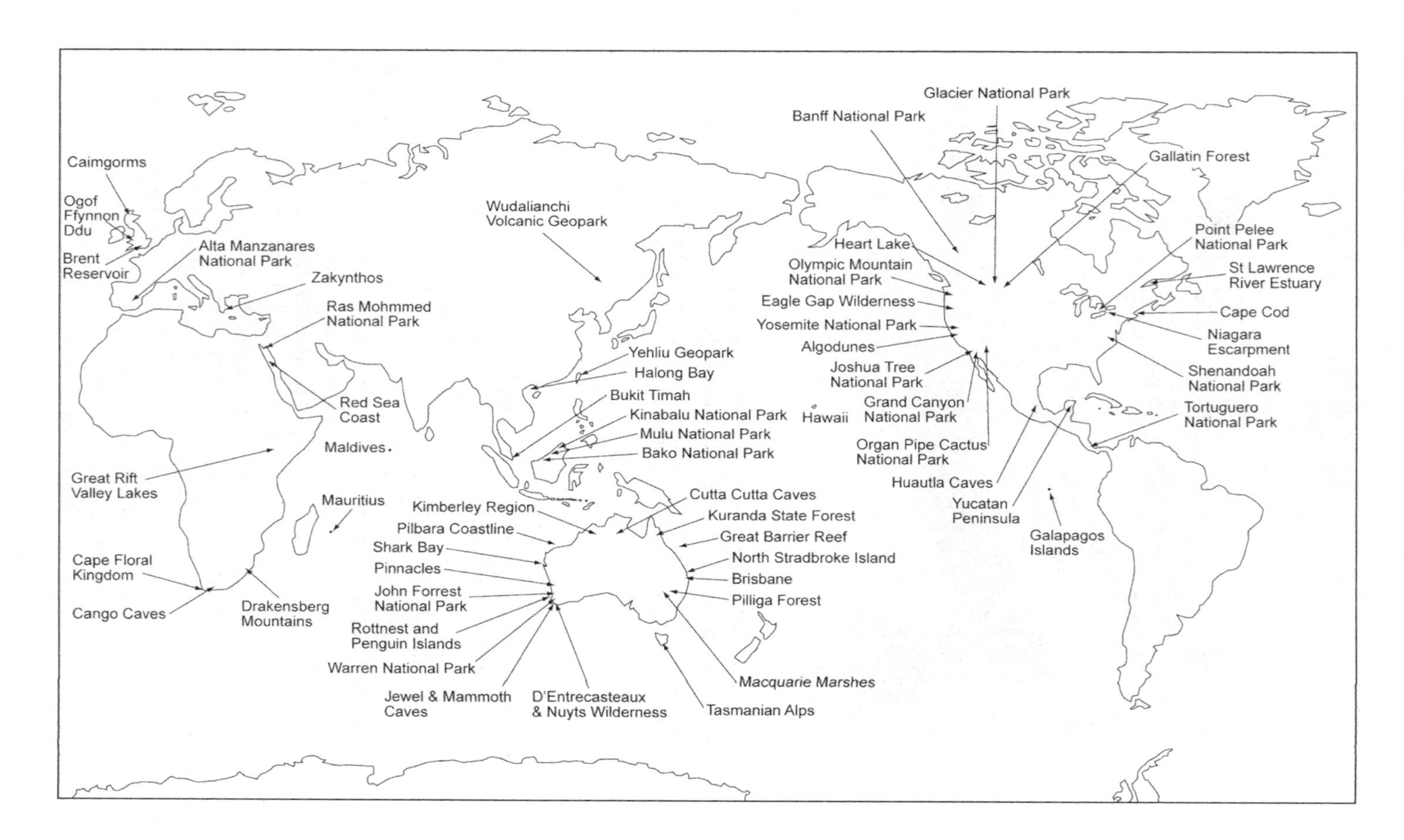

Figure 3.2 Location map of important nature-based tourism destinations referred to in this chapter

Scope of the chapter

In Chapter 2, as part of an exploration of the importance of understanding ecology, we considered the role that disturbance plays in creating landscape heterogeneity. Many of these disturbances constitute natural fluctuations within ecosystems and mostly result in a temporary change in ecological conditions. The disturbances brought about by recreation and tourism are multifaceted and reliant on a number of factors and, depending on their nature, can be permanent. Human disturbance from tourism/recreation within the natural environment constitutes a biophysical impact. Moreover, because aspects of the physical environment (abiotic factors) can be changed, there is the possibility of their translation into indirect ecological impacts.

Because of the importance and sensitivity of a large number of ecotourism destinations (see Figure 3.2 and also Figure 2.13) it is vitally important that the potential for, and nature of, any environmental impacts are properly understood and anticipated (Figure 3.3). As noted by Tyler and Dangerfield (1999), Hobbs *et al.* (2009) and Monz *et al.* (2010a), we must attempt to understand the ecosystems that we are dealing with. To this end we also need to appreciate the sources and nature of potential impacts on the tourism resource.

This chapter builds on the concepts explored in Chapter 2. A source-based approach is taken to examining and understanding environmental impacts in the context of valued ecosystems. The focus is thus on potential biophysical changes

Figure 3.3 Signs informing visitors of prohibitive activities at the Boulders Penguin Colony near Cape Town, South Africa. This signage clearly indicates that visitor impacts are real and need to be controlled (Adapted from National Parks Board Visitor Permit, the Boulders, Cape Peninsular National Park, South Africa)

Table 3.1 Examples of significant non-motorised outdoor recreational demand and activities taking place in natural areas

Recreational activity	*Characteristics*	*Impact potential*
Walking/hiking	Short-distance and long-distance day trips and walks that may occur over several days, either as a camping trip or utilising built accommodation. Walking usually occurs on formed and directional pathways/trails	Depends on trail planning and management and environmental conditions
Wildlife appreciation	Wide spectrum of activities, ranging from casual observations to highly organised touring. Specialist components include bird watching, safaris and watching marine mammals	Variable, depending on context and management
Sightseeing/day visits	Visits to scenic areas, landscapes and geotourism products such as caves, geotours and visitor centres	Usually low. High-visitation sites are hardened/managed
Camping	Can be planned and managed designated sites, or more informal sites in remote locations. Problems are more likely when uncontrolled, informal camping occurs	Variable, depending on planning and management
Tours to natural areas and wildlife destinations	Wide range of situations and experiences, from specialist excursions to mass tourism situations where people are highly managed by tour operators and infrastructure is installed	Usually low due to careful management and effective tour guiding
Rogaining/ orienteering	Long-distance cross-country navigation. Can involve large numbers of participants. Often a continuous 24-hour activity in which participants have to locate set check-points within a set time period. Involves day and night navigation and the use of a central base camp	Unknown/little specific information
Horse riding	Usually on approved bridle trails and not allowed in most protected areas	High impact potential, depending on context
Horse riding competitions and events	Competitive riding and endurance events. Events may take place over distances of up to 300 km and over 3 days. Control points and tethering sites may be employed. Prohibited in most protected areas	Unknown/little specific information
Mountain biking	Characterised by a complex spectrum of activity, ranging from passive use on formed pathways through to elite sporting activities. Complicated by users who create trails and modify pathways themselves. Mountain biking is sometimes a major component of many competitive sporting events, which can involve large numbers of participants and spectators	Limited data. Significant impacts have been demonstrated in peri-urban protected areas in Australia

Table continues opposite

Mountain runs/ funs runs	Involves a targeted course with the intention of completing the run in a personal best time or simply completing the run. Can involve large numbers of participants	Unknown/little specific information
Adventure racing	Commercially sponsored combinations of mountain biking, running, use of canoes, rock climbing, roping activities, use of a 'flying fox' and abseiling, possibly in association with orienteering and navigation. Entails transport of equipment and support crews. Can involve large numbers of participants and spectators. Increasing requests to access protected areas	Unknown/little specific information
Miscellaneous activities	Weddings, large group parties/social events, music events/orchestras. Can occur in caves and other sensitive environments	Unknown/no data

Derived from Newsome and Lacroix (2011) and Newsome *et al.* (2011).

that can be brought about by various recreation and tourism activities that may take place in the natural environment. Table 3.1 and Figure 3.3 indicate a range of non-motorised and motorised activities that can occur in different settings, such as on formed pathways, roads and other corridors, multiple-use trails, specific trails (bridle, mountain bike) and off-track. Specific recreational activities may not be sanctioned by management but still may take place. Various motorised vehicles – boats, all-terrain vehicle (ATVs), off-road vehicles (ORVs), quad bikes and motor/trail bikes – may also be used in natural areas, again with or without approval by management.

In addition to biophysical impacts there are also social and economic impacts associated with recreational activity and tourism development. Social impacts are briefly considered in this chapter as they relate to heavy use and crowding, and local community responses. Because of the specialised and diverse nature of economic, local community and cultural impacts these subjects are not covered in any detail in this book and the reader is guided to appropriate texts on page 200. Our main objectives are to provide on overview of biophysical impacts and to report some important recent work in this area. Biophysical impacts are examined according to major sources of biophysical impact and a global view is maintained. Many examples are drawn from the large database of research that has been carried out in the USA and more recently in Australia. For further details of methodology, assessment and the study of specific user impacts, the reader should consult the relevant cited work and further reading suggested at the end of this chapter.

Sources of Impact

There are various ways of usefully categorising the potential environmental impacts of tourism. Buckley and Pannell (1990) divided these into transport and travel, accommodation and shelter, and recreation and tourism activities in the natural environment. In terms of tourism and recreational activities, common sources of impact include boating, off-road vehicles (damage to tracks and landscapes), hiking, camping, mountain biking and horse riding. Particularly problematical is the presence of large numbers of people at a site such as a cave, waterfall or viewpoint. Sometimes the three main impact categories described by Buckley and Pannell (1990) occur together, with a cumulative impact. The degree of biophysical impact, however, will depend on the location and the type, diversity, intensity and duration of the activities themselves. Accordingly, the following major sources of environmental impact are examined (with increasingly complex assemblages of impacts): trampling and wear; access roads; permanent overnight dwellings; and focal points of activity such as rivers, lakes and wildlife habitats. Environmental responses will vary according to the resistance and resilience of individual species and ecosystems.

Tourism sites may be perceived to be at low levels of risk of biophysical damage where there are only few trails, little access and not many visitors. This situation often fits with the concept of wilderness tourism. Studies have nevertheless indicated

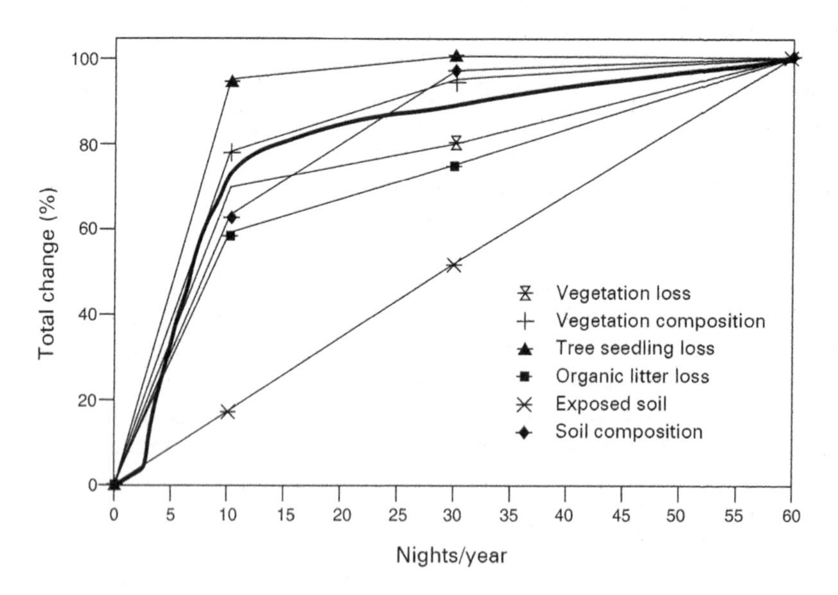

Figure 3.4 Change in campsite impact parameters under low to moderate levels of annual visitation, Boundary Waters Canoe Area Wilderness (From Leung & Marion, 1995; Hammitt & Cole, 1998)

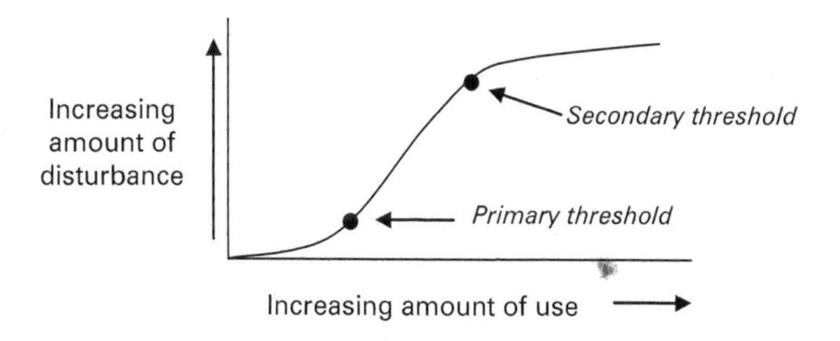

Figure 3.5 The proposed relationship between frequency of use and amount of disturbance, assuming two threshold points (Derived from Growcock, 2006)

that significant impacts can occur at recreation sites after only low levels of use (Hammitt & Cole, 1998; Cole, 2004). This work has shown that there is generally a rapid initial increase in impacts but this then slows even under moderate to higher levels of use (Figure 3.4), although Hammitt and Cole (1998) also point out that some impacts, such as campsite expansion and damage to trees at campsites, do not occur rapidly but instead show a gradual deterioration over time. The impact–response curve depicted in Figure 3.4, however, has been used to explain what is traditionally considered to be a common relationship in recreation ecology, where the response curve flattens out because the vegetation left behind is resistant to trampling or soil damage becomes limited due to the exposure of more resistant soil layers. Work by Growcock in the Australian Alps, nonetheless, indicated that other response curves are possible. Growcock (2006) found that an increasing amount of use resulted in continued deterioration (Figure 3.5), rather than resulting in a flattening out of the curve (Figure 3.4). A hierarchy of curves is therefore possible, reflecting different environmental conditions and scales of disturbance, as described by Growcock (2006) and Steven *et al.* (2011) (Figure 3.6). Steven *et al.* (2011) proposed that the hierarchy of different responses of birds to disturbance would be a function of the duration, intensity and extent of the disturbance. There would also be differences within each stage on the graph, depending on factors such as the species concerned, reproductive status, availability of alternative habitats and the nature of previous disturbances. Growcock (2006) suggests the hierarchy reflects loss of vegetation, with decreases in cover leading to increases in bare ground and soil erosion with increasing amounts of use (Figure 3.6).

A high and continuous impact potential may exist where there is high visitor pressure but also where there are many trails, roads, facilities and infrastructure. The latter situation comprises a spatial network of linkages and nodes. Networks include trails and roads, while nodes encompass car parks, campgrounds, river access points, day use areas and scenic lookouts (Hammitt & Cole, 1998). Concentrations of use mean that, although intense and severe impacts may occur at a node or along a

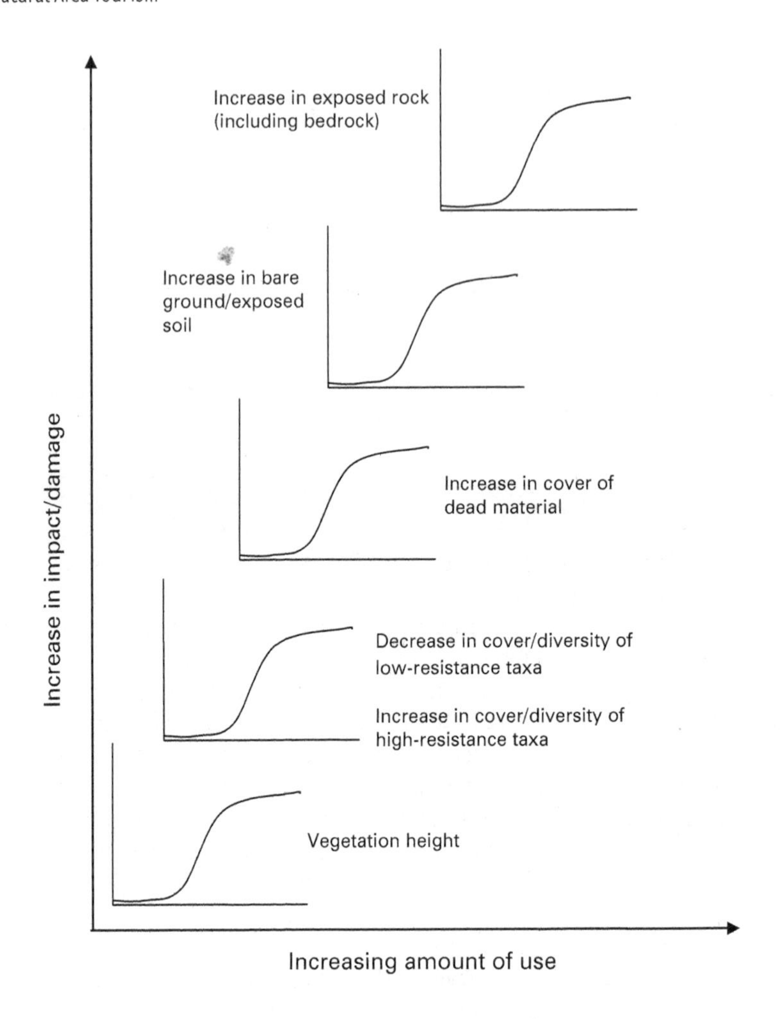

Figure 3.6 The hierarchy among environmental parameters with changing amounts of use (Derived from Growcock, 2006)

linkage, only small parts of large areas such as national parks are affected. In relation to this, Cole (1981a) estimated that only around 0.5% of two drainage basins in the Eagle Gap Wilderness Zone, USA (Figure 3.2), were significantly impacted by recreation. In contrast to this, small recreation areas such as some nature reserves, as in the case of Bukit Timah Nature Reserve in Singapore (Figure 3.2), are much more at risk of a larger percentage of the resource being damaged by recreational activity.

In determining the importance (severity or significance) of biophysical impacts Pickering (2010) provides a useful guide to the important criteria that need to be considered when evaluating the nature and degree of tourism/recreation-related

Table 3.2 Ten criteria for determining the significance of biophysical impacts

Criterion	*Situation contributing to the risk of high impact*
Conservation value of site	Where there are highly valued ecosystem components such as sensitive/ endangered flora and fauna and sites with significant biodiversity
Site resistance	Low (measured, for example, as the number of passes it takes to cause a 50% decline in vegetation cover – see Figure 3.9)
Capacity for recovery	Poor, for example plants that are slow growing are likely to have a lower resilience than rapidly growing plants
Susceptibility to erosion	High, as in the case of steeply sloping sites, erodible soils and heavy rainfall events as a feature of climate
Severity of direct impacts	In the case of high- impact activities such as horse riding and in-appropriate use of off-road vehicles. When large areas of bare ground are created, as in the case of informal trail networks made by mountain bikers
Severity of indirect impacts	Where there is the potential for colonisation by invasive weeds or for the spread of fungal pathogens
Amount of use	In the case of many visitors at all times/high usage, but this is dependent on how visitors use a site and the extent to which it is effectively managed. This relationship is complicated by the observation that a small percentage of 'bad' users can cause a lot of damage
Timing of use – social context	Periods of high usage such as weekends, holiday peak periods and public holidays resulting in overcrowding. Additionally depends on efficacy of management in the form of facility design and integrity and presence of staff and protocols to manage, supervise and control visitor behaviours
Timing of use – ecological context	The resistance and resilience of an ecosystem can vary over time. Seasonal aspects may contribute to the susceptibility of species during flowering periods for plants and breeding seasons for fauna. The suscepti-bility of soils to damage may be greater during wet seasonal conditions
Total area likely to be directly affected	Larger areas of damage will take longer to recover. Informal trail networks and areas of disturbance may become permanently impacted sites if they remain undetected and/or unmanaged

Adapted from Pickering (2010).

environmental impacts in natural areas (Table 3.2). She maintains that such criteria can be employed when a new trail is being planned or when management is considering upgrading an access road or recreational trail. Moreover, the criteria can be usefully employed when management is deliberating on whether a particular activity is appropriate for a particular trail or protected area. The criteria can also form part of the decision-making process in planning for natural area tourism, for example in the case of determining the recreation opportunity spectrum for a particular protected area (see Chapter 4).

Trampling

Vegetation

Trampling is a universal problem and damage to both soils and vegetation can take place as a result of visitors leaving established trails and pathways to traverse an area or take photographs, to follow the direction of a particular animal for a sighting or photography, or where certain users create informal trails to suit their own purposes. Trampling can also occur at sites of concentrated use or where visitor activity is not necessarily confined to trails. Consequently, the following common tourism/recreational activities are sources of trampling damage to vegetation: cross-country activities, camping and firewood collection, use of bush/informal toilets, horse riding, off-road vehicles, walking and hiking, wildlife viewing and photography, off-road bikes, trail bike motorcycling, access to riverbanks and viewing points, and boat launching. The possible effects of trampling are indicated in Figure 3.7. Liddle (1997) provides a comprehensive overview of the impacts of recreation on vegetation and notes that the type and distribution of visitor activity, amount and type of use, and the density and relative fragility of vegetation influence the degree of impact. Vegetated surfaces can be altered through the development of flattened vegetation or narrow trails, at one end of the scale, through to a situation where extensive areas are denuded of plant cover and dominant plants, such as trees, are lost.

Changes to vegetation as a result of wear consist of a reduction in plant cover, which is often measured as the proportion of the plant cover lost, the reduced height of vegetation or the reduced biomass of the original undisturbed vegetation. For example, Phillips (2000) investigated the trampling impacts of horse riding in D'Entrecasteaux National Park in south-west Western Australia and found significant reductions in vegetation height and cover (Figure 3.8). Such findings are in accordance with those of other trampling studies (e.g. Hylgaard & Liddle, 1981; Sun & Liddle, 1993a; Malmivaara-Lämsä *et al.*, 2008; Hamberg *et al.*, 2010).

Different vegetation types have been shown to respond differently to trampling. Liddle (1997) gives a useful summary of the resistance of different plant communities to trampling by walkers (Figure 3.9). The importance of these data are in their depiction of the contrasting sensitivities of different types of vegetation and the pertinence of its recreation ecology message (e.g. Hill & Pickering, 2009; Pickering & Growcock, 2009). For example, there is a marked contrast between two vegetation types that occur around Brisbane, Australia (Figure 3.2). While a subtropical grassland can tolerate 1412 passes before the vegetation is reduced by 50%, a eucalyptus woodland can tolerate only 12 passes before the same percentage of vegetation cover is lost. The reasons for such differences lie in the features of plants that confer resistance to trampling. Critical plant features are a low growth habit, growing points that are not easily damaged, flexibility and toughness. The latter structural aspects are conferred by the presence of lignified tissues, but too much lignin makes a plant inflexible and prone to snapping when trampled (Liddle, 1997; Newsome *et al.*, 2002). Grasslands are frequently more resistant because of their general low habit and growing points present just at the soil surface (e.g. Wiegmann

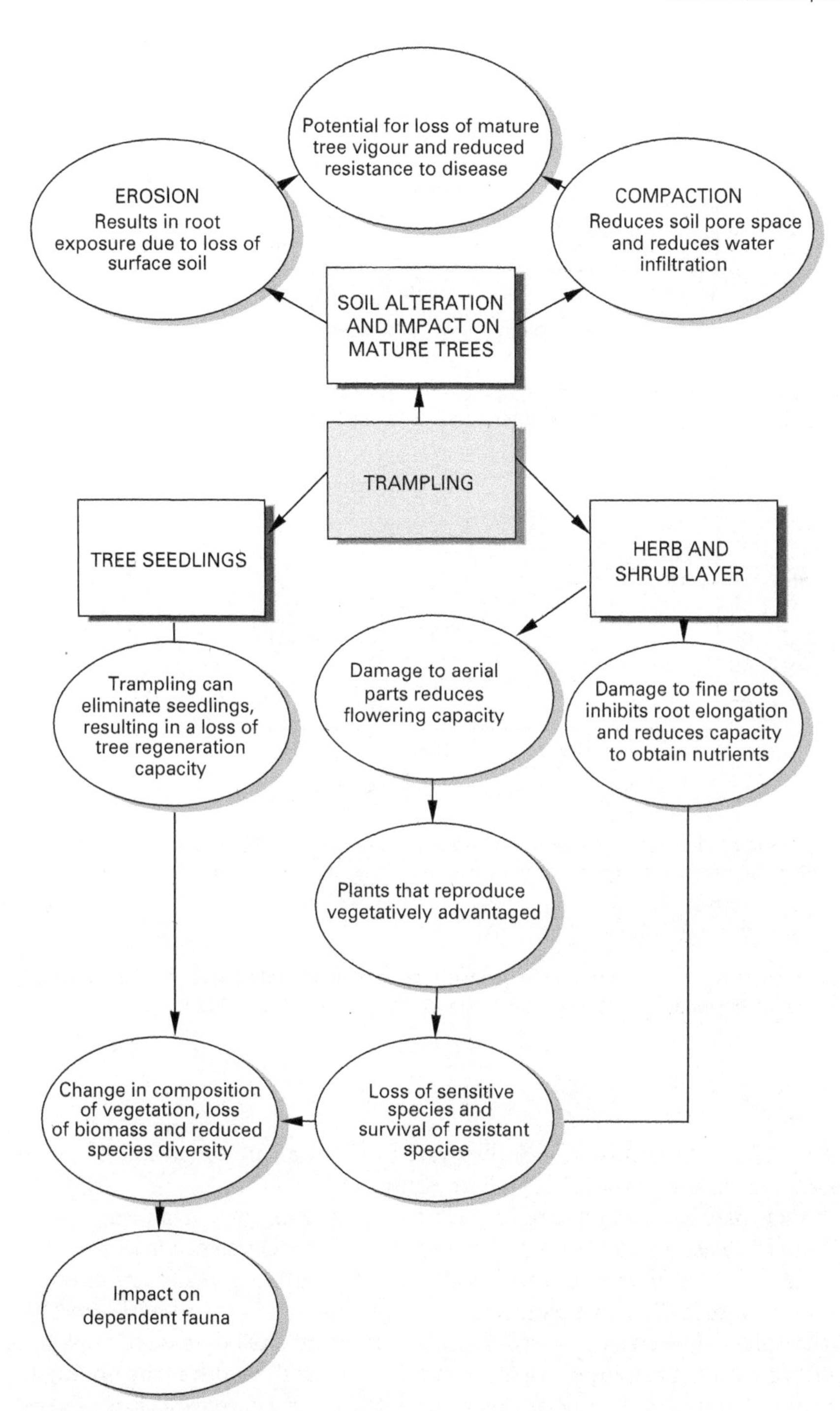

Figure 3.7 Impacts of trampling on vegetation and soils

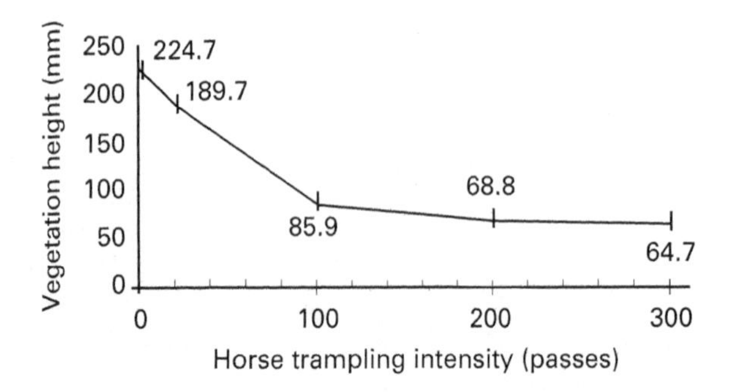

(a) Vegetation height averaged across the central 30–75 cm of a treatment transect cross-sectional profile after various intensities of horse trampling

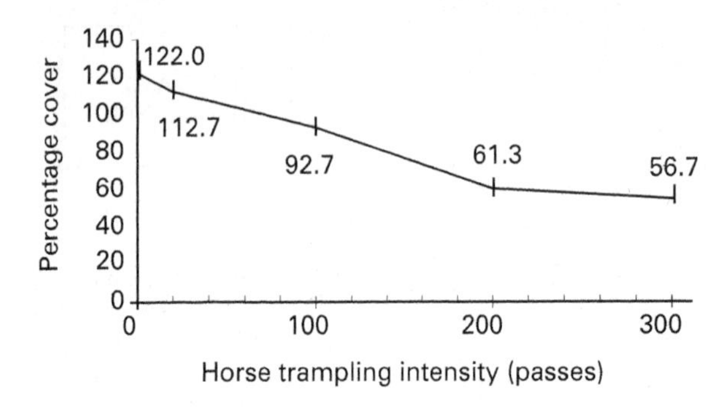

(b) Percentage of vegetation cover averaged across the central 30–75 cm of a treatment transect cross-sectional profile after various intensities of horse trampling

Figure 3.8 Result of a field experiment to quantify the environmental impact of horse riding in D'Entrecasteaux National Park, Western Australia (Derived from Phillips, 2000)

& Waller, 2006). Onyeanusi (1986) illustrated this in an investigation of off-road vehicle damage in the Masai Mara in East Africa (see p. 91).

Although many grasslands are, in terms of vegetation height and relative percentage cover, resistant to trampling, changes can nonetheless occur in the species composition of the plant community as a result of trampling. In a study in the Alta Manzanares Natural Park in Spain (Figure 3.2) Gomez-Limon and de Lucio (1995) found that plant diversity decreased at valley grassland sites that were popular for weekend recreational activities. An interesting aspect of the work is the finding that changes in soil and plant composition occurred before any detectable loss of vegetation cover. Loss of species diversity was linked to soil compaction. Gomez-Limon and

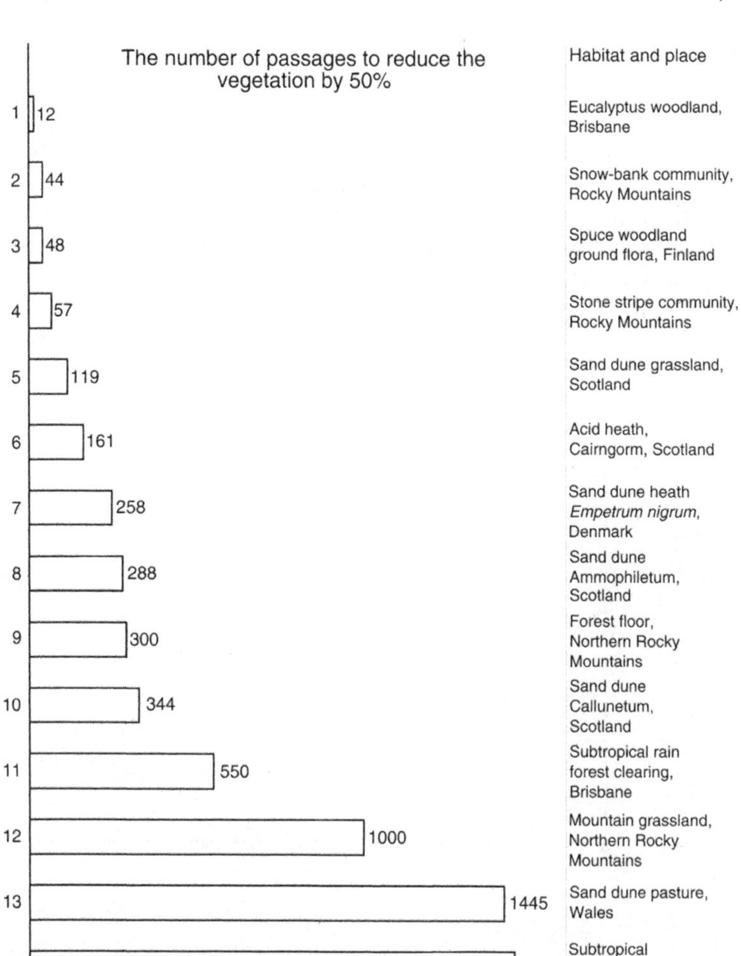

Figure 3.9 Resistance of different plant communities to trampling (Compiled by Liddle, 1997)

de Lucio (1995) suggested that a resistant species, *Spergularia rubra*, noted to increase as a result of trampling, could be used as an indicator of highly impacted areas.

Disturbances associated with the edges of trails and pathways frequently give rise to 'disturbance' communities. Changes in the species composition of a plant community are the result of sensitive species decreasing, more resistant species increasing and the possible arrival of invasive plants such as weeds (Liddle, 1997; Phillips & Newsome, 2002; Weigmann & Waller, 2006). Such changes can result in a decrease in plant diversity, with sensitive native and/or endemic species often being replaced with more aggressive native colonisers and/or exotic species. Some plant communities are particularly sensitive to trampling, as in the case of boreal forest under-storey, such as moss, lichen and dwarf shrub communities in Finland (Malmivaara-Lämsä *et al.*, 2008; Hamberg *et al.*, 2010). In contrast to these findings, increases in plant diversity have been recorded by Hall and Kuss (1989) in the Shenandoah National Park, Virginia, USA (Figure 3.2). In this case, plant cover and diversity were found to decrease away from the trails. Hall and Kuss (1989) considered that resistance to trampling and competition for light were the factors controlling the relationship. Trail-side plants were characterised by more trample-resistant growth forms (e.g. low habit), which would have benefited from the lack of competition for light from undisturbed vegetation.

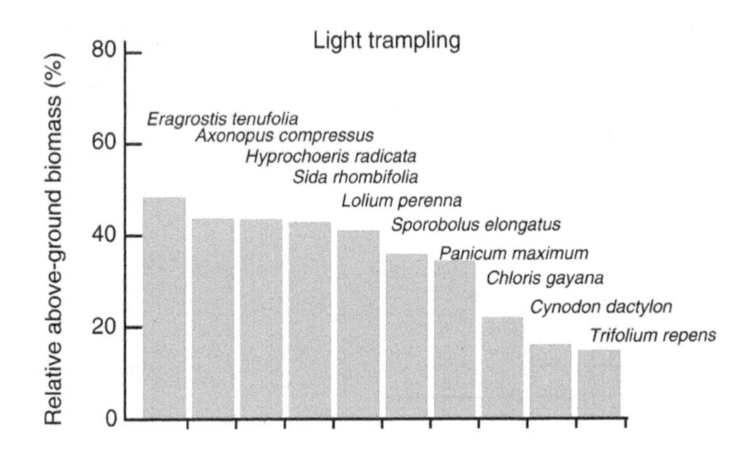

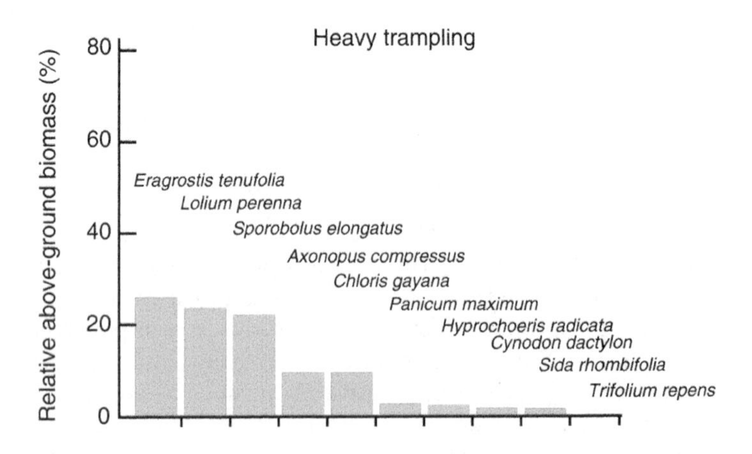

Figure 3.10 Change in relative biomass for different species of plant according to intensity of trampling (Adapted from Liddle, 1997; Sun & Liddle, 1993b)

The level of change also depends on the nature and intensity of disturbance. For example, Sun and Liddle (1993b) found that grassy tussock species were relatively resistant to trampling, while woody and erect herbs were more susceptible. They also showed that a reduction in root and above-ground biomass were both greater under a heavy, as compared with a light, regime of simulated trampling. Moreover, there was a change in relative biomass survival for different species, depending on whether they were heavily trampled or not (Figure 3.10). Liddle (1997) points out that the ultimate survival of vegetation depends on the morphological and physiological characteristics of individual species that comprise a plant community. This issue is readily illustrated by a comparison of two ecosystems. In the first case, reported by Boucher *et al.* (1991), the recovery of trail-side vegetation following trampling at a tropical rainforest site in Costa Rica was rapid, with substantial recovery after two years. Such recovery was related to the rapid growth of many tropical rainforest species under conditions where most of the important abiotic factors are not limiting. Boucher *et al.* (1991) suggested that a rotating system of trail access and closure would mimic the light-gap mosaic that is created when a large tree falls and allows light to penetrate the forest floor, stimulating the germination of pioneer species.

In contrast, environments where abiotic factors are limiting, as in the case of certain boreal forests, trampling damage can be significant and recovery slow (Malmivaara-Lämsä *et al.*, 2008; Hamberg *et al.*, 2010). In arctic alpine zones, plants are adapted to cold and dry conditions and the physiological response to this is slow growth, associated with poor recovery from disturbance (Liddle, 1975; Pickering & Growcock, 2009). Whinam and Chilcott (1999), working in the alpine zone of central Tasmania (Figure 3.2), found shrubs to be most susceptible to damage and most of the trampled vegetation had died within 42 days of the trampling period. The data show that vegetation continued to die even after the trampling had ceased. It was pointed out that the 'resting' of tracks (in contrast to the Costa Rica study) would not work, due to the very slow recovery rates of arctic alpine vegetation; other management strategies would need to be employed in such environments.

Biological crusts and microbes

Bacteria, fungi and related organisms play a key role in ecosystem function. They are instrumental in the functioning of key biogeochemical processes, as in the nitrogen and phosphorus cycles (see Chapter 2, p. 48). Moreover, soil microorganisms can determine the nature of plant communities by affecting the ability of plants to obtain nutrients (as in the case of soil mycorrhizae), or by giving certain species competitive advantage (via mutualistic relationships) in colonising and surviving on disturbed sites and new substrates.

An investigation into how trampling affected soil microbial communities at a subalpine campsite at Heart Lake, Montana, USA (Figure 3.2), was carried out by Zabinski and Gannon (1997). They found that there was a loss of microbial activity in the upper 6cm of soil in areas disturbed by camping activity. Such a decrease in microbial activity is likely to be due to a loss of vegetation, as that leads to a decreased availability of root exudates and organic matter as sources of energy for microbiota. As noted by Wardle (1992), any simultaneous loss of soil pore space that causes changes in aeration and water-holding capacity is also likely to have an adverse effect on soil microbes. The loss of microbiota, such as symbiotic mycorrhizae, can translate into less favourable soil conditions for plants and impact on nutrient cycling at the local scale. Newsome *et al.* (2002) flag recreational and tourism damage to ectomycorrhizal/proteoid root mats and also caution that the break-up and disruption of the fungal mycelial networks will reduce the vigour of susceptible plants. In addition to this, Waltert *et al.* (2002) report damage to mycorrhizal root mats due to trampling in *Fagus sylvatica* woodland in Switzerland.

In arid and semi-arid ecosystems, biological or microphytic crusts play a significant role in ecosystem processes. Based on a comprehensive review of the literature at the time, Eldridge and Rosentreter (1999) listed five important ecosystem functions of microphytic crusts:

(1) soil surface stabilisation and the reduction of erosion
(2) regulation of water flow into soils by absorbing and retaining water
(3) production of nitrogen and organic carbon at the soil surface
(4) sites for the establishment of higher plant seedlings
(5) refuge sites for soil invertebrates.

Cole (1990a) investigated the impacts of trampling on microphytic crusts in the Grand Canyon National Park, USA (Figure 3.2). Experimental work showed that after only 15 passes human trampling had reduced the microphytic crust cover by 50%. Cover was reduced to zero after 250 passes. Monitoring data showed that it took five years for the crust to establish itself to a level similar to the pre-trampled condition. Cole (1990a) also concluded that the crusts are highly susceptible to damage but are moderately resilient if trampling ceases and recovery can take place. In situations where random and dispersed trampling takes place, there is the potential for widespread and more permanent damage to the crusts.

Because of the role microphytic crusts have in stabilising the soil surface, their loss in arid landscapes could lead to increased soil erosion. The associated loss of soil nutrients, sites for the establishment of higher plants and attendant invertebrates is likely to lead to degradation of the biological condition of affected areas.

Soils

In terrestrial ecosystems, soils comprise a complex mix of abiotic and biotic components, provide a medium for the growth of vegetation and form the exchange pool for nutrient cycles. Soils thus exert a major control over the growth of plants and play a pivotal role in determining the nature of vegetation communities on land. Soil components, when eroded, can be moved through the landscape from one ecosystem to the next. Any loss of, or damage to, soil therefore has implications for the ecological integrity of nature-based tourism destinations.

Damage to soil arising from recreation and tourism is indicated in Figures 3.11 (trampling) and 3.12 (camping) (see also Figure 3.18, use of off-road vehicles, p. 140). Much of the research that reports soil erosion pertains to trails (p. 129); however, in

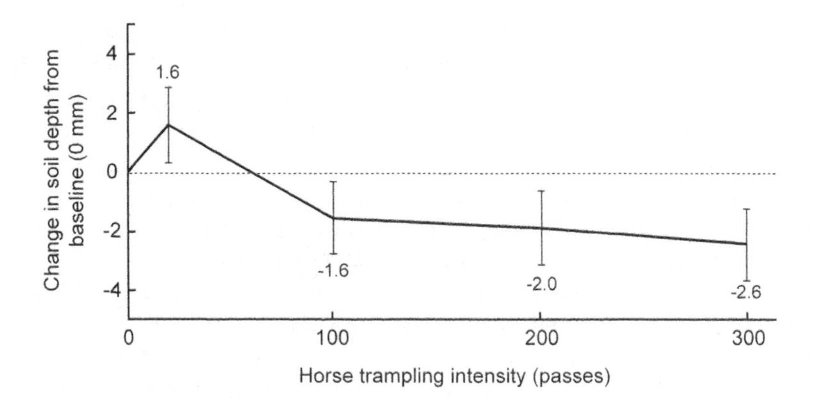

Figure 3.11 Change in soil depth from the baseline micro-topography averaged across the first 5–25 cm of the treatment transect cross-sectional profile after various intensities of horse trampling in D'Entrecasteaux National Park, Western Australia (Derived from Phillips, 2000)

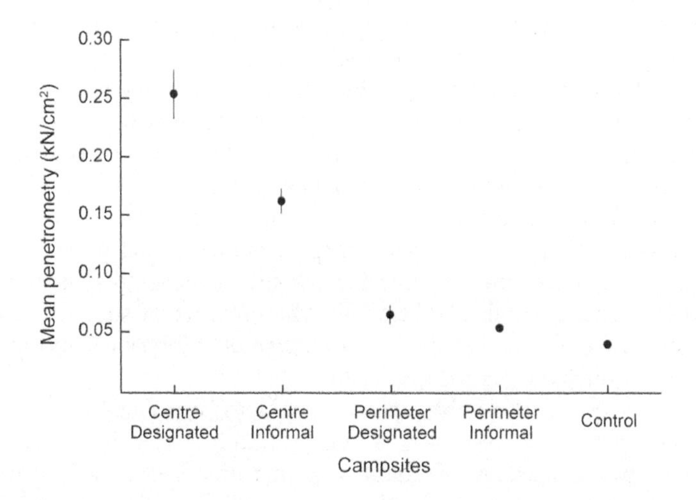

Figure 3.12 Mean penetrometry results for campsites in Warren National Park, Western Australia (Source: Smith, 1998)

off-trail situations soil damage is likely to take the form of changes in land surface micro-topography, loss of surficial organic layers, degraded soil structure, soil compaction and loss of permeability and porosity due to soil compaction (e.g. Liddle, 1997; Phillips & Newsome, 2002; Cole, 2004; Newsome *et al.*, 2004; Newsome & Davies, 2009; Pickering *et al.*, 2010b).

Soil compaction is a common problem and is measured according to bulk density and penetration resistance. For example, data from collected by Smith (1998) as part of a study on campsite impacts in Warren National Park, Western Australia (Figure 3.2), showed significant increases in bulk density and soil penetration resistance in the high-impact zones of campgrounds (Figure 3.12). Smith (1998) noted that the reduced infiltration capacity of the soil, caused by compaction, resulted in flat areas becoming saturated and boggy, and this in turn increased the need for visitors to excavate trenches around tents in order to facilitate drainage.

Of major importance is the degree of erosion that is occurring, and is likely to occur, in various recreational settings. Many soils can be inherently at risk from erosion and this is dependent on a number of important soil properties. Soil texture is a fundamental soil property and controls cohesion (the ease with which soil particles resist detachment), soil structure (the ability of the soil to form aggregates) and the infiltration of water into the soil profile. Soils dominated by sand and gravel are less cohesive than those with moderate to high clay content. The infiltration of water into the soil profile, on the other hand, is much higher in coarse-textured soils (coarse sand and gravel) than in fine-textured soils (those dominated by clay and silt) .

The formation of soil aggregates is dependent on the soil not being dominated by any particular size fraction. The presence of organic matter and 'cementing' agents such as calcium and magnesium also assists in the formation of stable soil aggregates. Such soil aggregates resist the splash impact of raindrops (see Figure 3.13), which, in the absence of soil structural aggregation, have the capacity to detach particles. Larger soil aggregates are more resistant to the detaching force of raindrops and their large size can increase the roughness of the ground; roughness acts to slow down any water that is moving down slope (Greeves & Leys, 2000).

Those factors that decrease the infiltration capacity of the soil promote the risk of surface run-off, which, in turn, can transport detached soil particles and erode soil. These factors include: fine-textured soils, poorly structured soils, hard surficial crusting and very shallow soils. Additionally, the presence of subsurface impermeable layers can slow down and reduce the drainage of soil profiles and increase the risk of surface run-off (Greeves & Leys, 2000).

Water erosion is a commonly reported problem, especially where raindrops strike bare soil. Aggregates are broken down by raindrop splash and detached soil particles can be transported by overland flow in thin sheets (sheet wash); heavy downfalls of rain can result in the development of rills (small channels) and gullies (large deep channels). Soil conditions that increase the risk of water erosion include low soil organic matter, poor soil structure (no or very little aggregate development), fine texture (especially silt and clay) and the impeded infiltration of water. Other factors include the length and steepness of slope and local climate. A critical aspect of climate is the amount, distribution, frequency and intensity of rainfall. Sudden heavy downpours on unvegetated, steep-sloping surfaces have considerable erosive potential. Vegetation protects the land surface from erosion. Initially, foliage dissipates the kinetic energy of raindrop impact and the interception of rainfall means that less water hits the ground. Second, the root networks of vegetation stabilises the ground surface by binding soil, enhancing the infiltration capacity and adding

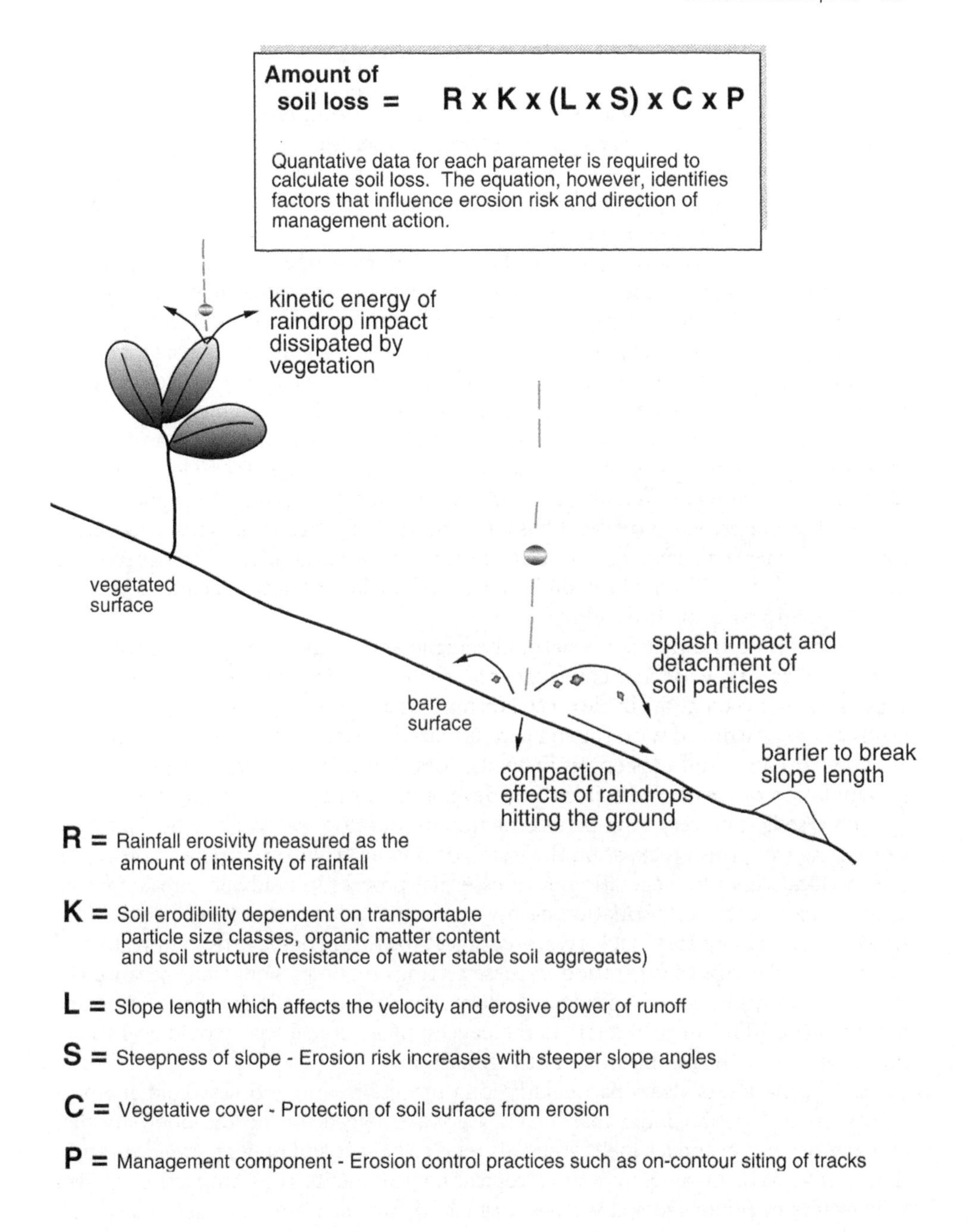

Figure 3.13 Components of the universal soil loss equation (equation developed by Wischmeir & Smith, 1978)

organic matter, which contributes to the development of soil structure. Different types of vegetation have different capacities to counter erosion. Kirkby (1980) demonstrated that forest was best at protecting soils from erosion, although Liddle (1997) indicated that the degree of protection from erosion is dependent on canopy continuity and height as well as the density of root mats and ground cover density. Rain falling through an open canopy or dripping off tall trees can erode soil in the absence of ground cover (Liddle, 1997).

The 'universal soil loss equation' (Figure 3.13), developed in the USA to assess water erosion on agricultural lands, has been applied in a recreational context by Kuss and Morgan (1984) and Morgan and Kuss (1986). The equation is used to predict soil loss, based on an assessment of rainfall characteristics (e.g. high-intensity rainfall), soil erodibilty (low organic matter, poor soil structure, sodic subsoils), slope characteristics (length and steepness), vegetation cover and land management factors (e.g. cleared land). Kuss and Morgan (1984) predicted that erosion, and thus a reduction in recreational carrying capacity, would occur if there was a decline to less than 70% of ground cover under different forests and woodlands. Kuss *et al.* (1990) pointed out that such an approach provides a basis for interpreting the sensitivity of different soil environments to degradation. Despite this, the applicability of this prediction model outside the USA has been limited because of a lack of detailed climate and soil data for many parts of the world.

Some soil environments are naturally fragile and easily damaged. Contributing factors are aridity, very cold environments, shallow rocky soils, wet soils and new substrates that become available for soil formation. For soil to develop, organic matter needs to become mixed with the soil parent material (e.g. sand, coral debris, volcanic ash). Disturbance and erosion result in the loss of organic matter and prevent the accumulation of weathered soil and the development of a deeper soil profile.

Some soils take very long periods of time to develop, especially when they are developing from fresh rock, as on the island of Hawaii (Figure 3.2). The colonisation of such landscapes by vegetation is an essential process in gradually breaking rock down as part of the soil formation pathway. Volcanic islands that have 'hostile' substrates, such as loose and friable lava, are subject to erosion damage from unrestricted hiking. Any damage to vegetation in these settings becomes significant because the thin and poorly developed soils do not allow the rapid colonisation of the landscape by vegetation. This in turn restricts the development of soil that would eventually further promote the establishment of vegetation.

Soil formation is also a particularly long process in arid and very cold environments. In the arctic tundra many soils are waterlogged during the brief summer period when snow and ice melt. Wet soils tend to be soft and undergo rapid compaction if trampled. In temperate and tropical environments the compaction of wet soils results in prolonged soil wetness, which, if further trampling takes place, can result in a permanent waterlogged trail or segment of pathway.

Arid landscapes are prone to wind erosion, which occurs in situations of low soil moisture, exposed surfaces and high wind speeds. Soils with a high silt content or fine sand are the most erodible. Where biological crusts are destroyed or where off-road vehicles erode large networks of tracks, wind erosion is more likely to lead to loss of both organic matter and soil nutrients and also to reduce the depth of soil. The wind erodibility of arid and semi-arid landscapes decreases as the size of the sand fraction and amount of clay in the soil increases.

Access Roads and Trails

Roads and traffic

Roads are an important means of access into and through natural areas and their ecological significance has been highlighted in Chapter 2. Moreover, many species benefit from the presence of roads, especially generalist animals that are able to exploit the edge effect and in particular those plants that colonise open areas such as road verges. Trombulak and Frissell (2000), however, state that wildlife impacts on roads commonly include changes in species composition and population size, the alteration and isolation of habitat and populations, deterrents to movement, extensive wildlife mortality and an alteration of hydrological and geomorphic processes. In addition, there is mortality from road construction, modification of animal behaviour, alteration of the chemical environment from exhaust, spread of exotic species and increased use by humans (see Figure 2.15, p. 68, and Table 3.3).

Mazerolle (2004) contends that the most immediate and direct effect of roads is mortality associated with vehicular collisions. According to Carr and Fahig (2001) life history traits and habitat requirements make many species vulnerable to road mortality and highly vagile species may cross roads with greater frequency and be susceptible to collisions. For example, according to Beaudry *et al.* (2008) road mortality is one of the biggest threats to semi-terrestrial freshwater turtles, due to their life history habits.

Table 3.3 Negative effects of road networks on wildlife

Impact	*Specific aspects*
Collisions with vehicles	Effects depend on wildlife traits and habitat requirements. For example, mortality can be significant if animals need to cross roads to access breeding sites
Pollution	Chemical run-off, especially of petroleum-based compounds; road salt attracts birds
Barrier effect (roads and roadside fencing)	Large gaps act as barriers
Noise	Can affect birds in particular: reduces density, abundance, nest site selection and disrupts communication
Artificial light	Attracts migrating birds, reduces habitat quality
Edge effects	Favour introduced species

Derived from Kociolek *et al.* (2010).

In terms of life history and habitat needs, snakes appear to avoid roads but suffer high mortality rates when they move across roads. Timber rattlesnakes (*Crotalus horridus*) cross roads slowly (10 cm/s) and stop moving and remain immobile for up to a minute or more in response to traffic noise (Andrews & Gibbons, 2005). Clark *et al.* (2010) report on the effect of roads on snake movement and habitat use and the implications for genetic structure, connectivity and gene flow among populations of timber rattlesnakes. Roads act as barriers to dispersal and natural population processes by isolating hibernacula (dens) and interrupting seasonal migration, resulting in fewer matings between individuals from hibernacula separated by roads. Rosen and Lowe (1994), in a study carried out in the Organ Pipe Cactus National Monument in the USA (Figure 3.2), found that up to 4000 snakes were killed per 22.5 km of road each year.

Jarvinen and Vaisanen (1977), in a study carried out in Finland, found that the edge effect created by roads favoured a number of species, resulting in an increase in the breeding populations of birds. Following on from these conclusions, Kuitunen *et al.* (1998) explored whether roads in central Finland impacted on the density of birds utilising roadside habitats. Their results showed that bird density was lower closer to highways. Kuitunen *et al.* (1998) considered that the reasons for such a decline could relate to traffic noise, gaseous pollutants from vehicles, some traffic-induced mortality and the road acting as a 'sink' for predators. The role of traffic noise in disrupting the defence of territories and habitat occupancy has also been reported to be a problem by Kociolek *et al.* (2010) and Goodwin and Shriver (2011). Kuitunen *et al.* (1998) pointed out that traffic-induced mortality (and road barrier effects) for birds are not significant when compared with the mortality suffered by other vertebrates. The situation, however, is complex, and specific to both situation and location, as evidenced by the Charles Darwin Research Station (CDRS, now the Charles Darwin Foundation), which reported that 4000 birds of 19 species were killed during a one-year period on a 40 km stretch of road on Santa Cruz Island, in the Galapagos (Figure 3.2) (CDRS, 2001).

The case with mammals is also a potential mix of positive and negative ecological impacts. For example, studies on the impacts of roads on the density, distribution and diversity of small mammals in the USA found that while five species did not prefer roadside habitats, more species actually benefited from the presence of roads (Adams & Geis, 1983). In this case more generalist species were attracted to the edge effect caused by the road (see p. 68) and any road kills were not detrimental to populations of small mammal.

The situation with larger mammals, however, is somewhat different, especially in the case of large carnivores. Gibeau and Heuer (1996) provided an account of the issues surrounding large-carnivore road mortality in the USA. Their work highlighted the role of roads in causing direct mortality, displacement of species and barrier effects, and as sinks for wildlife. Carnivores are particularly at risk because they have low reproduction rates, large territory and area requirements and low population densities. Table 3.4 shows road death data for six carnivores in and around Banff National Park, Canada (Figure 3.2). The large number of deaths of coyote (*Canis latrans*) resulted from the road acting as a sink, with coyotes being

Table 3.4 Highway mortality of large carnivores in the Bow River Valley, Alberta, 1985–95

Species	*Inside Banff National Park*	*Outside Banff National Park*
Coyote	117	39
Black bear	12	8
Cougar	1	2
Grizzly bear	1	0
Wolverine	2	0
Lynx	0	4

Source: Gibeau and Heuer (1996).

attracted to high densities of mice attracted to road edges. Gibeau and Heuer (1996) noted these mortality figures to be more typical of a harvested population rather than that of a population occurring in a national park. Coyotes were reported to be the fourth most frequently killed mammals in Yellowstone National Park, USA, after elk (*Cervus elaphus*), mule deer (*Odocoileus hemionus*) and bison (*Bison bison*) (Gunther *et al.*, 1998). Although the figures for bear deaths were relatively low (Table 3.4) they are, in fact, significant because of the low populations of these animals in the park. Similarly, the cougar (*Puma concolor*) occurs at low population densities and has large home ranges within the park. Gibeau and Heuer (1996) also pointed out that, with cougars, a road kill could result in the loss of the only male in a local breeding population. In contrast to Banff National Park, road kills appeared to be not as significant in Yellowstone National Park (Gunther *et al.*, 1998). Such differences may be due to a wide range of controlling features, such as the road siting and configuration, vehicle speeds, the pattern and extent of human visitation and ecological factors, illustrating the need to consider each case separately.

Road kills and the vegetation along quieter road edges are a source of food for grizzly bears (*Ursus arctos horribilis*). Roads, however, negatively impact on grizzly bears by acting as barriers, fragmenting habitats and reducing habitat effectiveness. Gibeau and Heuer (1996) reported on the low genetic diversity of grizzly bears in Banff National Park, Canada, which, in part, can be attributed to the barrier function of roads. Roadside fencing, which has been erected to reduce the road kill problem in Banff National Park, unfortunately was considered to enhance the barrier effect of roads and to fragment existing populations of bears.

Road kill is also a significant problem in Australia and according to Magnus *et al.* (2004) there are two major approaches to mitigating it: firstly, by altering driver behaviour (awareness through education or road modification); and secondly, by altering the behaviour of wildlife (discouragement, preventing access, providing safe crossings). Table 3.5 provides an account of various techniques, some of which have been trialled in Tasmania, Australia, for reducing the impacts of roads on wildlife. Underpasses, for example, have been employed widely around the world. Mata *et al.*

Table 3.5 Management context and evaluation of various strategies to reduce road kill in Australia

Mitigation strategy	*Evaluation in Tasmania*
Signage (wide variety of signs used)	Use sparingly to avoid habituation by divers
Escape routes to mitigate trapping effect of fences, banks and road cuttings	Very useful strategy, especially when applied along with new road construction
Ditch management (wildlife attracted to water and vegetation in ditches)	Objective is to reduce water accumulation and growth of tall herbaceous vegetation which is attractive to herbivores
Canopy crossings (closing of gaps created by roads in forested environments)	Cost-effective and relatively quick to install. Monitor to assess favourable conditions for success
Underpasses (various types and sizes)	Employ small underpasses where road kill of small animals is a problem
Odour repellents	Use of synthetic odour of canine urine to deter herbivores. Technique is currently undergoing trials
Ultrasonic whistle attached to the outside of a vehicle	Should not be considered as a road kill mitigation technique
Wildlife reflectors that reflect vehicle headlights onto the roadsides. Optical barrier created by reflected light	Effectiveness requires evaluation
Lighting to increase driver awareness and driver visibility	Expensive to employ. Effectiveness requires evaluation
Light-coloured road surfacing so that wildlife does not blend in with the colour of the road	Approach needs to be trialled and evaluated

Derived from Magnus *et al.* (2004).

(2008) investigated the use of motorway crossing structures in Spain and found that crossings were used by a variety of wildlife but overpasses were used more frequently by some species. Larger conduits were preferred by larger mammals but human use of the same conduits reduces their use by large wildlife. Jones and Bond (2010) found that the barrier effect of roads on forest-dwelling birds could be mitigated by the installation of well vegetated overpasses in tropical forest in Australia.

Many road kill problems occur on sealed roads but much recreation and tourism takes place where access is only on unsealed roads and with the use of off-road vehicles. For example, four-wheel-drive vehicles are often the only means of access into arctic alpine and most desert environments. Although off-road vehicles have the potential to cause damage in many different environments, their impact can be especially significant where plants and animals are at the limits of tolerance to harsh environmental conditions. Accordingly, the environmental impact of off-road vehicles in a range of environments is explored later in this chapter (see p. 139).

Trails

Trails in context

Nature tourism on land occurs mostly through corridors such as trails, pathways and sometimes informally along unsealed management tracks. Hiking trails form an important means of access and thus facilitate the recreational experience in many natural landscapes. Trails also serve to focus visitor attention, thereby helping to prevent more dispersed and randomised soil erosion and trampling of vegetation. There is of course a wide variety of trails and trail conditions. Trails differ in length, width and surface condition. Trail surfaces range from boardwalks and rubber or steel mesh or gravel pathways through to natural rock and soil. There may be engineered sections of trail, such as steps, staircases and viewing platforms. Trail networks may be highly managed and contain water bars, culverts, bridges, retaining walls and drainage cuts. Trails may or may not be maintained, as in the case of clearing obstructing vegetation and fallen trees, removing debris from sediment traps or by ensuring that wooden structures do not decay. Trails may have multiple uses and cater for walkers, cyclists and even vehicles on the same network. They may also be designed to be user specific, designated as being for hiking only or to cater for horse riding or mountain biking.

The degradation of trail systems through multiple treads, track widening, root exposure and soil erosion (track deepening) remain an ongoing problem worldwide (Figure 3.14) (Leung & Marion, 1999a; Marion & Leung, 2004a; Pickering *et al.*,

Figure 3.14 Severe trail erosion in Bako National Park, Sarawak, Malaysia (Photo: David Newsome)

2010b). Mountain areas, which attract many hikers, are particularly at risk of degradation due to steep slopes and harsh environmental conditions (Monz, 2000; Monz *et al.*, 2010b). Reviews of specific impacts associated with hiking carried out over the last decade include the work of Cole (2004), Buckley *et al.* (2004), Hill and Pickering (2006, 2009), Pickering and Mount (2010), Pickering *et al.* (2010b) and Steven *et al.* (2011). The condition of hiking trails and their effectiveness as access networks in natural areas is therefore a major management consideration and this aspect is explored further in Chapters 5 and 7. Here the focus is on the environmental impacts of various trail uses, especially in relation to trail damage, but the section also includes some consideration of impacts associated with moving off an established trail and the creation of new informal (illegal, unapproved, user-created) trails (Table 3.6).

Table 3.6 Comparison of actual and potential environmental impacts arising from three important recreational and tourism activities

Impact	*Hiking*	*Horse riding*	*Mountain biking*
On-trail impacts			
User conflict	✓	✓	✓
Degradation of existing trail network via trail incision, deepening, soil erosion and trail braiding	✓	✓	✓
Change in trail width	✓	✓	✓
Change to trail verge vegetation	✓	✓	✓
Tree damage and root exposure	✓	✓	✓
Transport of fungal pathogens	✓	✓	✓
Seed dispersal of introduced species	✓	✓	
Browsing of shrubs and vegetation		✓	
Nutrient enrichment from faeces and urine scolds		✓	
Disturbance to wildlife	✓	✓	✓
Off-trail impacts			
Creation and use of informal trail networks	✓	✓	✓
Loss of vegetation height and cover	✓	✓	✓
Damage to soils and creation of bare ground, leading to soil erosion	✓	✓	✓
Changes in plant species composition	✓	✓	✓
Transport of fungal pathogens	✓	✓	✓
Seed dispersal of introduced species	✓	✓	
Browsing of shrubs and vegetation		✓	
Nutrient enrichment from faeces and urine scolds	✓	✓	
Disturbance to wildlife	✓	✓	✓

Sources: Newsome *et al.* (2002, 2004, 2008) and Pickering *et al.* (2010b).

Trail degradation

Weaver and Dale (1978) surveyed the complex relationships that determine trail degradation. These included the amount and type of recreational activity, steepness and roughness of slope, the physical properties and moisture conditions of the soil, climate (rainfall characteristics) and vegetation type. Environmental controls on trail degradation have been investigated by Leung and Marion (1996), who presented a model containing climate, geology, user type and intensity of use as primary factors affecting trail degradation. Factors such as topography, soil, vegetation and user behaviour were identified as intermediate in importance. All these environmental factors need to be taken into consideration in predicting and assessing trail degradation, particularly the widespread problem of soil erosion.

Garland (1990) explored risk parameters as a means for assessing the erosion risk for mountain trails in the Drakensberg Mountains, South Africa (Figure 3.2). Access to the Drakensberg wilderness area is via thousands of kilometres of trails and due to steep terrain many of these trails remain at risk of degradation by soil erosion. Because of the increasing use of trails and the high cost of repairing damage Garland (1990) emphasised the importance of assessing the erosion risk before new trails are opened up (Box 3.1). Such approaches could be used in conjunction with the risk factors proposed by Pickering (2010) when planning new trail networks (Table 3.2).

Planning and management can reduce trail degradation (Marion & Leung, 2004a). Given that a constructed trail is a management footprint in need of maintenance to reduce impacts Randall and Newsome (2008) clearly demonstrated the benefits of planning, the application of management features and their subsequent checking and maintenance. In a comparison of two different sets of trails it was found that poorly planned (i.e. historical, user-created) trails (including the Peak Head and Bald Head trails) in Torndirrup National Park, Australia, which subsequently ended up being managed as formal trails, exhibited soil erosion, exposed roots and excessive width; trail proliferation was especially problematic where there were sections of indistinct trail or a viewpoint. In contrast, a section of the Bibbulmun Track in southern Western Australia, which has been subject to a higher level of planning and management intervention (alignment along natural contours, installation of maintenance features such as boardwalks, water bars and steps on sloped sections), was much less degraded (Figure 3.15). The presence of management features, however, does not always mean that they are effective, as demonstrated by Mende and Newsome (2006), who highlighted the importance of maintenance in preventing trail degradation in the Stirling Range National Park, Western Australia.

Multiple-use trails

Trails may be designed to cater for a variety of users and many trails around the world are currently shared by hikers, cyclists and even recreational vehicles (see Table 3.6). Wilson and Seney (1994) explored the relative impact of hikers, off-road bicycles, horse riding and motorcycles in the Gallatin Forest, Montana, USA (Figure 3.2). They found that horse riding and hiking caused more sediment loss than either off-road bicycles or motorcycles. In terms of environmental controls they

Box 3.1 Assessing the erosion risk in the Drakensberg Mountains, South Africa

Garland (1990) built on prior research that had shown that the most important erosion risk parameters in trail erosion were rainfall, soil type and slope. Previous work by Garland (1988) had shown that hourly rainfall intensity correlated with soil losses from trails. In addition, 'the long-term mean of the annual sum of all rain falling on the wettest day of each month' (MDR index) correlated strongly with levels of trail erosion.

The steepness of the slope is a well recognised factor in land stability and soil erosion. An assessment of the number of planned paths occupying various topographic slope classes (0–5°, 6–10°, 11–15°, etc.) provides an indication of erosion risk. In order to assess the soil factor in trail erosion, soil maps and analytical data are needed.

In the case of the Drakensberg Mountains, Garland *et al.* (1985) showed surface run-off to be one of the most important causes of trail erosion, in that it is the means by which detached particles can be moved. Soils which have different infiltration rates (see section on soils, p. 120), as in the case of soils derived from the Karoo sediments and Drakensberg basalts, will be at differing risk of surface run-off during rainfall events. Furthermore, those soils, like the Drakensberg basaltic soils, which exhibit aggregate stability are more able to resist the pounding effect of hikers. The major risk factors in trail erosion were divided into arbitrary classes and given scores (see below). Despite its partially subjective nature, Garland (1990) concluded that the method could be used in planning trail routes in the Drakensberg Mountains and flagged its applicability in other areas of extensive trail usage and development.

Scores for erosion risk parameters

Score	*MDR index (mm)*	*Lithology*	*Topographic slope*
1	<300	Basalt/alluvium	<6°
2	300–400	Sediments/dolerite	6–10°
3	401–500	–	11–15°
4	501–600	–	16–20°
5	>600	–	21–25°
6	–	–	>25°

Major land characteristics in each erosion risk class

Class	*Total score*	*Main characteristics*
1 low risk	<4	Sediments or dolerite only, sloping at <5°, MDR <300 mm
2 medium risk	4–6	Any rock type, MDR <600 mm and slopes <20°
3 high risk	7–12	Any rock type, MDR probably <400 mm and/or slopes probably >10°
4 very high risk	>12	Any rock type, MDR >500 mm

Source: Garland (1990)

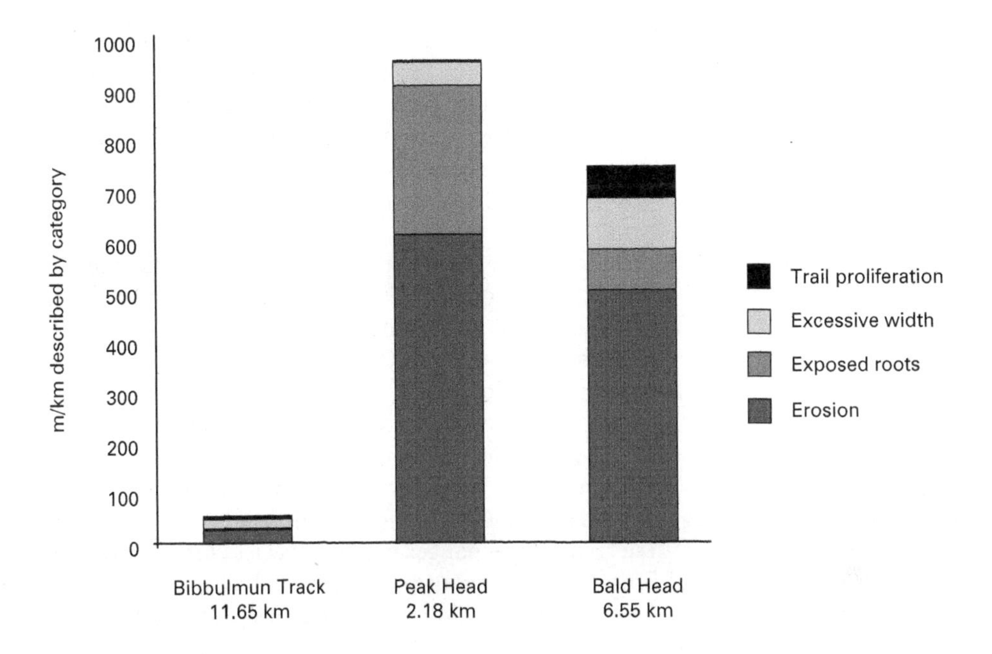

Figure 3.15 A comparison of degradation (m/km) between assessed trails in coastal south-western Australia (Derived from Randall & Newsome, 2008)

confirmed soil erosion to be positively correlated with steepness of slope and where trails followed slope, rather than contour, water was readily channelled downhill. It was also stressed that sediment yield, the precursor to erosion, is limited by detachment rather than transport. That is, the critical factor in initiating erosion is the ease with which soil particles are detached by the erosive force. Wilson and Seney (1994), however, acknowledged that the complex interplay of rainfall events, soils, vegetation type and geomorphology often makes it difficult to interpret which are the most significant environmental controls on trail erosion. Despite this difficulty they concluded that horses, by causing greater particle detachment, posed the greatest erosion risk and that sediment yield was greatest on pre-wetted trails (see p. 122). Moreover, of particular importance was the conclusion that these findings can be extrapolated to nearly all environments.

With increased visitation and usage of existing trail networks, and the public's requirement for a diversity of recreational experiences, user conflicts have arisen. For example, research carried out in the Sierra National Forests, California, USA, revealed that while 4% of horse riders disapproved of meeting hikers, some 36% of hikers did not like encounters with horses (Watson *et al.*, 1994). The use of trails by motorcyclists can create noise and air pollution problems that impact on the experience of other users (Newsome & Lacroix, 2011). Additionally, approved and non-approved use of trails by mountain bikers can cause conflict with hikers, especially over safety issues and the perceived impacts of mountain biking on trail networks (Davies & Newsome, 2009). Heavily used tracks are also likely to be subject to litter problems. This can be a particular problem for tourists visiting natural areas in those societies where the general public do not perceive littering to be of concern (see p. 157).

Horse trails/bridleways

The biophysical impacts of horse riding have been reviewed and summarised by Newsome *et al.* (2004, 2008) and Pickering *et al.* (2010b). Horses are important in the global recreation and tourism context. For example, there are 2.4 million horse riders in the UK alone and an estimated 20 million riders across the developed world (Newsome *et al.*, 2008). Horse riding may occur in a variety of recreational settings and when present in natural areas may be on multiple-use trails or dedicated bridleways, or off-track; it sometimes involves overnight camping. Newsome *et al.* (2004) explored the significance of the activity in the USA and Australia and identified as potential impacts trail degradation, multiple trail development, the loss of vegetation height and cover due to off-trail trampling, changes in plant species composition, the introduction of weeds and fungal pathogens, user conflict, fouling of water holes, collapse of wildlife burrows, disturbance of wildlife and local site degradation where horses are tethered at camp sites.

Many authors are in agreement that horse riding causes more damage to existing trails than do other uses (e.g. Dale & Weaver, 1974; De Luca *et al.*, 1998; Newsome *et al.*, 2004; Pickering *et al.*, 2010b). This can be attributed to the large magnitude of stress imposed on the soil surface. A horse and rider may weigh up to 500 kg and the weight-bearing surface is concentrated on four steel horseshoes. The forces applied by horse hooves are much greater than those of hikers or tyres and horse traffic readily penetrates and deepens tracks. For example, Lull (1959) reported that horses exert a ground pressure of up to 2.8 kg/cm^2, compared to a figure of 0.8 kg/cm^2 caused by hikers which was reported by Holmes and Dobson (1976).

The movement of horses along trails can also be particularly effective in introducing exotic weed species into pristine environments; horses may graze native species and can, in overnight stay and tethering areas, cause a reduction in the cover of vegetation (Pickering *et al.*, 2010b). Weed species readily germinate from horse faeces (in which there is a ready supply of nutrients) and colonise the untrampled, disturbed edges of trails. Indeed, a review of tourism-related weed vectors by Pickering and Mount (2010) concluded that weed seeds can be attached to the coat of horses and present in their dung and that horses are thereby capable of transporting weeds into protected areas. The establishment of weed communities can, in the longer term, provide a continuous source area for the invasion of exotic weeds along trail networks and into areas that experience regular natural disturbances, as in fire-prone ecosystems in Australia.

Newsome *et al.* (2002) wrote on the ecological significance of horse riding (e.g. commercial horse-back tours) in Australia and quantitative data have been collected on the biophysical impacts of horse riding in D'Entrecasteaux National Park in Western Australia (Phillips & Newsome, 2002). For example, trampling experiments showed that the percentage of bare ground increased from 5.2% at 0 passes to 31% following 300 passes. Horse riding changed the relative frequency of plant species by causing declines in herbaceous plants. Rapid reductions in percentage vegetation cover (e.g. 15.4% for 0–20 passes) and vegetation height (e.g. 56.5% at 0–20 passes) were also observed (see Figure 3.8).

Perhaps more significantly Newsome *et al.* (2002) reported that horses have never been a major component of Australian ecosystems and that the related or unrelated impact of weeds and introduced pathogens is particularly severe in such nutrient-poor ecosystems. Overall, Australian ecosystems are particularly vulnerable to the trampling impacts of horses because of slow growth associated with low nutrient levels, a long dry season, low soil organic matter, low soil resilience and easily disrupted surface mycorrhizal root mats. Susceptibility is further exacerbated by the brittleness of Australian vegetation and their propensity to stem snapping and slow recovery.

Cycling/mountain bike trails

Davies and Newsome (2009), Newsome and Davies (2009) and Newsome (2010) provide an account of the biophysical impacts of mountain biking. Their work describes the rise of mountain biking as a recreational activity and explains the complex nature of the mountain biking demographic in terms of rider styles and attitudes. Moreover, mountain biking comprises a spectrum of user activity, ranging from casual and family-oriented cycling through to dedicated thrill-seeking 'free riders' and 'dirt jumpers'. The latter two categories, because of the need for challenge, speed and technical skill, pose the greatest risk of environmental damage. Furthermore, mountain biking takes place in a range of outdoor settings, such as on approved and formed roads and pathways, multiple-use trails and on dedicated mountain bike trails, but also in non-approved circumstances, such as on hiking trails, informal trails or through areas where there are no trails.

Newsome and Davies (2009) recognise three major impacts associated with mountain biking:

(1) trail erosion and rutting caused by skidding and breaking (Figure 3.16) (often resulting in the formation of continuous linear ruts 1–8 m long)
(2) informal (user-created) trail development where mountain bikers seek greater challenge, shortcuts or to avoid trails with other recreational users
(3) unauthorised modification of both existing and informal trail networks – added features are referred to as technical trail features (TTFs), designed to add challenge, and comprise mounds, obstacles, ramps and even constructed seesaws, ladders and elevated cycle-ways (Figure 3.17; also see Newsome & Davies, 2009; Pickering *et al.*, 2010a).

Pickering *et al.* (2010b) have documented the specific impacts of mountain biking on biophysical conditions in the USA and Australia. Newsome and Davies (2009) report that informal trail networks can have significant impacts. For example, in John Forrest National Park in Western Australia, there are many areas affected by informal trails created by mountain bikers. In a snapshot study of just one area

Figure 3.16 Trail surface showing compaction and linear rutting caused by mountain bikes, John Forrest National Park, Western Australia (Photo: David Newsome)

Figure 3.17 Elevated cycle-way constructed off-trail for the purpose of mountain biking activity, John Forrest National Park, Western Australia (Photo: David Newsome)

of the park, Newsome and Davies (2009) found that 2540 m^2 of forest had been impacted in this way. Also in Australia, Pickering *et al.* (2010a) provide a detailed account of unauthorised TTFs in a 29 ha remnant woodland in Queensland. TTFs were made with soil, rocks and tree branches or constructed from materials such as cement and timber, transported to the site. Pickering *et al.* (2010a) found significant impacts in the form of 4010 m^2 of forest clearance and the creation of 1601 m^2 of bare soil in order to create 116 unauthorised TTFs.

Informal trails and their significance

Informal trails (illegal, unapproved, user-created) can develop where people go off-track to access shortcuts, wildlife and/or sites of interest such as riverbanks; they can also be deliberately created, as in the case of some mountain biking activities (Newsome & Davies, 2009; Pickering *et al.*, 2010a). Such unauthorised trail networks can damage ecologically sensitive areas, bring about changes in hydrology (e.g. increased channelling of water), degrade visual amenity, increase the disturbance of wildlife and necessitate potentially expensive management responses, such as site restoration (Newsome & Davies, 2009; Monz *et al.*, 2010b; Wimpey & Marion, 2011). Furthermore, the impacts of off-trail hiking and riding associated with informal trail development include damage to vegetation, such as loss of height, soil compaction, soil loss, reduced soil moisture, loss of organic matter, loss of ground cover vegetation, loss of native plant species, change in the composition of vegetation and the introduction of weeds and pathogens (Newsome *et al.*, 2002; Pickering *et al.*, 2010b, 2011; Wimpey & Marion, 2011).

Infection by pathogens such as *Phytophthora* is revisited here as a biophysical impact because there are a number of recreational activities that may take place off-track (e.g. mountain biking and orienteering – see Table 3.1), and it may be spread into natural vegetation as a result of informal trail development. As considered in Chapter 2 (see Box 2.2, p. 66) pathogens are a major threat to biodiverse communities in Australia (Newsome, 2003) and there is a clear association between the use of hiking trails and the spread of *Phytophthora* in Australia and the USA (Newsome, 2003; Cushman & Meentemeyer, 2008). The pathogen can be found on hiking trails in many parts of Australia and often spreads into adjacent native vegetation (Newsome, 2003; Turton, 2005; Daniel *et al.*, 2006; Boon *et al.*, 2008). Furthermore, *P. ramorum* is associated with hiking trails and is more common in regions with higher human visitation (Cushman & Meentemeyer, 2008). In terms of the introduction and spread of pathogens in protected areas Davidson *et al.* (2005) found that samples collected from 40% of the shoes of schoolchildren hiking on a short trail in a protected area in California showed the presence *P. ramorum.* With regard to pathogen viability Cushman *et al.* (2007) found *Phytophthora* to be viable for 24 hours if dry and up to 72 hours if shoes contained moist soil.

Once such pathogens are established they have the capacity to become a catastrophic ecological problem (Newsome, 2003; see also Box 2.2). For example, as discussed by the Dieback Working Group (2009), *Phytophthora* is able to survive within the roots of plants under dry conditions and is able to spread by root-to-root contact. Moreover, Shearer *et al.* (2004) caution that 2800 species of plant in south-west Western Australia are susceptible to infection and this represents a major visitor use management problem and risk to biodiversity in protected areas where the pathogen is present.

Use of trails for adventure racing and sporting events

During the last 10 years there has been an increasing trend for mountain biking to form the basis of organised action-based activities that take place in natural areas (Table 3.1). These activities focus on thrill seeking, competition, endurance and risk (Newsome *et al.*, 2011). Such approaches are likely to be in conflict with the contemplative and appreciative objectives of passive users of natural areas (Table 3.1; see also Newsome & Lacroix, 2011).

Newsome and Lacroix (2011) and Newsome *et al.* (2011) caution that the organisers of competitions and racing events are increasingly targeting natural and protected areas and note that there has been a rapid rise in public participation in such outdoor activities, as well as approval by the managers of protected areas. It is of concern that the trend is becoming institutionalised via promotion and marketing by tourism agencies and local government support in some countries (e.g. Australia). Moreover, Newsome and Lacroix (2011) and Newsome *et al.* (2011) maintain that the very nature of sporting activities and competitive events is inconsistent with the conservation and educative goals of protected areas, sustainable tourism and ecotourism (see Chapter 1 in relation to the dimensions of authentic ecotourism). Requests for, and the presence of, adventure, competitive and sporting activities such as rogaining/orienteering, horse-riding competitions and events, mountain runs and fun runs in ecologically important and protected areas should be viewed with caution. In particular, commercially sponsored combinations of mountain biking, running, canoeing, rock climbing, roping activities, use of a 'flying fox' and abseiling possibly in association with orienteering and navigation need environmental impact assessment, including the assessment of alternative sites, and a consideration of the capacity to manage such events in natural areas. This will be especially so when the activities involve the transport of equipment and support crews and feature large numbers of participants and spectators.

Nonetheless, we acknowledge that what comprises ecotourism and adventure tourism can be a subjective value judgement, as hiking to the top of a mountain, completing a long-distance walk or a gorilla-watching tour, which involves hiking and camping, may be considered by many to be adventure activities (see Chapter 1). Both hiking and the wildlife tourism, however, do have the opportunity for client engagement via interpretation (see Chapter 6), although this has been observed to be variable in the case of some forms of wildlife tourism (Newsome & Rodger, 2012a). But, as emphasised by Newsome and Lacroix (2011) and Newsome *et al.* (2011), the very nature of competitive sporting activity is more attuned to physical personal achievement rather than a passive, learning and natural-history engagement approach, such as bird watching or a physical but ostensibly appreciative activity such as hiking. Newsome and Lacroix (2011) therefore call for an increased dialogue on what are appropriate uses of a protected area and raise concerns about the message given to the general public if more and more sporting events are sanctioned in protected areas. They also reflect on the importance of preserving natural experiences. Has the push for increased use of protected areas (e.g. via the 'Healthy Parks, Healthy People' slogan) become corrupted, with some sectors of society failing to appreciate the impacts that their activities are having on the ecosystem, traditional park values and other users? Newsome *et al.* (2011) raise important research questions for the future, such as how well such events will be managed and what indicators of environmental damage can be used. Accordingly, various approaches to management and monitoring are discussed in Chapters 5 and 7.

Use of off-road vehicles (ORVs) as a recreational activity

Access to natural areas has been facilitated by the proliferation and widespread use of ORVs (motorcycles and four-wheel-drive vehicles) during the last three decades or so. Grant and Doherty (2009) estimate that in the USA nearly 50 million people had driven or ridden off-road at least once in the previous year and that in California alone the number of registered ORVs had increased by 108% since 1980 (also see p. 143). Accordingly, many natural areas that were once mostly accessible only through organised tours or expeditions are now available to a greater number of tourists due to rising levels of wealth, personal ownership of ORVs and an interest in visiting more remote locations. Increased access and visitation rates have occurred in the Australian and North American deserts, arctic tundra and mountainous areas, forests and beaches at many locations around the world.

ORVs are often used to view animals in savanna environments or for beach driving. Furthermore, ORV access allows people to penetrate remoter areas (e.g. Australian deserts) or sites that many other people cannot access. This creates the potential for the development of tracks, camping impacts, wildlife disturbance and pollution at more remote and relatively unused sites. In terms of track development Priskin (2004) found that along the central coastal region of Western Australia there were a total of 813 km of ORV tracks in 1998 compared with 517 km in 1965 and that the number of coastal access points had increased from 421 in 1965 to 908 in 1998. Furthermore, track density was greater in more remote areas where squatter shacks and bush camping sites were also evident.

The use of ORVs can also lead to social conflict. Bayfield (1986) found that in the Cairngorms (Figure 3.2) their impact, in particular their tracks, was one of visitor perception. He went on to state that their ecological impact is limited and that the impacts are social in nature. This is supported by a survey that showed that 72% of respondents thought vehicle tracks to be intrusive to their outdoor experience. It is also probably the case that associated biophysical impacts, such as noise, exhaust fumes and erosion scars, as a result of ORV activity also contribute to social discontent where hikers and ORVs use the same areas.

The use of ORVs has been found to impact on soils, vegetation and wildlife (Buckley, 2004a; Priskin, 2004). For example, Webb *et al.* (1978) studied the impacts of ORVs on soil properties in a heavily used designated vehicle recreation area in California. Changes in soil properties included increases in bulk density, decreased levels of soil moisture, a reduction in organic matter and plant nutrients and accelerated

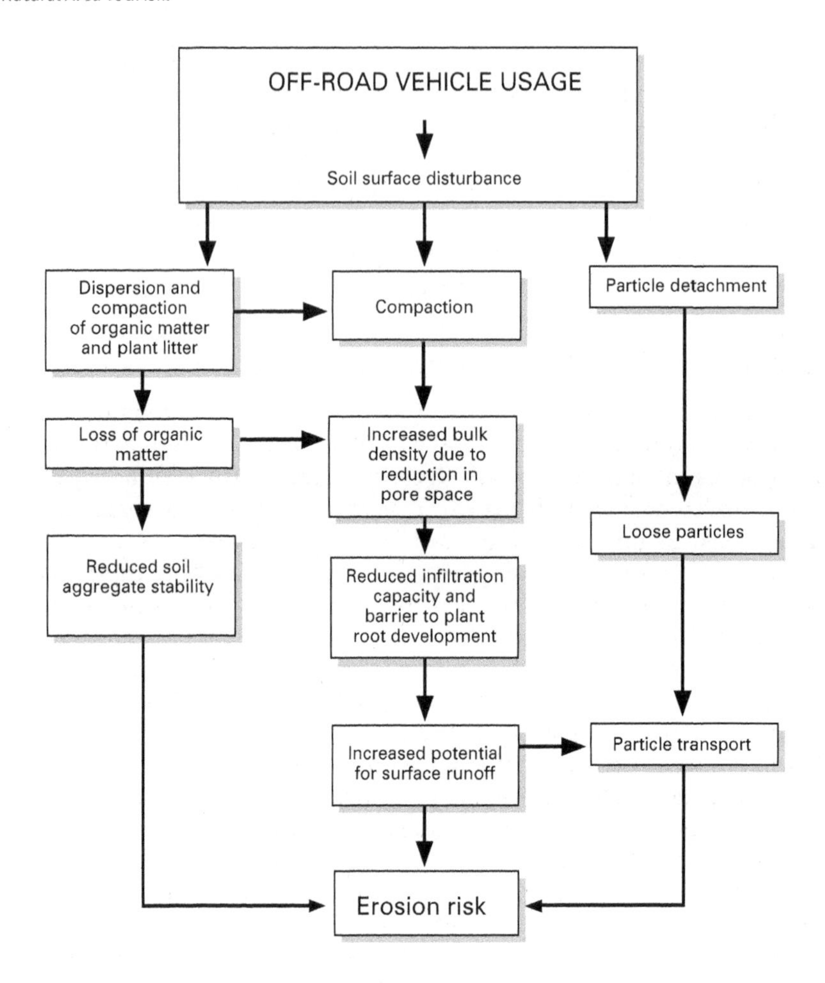

Figure 3.18 Impact of off-road vehicles on soils

erosion (Figure 3.18). More examples of the biophysical impacts of ORV access in different environments is provided below, with examples drawn from tundra, tropical, arid and beach environments.

Arctic alpine environments

Temperature is a major limiting factor and the growing season for plants is short. Low levels of biological activity are reflected in the low biological diversity that typifies arctic-alpine and tundra regions. Where the annual temperature regime is normally less than 0°C, sub-surface permafrost, overlain by a thin layer of soil, is a feature of the landscape. Rickard and Brown (1974) reported on the disruption of surficial organic layers and soils by ORVs, which then promote thawing of the

Box 3.2 Factors influencing the degree of impact caused by off-road vehicles

Environmental

- Surface soil characteristics
- Soil moisture content
- Vegetation type
- Slope
- Climate

Operational and vehicular factors

- Acceleration
- Speed
- Turning radius
- Wheel track pressure
- Wheel track configuration
- Skill and attitude of the operator
- Season in which the activity takes place

Source: Rickard and Brown (1974).

sub-surface permafrost. The change from ice to water results in subsidence (partly from the volume difference between ice and water) and the creation of ponded areas, which are then susceptible to erosion. Rickard and Brown (1974) also considered aspects of environmental sensitivity and the factors influencing the degree of bio-physical impact caused by ORVs on arctic tundra ecosystems (Box 3.2).

Rickard and Brown (1974) additionally considered a number of impacts in terms of increasing severity of disturbance in the arctic tundra landscape: aesthetic impacts; disturbance and damage to vegetation; destruction of plant cover and soil compaction, leading to soil erosion; and, finally, surficial peat disruption, subsidence of frozen ground and the ponding of water. They noted that the latter two impacts are dependent on higher intensities of ORV activity taking place during the summer.

Various other authors have also reported on the slow recovery of tundra vegetation following disturbance by ORVs (e.g. Greller *et al.*, 1974; Forbes, 1992; Kevan *et al.*, 1995). Furthermore, it has been shown that different plant communities within the tundra ecosystem show different susceptibilities to damage. Greller *et al.* (1974), for example, showed that the impact of snowmobiles was greatest on soil and rock lichens and rigid cushion plants. Those plants that resisted and tolerated the damage were found to have less height and a reduced woody biomass, such as grassy species. Other studies of vehicle damage to tundra vegetation reiterate the findings of earlier researchers that the passage of vehicles in tundra ecosystems results in a reduction of woody species, loss of vegetation cover and subsidence of vehicle tracks, and that recovery from damage is slow (e.g. Forbes, 1992; Kevan *et al.*, 1995).

Tropical environments

As discussed previously, a common impact of off-road driving is damage to vegetation and the increased potential for soil erosion. While arctic alpine tundra is particularly sensitive to damage, the savanna vegetation in Africa has been shown to be relatively resilient and to recover quickly (Onyeanusi, 1986). Other impacts, however, may be apparent, as in the case of the aesthetic impact of numerous track lines, which detract from the wilderness experience of visitors, for example in the Masai Mara in Kenya (see p. 91).

Wildlife disturbance is a problem associated with ORVs in various settings around the world (Buckley, 2004c). It can occur directly – as a result of road mortality – or indirectly – as a result of habitat alteration. Goosem (2000) provides an account of how roads can lead to changes in the faunal composition of a tropical rainforest in Queensland, Australia. Roads that penetrate forested environments allow increased levels of light to reach the forest floor and this in turn reduces the normally higher levels of humidity on the forest floor. The resultant microclimatic changes lead to a change in vegetation, particularly towards disturbance-adapted species. Such changes in the plant community alter the prevailing conditions for resident fauna. Goosem (2000) found that the community of small mammals along an unsealed road with a low traffic volume in the Kuranda State Forest (Figure 3.2) was different to that in a non-roaded area in the same forest. For example, *Rattus sordidus* and *Melomys burtoni*, non-rainforest mammals, had penetrated the forest by utilising the open habitat created by the presence of the road. Studies such as this show the potential for the penetration of non-rainforest and even exotic mammals into areas of forest, even along relatively unused and unsealed roads. Such changes, over time, could lead to an alteration in the abundance and diversity of small mammals. This is an issue that probably applies to most forest environments that are dissected by roads.

Arid environments

ORV activity in arid environments has been particularly investigated in the USA (e.g. Bury *et al.*, 1977; Webb, 1982; Webb & Wiltshire, 1983). The wider implications of soil compaction have been discussed by Adams *et al.* (1982), who reported that a reduction in the cover of annual plants was related only to low levels of ORV-induced soil compaction. As also observed in other environments, the response of vegetation to disturbance varied according to plant structure and growth habit. For example, Adams *et al.* (1982) found that large plants like *Erodium cicutarium* showed a greater loss of cover than grasses such as *Schismus barbatus*.

Cases of reduced biota – impacts on plants, invertebrates, reptiles and mammals – have been reported from the Algodunes in California, USA (Figure 3.2), a popular recreation area for users of dune buggies (Luckenbach & Bury, 1983). Although such human use of a natural environment can, strictly speaking, be distinguished as recreation, as opposed to tourism, such studies are directly applicable to the tourism situation. This is because, as shown by Luckenbach and Bury (1983) and others, these impacts can be brought about even by low levels of access and activity.

Impacts on wildlife in arid areas are of particular importance because those animals that are adapted to arid zones and sand dunes will be restricted to them. Furthermore, as in the arctic tundra, there are environmental limitations that give rise to slow rates of recovery from disturbance.

Edington and Edington (1986) summarised the ecological consequences of ORVs in desert ecosystems. Figure 3.19 depicts the impacts of ORVs on ecosystem structure and function that lead to reduced biodiversity in arid environments. Edington and Edington (1986) pointed out that there are many potential indirect impacts on animals. An example of such an indirect impact on fauna is the loss of food supply, such as desert annual plants. This is because different species of rodent are dependent on the different-sized seeds of various annual plants. In addition to this, the loss of shrubs means a loss of ecosystem structure, which translates into reduced cover. This in turn reduces the scope for: prey species avoiding predators; predatory species successfully ambushing prey; and animals readily gaining shade under high-temperature conditions. It also constitutes a reduction in the availability of breeding sites. Because vegetation is sparse and well spaced in desert ecosystems, such losses can rapidly lead to a reduction in faunal populations.

The need for desert reptiles and mammals to gain shelter from the excessive heat and predators in a sparsely vegetated environment has led to the behavioural habits of small reptiles burying themselves and the widespread use of burrows by larger reptiles and mammals in sandy deserts. Kinlaw (1999) stressed that open burrows are also a major resource for many other species and their presence in semi-arid and arid environments contributes to increased species richness. Kinlaw (1999) cited various examples of this from the USA and from southern Africa. In New Mexico, for example, burrows of the kangaroo rat (*Dipodomys ordii*) have been found to house 14 species of reptiles, 22 families of insects and 6 orders of other invertebrates associated with them. The 45% or more losses of mammals reported by Bury *et al.* (1977) and Luckenbach and Bury (1983) at heavily used locations is thus likely to translate into a wider ecological impact, depending on how the associated burrowing animals, as indicated above, are also affected.

Sandy beaches

Recreation on beaches comprises beach combing, swimming, surfing, wind and kite surfing, fishing, camping and beach driving. Nature-based tourism activities include bird watching, walking and viewing the seascape and coastal landforms (beach tourism-related impacts are considered further on p. 166). These activities can be in conflict with the use of ORVs, especially as the noise and visual aspects of beach driving can impact upon the natural experiential value that a beach can provide. According to Groom *et al.* (2007) there has been a massive increase in the use of ORVs in the USA alone since 1999. In the USA, ORV usage increased by 42% between 1999 and 2004, to 51 million participants. Schlacher *et al.* (2008) wrote that a large proportion of beach recreation in Australia involves the use of ORVs and recorded peak-season beach traffic of 30,000 vehicles per month and at times 5000 cars a day at North Stradbroke Island in eastern Australia (Figure 3.2). Schlacher *et al.* (2008) also made the observation that any seasonal recovery was not likely

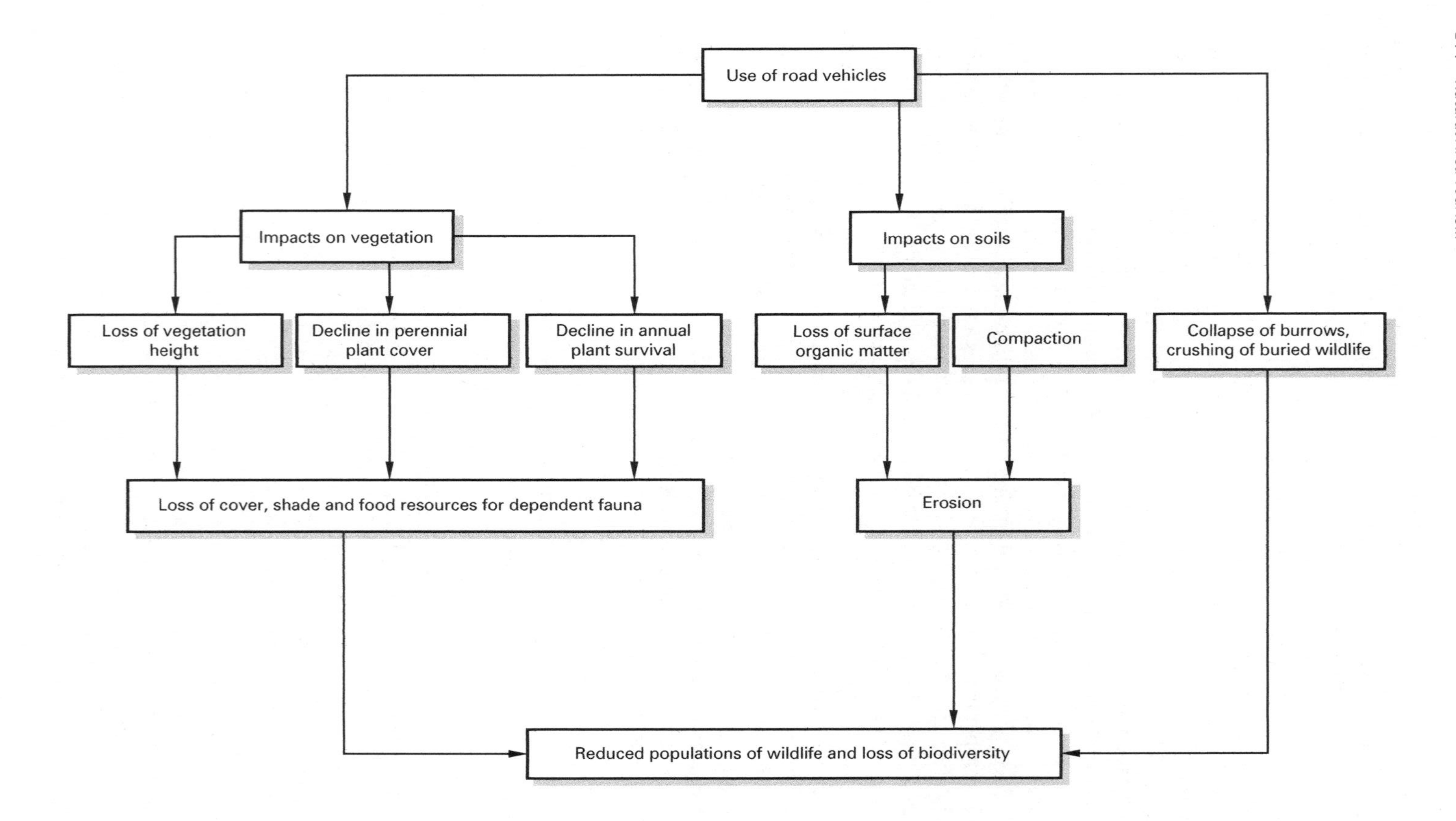

Figure 3.19 Environmental impacts of off-road vehicles in semi-arid and arid ecosystems (Based on Edington & Edington, 1986)

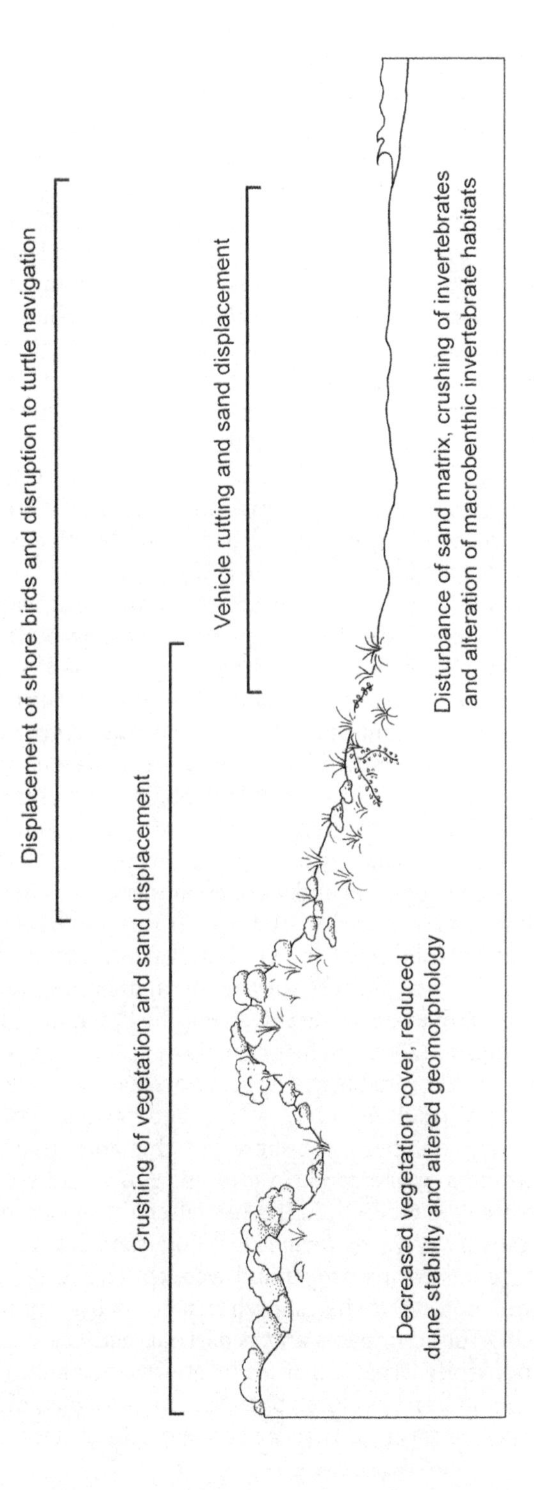

Figure 3.20 Environmental impacts of off-road vehicles on beach environments

to be possible due to the occurrence of 15,000 vehicles a month on the beach even during the low season. They also cautioned that areas impacted by ORV activity (Figure 3.20) include national parks and World Heritage sites.

Various studies have also reported damage to vegetation as a result of ORV use in beach environments. Broadhead and Godfrey (1977) observed the total destruction of above-ground vegetation occupying dune systems at Cape Cod in the USA. ORVs present a risk to rare species where beach driving takes place (Groom *et al.*, 2007). Such loss of vegetation increases the risk of wind erosion (Anders & Leatherman, 1987). Furbank (2010) reports on the impacts of beach driving on shore-nesting birds, including vehicle access to remote areas, vehicles crushing nests and eggs, disturbance of adults incubating eggs, disruption of feeding activity and the negative impacts of the energetic costs of birds having to escape from vehicles.

In an Australian case study (in south-east Queensland) Schlacher *et al.* (2008) demonstrated that the assemblage of macrobenthic invertebrates was much reduced on ORV-impacted beaches. Worms, clams and crustaceans can be killed directly. Given the ecological importance of these animals in sandy-beach food chains, species that prey on them, such as birds, crabs and fish, may be affected by changes in macrobenthic invertebrate populations (Figure 3.20).

Schlacher and Thompson (2008) quantified the physical impacts of ORVs on sandy beaches on North Stradbroke Island, Australia. During peak vehicle activity the density of tyre tracks per linear metre of beach was 2.69–6.35 on Flinders Beach (61% of beach affected by vehicle tracks) and 2.38–8.06 on Main Beach (54% of beach affected) (up to 90% in some parts). They recorded corrugated sand up to 28 cm in depth (mean 5.86 cm), with 57% of vehicle ruts deeper than 5 cm. Large volumes of sand were found to be disrupted (e g. 38,018 m^2 of sand disrupted in a single day on Main Beach) by compaction and displacement. When the top 30 cm of the faunal habitat zone of the beach was taken into account ORVs were estimated to have disrupted 5.8% of Main Beach and 9.4% of Flinders Beach habitat matrix in a single day. Vehicles were also seen to disturb the drift line and base of the fore-dune area. There are also likely to be fewer newly hatched turtles reaching the sea on beaches with many vehicle ruts, because they render hatchlings more vulnerable to predation as they are slowed down while traversing the beach in sight of predators.

Schlacher and Thompson (2008) predict that those species (ghost crabs, isopods, some polychaete worms) that inhabit the upper shore will be subject to significant habitat modification as a result of ORV traffic and residual corrugations. The corrugations are shallower on the lower shore but this zone receives more traffic due to the relative firmness of the sand. In a recent Australian study, Lucrezi and Schlacher (2010) reported that beach traffic had altered the size distribution and behaviour of ghost crabs (*Ocypode cordimanus*). Where sand was affected by traffic, crabs were smaller, were burrowing deeper and were constructing simpler burrows. Earlier, Neves and Bemuenuti (2006) had shown that the density of burrows is lower on beaches experiencing traffic impacts and pedestrian activity. Moss and McPhee (2006) have also demonstrated that densities of ghost crabs are lower on beaches with ORV impacts than on beaches closed to traffic. In particular, night driving may result in the death of nocturnal crabs that are active on the surface after dark.

Use of Built Facilities and Camping Areas

Environmental impacts associated with the construction and operation of built facilities

In many natural areas there is often a node of intense visitation (with its own potential impacts), such as a tourist resort, campground, day-use area, picnic area or car park. Accommodation and shelter, in particular, provide consistent focal points of activity that range from simple overnight huts, campsites and caravan parks through to resort and hotel development (Buckley & Pannell, 1990). The site clearance for a hotel complex may be substantial, resulting in a complex node with corridors acting as sources of disturbance. This is particularly so if the development is situated entirely in a natural setting. More commonly, however, such developments are on the edge of a natural area/national park and form part of a modified/non-natural landscape matrix. The widespread interest in beaches has put coastal areas particularly at risk and mangrove systems, rocky shorelines and seagrass beds are under threat from dredging, reclamation and the construction of boating facilities. One example of negative impacts associated with facility development is where hotel providers like to create sandy swimming areas for their clients. In Mauritius, seagrass beds are removed because they are deemed unsightly and unhealthy by certain hotel businesses. The practice has resulted in destabilisation of the seabed and disruption of ecological processes via the loss of fauna associated with the seagrass. Longer-term consequences are likely to include erosion, further ecological impoverishment and increased turbidity, which degrade the experience of nature at the site (Daby, 2003).

In many instances such developments and actions would be subject to formal environmental impact assessment procedures, as highlighted by Buckley *et al.* (2000). Ski resort development is a major issue in mountainous areas and the types of effects listed by Buckley *et al.* (2000) and others serve as a guide to the potential environmental impacts of tourist accommodation and resorts as well as their associated structures and facilities (Table 3.7). The degree and extent of any negative impacts, however, will depend on where the development is located, building design and adaptation to existing natural conditions, waste treatment systems, recycling and pattern of resource consumption as well as approaches to the recreational activities that take place in association with the tourism development (see Box 3.3). Accordingly, Chapter 5 provides an array of approaches to the management of built accommodation in natural areas.

Hotels and resorts clearly need access roads, water and energy supply plus waste disposal facilities. Infrastructure supply corridors frequently require regular maintenance and these, because of their open nature, constitute disturbance corridors, especially where they occur in a natural landscape matrix. Electricity generators can be semipermanent or even permanent sources of noise and the use of wood for fuel can result in coarse woody debris being taken from the surrounding natural environment. The latter issue can be particularly acute in mountainous environments where woody vegetation is already being removed for fuel, and the provision of mountain

Table 3.7 Environmental impacts of infrastructure and support facilities in the development of tourism

Activity	*Possible impact*
Land clearing	
Noise	Disturbance to wildlife
Light pollution at night	Disturbance to wildlife
Removal of vegetation	Loss of habitats Shift in species composition of area Smaller population of plants and animals Weed invasion Increased fragmentation of habitats
Soil erosion	Stream sedimentation and reduced water quality Sedimentation of coral reefs
Energy supply	Noise from generators Pollution from fumes and oil/reduced air quality Disturbance corridors
Water supply	Disturbance corridors Ground water abstraction/reduced water tables Construction of dams/disrupted stream flow
Waste disposal	Need for solid landfill or removal of waste off-site Liquid treatment facilities/odour, litter
Transportation infrastructure	
Roads	Nutrient, fertilizer, pesticide and oil run-off Road corridor impacts and noise from vehicles Barriers to animal movement
Airstrips	Noise
Boat landings	Damage to water margins

Adapted from Buckley *et al.* (2000).

village accommodation for travellers can put increasing stress on already depleted mountain forest resources.

Tourist facilities built in remote locations may also use local groundwater instead of water piped from dams. The intensive use of groundwater supplies in more arid environments can result in draw-down of the water table, which in turn can result in the death of vegetation. Any loss of vegetation, such as trees, constitutes a major impact on the biotic environment. Water and power supply can also be a critical issue in island settings (Gajraj, 1988). Jackson (1986) reported that, historically, several Caribbean islands experienced water and power shortages that were directly attributable to tourist demand exceeding capacity. At the time it was estimated that the average tourist used twice the amount of water as local people.

Box 3.3 Case study: Sukau Rainforest Lodge, a tour and accommodation facility situated on the banks of the Kinbatangan River, Sabah, Malaysia

In recent years tour operators and tourism facilities have made substantial progress in developing bona fide ecotourism practices and minimising their footprint via building design, education and sustainability awareness. There are now many examples of serious attempts at sustainable tourism through 'green' design that focuses on solar energy, recycling, waste minimisation, water conservation and waste disposal. One example is Sukau Rainforest Lodge, located in Sabah, Borneo, an ecotourism facility that offers day and night guided boat tours, and interpretive walks on and around the Kinabatangan River. Single boat trips are limited to eight persons and during wildlife observation electric motors and oars are used to minimise noise and exhaust from engines (Figure 3.21). The accommodation facility is built on stilts and is located 30 m from the riverbank. Nearby native vegetation has been retained. A significant contribution is made towards reducing the footprint of the lodge through the promotion of environmental awareness, environmental waste disposal, employment of recycling and the promotion of the sustainable use of energy. In terms of reducing the impacts of resource use, rainwater tanks are used and the power generator is noise insulated and solar supplied. Solid waste is transported out and septic tanks have been installed to deal with toilet wastes. Recycling is in place and clients are encouraged to participate. Lodge management provides opportunities for visitors to become involved in local conservation projects. Accordingly, clients are provided with the opportunity to plant trees on the edge of oil palm plantations, along the riverbank where riparian vegetation has been lost and in previously logged areas.

Figure 3.21 Boat trip from Sukau Rainforest Lodge with wildlife observation (Borneo Eco Tours use electric power in order to reduce noise impacts when approaching wildlife) (Photo: Albert Teo)

As already noted (Table 3.7) various forms of environmental pollution can arise from the operation of permanent built facilities. This can range from visual pollution in the form of badly sited or designed hotels or ski complexes, solid and liquid waste disposal through to the ill-considered use of pesticides to control insect nuisance. The use of insecticides may be substantial in areas liable to termite damage or malaria and can particularly impact on bird populations through direct acute toxicity or by the concentration of toxic compounds along food chains.

A major issue, however, is that of waste disposal and its effects and significance depend on the volumes produced, the application of recycling, waste prevention strategies and the nature of the receiving environment. The problem of waste disposal falls into the two main categories of solid and liquid wastes. Solid waste tips, depending on their location, typically attract opportunistic and scavenging birds such as gulls, crows, kites and vultures. Increased numbers of aggressive species like gulls have the capacity to displace and predate local populations of resident species. This has been reported from islands along the Great Barrier Reef, where silver gulls (*Larus novaehollandiae*) predate the eggs and chicks of the crested tern (*Sterna bergia*); the breeding population of the terns is impacted upon by gulls attracted to tourist sites (Edington & Edington, 1986).

In a survey of the activity of animals around solid waste dumps Edington and Edington (1986) found that the presence of wildlife constituted two main problems for humans. The first comprises a direct threat to tourists from large and powerful species, such the black bear (*Ursus americanus*) and grizzly bear (*Ursus arctos horribilis*), which are attracted to refuse in the USA park system. Polar bears (*Ursus maritimus*) are also considered to be a threat where tourism occurs in the arctic region. The second issue is that of disease transmission through the activity of gulls, rats and flies. Gulls in particular are attracted to refuse tips and can transmit disease when they also congregate at tourist sites such as picnic and bathing areas. Iveson and Hart (1983) demonstrated that wildlife can be a reservoir for diseases and reported the occurrence of salmonella in mammals, birds, reptiles and frogs on Rottnest Island, a tourist destination in Western Australia (Figure 3.2).

Edington and Edington (1986) have described the role of rats in transmitting Weil's disease. Rats transmit the pathogen *Leptospira icterohaemorrhagiae*, which can enter water through rat urine. They also note that this becomes an issue where refuse tips are in close proximity to swimming areas. Refuse tips can also become breeding sites for flies and mosquitoes and thereby increase the risk of insect-borne disease.

Liddle (1987) has reported on the disposal of liquid wastes in recreational settings. Sewage enters natural areas from tourist developments, boats, chemical toilets and buried faeces (see p. 158). The general effects of sewage pollution in freshwater ecosystems are summarised in Figure 3.22. The relative impact of sewage disposal depends on volumes discharged, the degree of treatment, dilution factor and flushing regime of the water body. Those areas with poor flushing regimes, enclosed lakes and small bodies of water will be more susceptible. Moreover, oligotrophic lakes, which have naturally low concentrations of nitrogen and phosphorus, are particularly at risk of ecological damage from organic effluent.

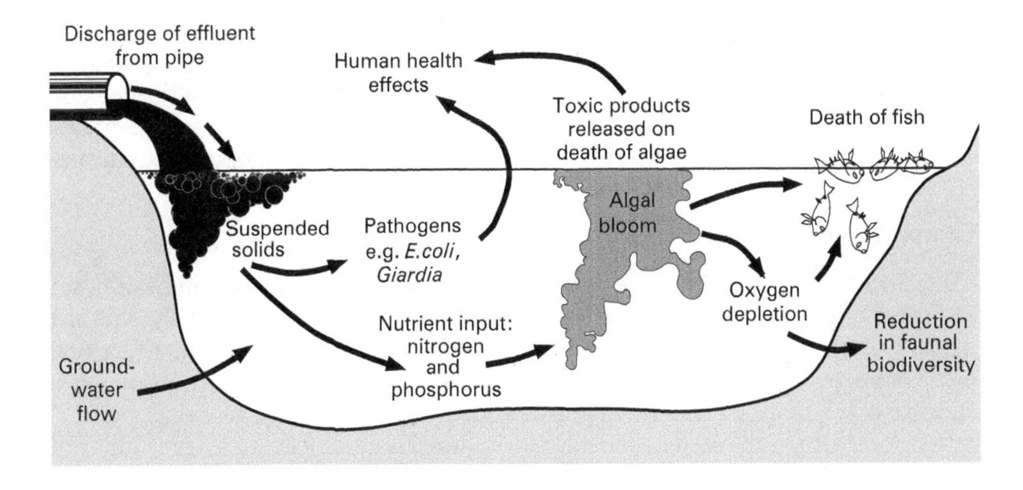

Figure 3.22 Conceptual diagram of the effects of sewage pollution in freshwater ecosystems

The changes that occur as a result of sewage entering rivers and lakes include increased levels of suspended solids and nutrients and decreases in the oxygen content of the water. These changes in abiotic factors provide different conditions for microorganisms and invertebrates. The increased levels of nutrients give rise to dramatic increases in algal biomass and the resultant intense metabolic activity reduces the oxygen content of the water. De-oxygenation of rivers and lakes results in the death of fish and those invertebrates that are unable to tolerate low oxygen conditions. In severe cases of pollution only tolerant species, such as sewage fungus, tubifex worms and chironimid larvae can survive.

In marine settings the major risk is the contamination of bathing areas by pathogens and the degradation of coral reef systems (see Table 3.12, p. 170). Liddle (1997) pointed out that marine systems differ from freshwater environments in terms of water chemistry (often naturally low levels of phosphorus and nitrogen), the larger volume of water and in the presence of tidal cycles and wave action. Sewage impacts are much more pronounced where there is reduced flushing and the residence time of nutrients is measured in days or weeks. Various authors have documented organic pollutant damage to coral reefs, reflected in declines in species diversity, a reduction in the size of coral colonies and significant decreases in coral cover (see Liddle, 1997).

As reported by Morrison and Munroe (1999), the management of wastes is a critical issue in many island settings. Many islands around the world have already established tourism industries and others are developing. These same countries are faced with aesthetic impacts as well as land and water contamination problems arising from waste generation and its disposal. Landfill can give rise to odour, groundwater contamination, wind-blown debris and dust, while the burning of refuse and rubbish results in smoke and odour. The management of wastewater varies from country to country, from effective facilities through to overloaded and outdated systems such as septic tanks and open latrines that pose the risk of disease and inefficient incineration of solid wastes, which can result in air-borne pollution.

Camping and campsites

Camping is one of the most popular of all recreational activities and can occur in formal (approved) and informal (non-approved) situations. There are also varying degrees of management, ranging from extensive, hardened sites to cleared natural areas through to unmanaged situations where campers simply pitch a tent on a suitable patch of ground. Camping in the USA may be described as 'front-country' or 'back-country', the latter occurring in minimally developed areas generally not easily reached by roads. Such camping areas may have groups of campsites or occur as individual pitch sites located at some distance from each other. Front-country campsites, on the other hand, can be readily accessed by road in two-wheel-drive motor vehicles. In Australia, front-country sites (as they are known in the USA) are often specially designed as camping areas with tent-pads (permanent hardened areas for pitching tents), on-site toilets, facilities such as constructed shelters and dining areas; car bays are often included and rangers or campsite hosts may service the site.

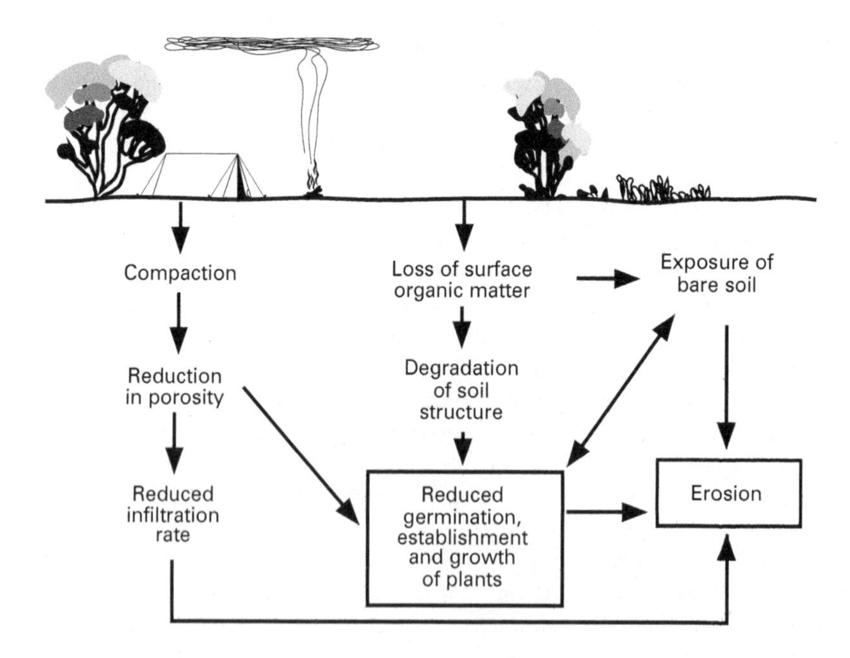

Figure 3.23 Campsite impacts on soils and ultimate effects on vegetation

Various workers have shown that, because of its highly concentrated use, camping has the capacity to result in significant localised effects, which include impacts on soils, vegetation, wildlife and riparian zones, as well as social impacts (Huxtable, 1987; Cole, 1990b; Kuss *et al.*, 1990; Hammitt & Cole, 1998; Smith & Newsome, 2002; Cole, 2004; Cole *et al.*, 2008). The more obvious impacts on soil and vegetation are summarised in Figure 3.23. The overall biophysical problems occurring as a result of camping under non-managed conditions is soil compaction and damage to vegetation, resulting in the proliferation of bare areas and changes in the composition of vegetation in and around campsites (e.g. Smith & Newsome, 2002; Cole, 2004; Reid & Marion, 2004). Clearly, where formal campsites occur there will be a permanent cleared area, known as a management footprint. In many parts of the world such areas are often hardened and contain facilities such as fire rings and other facilities, as in the case of Australian front-country campsites. Under these conditions the zone of intensive use comprises a permanent camping area of cleared vegetation. This can be contrasted with less frequently used, non-hardened back-country and informal (non-approved) campsites. In these latter cases, the impacts of camping beyond designated camping areas on soils and vegetation become much more of an issue, especially where the camping takes place in 'pristine' riparian or alpine and other fragile natural environments.

Firewood collection

Although many formal (permanent) campsites have toilets, designated tracks and wood supplies provided to reduce the zone of impact, informal firewood collection from the area surrounding a campsite can result in the trampling of vegetation and the disturbance of ground-dwelling and tree-hole-nesting wildlife. Additionally, wildlife can be disturbed indirectly by camping activities, through the collection of dead wood or coarse woody debris, as this is an important structural component and functional element in forested ecosystems and wholly or partly forms the habitat for a range of forest organisms (Harmon *et al.*, 1986; Freedman *et al.*, 1996; Lindenmayer *et al.*, 2002; Woldendorp & Keenan, 2005; Smith *et al.*, 2012). Thus, while soil compaction and damage to vegetation as a direct result of camping may not be significant at permanent, hardened sites, the collection of firewood can be. This is because firewood collection and the associated formation of informal access pathways occur where firewood is not carried by campers or not provided by management. Campers are then likely to search the area around the campsite for wood, with small pieces being sought for initiating the fire and larger pieces (dead standing trees may be cut) then harvested in order to sustain the campfire (Smith *et al.*, 2012).

Cole (1990b), Liddle (1997) and Smith and Newsome (2002) have reported that where campfires are allowed, there is a gradual depletion of coarse woody debris (CWD) in the surrounding area. Losses of CWD can extend well beyond the immediate camping area. The ecological importance of CWD in forested ecosystems is well established, as indicated by the work of Harmon *et al.* (1986), Christensen *et al.* (1996), Hecnar and M'Closkey (1998), Bowman *et al.* (2000), Driscoll *et al.* (2000), MacNally *et al.* (2001), Grove *et al.* (2002), Lindenmayer *et al.* (2002) and Woldendorp and Keenan (2005). Harmon *et al.* (1986), for example, provided a comprehensive

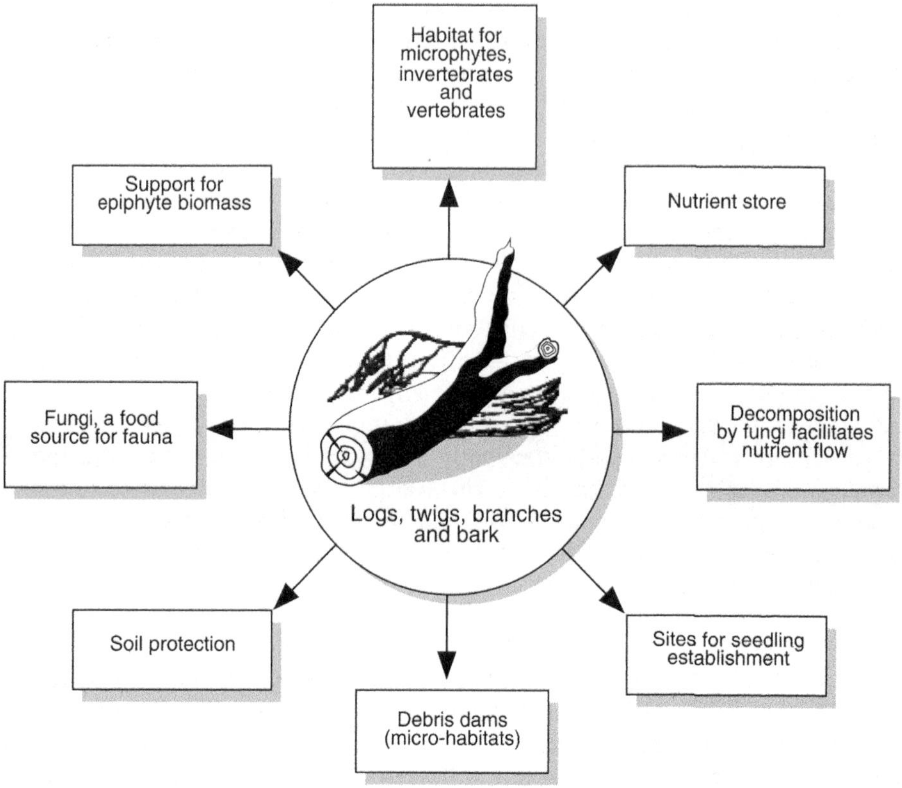

Habitat for microphytes, invertebrates and vertebrates

Support for epiphyte biomass

Nutrient store

Fungi, a food source for fauna

Logs, twigs, branches and bark

Decomposition by fungi facilitates nutrient flow

Soil protection

Sites for seedling establishment

Debris dams (micro-habitats)

Figure 3.24 Ecological significance of coarse woody debris in forested ecosystems

overview of the ecological significance of CWD in temperate ecosystems. Their extensive work on the subject includes amounts, decomposition and distribution of CWD. The ecological role of CWD is then described in terms of plant and animal habitats, nutrient cycling and geomorphic function. Some of the most significant functions that can be impacted by collecting firewood are depicted in Figure 3.24. Accordingly Harmon *et al.* (1986) point out that the removal of CWD is likely to result in ecosystem simplification, leading in turn to a reduction in biotic structure and ecosystem function. Loss of CWD also results in the loss of habitats (shelter and food), which in turn may lead to a reduction in local area biodiversity.

In the face of sustained and increased levels of camping activity, managers of natural area tourism need to understand the degree of firewood gathering and assess its significance (Box 3.4). Firewood collection for campfires and barbecues is widespread and a particular problem in Australia, despite the fact that many recreation sites have electric or gas barbecue facilities. Huxtable (1987), in a study of

Box 3.4 Assessing the loss of coarse woody debris (CWD) around campsites in Warren National Park, Western Australia

As part of the campsite study carried out in Warren National Park, Western Australia, Smith (1998) also set out to obtain an overview of the firewood-gathering problem. The amount of CWD at each campsite was assessed using an adapted form of the line intersect technique (Figure 3.25) of Van Wagner (1968) and Smith and Neal (1993). CWD was categorised according to diameter classes of 25–70 mm, 70–300 mm, 300–600 mm and >600 mm. Quantities of CWD were determined at both formal and informal campsites and in adjacent control areas. For CWD <70 mm diameter, formal campsites recorded 93% less CWD than the control sites. Informal sites had 58% less CWD (Figure 3.26). Formal campsites recorded CWD of >70 mm diameter at a level 64% lower than the control sites. Informal sites had 27% less CWD (Figure 3.26), despite the *gratis* provision of firewood by park management and the possible complicating factor of natural/prescribed fire regimes. Smith (1998) concluded that the removal of CWD in the size class <70 mm diameter constitutes less of an ecological impact than the removal of larger classes. This is likely to be due to the reduced water-holding capacity, smaller nutrient store and reduced habitat potential relative to that afforded by CWD >70 mm diameter.

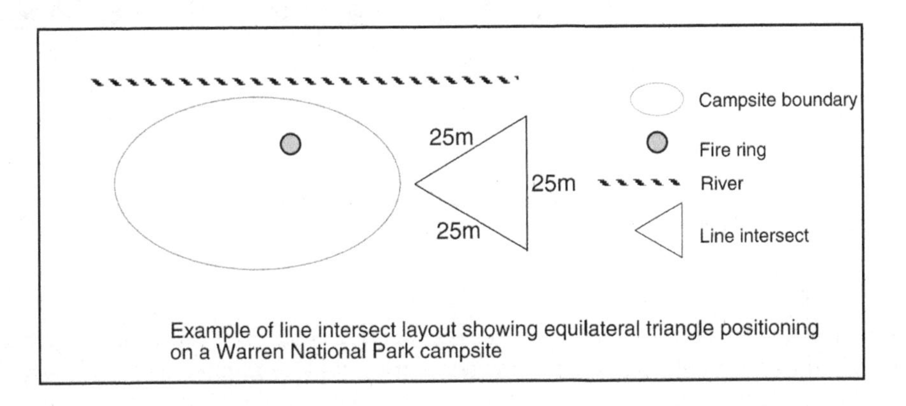

Figure 3.25 Methodology applied in assessing the loss of coarse woody debris from campsites in Warren National Park, Western Australia: The line intersect triangle defines the survey lines for recording the presence of CWD (Derived from Smith, 1998)

firewood-gathering at campsites in Innes National Park, South Australia, found that impacts extended well into the adjoining vegetation, where there was a reduction in CWD and increased levels of tree damage, although the provision of firewood by management can reduce the problem (Smith *et al.*, 2012).

In a study of the importance of CWD as fauna habitat, Hecnar & M'Closkey (1998) investigated the effects of human disturbance on the five-lined skink (*Eumeces fasciatus*) at Point Pelee National Park, Canada (Figure 3.2). It was found that in areas of heavy human usage there was a corresponding lack of CWD and a reduced abundance of skinks. Because the skinks prefer to use large, but moderately decayed

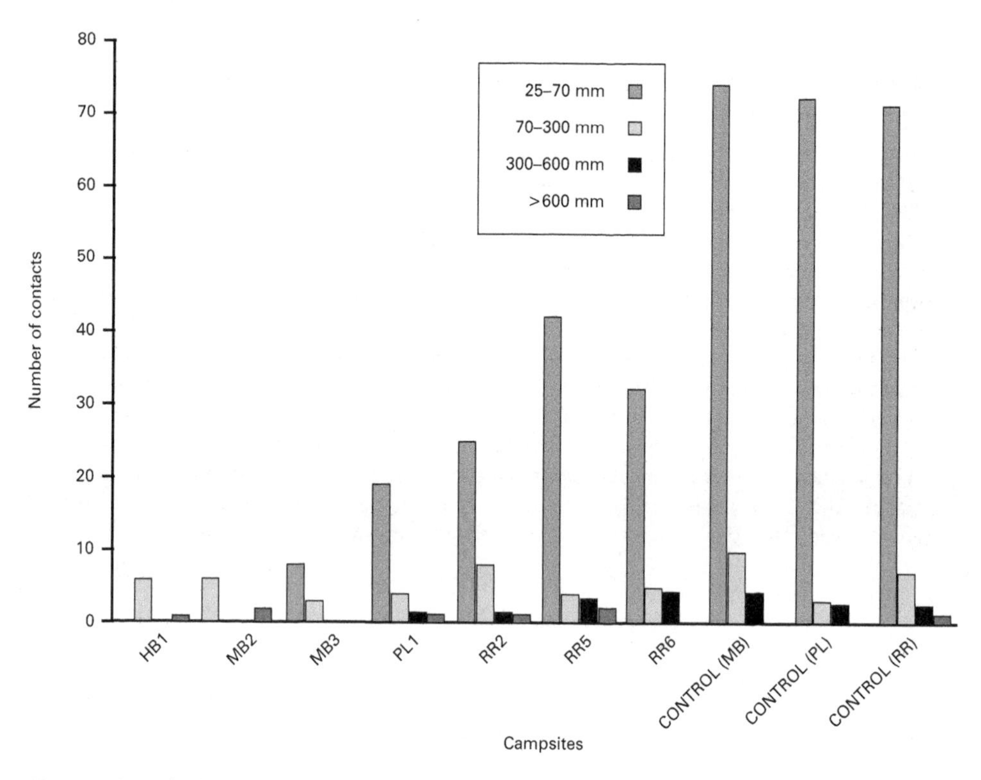

Figure 3.26 Relative abundance of coarse woody debris at campsites in Warren National Park, Western Australia: Number of 'contacts' by diameter class (Source: Smith, 1988)

logs as habitat, the removal or accelerated disintegration of this CWD impacts on the skinks. An experiment in microhabitat restoration was conducted in which CWD was placed in areas in which previously there had been none. Despite the presence of human disturbance the reptiles colonised the 'new' CWD. The experiment showed that the larger CWD constituted a critical habitat for the skinks. Hecnar and M'Closkey (1998) suggested that such CWD is also likely to be of benefit to at least eight other small vertebrates and a much larger number of invertebrates.

The work of Hecnar and M'Closkey (1998) is important in demonstrating that small vertebrates are dependent on abiotic structure and invertebrate food sources (biotic factors) afforded by the microhabitat qualities of larger CWD. The extent to which rare and endangered invertebrates, other vertebrates, such as birds, and larger vertebrates (as in food chain relationships) are dependent on CWD in various different ecosystems remains an important avenue of research. MacNally *et al.* (2001), for example, indicated the importance of CWD for the maintenance of bird diversity in floodplain forests in south-eastern Australia. They found that ground-dwelling and CWD-using birds were more prevalent in areas that had accumulations of CWD that provided food and foraging opportunities.

Litter and human waste

Although not confined to campsite situations, litter and the disposal of human waste are particular problems at camping locations. There can be social and health issues and litter can impact on wildlife. Discarded items in the form of cigarette butts, food wrappings, drinks cans, plastic cups and food debris are often a feature of moderately to heavily used trails and camping areas. Wild animals may eat or become entangled or trapped in discarded materials and subsequently die (Table 3.8).

Because natural area tourism is clearly about appreciating nature under natural conditions, visitor experience can be spoilt by the presence of litter (e.g. Lucas, 1990a, 1990b; Dixit & Narula, 2010). Two studies clearly demonstrate this. For example, 34% of visitors to Bako National Park, Sarawak (Figure 3.2), rated litter around the park as a serious problem (Chin *et al.*, 2000). In a study of recreation impacts in the Nuyts Wilderness in Western Australia (Figure 3.2) 71% of respondents rated the amount of litter as an extremely to very important factor in influencing the quality of their experience (Morin *et al.*, 1997). The findings of Morin *et al.* (1997) and Chin *et al.* (2000) are supported by Cochrane (2006), who found that litter was a significant detraction from appreciating the natural environment at Mt Bromo in Indonesia. Newsome and Lacroix (2011) are of the view that litter constitutes a visual impact that reduces the naturalness of an area and might also be interpreted as a lack of concern for the environment, as well as giving the visitor a poor impression of the management's capacity to deal with the problem.

Table 3.8 Impacts of littering: Items commonly found along walk trails, at campsites and at day-use areas, and their likely social and biophysical impacts

Impact	*Cigarette butts*	*Glass*	*Plastic**	*Food packaging**	*Food waste*	*Toilet paper*
Reduced aesthetics	✓	✓	✓	✓	✓	✓
Soil pollution	✓				✓	✓
Spread of weeds from discarded fruit items					✓	
Attraction of wildlife				✓	✓	
Attraction of undesirable species				✓	✓	✓
Injury to wildlife	✓	✓	✓		✓	
Threat to human safety		✓				✓
Increased risk of wildfire		✓				
Reduced water quality	✓				✓	✓
Negative public perception of management effectiveness	✓	✓	✓	✓	✓	✓

*The accumulation of broken-down and tiny fragments of plastic in the world's oceans is an emerging pollution problem, as they can be ingested by marine wildlife.

Morin *et al.* (1997) also found that the inadequate disposal of human waste was rated to be extremely to very important (68% of respondents) or moderately important (15% of respondents) in affecting the quality of visitor experience. Human waste also poses a health risk, as shown by the work of Temple *et al.* (1982). In their study, in the absence of pit toilets, campers buried their faeces in shallow pits. Temple *et al.* (1982) demonstrated that pathogenic bacteria were able to survive in large numbers for several months and that *Salmonella* survived the winter period. In addition, von Platen (2003) found that faecal bacteria persist in soils for up to 12 months. Bridle and Kirkpatrick (2003), however, noted that the decay rates of toilet paper and faeces depend on the receiving environment. They found that in Tasmania toilet paper and faeces buried to a depth of 15 cm were visible 12 months after burial in waterlogged peat soils but bacteria counts were low. In contrast, such items buried in coastal eucalyptus forest were not detectable within months of being buried but faecal bacteria were present in the sandy soils and would have comprised more of a health risk than at the waterlogged site. Furthermore, and continuing the work from nine different environments in Tasmania, Bridle and Kirkpatrick (2005) reported that toilet paper and tissues decay more rapidly than tampons and that depth of burial was not significant in relation to the speed at which these materials decayed.

The long-term survival of bacteria in some soils also poses the threat of water contamination if waste product burial pits are located close to a water body. In relation to this, Bridle *et al.* (2007) discovered that even the minimal impact guidelines for day and overnight walkers in the back-country of Tasmania, Australia, were not being properly followed and that human faeces could be found only 50 m away from an overnight hut. Bridle *et al.* (2007) also posited that poorly buried faeces were the most likely cause of faecal bacteria being detected in a nearby alpine lake (also see Chapter 5, Box 5.5).

Ongerth *et al.* (1995) examined the contamination of water by pathogens originating from human faeces in the Olympic Mountains National Park, USA (Figure 3.2). The study areas comprised two remote pristine watersheds in which hiking and riverbank camping occurred. Ongerth *et al.* (1995) found that the more heavily used Hoh River had one to three *Giardia* cysts per 100 ml of water, whereas the much less intensively used Queets River had 0.2–1 *Giardia* cysts per 100 ml of water. The presence of coliforms also peaked in accordance with human activity. Coliform bacteria were most abundant following the summer peak of tourist activity. The data also illustrate the importance of regional environmental conditions in influencing the numbers of bacteria. For example, the Queets River only had about five people per kilometre of river per month, but the faecal coliform count was as high as in the Hoh River, which had around 70 people per kilometre of river per month. The latter river having lower water temperatures and higher stream flow than the Queets River explains this relationship.

Significance of camping activity

Given that recreational demand for campsites continues to grow, the environmental impacts of camping are likely to increase in the future (Smith & Newsome,

2002; Cole, 2004). Such a trend places importance on the assessment and monitoring of these impacts (Leung & Marion, 2000; Farrell & Marion, 2001; Daniels & Marion, 2006; Monz & Twardock, 2010). The earlier work of Cole (1992) and Marion (1995) has especially set the scene for investigating campsite impacts in the heavily used and popular recreation/tourist sites in the USA. The major concerns for the future will involve expansion of formal campsites and the degradation of soils and vegetation at informal campsites in natural areas around the world.

Data need to be collected that provide resource managers with information relating to design, location, management, camp ground rehabilitation requirements and visitor attitudes. For example, Smith (1998) found that although only about 1% of Warren National Park, Australia (Figure 3.2), had been impacted by camping this was a significant influence on visitor experience. The cumulative impacts of camping, associated walk trails and access tracks are ecologically and socially relevant, especially at the severely impacted designated campsites. Approaches to investigating campsite impacts (using examples from Africa, Australia and the USA) are therefore considered in more detail in Chapter 7.

Use of Water Edges

Overview

Water-edge environments comprise streams, lakes, rivers, estuaries, marine beaches, rocky coastlines, mangroves, estuarine environments, salt marshes, lake shores, small-island coastal habitats, various wetland fringing habitats, riparian vegetation and stream edges. Visitors tend to find all these captivating components of the landscape (Cole *et al.*, 2005; Hall & Härkönen, 2006). Hadwen *et al.* (2008) have presented the first complete attempt to relate terrestrial and aquatic activities to the ecological function and status of aquatic ecosystems, particularly those devoid of other uses (Table 3.9). Many studies have focused solely on terrestrial environments (Whinam & Chilcott, 1999; Cole & Monz, 2004; Dixon *et al.*, 2004), or on activities within aquatic ecosystems (Butler *et al.*, 1996; Hadwen *et al.*, 2005), yet the greatest capacity for visitor impacts on aquatic ecosystems probably occurs at the ecotone between terrestrial and aquatic environments (Table 3.9).

When utilising aquatic ecosystems, such as rivers and lakes, tourism and recreational activity can be more dispersed than in the well defined 'confined' situation of a campground or walking trail. Despite this, there remains the tendency for visitor activity to be concentrated at specific sites or edges, especially beaches and the most accessible parts of lakes and rivers. With rivers and lakes, the riparian zone, frequently consisting of specific plant communities, marks the interface or ecotone between land and water. The importance of riparian vegetation lies in its protective functions; fringing vegetation will have influences on aquatic flora, fauna and ecological processes.

In addition to impacts on ecological components, the issue of some recreational pursuits intruding on natural experiences are especially pertinent in coastal and

Table 3.9 Ecological implications of visitor activities in and around aquatic ecosystems

Major activity	*Specific aspects*
Trampling	Track erosion Trampling of riparian zone vegetation Trampling of littoral zone vegetation Bank stability at risk Sediment transport at access points
Terrestrial weed dispersal	Changed riparian zone structure and composition Alteration to in-stream habitats due to loss of coarse woody debris
Aquatic weed dispersal	Changed in-stream habitat and resources Changes in flow and hydraulic characteristics
Camping	Compaction of soil Littering and unburned matter Human wastes and microbial contamination Use of detergents and soaps Run-off from fire pits and campfire ash delivery to waterways Removal of wood and riparian vegetation
Swimming	Sediment re-suspension Nutrient additions Microbial contamination Water-level fluctuations and wave generation Trampling and scraping of littoral surfaces Removal of coarse woody debris Modification of pool habitat by piling of rocks to create dams
Picnicking	Clearing of picnic spots Compaction of soil Littering and food wastes Interactions with wildlife
Boating	Noise pollution Oil pollution Contamination with TBT (tributyltin, used in marine paint) Anchor damage Wake creations and its effects Shoreline mooring and access points Sediment re-suspension and turbidity
Hunting	Trampling and site access Increased mortality of target species and food web implications Lead contamination Recreational stocking of introduced fish Removal of fish as top predators Mortality from handling and capture of target species Littering
Wildlife disturbance	Displacement of wildlife Feeding wildlife

Derived from Hadwen *et al.* (2008).

water-edge settings where visitors with different needs and attitudes may come together. For example, where there are large numbers of people and the use of motorised watercraft there is the potential for noise and disturbance, which confer social impacts relating to perceptions of tranquillity, visual amenity and soundscape (see Newsome & Lacroix, 2011). The impacts of tourism and recreation on water-edge environments, accordingly, are explored under the headings of riverbanks, rivers, lakes and reservoirs, coastal areas and coral reefs.

Riverbanks

Tourism and recreation activities that take place along riverbanks and among riparian vegetation include: accessing viewpoints, fishing, boat launching, camping and access for wading, tube floating and swimming (Hadwen *et al.*, 2008). The major impacts arising from these activities are trampling and destruction of the zone of riparian vegetation, loss of vegetation, bank erosion, pollution, habitat loss and disruption of aquatic food webs.

Natural river channels demonstrate a balance between bank erosion and the deposition of sediment where the velocity of water flow decreases. Malanson (1993) reported on the protective function of riparian vegetation. The riparian fringe confers channel stability, especially under the erosive potential of flood conditions and through root systems binding riverbank sediments. The presence of riverbank vegetation reduces the tractive force of water through sediment trapping and thus facilitates sedimentation. Malanson (1993) also considered the role of riparian vegetation in attenuating pollutants by trapping sediments with attached pollutants and intercepting nutrients, such as nitrogen, which may be transported by a river or stream and promote excessive growth of algae and aquatic plants.

The riparian fringe always sits within a broader ecosystem context. Contrasting examples include river channels in the forested ecosystems of the USA, rivers running through savanna ecosystems in Africa and the vast intermittent river and floodplain networks that occur in semi-arid Australia. In more arid settings the presence of fringing vegetation often constitutes important habitat, due to the presence of large trees in an otherwise essentially treeless environment, in association with permanent pools of water that allow both aquatic and terrestrial organisms to survive the dry season. Such refuge pools remain cooler and receive additional inputs of organic matter when shaded by large trees. The shading, litter deposition, provision of CWD, river and terrestrial habitat functions of riparian vegetation have been reported by Kuss *et al.* (1990) and Hammitt and Cole (1998), and these ecological functions of riparian vegetation have particular relevance for fish and other members of the aquatic food web (see Pusey & Arthington, 2003).

A consistent problem is trampling along the banks of rivers resulting in changes to and loss of vegetation. Depletion of the protective function of vegetation can lead to bank erosion. Such problems have been reported from Yosemite National Park (Figure 3.2) by Madej *et al.* (1994), who investigated bank erosion on the Merced River using historical photographs, air photographs and measurements of channel width. There was a strong association between poor bank stability and recreational

Figure 3.27 Bank erosion: 'Scalloping', root exposure and loss of riparian vegetation adjacent to campsites in Warren National Park, Western Australia (Photo: Amanda Smith)

activity. For example, bank erosion was significant in campground areas. One study area had more than 1000 camping sites located within 500 m of the river.

Smith (1998) also demonstrated impacts on riparian vegetation in turn resulting in erosion associated with riverbank campsites (Figure 3.27). Trail networks leading from campsites to and along the Warren River showed severe bank erosion as a result of the loss of riparian vegetation (Table 3.10). Study sites HB1, MB2 and MB3, for example, are advertised formal campsites and the indicators of degradation (root exposure, bank collapse and gully development) reflect easy access, heavy use and waterside activities. In contrast, other sites, such as PL1 and RR2, are informal sites and not so heavily used, due to difficulty of access, and as a consequence not so impacted (Table 3.10).

The loss of riparian vegetation has ecological impacts through diminished shading, leading to increased water temperatures, reduced inputs of organic matter as sources of energy for aquatic organisms and increases in turbidity, which may lead to changes in the populations of invertebrates and fish (Cole & Landres, 1995; Hammitt & Cole, 1998). Increases in turbidity have implications for algal and plant growth, primary productivity and food web structure.

Pollution from human wastes has already been dealt with, in the context of terrestrial environments (pp. 157–158); additionally, Warnken and Buckley (2004)

Table 3.10 Mean width, depth, distance and erosion of main riverbank access trails from campsites at Warren National Park

Campsite	*Mean width (cm)*	*Mean depth of incised trails (cm)*	*Distance (m)*	*Root exposure*	*Bank collapse*	*Gully development*
HB1	191.0	55.0	6.2	Severe	Severe	Severe
MB2		Not applicable		Severe	Severe	Severe
MB3	209.4	86.0	15.5	Severe	Severe	Severe
PL1	69.7	0.7	37.9	Low	Negligible	Negligible
PL2	61.0	Flat	19.3	Low	Negligible	Negligible
PL3	63.0	6.8	15.5	Low	Low	Low
RR1	111.7	Flat	13.7	Negligible	Negligible	Negligible
RR2	180.1	Flat	14.3	Negligible	Negligible	Negligible
RR3	66.8	20.0	3.3	Moderate	Mild	Moderate
RR4	96.0	6.8	7.5	Low	Mild	Low
RR5	96.1	1.0	7.5	Moderate	Mild	Low
RR6	120.0	20.0	4.1	Low	Mild	Moderate

Source: Smith (1998).

provide an account of the risks associated with tourist use of rivers, streams and swimming holes in protected areas. They note that water quality can be affected by the rinsing out of food containers, swimming (e.g. babies in nappies), disturbance of stream-bed sediment, and urination and defaecation in the catchment. Of particular significance is the observation that increases in *E. coli* were found in a popular swimming hole and that the concentration of coliform bacteria was 1000 times higher in stream-bed sediment than in the overlying water column.

Rivers, lakes and reservoirs

Recreation and tourism that focus on rivers and lakes include boating, sailing, swimming and bank/shoreline activities. Impacts on the riparian zone have already been considered and largely involve the destruction of fringing vegetation and bank erosion. Even the dry lake-beds of inland salt lakes such as those that occur in semi-arid and arid Australia are not free of potential damage. When dry, visitors can cross them in vehicles, leaving tyre scars and damaged shoreline vegetation at entry and exit points. This section, however, focuses mainly on the effects of boating activities on bodies of freshwater and draws from comprehensive reviews undertaken by Mosisch and Arthington (1998, 2004), who provide a systematic discussion of the impacts of motorised recreational activities, especially power boating, on lakes and reservoirs. Mosisch and Arthington (1998, 2004) divide the impacts of power boating into physical disturbances (wave action, turbidity, direct boat contact, noise and visual disturbance), chemical effects (various forms of pollution) and ecological impacts. Major physical impacts include the erosional effects of wave action on riverbanks and lake shorelines. According to Liddle and Scorgie (1980), waves and ripples generated by boats wash out the roots of emergent macrophytes and riparian vegetation. Many authors agree that this destabilises the bank and leads to bank erosion, which then leads to a further decrease in vegetation.

Liddle (1997) noted that aquatic plants differ in their sensitivity to erosion. Data collected mostly in Europe showed that soft-leaved species such as *Elodea canadensis* and *Rorippa amphibia* were most easily eroded; species such as *Potamogeton crispus* and *Sparganium emersum* were more difficult to erode, with larger, tougher, more robust *Phalaris arundinacea* and *Phragmites communis* being the most difficult to erode. Poorly vegetated, exposed banks are particularly subject to direct erosional effects. These effects can also be caused by the destructive action of direct contact with shoreline and riverbank vegetation by boats, trailers and people.

Direct contact also takes place in the form of propeller action, which can cut and damage macrophytes and disturb benthic organisms by stirring up sediment. Turbidity is caused by the stirring up and suspension of sediments in the water column. This can introduce nutrients stored on sedimentary particles into the water column and, depending on other limiting factors, give rise to an algal bloom. Continuous turbidity, nevertheless, obstructs light from reaching phytoplankton and can reduce algal productivity and limit the growth of submerged macrophytes. High levels of turbidity can also clog the gills of invertebrates and fish and reduce the feeding efficiency and food resources of fish-eating birds (Murphy *et al.*, 1995).

In their account of the biological impacts of power boating, Mosisch and Arthington (1998) note that species of fish appear to differ in their tolerance of the suspension and redeposition of fine particles. Power boating has been shown to cause some fish to abandon nest-guard duties, leaving eggs at risk of predation. The stirring up of river and lake-bed sediments can also kill fish eggs. Murphy *et al.* (1995) report that fish populations can be reduced as a result of disrupted courtship and spawning and declines in the viability of fish eggs.

Recreational boating can also result in the spread of aquatic weed species as a result of propagules and fragments of vegetation being held on boat attachments such as propellers. In this way weed species can be transported from one lake to another. Johnstone *et al.* (1985), who investigated 107 lakes in New Zealand, found that the occurrence of five aquatic weeds could be traced back to recreational boating. The proliferation of weed species has the potential to alter ecological conditions and impact on the aesthetic and recreational values of lakes.

A number of studies have shown that the noise generated by boating causes water birds to seek refuge on quieter stretches of water. Visual effects and boat movement can also be important. In a study of the impacts of sailing on water birds utilising the Brent Reservoir in England (Figure 3.2) Batten (1977) observed that birds either temporarily left the site or congregated in refuge areas where sailing was not taking place. The importance of water bodies for over-wintering birds is also highlighted by Batten (1977), who found that teal (*Anas crecca*) and wigeon (*Anas penelope*) had stopped using the site as a winter refuge. The significance of such disturbance depends on the availability of other water bodies and quiet refuge areas and the intensity of recreational use. Tuite *et al.* (1984) reported that local and short-term disturbance may cause changes in the distribution of birds and even more significant changes may occur at intensively used recreation sites (especially for power boating and water skiing). The situation, however, can be ameliorated by the widespread occurrence of inland water bodies and reserved areas. In the UK the mobility of ducks and the fact that only low levels of recreational activity generally take place during the important over-wintering period also serve to reduce the impact.

Activities that lead to the disturbance of wildlife ranges from sailing and windsurfing, power boating and the use of jet skis through to tour boat visits to wildlife areas. In relation to tour boat visitation, Galicia and Baldassarre (1997) studied the effects on American flamingos (*Phoenicopterus ruber ruber*) in the Celestun Estuary, a coastal lagoon which lies parallel to the coastline in the north-western part of the Yucatan Peninsula in Mexico (Figure 3.2). It is a biosphere reserve and a major site for one of the non-breeding populations of the American flamingo. Tour boats carried 7488 tourists in 1992–93 and averaged 16 boat-viewing trips per day (Galicia & Baldassarre, 1997). Activity budgets of both disturbed and undisturbed birds were assessed. Tour boat visits were seen to significantly increase alert time, but reduce the feeding time from 40% to 24%, which translated into a loss of some 30 minutes of feeding time per individual per day. Galicia and Baldassarre (1997) noted that on days of exceptional numbers of tourist visits (97 tour boats) feeding activity may be halted altogether. The biological significance of these findings relates to the possible impacts of reduced feeding time on the breeding activity of the flamingos. For example, if time spent in courtship reduces feeding time then flamingos need to acquire sufficient food reserves beforehand. An additional aspect of this study was the distribution of flamingos according to food supply in the estuary and that boat traffic could reduce their access to optimal feeding areas. Such impacts are additional sources of stress and may cause the decline of an important population of flamingos, as well as threaten the viability of an important tourism resource.

The pollution of rivers and lakes as a consequence of boating activities occurs mainly in the form of contamination from oils and fuel combustion products and in the liberation of motor exhaust fumes. Mosisch and Arthington (1998) reported that 380–600 million litres of outboard motor fuel are discharged into waters each year; the details of how oils and fuel enter the water and their chemistry are detailed in their reviews (e.g. Mosisch & Arthington, 2004). Environmental impacts consist of reduced water quality through hydrocarbon contamination, the accumulation of pollutants (e.g. lead) in sediments, slicks of lubricating oils on the surface of the water and the deposition of unburnt oil on algae.

The presence of potentially carcinogenic hydrocarbons such as polynuclear aromatic hydrocarbons (PAHs) has been reported to correlate with peak power boating activity by Mastran *et al.* (1994). After entering the water, these compounds quickly become bound to suspended organic and inorganic particulate matter, which in turn, is gradually deposited in the bottom sediments, carrying the bound PAH with it (Neff, 1979). Once contained within the sediment layer, these compounds are extremely resistant to degradation and may persist for long periods, accumulating to high concentrations, particularly if the receiving sediments are anoxic (Neff, 1979; Mastran *et al.*, 1994). A wide range of microscopic and macroscopic animals are associated with the bottom sediments of water bodies and many of these organisms ingest organic material from these sediments (Harvey, 1997; Neff 1979). Contamination by these and other hydrocarbons has been found to be toxic to invertebrates and fish (see Mosisch & Arthington, 1998, 2004).

Coastal areas

Coastlines, consisting of beaches, mangroves, salt marshes, mud flats and rock pools, attract many people; they engage in a variety of activities, ranging from the typical 'beach holiday' through to activities more centred on natural history, such as searching rock pools and bird watching on estuaries. Hardiman and Burgin (2010) provide an excellent review of the impacts of coastal tourism, recreational fishing, physical disturbance as a result of recreation, impacts of recreation on water quality and disturbance as a result of coastal marine tourism activities in the Australian context (Table 3.11). Warnken and Byrnes (2004) specifically focus on the impacts of tour boats in the marine environment.

Recreational fishing and the excessive collection of shells and organisms from rock pools can lead to a local depletion in fauna and faunal resources. Remote coastlines that have only ORV access can attract the more adventurous tourist but this can also lead to the damage of important coastal ecosystems. This has been reported for a unique tropical arid area, the mangrove systems which occur along the north-west (Pilbarra) coast of Western Australia (Figure 3.2), where Gordon (1987) reported on the erosion of salt flats, the destruction of algal mats and the restricted movement of tidal water as a result of recreational ORV activity.

The most significant issue in coastal tourism, however, is access to and use of beaches. Beaches are naturally dynamic systems and change according to tidal fluctuation. They are also subject to seasonal variations in sediment removal, transport and deposition. Beach sand is especially mobilised during stormy periods, when the sea moves the sand offshore and strong winds can blow it landward. During calm conditions sand is mostly deposited on the beach, where onshore winds blow it inland to be trapped by vegetation to form sand dunes.

Sand dunes do not form where onshore winds are relatively weak, as in the case of equatorial tropical beaches. In many parts of the world, however, strong onshore winds are a feature of the coastal environment and in some places extensive dune systems can develop, as in south-west Western Australia. In these cases the fore-dune systems act as a reservoir of sand and buffer against the penetration of erosive activity further inland. Coastal dune systems, however, are easily disrupted. Any disturbance to the natural cycling of sand has the capacity to alter the shape and position of a sandy shoreline. In particular, the removal of vegetation can lead to the development of localised pockets of fore-dune erosion called blowouts, which reduces buffering capacity and can lead to the erosion of and transportation of sand further inland.

Table 3.11 Recreational impacts on coastal marine ecosystems and water-edge environments

Major category	*Impacts*
Tourism and recreation infrastructure	Eutrophication, clearing of coastal vegetation, invasion of weeds, chemical pollution, impact of shark nets and the use of artificial structures
Recreational fishing	Collecting bait, depletion of shellfish, capture of non-target species and use of introduced species
Physical disturbances	Off-road vehicle access Use of power boats Walking on rock platforms Trampling of intertidal and marine algal and invertebrate communities Vessels striking wildlife Damage to coastal vegetation
Water quality	Fuel, oil and chemical discharge from boats Contamination from anti-fouling paints Sewerage
Disturbance of wildlife	Displacement of wildlife from feeding and breeding areas Habituation of wildlife Food-provisioned wildlife attracted to humans, with consequent disruption of their normal foraging behaviour Aggressive responses from wildlife
Quality of visitor experience	Exhaust fumes from boats Noise and intrusion from power boats and jet skis Sewerage and litter Damage to vegetation Disturbance and displacement of wildlife

Modified from Hardiman and Burgin (2010).

The recreational use of motorcycles and beach buggies and pedestrian access can destabilise coastal dune systems, especially where access is uncontrolled and a network of multiple access tracks develops. The resultant bare areas become focal points of erosion and blowouts can develop. Headlands can be especially susceptible, due to the continuous presence of strong winds. Here the loss of sand can result in the exposure of underlying rocks.

Beach erosion that leads to the subsequent burial of once stable vegetation communities constitutes a major change. Extensive burial of fixed dune vegetation can re-set the beach successional process, in that only those species tolerant of burial will be able to survive and the new vegetation will approximate that of a fore-dune community. Such a loss of the fixed dune vegetation becomes especially significant if it contains rare and endangered species or critical habitat.

Sand dune systems are only poorly developed or are absent from most tropical environments. In many of these areas, however, the beach itself is a critical habitat as it is used as a breeding site for turtles and seabirds. A combination of hunting for food and habitat degradation has decimated many turtle populations and viable nesting habitat is essential for their survival. There has been a widespread loss of turtle breeding beaches in the tropics and Mediterranean region (encompassing North Africa and southern Europe). Poland *et al.* (1996) noted that tourism comprising hotel sprawl, light pollution, human activity and obstacles on the beach and even building sand castles is a major threat to already stressed and reduced turtle populations on the Greek island of Zakynthos (Figure 3.2).

Many turtle egg-laying beaches have become tourist attractions in their own right. For example, Tortuguero National Park in Costa Rica (Figure 3.2) hosts one of the largest nesting areas for the green turtle (*Chelonia mydas*) in the world. The impacts of tourism, as observed by Jacobson and Lopez (1994), on turtles at Tortuguero comprised light disturbance from torches and flash photography, touching and blocking the progress of turtles, digging and moving around nests and the trampling and handling of hatchlings. The interruption of turtle activity also caused turtles to return to the sea without laying, with the risk that the eggs are wasted as they are discharged at sea. Jacobson and Lopez (1994) collected data on the number of turtle arrivals and nesting attempts along 7 km of beach. The amount of successful nesting activity was recorded and the data were then compared between zones of low and high tourist activity. It was found that at times of high tourist concentration (weekends) 30% fewer turtles visited the beach. Jacobson and Lopez (1994) also found, though, that specific turtle behaviour such as successful nesting and females returning to the sea without nesting did not differ between the zones. A reduction in total nesting activity over time, however, could have repercussions for this turtle population. Codes of conduct may help to manage turtle tourism. Waayers *et al.* (2006), however, found at an important turtle nesting area in north-west Australia that 77% of tourist groups breached the code by shining lights on turtles, positioning themselves too close to the turtles and not staying behind turtles during their egg laying.

Bathing, swimming and boating comprise major activities in coastal environments. Studies have shown that these also have potential impacts and can pose a risk for recreationists and tourists (Liddle, 1997). A combination of sunbathing, picnicking and swimming can result in litter and human wastes, which in turn can lead to water contamination. Contamination of water also takes place in the form of suntan oils, sun-screens, soaps and bacteria derived from human skin. Liddle (1997) cites several studies (e.g. Cabelli *et al.*, 1982) that demonstrate increased levels of pathogenic organisms and the risk of water-borne disease resulting from water-based recreation.

As discussed above, in the context of lakes, various boating activities and water sports can disturb wildlife. In the marine environment injuries to sea mammals can occur when an animal is suddenly surprised at the surface of the water by boats. Power boats and water sports can increase the mortality of resident fauna, as demonstrated by injuries to turtles in the Greek islands and to manatees in Puerto Rico (Poland *et al.*, 1996; Mignucci-Giannoni *et al.*, 2000). As already noted, motorised water sports produce noise and exhaust fumes. Furthermore, there is also the potential for the accidental release of oil and petrol and chemical pollution emanating from the weathering of anti-fouling paints.

Indirect impacts on marine wildlife can occur as a result of beach recreation and tourism. For example, the protection of popular Australian and South African swimming beaches with shark nets has resulted in the death of many species of sharks and other vertebrate species. Most shark species are declining worldwide and because of low fecundity they have a slow capacity for recovery. Liddle (1997) shows data from various sources on the entanglement and mortality of large sharks, dolphins and small whales, birds, six species of turtle and large fish such as rays. In a number of cases such deaths are decreasing already depleted and stressed populations.

Coral reefs

Some background to coral reef ecosystems and an account of some early work has already been given in Chapter 2. This section follows on from this and provides further details on how recreation and tourism can damage coral reef ecosystems. Table 3.12 provides a summary of the most common tourism impacts on coral reefs. Impact research has continued to grow, especially in relation to the major tourism and recreational activities of snorkelling (e.g. Plathong *et al.*, 2000; Harriott, 2002; Leujak & Ormond, 2008; Hannak *et al.*, 2011) and diving (e.g. Tratalos & Austin, 2001; Zakai & Chadwick-Furman, 2002; Barker & Roberts, 2004). Earlier works, by Liddle and Kay (1987), Hawkins and Roberts (1993) and Allison (1996), however, remain useful case studies of the environmental impacts of snorkelling and diving on coral reef systems.

Liddle and Kay (1987) conducted several experiments in order to study the relative susceptibility to damage, survival and recovery of corals to reef walking on the Great Barrier Reef, Australia (Figure 3.2). It was found that the branching coral *Acropora millepora* was the least resistant to breakage and most resistance was offered by *Acropora palifera*. A study of trampling on the non-branching massive coral *Porites lutea* showed that damage accumulated over time, depending on how much trampling took place. Liddle and Kay (1987) also reported that corals differ in terms of survival and recovery from trampling. The highest recovery rates were reported for *Acropora millepora*. Their work clearly established that the branching corals are more susceptible to damage and that some species are slower to recover than others.

Allison's (1996) work carried out on Kaafu Atoll, a resort in the Maldives (Figure 3.2), found that remote parts of the reef were in good condition but coral breakage was apparent in the vicinity of a well used snorkelling channel. There was a positive correlation between snorkelling and the presence of broken coral. A useful aspect of Allison's study was observations of snorkeller behaviour. It was observed that snorkellers would kick and stand on the coral, especially if the snorkeller was not experienced or was being dragged about by waves.

Hawkins and Roberts (1993) examined the impacts of snorkellers and scuba divers at the Sharm el Sheikh resort on the Red Sea coast, Egypt (Figure 3.2). Two different sites were investigated. The Tower site showed that trampling caused an increase in the amount of broken coral colonies, rubble and live loose coral fragments

Table 3.12 Major tourism-related sources of damage to coral reef ecosystems

Recreation/tourism activity	*Impacts*	*Control measures*
Snorkelling	Local damage to susceptible corals Sediment disturbance Abrasion of corals Direct breakage	Education Code of practice Briefings/direct supervision Provision of resting buoys and flotation platforms
Diving	Local damage to susceptible corals Sediment disturbance Abrasion of corals Direct breakage	Education Code of practice Briefings/direct supervision Site selection for inexperienced divers
Reef walking	Coral breakage Abrasion of corals	Education Code of practice Designated trails to focus damage
Anchoring	Local/site coral damage but can be extensive if many boats use the reef system	Private and public moorings/zoning Education Code of practice Use of sandy areas
Waste/pollution from boats	Impacts on aesthetics Nutrient enrichment	Education Penalties Holding tanks
Uncontrolled fish feeding	Disease risk to fish Enhanced capture risk Dependency on feeding Risk of humans being bitten	Education Code of practice/guidelines Feeding permit Controlled/best-practice operations at designated locations
Tourism infrastructure to facilitate activities: moorings, pontoons	Local damage to benthic communities	Permit Design Site selection Education
Coastal development	Water quality (pollutants, increased turbidity and nutrient enrichment) Introduced species Sediment runoff Social impacts	Environmental impact assessment and management plans Best-practice waste treatment/disposal Water quality monitoring

Modified from Harriott (2002)

(Table 3.13a). The impacts of trampling were also assessed at the Ras Umm Sidd site and it was found that mean coral colony height was reduced where trampling occurred (Table 3.13b).

Also in Egypt, Leujak and Ormond (2007) investigated visitor perceptions of reef condition as compared with social conditions at coral reef sites and cautioned that the social carrying capacity of snorkelling sites may be above that of their ecological carrying capacity. Of particular interest was the finding that the snorkelling

Table 3.13 Impacts of trampling and snorkelling on coral reefs

(a) Impacts of trampling at the Tower site, Sharm el Sheikh, Egypt

Parameter	*Trampled*	*Untrampled*
Number of broken coral colonies	2.6	0.5
Number of live, loose coral fragments	1.0	0.1
Number of reattached fragments of hard coral	0.1	0.0
Number of clams	0.3	0.2
Number of hard coral colonies	28.6	42.4
Proportions of species		
% hard coral	10.4	26.6
% soft coral	0.7	2.7
% bare substrate	87.4	70.5
% rubble	1.4	0.1

(b) Impacts of trampling and snorkelling on coral reef at the Ras Umm Sidd site, Sharm el Sheikh, Egypt: Mean coral colony heights and diameters on the reef flat according to tourist activity

Zone and treatment	*Mean colony height (cm)*	*Mean colony diameter (cm)*
Middle, trampled	1.8	6.5
Middle, snorkelled	1.9	6.6
Middle, untrampled	2.6	9.0
Outer, trampled	1.7	7.4
Outer, snorkelled	2.2	8.2
Outer, untrampled	3.1	13.1

Source: Hawkins and Roberts (1993)

demographic has changed over time, with more inexperienced and less environmentally aware snorkellers, who were also less concerned about crowding when visiting the Ras Mohammed National Park and Sharm El Sheikh reef sites in Egypt. The implication of this trend is the risk of further decreases in the quality of reef ecosystems. Leujak and Ormond (2007) therefore highlight the importance of determining and monitoring ecological carrying capacity, such as changes in the abundance of fish, amount of coral cover and damage to coral.

Studies on fish feeding in marine tourism settings by Milazzo *et al.* (2005, 2006) have indicated that provisioning by snorkellers can cause increases in the abundance of some species of fish, bring about increased predation of some species and result in fish being attracted to humans when they are in the water. Milazzo *et al.* (2006) went on to caution that while it is a popular tourism activity that facilitates close access and a visual spectacle for snorkellers, it is not a natural experience and is likely to conflict with conservation objectives when such tourism occurs in marine protected areas.

Recreation and Tourism in Mountainous Areas

Mountainous environments comprise many of the 'classic' natural area tourism destinations around the world. Tourism can focus on single mountains such as Mt Kinabalu (Sabah) and Kilamanjaro (Tanzania) or occur in entire regions such as the Cairngorms (Scotland) and the Himalayas (Nepal). Depending on the site and season there may be a range of recreational and tourism activities, such as hiking, camping, skiing, snowboarding, abseiling, paragliding, rock climbing and mountaineering. There may also be the development and operation of tourist-dedicated built facilities on lower slopes and at the base of mountains.

Mountain environments are susceptible to disturbance due to steep slopes and thin soils and this is especially so in the high-rainfall environments that span the tropics (Ahmad, 1993). The risk of negative impact also tends to be greater in mountainous areas because of the presence of slow-growing and fragile arctic alpine plant communities at high altitudes and in the colder temperate zones (Pickering *et al.*, 2003). Monz (2000) points out that this is of particular concern when rare, endemic plant species of restricted distribution are involved, as is the case with potential recreational climbing damage to remnant arctic alpine vegetation in Snowdonia National Park, Wales (Edington & Edington, 1986).

Singh (1992) highlighted the wide spectrum of impacts arising from tourism and recreation in mountains. These embrace a complex array of ecological, socio-cultural and economic effects. It is noted that the increasing demand for tourism in the less-developed world can result in an increased level of deforestation and loss of wildlife due to hunting, as more people are attracted to tourist routes in search of employment (see for example Table 3.17, p. 197). In global terms, significant bio-physical effects include: disturbance to wildlife; camping impacts; trail degradation and erosion; damage to vegetation; water, air and noise pollution; litter; human waste; and the negative social conditions of crowding and congestion on popular routes (e.g. Pickering & Barros, 2012). On the other hand, Singh (1992) also pointed out that a number of positive impacts can flow from mountain tourism, including increased conservation efforts, changes from marginal agriculture, cultural preservation, upgrading of facilities, improved infrastructure and economic opportunities.

Pickering and Barros (2012) note that mountains can attract different user groups, ranging from adventurous mountaineers through to nature tourists in search of natural experiences and wildlife. Pickering and Barros (2012) also identify the recreational focus on snow in mountain environments and consider the importance of winter-based tourism, such as skiing, cross-country hiking and riding snow-mobiles. These activities can result in noise disturbance, pollution by human waste, littering and compaction of the subnival space, which, if extensive, would disrupt the movement of animals under the snow.

Many of the issues pertaining to accessing mountain environments during the snow-free summer period are covered under trampling, access roads and trails and built facilities and campgrounds and the cumulative nature of mountain tourism alongside other uses is explored in later in this chapter (p. 191). Included here, however, is a brief account of some of the less obvious ecological effects of recreation and tourism in mountainous areas and where cliffs occur, namely that of the increasingly popular activity of rock climbing.

The human impact on cliff ecosystems is increasing, with associated impacts such as damage to rock faces, displacement of animals from feeding and breeding areas, and damage to vegetation communities (Cater & Hales, 2008; Vogler & Reisch, 2011). Rock ledges were reported by Giuliano (1994) to be an important habitat in Europe for cliff-nesting birds such as the buzzard (*Buteo buteo*), peregrine falcon (*Falco peregrinus*) and golden eagle (*Aquila chrysaetos*). Birds that are typical of upland environments utilise rock ledges during the breeding season for nesting sites and fledging their young. Giuliano (1994) noted that disturbance by climbers may cause birds to desert their nests and the resultant exposure of eggs and young can lead to mortality. Similarly, the disturbance of unfledged birds can cause them to fall from the nest area, risking injury and predation.

Other species of bird are also affected by rock climbing, as shown by the work of Camp and Knight (1998) at Joshua Tree National Park, California, USA (Figure 3.2). In this study, cliff and rock ledge bird communities were studied in the context of different levels of rock climbing. Camp and Knight (1998) found that birds at popular rock climbing sites, which had up to nine climbing routes on the cliff, responded by flying rather than staying at the site. When rock climbers were present, birds tended to avoid the site. Birds were more likely to stay at unclimbed cliffs, as evidenced by their perching behaviour. Camp and Knight (1998) also observed that rock climbers as they prepared for the climb spent more time at the base of cliffs than did hikers. Varying lengths of time are spent on the cliff faces themselves. In some areas this can be as much as two days, as frequently occurs in Yosemite National Park during the peak climbing season, which lasts from June to September.

In the Joshua Tree National Park study it was concluded that birds appeared to be responding to rock climbing disturbance by adjusting their daily activities, habitat usage and spatial occurrence on the cliff faces. Such changes to preferred activity could increase stress on the affected birds, through the disruption of feeding behaviour, breeding activity and increased avoidance of predators. The observed presence of aggressive, invasive species such as the European starling (*Sturnus vulgaris*) at climbed sites could also lead to the displacement of native bird species as a result of competition for nest sites.

In relation to disturbance to mammals, White *et al.* (1999) studied the effects of mountain climbers on grizzly bears (*Ursus arctos horribilis*) in Glacier National Park, Montana (Figure 3.2). Grizzly bears forage for cutworm moths (*Euxoa auxiliaris*) in the alpine zone during the summer period, a time when they are also likely to encounter mountain climbers. White *et al.* (1999) highlighted the potential disturbance to bears because of the many access routes through the mountains and in particular in the vicinity of moth sites. They noted that bears disturbed by climbers spent 53% less time foraging and 23% more time engaging in aggressive behaviour. It is suggested that such a reduction in feeding time and increased energy consumption could impact on the physiological status of the bears and reduce their reproductive success.

The relative sensitivity of mountain floras has already been considered. Studies of the impacts of rock climbing on cliff plant communities provide further insight into the potential damage that climbing can do in mountain environments. Kelly and Larson (1997), in a study of the impact of rock climbing on eastern white cedar (*Thuja occidentalis*) on the Niagara Escarpment, Canada (Figure 3.2), showed that the density of trees on cliff faces is lower where climbing occurs. In addition, there was damage to trees in the form of rope abrasion, sawn branches and trees being cut down.

Farris (1998) emphasised the problems of determining the effects of rock climbing on cliff vegetation. Importance is placed on comparison with suitable controls, which can be difficult to find due to variation in the proportion of bare rock face, ledges and vegetation cover between sites. Of interest is the observation that degree of slope, micro-surface features and aspect are likely to influence the nature of vegetation and amount of disturbance that takes place. Farris (1998) found that complex upstanding lichens were readily damaged by rock climbers. Total plant cover was also found to be lower where climbing took place. This can be particularly significant where cliff-edge vegetation exerts a control over run-off and soil erosion, and if weed invasion is taking place. Additionally, if a patch of cliff vegetation is an important source of seed and spores for the establishment of new vegetation in the area, then its loss, and or replacement with weeds, also reduces site regeneration potential.

Recreation and Tourism in and around Caves

Cave tourism

Unique features, fossils, archaeological remains and unusual wildlife mean that caves are very important tourism resources and have a high economic value. According to Anderson (2010), caves are likely to represent one of the oldest forms of natural area tourism. Today, cave recreation and tourism ranges from relatively easily accessible show caves (e.g. the Jewel and Mammoth Caves in south-west Australia and the Cango Caves in South Africa) and wildlife-centred cave viewing (e.g. Niah and Mulu Caves, Sarawak) through to the more adventurous exploration of caves and cave diving (Huautla Caves, Mexico). It spans new aspects of tourism such as geotourism and niche areas like ecotourism and adventure tourism as well as more traditional forms such as cultural and spa tourism. In many cases one cave system may cater for all of these forms of tourism.

Visitation varies greatly around the world. Baker and Gentry (1998) reported around 40,000 visitors a year more than a decade ago at a British show cave, Poole's Cavern, Derbyshire. Gillieson (1996) indicated that Mammoth Cave in Kentucky, USA, received more than 2 million visitors per annum and also noted that, at the time, there were some 650 tourist caves with lighting systems and an estimated worldwide total of 20 million visitors a year. The Asian region is rich in carbonate rocks and extensive cave systems occur in Laos, Vietnam and China. The World Heritage site of Halong Bay in Vietnam receives over 1 million visitors per annum, many of whom take a cave tour. In China 3319 caves had been identified by 2000 and in southern China 60% of identified caves had been developed. In contrast to these carbonate cave systems, the Wudalianchi Volcanic Geopark in north-east China, which receives 1–5 million visitors a year, has rare lava tube ice caves that are a significant attraction.

Sources of impact

Although deliberate damage and graffiti tend to be contained due to education (see Chapter 6) and environmental protection measures such as controlled access, the touching of cave features (stalactites and stalagmites) during a tour can result in breakage and discolouration. Deliberate vandalism can, though, occur where there is high visitation and overcrowding and where management finds it difficult to supervise large groups of visitors. Inappropriate management footprinting is a common situation, where natural cave conditions are modified to 'appeal' to tourists and/or to cater for heavy visitation. This is a particular issue in Asia, where many show caves are modified by coloured lighting. The lava tube caves at Wudalianchi, for example, have been extensively modified by infrastructure, lighting and ice carvings and the focus of tourism is on human-made features rather than the permanently frozen caves. Some cave managers may wish to alter natural conditions by the use of music and concerts are sometimes held in cave systems. Caves may also be used for cultural programmes and religious ceremonies. Given the importance of caves as unique ecosystems, the focus here is to recommend maintenance of caves as natural systems where the true nature of the cave (darkness, natural soundscape) can be appreciated under low light conditions (see Newsome & Lacroix, 2011). Good examples of cave tourism can be seen at the caves in Mulu National Park, Sarawak, Malaysia, where there is unobtrusive infrastructure to manage access, guiding and interpretation to manage behaviour and visitor satisfaction and remote-camera facilities and a managed area for viewing bats (Figure 3.28).

Artificial lighting has been installed in many caves to facilitate access and the viewing of speleothems. Such lighting, however, may allow normally absent algae, mosses and lichens (lampenflora) to grow on and discolour cave features. The world-famous Lascaux cave paintings in France were discoloured by algae as a result of artificial lighting and increased carbon dioxide levels derived from numerous visitors (Dellue & Dellue, 1984). Many cave managers are now reviewing their lighting systems with a view to their replacement with low-intensity lights.

Respired carbon dioxide and heat derived from visitors and artificial lighting have the capacity to change the microclimatic conditions in a cave (Cigna, 1993). This can become significant where the cave system has limited air exchange with the outside environment. Those caves that experience naturally rapid air movements or have constant rapid water transit are less likely to be negatively affected by a fluctuating cave climate and increased levels of carbon dioxide. Increased levels of carbon dioxide associated with localised increases in temperature and reduced humidity under conditions of poor ventilation have been reported to increase the dissolution of speleothems (Craven, 1996, 1999). Kiernan (1987) notes that, under conditions of high humidity, a temperature increase of 1°C can increase the vapour capacity of the air by as much as 8%, in turn increasing the evaporation rate and potentially resulting in the desiccation of speleothems.

Baker and Gentry (1998) suggested that the calcium ion content of drip water was another critical factor. The calcium content of dripping water is pivotal in speleothem growth and any change in concentration will influence the growth of

Figure 3.28 Bat observation platform Mulu National Park, Sarawak, Malaysia (Photo: David Newsome)

stalactites and stalagmites. Baker and Gentry (1998) advised that cave features may be at risk of damage where the calcium content of drip water become naturally low or where land use change alters the chemistry of cave system drip water.

Kiernan (1987) reported that clearing vegetation from above cave systems may result in increased soil surface temperatures, which could then facilitate biological activity and result in the increased generation of organic acids. More acidic waters, which dissolve and mobilise calcium ions, percolating into the cave system could result in the dissolution of speleothems rather than incremental growth. Above-ground impacts can cause water pollution and changes in water flow into the cave, and soil erosion can lead to the sedimentation of cave passages.

Calaforra *et al.* (2003) indicated the importance of defining baseline conditions and conducting experimental work in understanding human visitation impacts on caves that are yet to be opened up for tourism. In a study of the thermal modification of the cave environment, they demonstrated that the rise in air temperature due to the presence of visitors was rapid and, depending on the size of the cave chamber, the thermal perturbation could last up to six hours following a visit. The inner cave zone was noted to take longer to recover. In order to avoid long-term disruption of the natural thermal regime Calaforra *et al.* (2003) recommended small group sizes in sensitive caves such as Cueva del Agua de Iznalloz (Spain) if opened up for tourism.

Sensitivity of cave fauna

The dark conditions and relative constancy of temperature and humidity have given rise to the evolution of distinctive cave faunas. Many organisms are adapted to the dark, as shown by the loss of pigment and blindness. Cave faunas ('troglobites') occupy a very narrow ecological niche and are thus are very sensitive to disturbance. Examples of these animals from Australian cave systems include the giant blind cockroach (*Trogloblattella nullarborensis*), cave spiders (e.g. *Tartarus mullamullangensis*) and blind fish (e.g. *Anommatophasma candidum*).

Besides these specialised cave faunas a range of other animals use caves and cave entrances. These include various invertebrates, snakes, some birds and bats. Hamilton-Smith (1987) cautions that some cave-entrance-dwelling species are highly susceptible to disturbance, as seen by the elimination of cave-dwelling insects from the Alexandra Cave in south-eastern Australia following the insertion of a draught-proof door to control desiccation. Besides the intermediate habitat of the cave entrance area, the roof and cave walls provide resting areas for birds and bats. Various aquatic habitats can also be recognised, comprising streams, flood pools, water percolation pools and flow stone pools (Figure 3.29).

The ecology of cave systems is dependent on inputs from outside the cave. Cave drainage systems such as streams and pathways made by large trees are natural corridors and a major source of organic debris. Wet surfaces, such as flow stone areas, trap organic debris which supports bacterial mats. These, in turn, support colonies of crustaceans. Organic matter, derived from ecosystems outside, enters the cave system in the form of guano deposited by birds and bats that roost in caves. The reliance on energy derived from outside the cave system makes cave fauna susceptible to changes that occur at the surface and beyond the cave system itself.

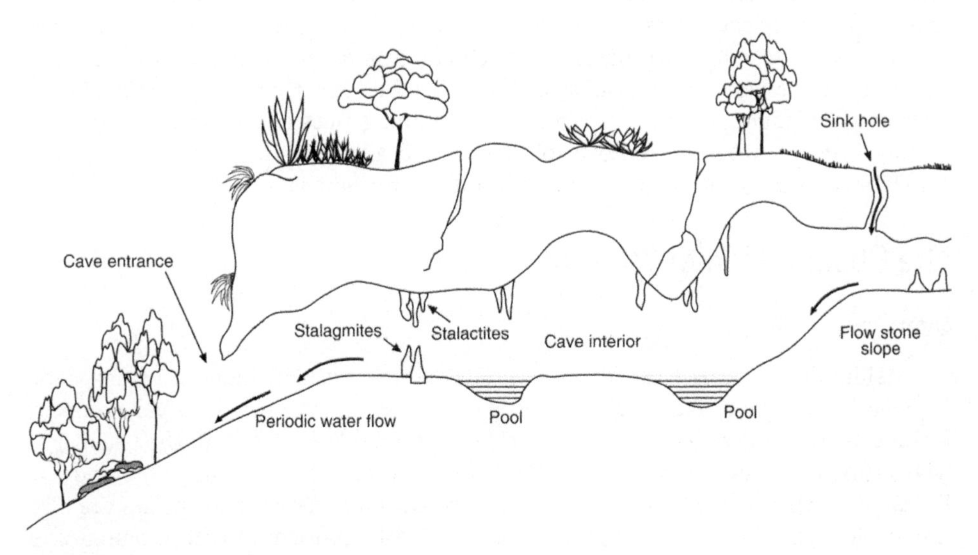

Figure 3.29 Conceptual diagram of cave habitats

Bats may decline as a result of increased human visitation to caves (Craven, 1999). Caves that receive large numbers of visitors are also subject to track degradation and pathways can become muddy when wet or liberate dust when dry. The habitats of cave faunas can be damaged by constant trampling and the stirring of sediment. Significant depletion of aquatic crustacean cave fauna has occurred in the Ogof Ffynnon Ddu cave system in Wales (Figure 3.2) as a result of caving activity. Edington and Edington (1977) reported that aquatic crustacean fauna were intact only in pockets outside the main caving route. Furthermore, entire populations of aquatic crustaceans can be lost due to chemical pollution, for example that caused by the careless dumping of torch batteries in pools.

Tourists also visit caves to see wildlife because caves are used as roosts and maternity sites by bats in many parts of the world and by swiflets (e.g. South-East Asia) and oilbirds (South America) in the tropics. Bats also use caves as rest areas when their normal food supply becomes scarce or during adverse weather conditions. Hibernating or over-wintering bats may be particularly prone to disturbance from large groups of people searching the roof-line of caves with torchlight. In seasonal environments over-wintering bats rely on stores of fat built up from the previous summer. Disturbance can cause them to utilise important reserves of energy and repeated disturbance can threaten their survival into the next season. Similarly, during the breeding season many species of bat leave the young in groups clinging to rock walls in the cave. Disturbance may cause the young to fall and frequent disturbance may increase the mortality of a population.

Techniques to manage visitor impact in caves have also resulted in population declines in bats. For example, Churchill (1987) found that the application of a steel mesh grille to prevent illegal cave entry at Cutta Cutta Caves in northern Australia (Figure 3.2) caused a decline from a bat population of 5000 to just 350 bats within 12 months. The population subsequently recovered when the grille was removed. Five species of bat use the Cutta Cutta Caves and one species, the rare orange horseshoe bat (*Rhinonicteris aurantius*), has strict microclimatic requirements in needing high humidity levels and temperatures of 28–32°C at its roost sites (Churchill, 1987). This example highlights the importance of retaining trees and vegetation adjacent to cave entrances, as the removal of trees could alter temperature and humidity conditions inside the cave, making it less suitable for fauna.

The Observation of Wildlife

Overview

Major wildlife experience destinations around the world include the Masai Mara, Serengeti and Kruger National Parks in Africa; the East African Rift Valley Lakes; Kakadu National Park in Australia; Chitwan National Park in Nepal; the Galapagos Islands in the Pacific; Antarctica; Yellowstone National Park in the USA; and the Pantanal Wetlands in South America. Spectacular and charismatic species are also the focus of specific tourism activity as seen in gorilla tourism in Africa, orang-utan viewing in Indonesia and Malaysia, watching lemurs and birds in Madagascar, swimming with whale sharks in Western Australia, observing elephant seals in the USA and Argentina, and whale watching in Australia, New Zealand, South Africa and North America.

Visitation to sites of wildlife interest has continued to grow (Newsome *et al.*, 2005; Rodger *et al.*, 2007). Mintel (2008) reported that, at the global scale, the market for wildlife tourism is estimated to be 12 million trips annually, with a growth rate of around 10% per annum. Mintel (2008) also noted that wildlife tourism is worth about £30 billion globally and that each year around 3 million people take a holiday specifically to view wildlife. In accordance with this trend, there has been a significant growth in the number of businesses offering wildlife tourism trips and tours (Curtin, 2010).

In a recent overview of wildlife tourism Newsome and Rodger (2012a) state that wildlife tourism is delivered and experienced in the wild according to a wide range of viewing expectations, involving many different species, and at a wide range of destinations and management conditions (Table 3.14). Newsome and Rodger (2012a) divide wildlife tourism into general nature-based tourism tours, wildlife experience destinations such as the Kruger National Park in South Africa and specialised wildlife tours that target certain groups such as birds (Monteverde, Costa Rica) or primates (e.g. Sulawesi macaque) and apes (e.g. the proboscis monkey in Malaysia).

Besides those people who seek out specific species and concentrations of wildlife, the presence and observation of wildlife play an important part in the recreational experience of hikers, campers and other users of natural areas. In surveys conducted in the USA, 96% of campers stated that the opportunity to observe wildlife in natural settings added to their outdoor experience (Hendee & Schoenfeld, 1990). Further, in Western Australia, 70% of visitors to Warren National Park stated that viewing wildlife was an important reason for their visit (Smith, 1998). Additionally, much of the scuba diving that occurs in kelp beds and off rocky shores and around coral reef systems is focused on viewing animal life.

Given this increasing interest in seeing animals in the wild and the large number of species involved, there is a risk of negative impacts occurring (see Higginbottom, 2004; Newsome *et al.*, 2005; Newsome & Rodger, 2008a; Higham & Liick, 2008; Newsome & Rodger, 2012a). In this section, therefore, we provide a brief account of some important principles relating to wildlife response to disturbance. Because other texts (e.g. Newsome *et al.*, 2005) deal with wildlife tourism in detail, our intention here is to provide an overview. Some examples of disturbance to wildlife have already been covered in Chapter 2 and these serve as additional content with regard to the vulnerability, responses to disturbance and assessment of wildlife tourism.

Vulnerability to disturbance

The vulnerability of an animal to disturbance depends on its life history traits and evolutionary strategies such as longevity, degree of parental care and reproductive effort (Hammit & Cole, 1998; Newsome *et al.*, 2005). Some animals, for example bears in the USA and elephants in Africa, are known to produce a more dramatic response to disturbance when caring for very young offspring. Tolerance levels can

Table 3.14 Examples of wildlife sighting opportunities and activities

Target species	*Country*	*Viewing conditions*
Gorilla	Rwanda, Uganda	Camping and trekking to locate habituated animals
Orang-utan	Sabah	Boat trip
Polar bear	Canada	From all-terrain vehicle
Rhinoceros	India, Nepal	From vehicle or on elephant back
Sperm whale	New Zealand	Viewing from helicopter
Minke whale	Australia	Boat trip and swim with activity
Bottlenose dolphin	Australia	Shore-based hand feeding and boat excursions
Whale sharks	Australia	Detection via spotter plane, boat trip and swim with interactions
Great white sharks	South Africa	Boat trip, attraction with food and cage diving
Sting rays	Cayman Islands	Boat-based access and feeding activity
Turtles	Australia	Guide-led excursions in restricted area to view egg laying; independent viewing of turtles emerging from sea to lay eggs
Chameleons	Madagascar	Guided walks
African savanna species	East and southern Africa	Accommodation at artificially created water holes; coach tours; private vehicles; tour operator vehicles; guided walks and camping; hot-air balloon trips
Birds	New Zealand	Boat trips
	UK, USA, Europe, Australia	Independent visits to a number of sites; birding trails
	Canada, Iceland	Bird watching hot-spot sites
	India, Costa Rica, Australia, UK	Water bodies with access trails and hides
	Bulgaria	Vulture feeding station
Frogs and toads	Tabin Reserve, Sabah	Guided walks
Freshwater fish	Bonito, Brazil	Snorkelling and scuba diving
Coral reef fish	Australia, Sulawesi	Shore- and boat-based access to reef; snorkelling and scuba diving
Glow worms	Australia, New Zealand	Constructed viewing area; visit to cave habitats
Nocturnal species		
Fireflies	Malaysia	Night boat trip
Possums and gliders	Australia	Guide-led spotlighting
Tasmanian devil	Australia	Devil 'kitchen' (feeding site with observation hide)
Bats	Sarawak, Malaysia	Constructed viewing area; visit to cave habitats; remote video technology

Adapted from Newsome and Rodger (2012a).

also vary with age, breeding season, time of year and type of habitat. Species with specialised food or habitat requirements are more vulnerable than generalist animals, and species that live in large groups tend to respond less to disturbances than do solitary animals (Hammitt & Cole, 1998; Newsome *et al.*, 2005).

Species vary in their degree of tolerance to human intrusion. Some, for example, are very shy and move away at the slightest detection of a human by sound, smell or sight. Previous experience, however, plays an important part in determining the response of a species to disturbance. For example, in Australia, red-necked wallabies (*Macropus rufogriseus*) will flee at the presence of humans in certain rural areas but those at picnic sites or golf courses will tolerate a much closer approach by humans or may actually seek out humans, in search of food (Green & Higginbottom, 2001).

Table 3.15 An approach to understanding human interest in wildlife

Human interest in wildlife	*Potential impact*
Accessing natural areas to see wildlife	Construction of roads Noise and disturbance Road kill Barriers Pollution Facilities constructed in natural areas and close to wildlife habitat
Locating wildlife	Focus on breeding colonies and islands Locating migratory movements Targeting of rare and endangered species
Observing wildlife	Animal response to intrusion varies according to species, breeding status, sex, age, habitat and experience Focal activities such as spotlighting Can occur in sensitive environments (e.g. cave habitats)
Photographing wild animals	Potential for extended visitor–wildlife interaction Risk of close approach Disturbance to feeding animals (e.g. encirclement of predators by vehicles in African national parks) Animals may be followed May be attempts to manipulate and feed Focus on breeding birds and/or adults with young Small animals may be captured
Feeding wildlife	Disruption of normal feeding activities Nutritional problems for target species Attraction of scavenging species Habituated/attracted species a potential danger to visitors Risky and dangerous animals may have to be destroyed
Touching and close interaction	Risk of disease transmission Visitor conflict in relation to differing values regarding appreciation of wildlife (manipulative versus authentic) Habituated/attracted species a potential danger to visitors Risky and dangerous animals may have to be destroyed

From Newsome *et al.* (2005).

Birds in urban environments are generally much less wary than those in more natural settings. Moreover, when disturbance does not result in negative effects birds will often cease reacting to humans, as it is important for them to conserve energy for their normal daily activities (Burger *et al.*, 1995). Overall, it appears that larger animals are affected more by the direct presence and activity of humans, while smaller animals are more vulnerable to habitat modification or indirect impacts (Hammitt & Cole, 1998).

Many researchers have divided the effects of recreation and tourism on wild animals into direct and indirect impacts. Indirect impacts include habitat modification and impacts associated with infrastructure (Roe *et al.*, 1997; Green & Higginbottom, 2000). Liddle (1997) divided direct impacts into type 1 and type 3 disturbances. Type 1 disturbance is defined as 'an interruption of tranquillity' and a type 3 disturbance is where the animal is wounded or killed. Liddle (1997) defined indirect impacts as a type 2 disturbance incorporating those impacts that result from changes to habitat. The focus, however, has mainly been on direct impacts, that is, a response to humans being present, as in walking, human-created sound and driving vehicles in a natural setting. Newsome *et al.* (2005) went further in describing an approach to understanding the nature of human interest in wildlife, and the complex nature of wildlife tourism is described according to modes of access and various types of disturbance in Table 3.15.

Behavioural responses of wild animals to humans

Tourists can have various attitudes towards wild animals. While some people enjoy and lobby operators for close sightings and even direct contact, members of the same tourist group may become alarmed if approached too closely by certain animals such as large bird, bats or primates. Some species appear to be universally acceptable and close contact with them is often desired, for example dolphins, while others, such as reptiles, invoke mixed feelings and are usually viewed from a distance. Tourists may be completely unaware of the effects they are having on a wild animal whether the contact is close, as in handling chameleons in Madagascar, or viewing from vehicles, as in the case of African safari tourism.

How a wild animal responds to the presence of a human depends on the sensitivity of the species (sometimes dependent on nutritional and reproductive condition), the age and sex of the animal, the animal's past experience and characteristics of the habitat in which it occurs. In addition, animal response also depends on the frequency, magnitude, timing and location of the disturbance (Vaske *et al.*, 1995; Liddle, 1997; Hammitt & Cole, 1998; Newsome *et al.*, 2005; Bejder *et al.*, 2009).

Three different types of behavioural reaction, namely avoidance, attraction and habituation, have been recognised as being fundamental in understanding wildlife responses to humans (e.g. Whittaker & Knight, 1998; Newsome *et al.*, 2005). Underpinning these reactions are thought to be the behavioural strategies that various animals employ in order to survive in the wild. For example, with attraction, an animal may associate humans with sources of food or shelter. Such behaviour under natural conditions increases the chances of survival in the wild and has thus evolved as a natural strategy. Whittaker and Knight (1998), however, caution that the attraction response is often confused with habituation. They see the difference between the two as being that in attraction there is a positive reinforcement of stimuli while habituation refers to a 'waning of response to repeated neutral stimuli'. This latter perspective is exemplified by the case of gorillas that tolerate the presence of humans once the process of gradual deliberate habituation is complete. Attraction, as Whittaker and Knight (1998) posited, would be where animals deliberately associate with humans in order to gain food. An example of this includes the many species of birds that are attracted to humans eating food, as in the case of glossy starling attendance at dining and picnic areas in the Kruger National Park, South Africa. Many tourists see such close contact with colourful birds as an addition to the wildlife experience. Attraction to food can, however, cause problems, as in the case where monkeys have totally lost their fear of humans and exhibit aggressive behaviour or elicit unwelcome close contact when they attempt to obtain food items from people.

According to Whittaker and Knight (1998), wild animals can become habituated to a wide range of human stimuli. This can have a negative impact on wildlife, for example where wild animals are habituated to the sight and sound of traffic. Such animals do not move away from fast-moving vehicles when crossing roads and are at risk of being hit, resulting in a Liddle (1997) type 3 disturbance (Green & Higginbottom, 2000). Shackley (1996) also warned that habituation has the potential to cause problems where it disrupts normal foraging and daily activity patterns.

Despite such potential problems, Shackley (1996) is of the view that habituation can be perceived as neutral to the animals concerned and a benefit for tourism. For example, wild apes are habituated to humans in Africa for the purposes of tourism. Gorillas, through repeated neutral contact, will gradually lose their fear of humans so that they can be viewed in the wild. This process may take 1–3 years for mountain gorillas and up to 5–10 years for lowland gorillas and chimpanzees (Robbins & Boesch, 2011). Nevertheless, Robbins and Boesch (2011) also caution that the habituation process causes wild apes considerable stress and fear and the process of habituation may expose the apes to disease (see p. 320) and also render them more susceptible to being successfully hunted by poachers. However, Robbins and Boesch (2011) further their view that the benefits outweigh the costs, as there is evidence that ape tourism reduces poaching and other illegal activity and fosters support for the conservation of wild apes in Africa. Following on from the issues raised by Robbins and Boesch (2011), concern as to how habituation might be understood and utilised in human–wildlife interactions has been raised by Bejder *et al.* (2009). The key contention brought forward by Bejder *et al.* (2009) is that the widely understood perception of habituation is that the response implies an absence of detrimental consequences and that there could be physiological responses even when there is little or no sign of behavioural reaction or sign of disturbance. In some cases, and where appropriate, it may be prudent therefore to recommend the assessment and monitoring of physiological condition such as heart rate, body temperature and hormone levels in attempting to understand whether wildlife is impacted or otherwise in a 'habituated' situation.

The measurement and assessment of recreational disturbance to wildlife

Although wildlife tourism is a widespread and increasing activity, there is dearth of 'hard' data on the nature and significance of its impacts on animals. The lack of data stems from the difficulties of researching animal behavioural responses to disturbance and relating this to tourism activity. Judgements regarding recreational impact on wild animals need first to consider total numbers in the animal population being studied, habitat requirements and the natural distribution of the species of interest. Secondly, a profile of tourism activity and pressures needs to be catalogued. Measures of impact can then be judged according to changes in the population, alterations in distribution and behavioural and physiological changes in the target species. A comparison between tourist/recreational sites with control (non-recreational) sites provides scope for the detection of impact. Various researchers have pointed out that impacts detected at the population level are likely to be significant in terms of reproductive success and long-term survival. Impacts on individuals or groups may be significant where local distributions are displaced or lost (e.g. IFAW, 1996; Newsome *et al.*, 2005; Rodger *et al.*, 2011)

Rodger *et al.* (2011) developed a framework to assess the sustainability of marine wildlife tourism operations. The framework was designed to collect ecological and environmental data and requires data inputs on the group dynamics of the target species, known behaviour, population biology, feeding and habitat requirements, nature of the tourism activity, known responses to tourism disturbance and the nature of any monitoring that might be in place. The framework also requires data inputs relating to operational and social conditions, such as operator licensing, number and type of tours, management strategies, knowledge of tourist compliance with regulations, nature of guiding, interpretation and visitor satisfaction. The framework was tested against the whale shark tourism industry at Exmouth in Western Australia and was found to be useful in identifying existing information, knowledge gaps and issues of concern.

Wildflower Tourism

Alongside the plethora of field guides for birds and other species of animal is the vast range of books available to assist in the identification of flowering plants. Such plant identification guides are widely available for tourists visiting protected areas in the USA, Australia, Europe and South Africa. Some locations like the Cape Province in South Africa and south-western Australia are wildflower tourism hot spots, with tours targeting specific geographical and protected areas (e.g. Figure 3.30). Various sites in Western Australia, for example, have trails specifically designed to provide direction and access for wildflower enthusiasts. Globally, a wide range of natural environments provide opportunities for wildflower tourism. Groups of plants of specific tourism interest include orchids, carnivorous plants, heathland species and large showy species such as *Amorphophallus* and *Rafflesia* (Table 3.16). Plant communities such as forests, bogs and wetland areas (e.g. *Victoria amazonica* wetlands in Brazil) are also tourism attractions.

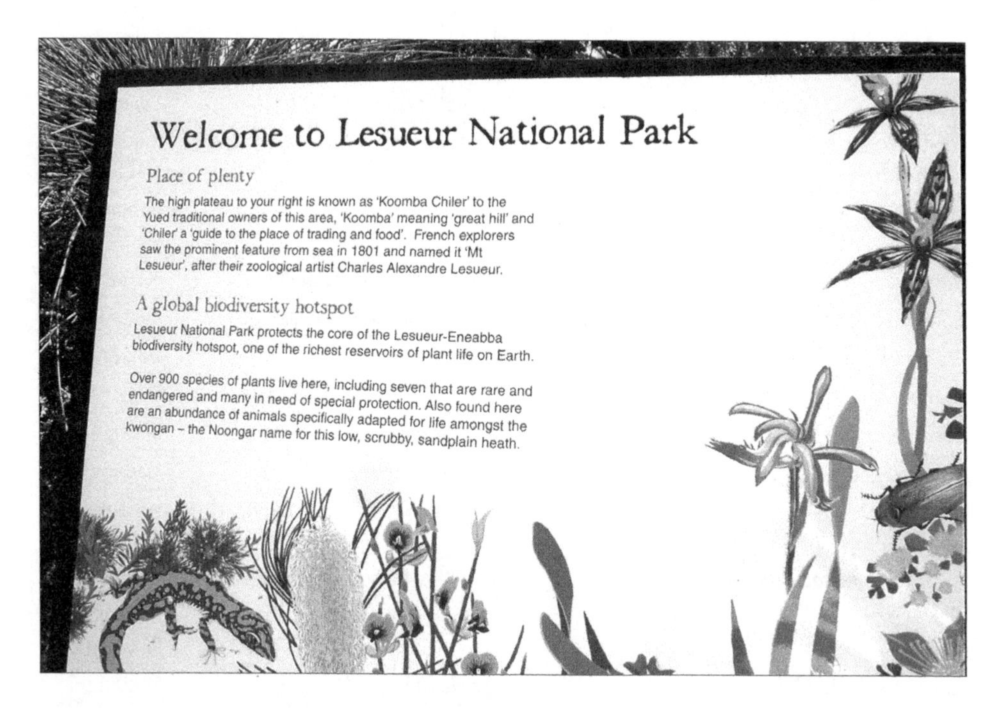

Figure 3.30 Information panel at Lesueur National Park, Western Australia, an international wildflower tourism destination (Photo: David Newsome)

Pickering and Hill (2007) reviewed the impacts of recreation and tourism on plant diversity in Australia and noted that despite the international significance of Australia's flora more research is needed. Accordingly, the environmental impacts of wildflower tourism per se have not been extensively researched and comprise an important area of recreation ecology that remains to be investigated. Pickering and Hill (2007) also make the point that the impacts of tourism on rare and threatened plants is particularly serious, especially in the case of infection with plant pathogens (see Box 2.2).

Pickering and Ballantyne (2012) describe orchid-focused ecotourism as travel to natural areas to view, photograph and learn about orchids. They describe the case of the Phu Khao Khoay natural area in Laos, where local guides lead treks for orchid enthusiasts. The Nepenthes Rajah Nature Trail at the Mesilau Nature Resort, located in the vicinity of Kinabalu National Park, Sabah, is at present the only place where tourists can reliably see one of the world's largest pitcher plants, *Nepenthes rajah*, growing in the wild.

Ballantyne and Pickering (2012) report on the problems associated with orchid tourism. These include habitat loss as a result of resort and infrastructure development, site disturbance by trampling, weed invasion, infection with *Phytophthora cinnamomi*, and the collection of orchids. This list of impacts will also be applicable to other forms of wildflower tourism. To this end, five terrestrial orchids are under

Table 3.16 Examples of tourism centred on wildflowers and specific types of vegetation

Focus	*Example of location*	*Country*
Heath forest/Kerangas (e.g. *Drosera spatulata var. bakoensis*)	Bako National Park	Sarawak, Malaysia
Pitcher plants (*Nepenthes*)	Mt Kinabalu	Sabah, Malaysia
Rafflesia	Taman Negara National Park	Malaysia
Epiphytic orchids (*Warszewiczella discolor*)	Bosque De Paz Biological Reserve	Costa Rica
Giant senecio (*Dendrosenecio kilimanjari*)	Mt Kilamanjaro	Tanzania
Heathlands (e.g. *Leucodendron*)	Fernkloof	South Africa
Succulents (e.g. *Pachipodium namaquanum*)	Garies, Namaqualand	South Africa
Spiny forest (*Alluaudia procera*)	Berenty Reserve	Madagascar
Heathlands and terrestrial orchids (e.g. *Banksia*)	Stirling Range National Park	Western Australia
Cacti (e.g. *Stenocereus thurberi*)	Organ Pipe National Monument	USA
Alpine flora (e.g. *Gentiana*)	European Alps	Austria, Switzerland
Ancient woodland (e.g. *Quercus*)	Eve's Wood	England
Giant redwood forest (*Sequoiadendron giganteum*)	Muir Woods	USA
Bog environments with a focus on mosses and lichens	Jostadalen	Norway

threat from illegal collecting in South Australia (SAG, 2006). Flower picking and the collection of whole plants are threats to wild orchids in several protected areas in Western Australia (Ballantyne & Pickering, 2012). Trampling, as discussed previously (pp. 114–125), is major cause of damage to vegetation and ground orchids are especially prone to such damage (SAG, 2006). Orchids in Western Australia are under threat as a result of hiking and illegal access by mountain bikers and trail (motor) bike riders. Ballantyne and Pickering (2012) report that wildflower tourists can damage orchids via the creation of informal trails. Existing trails may be widened as a result of tourists wishing to access flowering plants and such trails also provide for greater access to the wildflowers themselves.

Areas of disturbed ground may facilitate the invasion of weeds via the creation of bare soil, causing changes in the amount of soil nutrients and increasing the amount of available light. Pickering and Hill (2007) note that weeds are able to out-compete native species and can also alter resources for native wildlife, enhance the fire risk and bring about changes in soil chemistry. Wildflowers of tourism interest are also susceptible to climate change and particularly to reductions in precipitation, which cause desiccation of soils, reduced beneficial fungal activity, changes in the activity of pollinators and increase the risk of fire (Liu *et al.*, 2010; Seaton *et al.*, 2010).

Recreation and Tourism Focused on Sites of Geological Interest

Tourism focused on landscapes and geosites has always been part of nature appreciation and ecotourism, but the development of this type of visitation into a sub-sector of nature tourism with its own issues, impacts and management is a very recent development (Dowling & Newsome, 2006, 2010; Newsome & Dowling, 2010; Newsome *et al.*, 2012). A detailed account of geotourism is, however, beyond the scope of this chapter and the emphasis here is on some of the impacts that may be specific to sites where the focus is on landforms and rock exposures.

Very little has been published in this area (an exception is Dowling & Newsome, 2006) but a recent paper by Newsome *et al.* (2012) highlights some of the major impacts and provides guidance for management. Their work describes two case studies of iconic geotourism destinations: Yehliu Geopark in Taiwan and the Pinnacles Desert in Western Australia. Both sites are located in coastal settings and comprise upstanding landforms expressed as mushroom-shaped sandstone rocks at Yehliu and limestone pillars in the Pinnacles Desert. Yehliu is a mass tourism precinct (1.7 million visitors per annum) and the Pinnacles (200,000 visitors per annum) is situated in a remote location. Environmental impacts observed at Yehliu include touching and climbing on landforms, congestion, accessing of restricted areas and the creation of informal trails. The Pinnacles are also subject to climbing on the landforms and touching, the latter being particularly significant where this occurs at exposures of fragile aeolianite. Both sites are managed in an attempt to prevent damage in the form of scraping, graffiti and accelerated erosion, as well as to encourage visitor safety.

Perhaps the most important message that can be derived from this study is that the management of a site can be ineffective and even lead to a management footprint that carries a significant impact in its own right. For example, at Yehliu management actions have resulted in the development of hardened walkways, viewing platforms, installation of educational materials, life-saving equipment points, a boardwalk, barriers, extensive signage, security camera points and signed restricted areas (Figure 3.31). Newsome *et al.* (2012) view this as a case of overdevelopment in the form of a substantial management footprint which impacts on the natural values of the site. Moreover, even with the presence of wardens, management was not able to prevent visitors climbing over the landforms and with an estimated 10,000 visitors in a single day the congestion resulted in people straying from paths, not using the boardwalk and not adhering to park regulations specifying no-go areas. Overall, despite the management effort, it was ineffective at containing visitors and is unlikely to be entirely successful in containing impacts. These case studies have raised two important points: (1) when there are large numbers of tourists it is extremely difficult to manage them effectively; (2) there is the possibility that those people who are used to such 'free access' may exhibit non-conformist behaviours when travelling overseas, or to sites which are particularly fragile, only conservatively managed and/or where there is no management presence.

Figure 3.31 Extensive management presence and overdevelopment at Yehliu Geopark in Taiwan. Note that the boardwalk fails to contain visitors (Photo: David Newsome)

Social Impacts

Local communities

Natural area tourism offers new development and social and economic benefits for populations in regions within natural areas. But natural area tourism, like other forms of tourism, can generate both negative and positive impacts. Examples of the social drawbacks include the loss of community coherence, degradation of local culture, growth in crime and other social ills, and a loss of access to facilities for local people. Positive social benefits include the empowerment of local communities, community organisation, improved education and facilities, and the promotion of local culture (Koens *et al.*, 2009).

One strategy to maximise the ecological, social and economic benefits and to minimise any adverse costs is to engage the stakeholders in the development, planning and management of any nature tourism opportunity. Net benefits from tourism accrue from the balance of economic, social and environmental interactions of tourists with a destination (Greiner *et al.*, 2004). Any nature tourism ventures should be considered successful only if local communities have some measure of control over them and if they share equitably in the benefits (Scheyvens, 2002).

The proponents of natural area tourism have often presented it as a market-based activity that will provide income and empowerment to local communities while promoting environmental conservation. This is not always the case, however. For example, Costa Rica's Osa Peninsula, which hosts a national park and conservation area, is relatively isolated, and this has limited the employment of transnational corporate capital for the development of ecotourism. Although local, community-based ecotourism has been small in scale, there have nonetheless been struggles over access to land and natural resources, as well as the economic benefits of tourism, and representations of the environment (Horton, 2009). While natural area tourism offers potential economic opportunities for the Peninsula, it is also viewed as reproducing pre-existing patterns of stratification, particularly as state policies favour larger and foreign ecotourism enterprises. Similarly, research conducted in Lombok, Indonesia, into the social and socio-economic outcomes of ecotourism indicates that few economic benefits are gained by the local community (Schellhorn, 2010). Schellhorn (2010: 132) concluded that there is a need for a thorough scrutiny of the social effects of ecotourism, 'especially in the light of its increasing promotion by international donor agencies'.

Local residents almost always pay the social and environmental costs of conventional forms of tourism but seldom partake fairly in the benefits (West & Carrier, 2004). By contrast, nature tourism is designed to confer benefits directly on local communities. The positive benefits of nature tourism have been defined as economic – measurable as new employment or cash income (Walpole & Goodwin, 2001; Agrawal & Redford, 2006) – or social, for example in terms of community empowerment – measurable by the development of new skills, broader experiences in managing people and projects, strengthened abilities to negotiate with outsiders, and expanded circles of contacts and support for community efforts (Scheyvens, 1999).

In a survey of community views of ecotourism development in three Amazon communities in Ecuador, Peru and Bolivia, the local people indicated that economic benefits as well as increased self-esteem and greater community organisation were important (Stronza & Gordillo, 2008). Tourism development in the Maya Biosphere Reserve, Guatemala, was the catalyst for women to build alliances and work outside their immediate family, thus broadening their social networks as well as self-esteem (Sundberg, 2004). Similarly, women working in handicraft production in ecotourism projects in Monte Verde, Costa Rica, were given social and economic power that they did not have previously (Vivianco, 2001; Koens *et al.*, 2009). Such shifts should be considered in relation to conservation, as they affect the stability of local institutions and the prospects for long-term collective action for resource management. In Kermanshah Province, western Iran, community attitudes to the development of tourism were mostly positive, with residents saying they were happy to meet tourists, especially from other countries; they enjoyed the increase in local recreational facilities and viewed tourism as providing an incentive for heritage restoration (Mohammadi *et al.*, 2010). Social benefits for the people of the Masai Mara and Amboseli National Parks in Kenya have been that large numbers of Maasai have been hired as guides, drivers, hotel staff, artisans and cultural performers (Honey, 2009).

Crowding

Any examination of the use of natural areas should be centred on three elements: the natural or resource environment; the social or people (in this case tourist) environment; and the management context. A key intersection of the three elements of this framework is that, as well as the impacts of tourism development on people living in natural areas, there are also impacts on the other tourists visiting the areas as well. Recreational use can cause social impacts in the form degrading the quality of the visitor experience through crowding, conflicting uses and the social and aesthetic implications of inappropriate behaviour by some tourists.

One well identified impact is that of crowding, whether perceived or real, and is often referred to as overcrowding. The idea that there is a visitor use threshold, beyond which the visitation experience is diminished, has been the topic of much research and was instrumental in spawning the 'carrying capacity' approach to managing natural areas in relation to both the natural resources and the social environment (e.g. Wagar, 1964; Stankey & Lime, 1973). This led to a number of management frameworks, including: the Limits of Acceptable Change (e.g. Stankey *et al.*, 1985); Visitor Impact Management (e.g. Graefe *et al.*, 1990); Visitor Experience and Resource Protection (e.g. Hof & Lime, 1997); Carrying Capacity Assessment (e.g. Shelby & Heberlein, 1984); and Visitor Activity Management (e.g. Environment Canada & Park Service, 1991) (see Chapter 4).

The frameworks have been compared and reviewed by Manning (2011: 110), who argues that 'crowding appears to be a normative concept, dependent upon a variety of circumstances'. Key elements in the discussion of crowding in natural areas includes visitor characteristics, perceptions and coping behaviour, contacts with other groups encountered, and other vicarious situational variables.

Crowding can also occur in marine settings and a recent study in Molokini Shoal Marine Life Conservation District in Hawaii, USA, of encounters, norms, crowding and support for management found that the majority of boat users expected to escape crowds at Molokini, but more than 65% felt crowded (Bell *et al.*, 2011). Similarly, Szuster *et al.* (2011) found that the number and proximity of divers were strong determinants of scuba diver perceptions of crowding.

The view that crowding alone is a problem is a simplistic perspective (Arnberger & Haider, 2007) and, of course, crowding is not a problem in all countries and at all sites. Research undertaken in two Western Australian national parks found that while overcrowding was an issue at one (Kalbarri National Park), it was not at the other (Nambung National Park) (Moore & Walker, 2008).

The Impacts of Natural Area Tourism in the Context of Wider Environmental Issues

In this chapter a range of potential negative impacts that can arise as a result of recreation and tourism have been surveyed. Such impacts, however, can be anticipated and minimised via tourism planning (Chapter 4), the employment of various management strategies (Chapter 5) and the application of interpretation (Chapter 6).

Ultimately for natural area tourism to be successful we need intact and undisturbed natural areas that are adequately protected and properly managed according to the principles of ecologically sustainable development. Moreover, Weaver (2000) stated more than a decade ago that, with the pressures of mining, logging and agriculture, tourism was seen to provide an incentive to conserve natural areas. More recently, Carlsen and Wood (2004) and Hughes (2011) have written that tourists can bring economic benefits to a region and that natural and protected areas can generate significant income on a sustainable basis, resulting in additional investment by governments for environmental management and conservation. Accordingly, this section examines recreation and tourism in the wider landscape context, and considers threats such as urban encroachment and the human need to develop land for housing, infrastructure and resource exploitation. This section also explores the implications of an absence of tourism in natural areas and problems associated with inadequate security for protected areas.

Cumulative impacts

Small impacts can combine together to result in a much larger and significant impact. For example, a national park may face the combined impacts of tourist accommodation, infrastructure such as road linkages with attendant traffic, high levels of resource consumption, recreation/tourism infrastructure such as campsites, trails and car parks, as well as from the nature, location and intensity of various activities such as horse riding or rock climbing. Clearly, the extent and significance of a cumulative impact will depend on the sensitivity of the environment, the scale at which the sources of the impacts are developed and applied, and the effectiveness of prevailing management systems. The larger the natural area is, the more likely it will be able to 'absorb' various impacts. This is illustrated by the Kruger National Park in South Africa (p. 91), which, despite an extensive road network and some 25 camps, remains largely a wilderness area. The park receives up to 800,000 visitors a year but, at 1.9 million hectares, it is able to accommodate such tourist pressure because of its size, the management systems in place and the fact that the major activity is viewing wildlife from the safety of vehicles. The principal threats to the integrity of the Kruger National Park in fact reside in the landscape matrix in which it sits (p. 93).

Cumulative impacts are more likely to occur when both tourism and other factors come together. This has been observed at a prime natural area tourism location, the Cairngorm Mountains in the UK. This 500,000-hectare mountainous region is a major recreation/tourism site in Scotland, with infrastructure such as ski areas, mountain resorts, access roads and hotels. Recreational activities include hiking, mountain biking, ORV driving and self-drive touring (Crabtree & Bayfield, 1998). The negative environmental impacts associated with the infrastructure and activities include changes in site hydrology, visual intrusion, disturbance to wildlife, erosion, air pollution and the generation of solid wastes. It could be argued that these impacts themselves, when combined together, have a larger cumulative impact. However, as mentioned above, it is when some of the wider environmental issues are added to this list that the cumulative impact becomes very apparent.

Crabtree and Bayfield (1998) stated that the ecological sustainability of such a fragile mountain area is connected with the entire spectrum of land uses in the locality. They report that agriculture, forestry and estate management also generate significant environmental impacts. Examples include high stocking rates on grazing land, farm intensification reducing habitat diversity, drainage of wetlands and pollution. Superimposed on this are pollutants, such as acid deposition, derived from industrial landscapes beyond the Cairngorms. There are also positive economic impacts but, as Crabtree and Bayfield (1998) pointed out, it is mostly economic activity that has put the Cairngorms environment under considerable pressure. These problems will require wider social and economic solutions and a detailed consideration of this in the context of the Cairngorms is beyond the scope of this book. Nevertheless, such a complex set of land uses functioning on a regional basis is clearly governed by social and economic factors as well as aspects of land use policy.

The landscape matrix

Although this chapter is primarily concerned with impacts originating from recreation and tourist activities, it is important to demonstrate that tourism resources can also be damaged by adjacent land use and activities. Furthermore, it needs to be appreciated that such situations can lead to a combination of stressors affecting natural areas, creating a situation which may require additional management of a particular reserved area and/or attention given to source of impacts emanating from the landscape matrix. Clearly, where natural areas such as reserves, national parks and other protected areas exist as a patch, within a matrix of agricultural or other land uses, there is capacity for the matrix to influence this patch. Such influences are likely to be more significant if the natural patch is small and industrial or urban land uses occur as part of the matrix (Figure 3.32).

Cole and Landres (1996) pointed out that the impacts of adjacent land use include regulation of water flow, diversions of water and the location of dams on watercourses, air pollution and the effects of particular land management regimes (e.g. intensive agriculture). Impacts derived from regulated river flows and the impoundment of water upstream highlight the role of river corridors as sources of impact on natural areas. The role of rivers, sourced beyond a reserved area, in influencing the ecology of a particular region is illustrated by the impacts of water abstraction, turbidity and pollution problems on segments of the Kruger National Park in South Africa (see Box 2.3, p. 93).

Rivers systems in the matrix

Petts (1984) estimated that up to 75% of the world's rivers have their flow controlled by dams or their water flow otherwise regulated. Such a scale of river system modification has affected a number of important natural areas, as in the case of reduced water supply to the Everglades and Grand Canyon National Parks in the USA. The impacts include the degradation of riparian landscapes, altered sedimentation regimes and changes in natural fish communities. These alterations are brought about by altered flow patterns, the introduction of exotic species and disrupted barriers to fish migration (Cole & Landres, 1996).

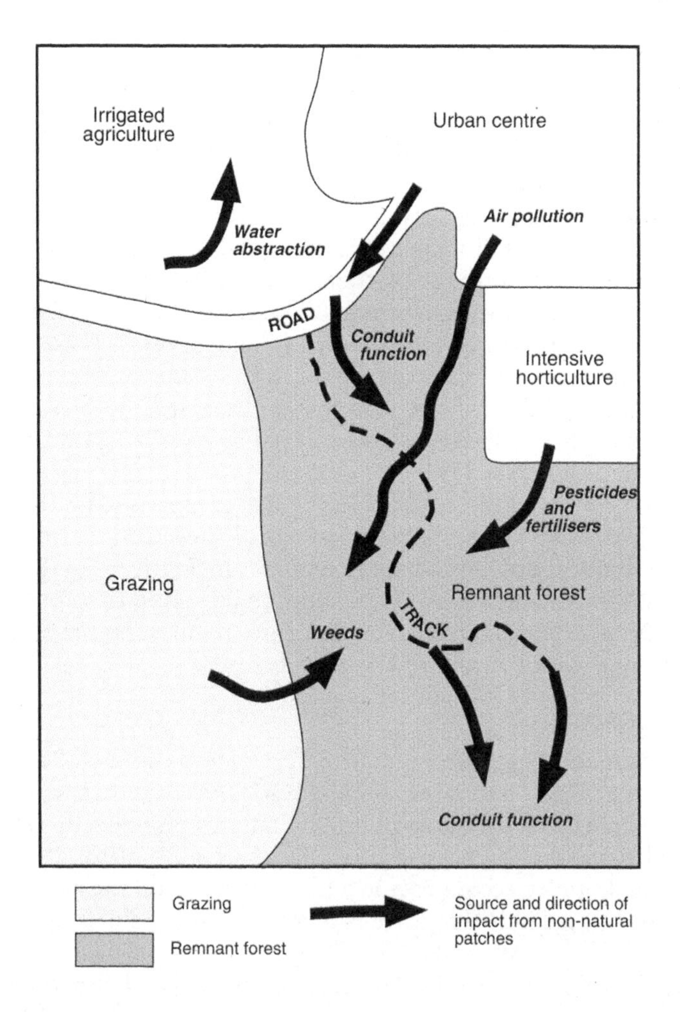

Figure 3.32 Potential impacts from non-natural patches

Kingsford (2000) in Australia demonstrated the significance of river modification to actual and potential natural area tourism resources. Citing various authors Kingsford (2000) reported on the decline of wetland ecosystems around the world. Wetlands are indeed significant tourism destinations globally, as indicated by important ecotourism resources such as the Pantanal in southern Brazil, the Everglades in Florida, USA and Lake Tempe in Sulawesi. Kingsford (2000) noted the high biological diversity of Australian floodplain wetlands, as particularly reflected by the diversity and large populations of waterbirds. These floodplain wetlands consist of swamps, floodplain marshes and tributaries, river overflows, lagoons, lakes and water holes, all of which are dependent on flood water from rivers. Many ecological processes in these wetlands are dependent on the arrival of floodwaters. Examples of theses include: organic inputs providing energy for microorganisms; stimulation of drought-resistant zooplankton eggs to hatch; germination of plants from seed banks; and the stimulation of breeding activity in frogs and birds (various authors cited by Kingsford, 2000).

Dams can either submerge a wetland or substitute a variable flooding regime with a permanent one so that some wetlands never dry out, with resultant ecological changes. Kingsford (2000) uses the Macquarie Marshes (Figure 3.2) as a case study in demonstrating the impacts of dam construction, water diversion schemes and water abstraction on an internationally important wetland for birds. These landscape-level activities have reduced the extent of the marshes by up to 50%, caused a decline in the abundance and diversity of waterbirds, decreased the breeding activity of colonial waterbirds and reduced the area covered by river red gum trees (*Eucalyptus camaldulensis*) by as much as 15%.

The natural corridor function of rivers is also emphasised by their capacity to move materials through landscape. This has already been considered in the context of the Kruger National Park and adjacent land uses in South Africa. Eroded soils and toxic chemical components such as pesticides can be moved from one ecosystem to another in this way. Toxic materials can also enter natural areas at their boundaries or be moved along disturbance corridors such as roads.

The agricultural matrix

In a predominantly agricultural matrix, the widespread use of pesticides could affect adjacent patches of natural ecosystem. Pesticides are known to impact on birds through direct mortality, by causing breeding failures or indirectly by the removal of insect food supplies (e.g. Newton, 1995; Burchart *et al.*, 2010). It is also possible for toxic materials which accumulate in a food chain within the matrix to cause the breeding failure of predators that leave patches of natural habitat in order to feed in the matrix.

As pointed out by Cole and Landres (1996), the flow and dispersal of organisms from natural areas can be detrimental, with losses occurring as a result of accidental road mortality or deliberate shooting, as when wild animals kill domestic stock or when crops are grazed and damaged by herbivores. This can reduce the abundance and long-term viability of natural populations of animals confined to patches within a landscape mosaic. The problem also becomes much more significant where natural areas are small and the scope for in-migration from other source areas is limited, due to the presence of an extensive modified matrix and lack of connectivity between separated patches of natural ecosystem.

In contrast to the dispersal of organisms out of patches, predators, parasites, weeds and pathogens can enter patches of natural area from the adjacent matrix (Cole & Landres, 1996). An example of this is the spread of the fox (*Vulpes vulpes*), feral herbivores and weeds from adjacent agricultural land across the boundary of, and into, the Stirling Range National Park in south-western Australia (Box 2.2, p. 66, looks at the spread of *Phytophthera* within that park). Moreover, in Western Australia isolated reserves of natural vegetation, riparian landscapes and attendant wildlife, which occur in the extensive agricultural zone, are under threat from widespread landscape salinity. This problem is occurring as a result of rising saline water tables induced by the extensive removal of deep-rooted woody vegetation which originally kept the water table at greater depths.

Urban encroachment and the need for land

Urban encroachment will be an increasing source of landscape-level impacts globally in the future. In Australia, for example, some of the most popular natural areas, which have increasing levels of recreation and tourism pressure, lie in the proximity of expanding urban areas. Housing development, for instance, is expanding up to and alongside the borders of John Forrest National Park on the outskirts of the city of Perth, Western Australia (Figure 3.2). The proximity of this urbanised landscape to the park increases the chance of fauna mortality should any out-dispersal take place. Furthermore, in a European study, Van der Zande *et al.* (1984) demonstrated that the density of breeding birds declines in association with high recreation intensities emanating from adjacent residential areas. The close proximity of urban areas means that there is a much greater chance of more people utilising the area for a variety of reasons (e.g. walking their dogs) and damaging natural resources. Nearby urban areas can also act as sources of weeds and feral animals that may compete with or prey on native species. In the case of the mountain gorilla in Rwanda agricultural encroachment has driven gorillas further and further up into the mountains and into less favourable habitat. With more people in the vicinity there is also a greater risk of disturbance and, where poverty is a factor, an increased threat of hunting and poaching.

Air pollution, recognised by Cole and Landres (1996) as a potentially significant impact originating from the landscape matrix, can take the form of acid rain, heavy-metal contamination and photochemical smog. These forms of air pollution can be transported from urban areas into natural areas by local winds. Sigal and Nash (1983) reported premature leaf senescence, reduced growth and an increased susceptibility to disease in vegetation caused by air pollution in the Cascade Mountains in the western USA. The urban encroachment around John Forrest National Park, described above, could lead to pollution-induced physiological stress on the vegetation. Pollutants flowing into the park could increase disease susceptibility, which is a critical issue given that *Phytophthora cinnamomi* that has already been accidentally introduced into the park.

Globalisation has meant that, in many cases, foreign investors now drive local changes in land use, with a specific emphasis on the expansion of agricultural activity (Friis & Reenberg, 2010). According to Nayar (2012), 203 million ha of such land deals were conducted in the period 2000–10. A particular focus has been on Africa, where Liberian forests have been sold to land investors for oil palm development. The 'Africa land grabs', as they are known, may be at the expense of as yet unrecognised tourism resources, with 'unused' natural areas in more marginal areas being targeted for food security, especially with the aid of technology to assist in agricultural development. Ananthaswamy (2011) reports that China, India and Saudi Arabia have leased large areas of land in sub-Saharan Africa and in the period 2004–09 Saudi Arabia leased 376,000 ha of land in Sudan to grow wheat and rice. Such practices can also lead to wider problems and land degradation over time.

Natural areas with no tourism

Natural area tourism is a land use that provides people with the opportunity to experience and learn about landscapes, natural ecosystems, plants, animals and geology. In the period 2002–12 the world continued to lose natural areas and wild places and there was a trend towards further reductions in the populations of wild plants and animals. For example, it is believed that 2078 bird species (out of an estimated world total of 9895) are in decline (Burchart *et al.*, 2010). A positive aspect, however, is that over the last two decades or so many countries have increased their networks of protected areas (Lockwood *et al.*, 2006). Nevertheless, not all important nature-based tourism resources are safe from damage or loss. For example, viable ecotourism resources can be converted into plantations (e.g. oil palm in Malaysia and Indonesia) or agricultural land (e.g. slash-and-burn land use in Madagascar). Furthermore, traditional economic imperatives, such as mining, may seem more attractive to governments if natural areas are seen to have no or little economic value. A case in point is Pilliga Forest in New South Wales, Australia (Figure 3.2), a large remnant of semi-arid woodland containing significant biodiversity, currently at risk of clearance (2400 ha) and fragmentation with the development of an onshore gas reserve (Birds Australia, 2011). Norman (2011) expresses concern over large-scale industrial projects such as mining and the processing of liquefied natural gas where there has been a highly significant recent discovery of dinosaur footprints. The find, located in the remote Kimberley region of northern Western Australia (Figure 3.2), represents as many as 15 species of dinosaur and is possibly the largest stretch of dinosaur footprints in the world. Norman (2011) also points out that hundreds of mining exploration licences (uranium and bauxite) have already been granted, which poses an additional threat to existing nature-based tourism attractions in the region.

Inadequate security for protected areas

Table 3.17, although focusing on Asia, provides an overview of threats to nature-based tourism resources. These threats include roads and infrastructure, energy supply, hydro-electric power development, illegal logging, expansion of plantations and agricultural land and resource extraction. For example, UNESCO has expressed concerns (*Brisbane Times*, 2011) about a proposal for a plant for processing liquefied natural gas and a port facility that has the potential to impact on the Great Barrier Reef in Australia, and approval has also been granted for oil and gas exploration close to the Ningaloo Reef in Western Australia. Both reef systems are highly significant conservation reserves (and are World Heritage sites) and tourism destinations and could be impacted if such development projects proceed, especially in the context of an industrial accident such as an oil spill.

Table 3.17 Threats and obstacles to conservation of natural areas and wildlife, protected areas and sustainable tourism in South and East Asia

Problem	*Selected examples from across the region*
Logging and habitat loss	World demand for palm oil is resulting in the expansion of oil palm plantations in Indonesia and Malaysia. Illegal exploitation of national parks, for example in Indonesia
Fuel-wood consumption	An estimated 21 million tons of fuel-wood consumed per annum in Vietnam
Harvesting of forest products	Over-exploitation and environmental damage caused by access and harvesting of valuable timbers, plant roots and wildlife
Hunting wildlife	Increased demand for animal products in Vietnam, Laos and Thailand. In Laos much protein is derived from natural areas
Traditional shifting cultivation	A cause of forest loss and land degradation in Vietnam and Laos
Poaching and illegal trade in wildlife	Soaring world demand for traditional medicines
Forest fires	In Vietnam 56% of remaining forest prone to fire in the dry season. In 1982–83 about 5 million ha of forest was burnt in logged and drought-affected primary forest in Borneo
Population increase	Global population reached 7 billion in November 2011, much of the increase sourced in developing countries
Urban expansion and infrastructure development	Connected with increased population and economic activity promoting increases in the demand for basic raw materials, transportation, water and energy. Road corridors planned to cut through protected areas in Vietnam
Hydro-electric dams	Additional dam construction projects planned for river systems in Vietnam, Laos Thailand. Cat Tien National Park in Vietnam is under threat from a hydro-electric project
Migration policies	Migration policies have led to demographic imbalance, changes in cultivation practices and increased demand for natural resources in Indonesia and Vietnam
Poverty	Many indigenous people suffer poor health and reduced life spans. Large numbers of people in Vietnam, Laos and Indonesia live in rural areas and are dependent on natural resources
Economic policies	Expansion of export markets for agricultural products and increased demand for agricultural land in Vietnam, Thailand and Indonesia
Mass domestic tourism and unmanaged tourism	Malaysia, Vietnam, Thailand and Indonesia. As much as 80% of original forest has been lost in Thailand. There are tourism pressures on remaining forests (e.g. resort and road construction around Khao Sam Yot National Park in Thailand). Visitation in Thailand is at around 10 million, 90% of which is of domestic origin. In Vietnam 50% of the domestic tourism is focused on natural areas. In both Thailand and Vietnam there are low levels of environmental awareness, no codes of conduct, lack of park management and over-development

Outstanding natural areas can be significantly influenced by the surrounding landscape, as with the encroachment of oil palm plantations along the fringe of the Kinbatangan River in Sabah (Newsome & Rodger, 2012b). The Great Rift Valley Lakes Elementaita, Nakuru and Bogoria in Kenya have been affected by the clearing of natural vegetation in their catchments, pesticide and fertiliser inputs into rivers that drain into them, water abstraction and mineral exploitation. Human pressures combined with the characteristics of a semi-arid climate that is susceptible to drought and climate change pose an ongoing risk of land degradation. Farming activities have resulted in soil erosion and reduced water flows into Lake Elementaita and extractive practices constitute a threat to this lake ecosystem.

Even where Protected Area Management Plans exist, their use and application may not be effective. In a global analysis of management effectiveness for protected areas, Leverington *et al.* (2010) found that many (40% of assessments) protected areas show management deficiencies in the form of inadequate resourcing (funding, staff numbers and facilities), communication and community relations problems and problems with management planning. Management plans need to be effective in terms of mitigating human impacts, in ensuring biodiversity conservation and in facilitating sustainable tourism management (see Chapter 5).

Human pressures and poverty remain significant problems. Indeed, the role of indigenous people, in the form of guides and rangers, is potentially very exciting in terms of opportunities for indigenous and local people to share their knowledge of the natural environment with tourists. But in many parts of the world there are problems with traditional practices and hunting impacting on protected areas; furthermore, local people may have poor environmental practices regarding waste disposal and littering. One example of potential conflict is in Australia, where, under joint management agreements, park management agencies may allow traditional practices in protected areas. These may involve burning ceremonies, hunting native wildlife (sometimes rare species, using vehicles, dogs and guns) and collecting plant materials. Such activities are likely to conflict with tourist expectations of natural area management and conservation.

Aiyadurai *et al.* (2010) state that hunting for food, trade, leisure and festivals is a serious threat to wildlife worldwide. Large carnivores are declining in South America due to hunting activity (Carvalho & Pezzuti, 2010). Woodroffe and Ginsberg (1998) note that hunting may occur in the absence of habitat loss and inside protected areas. The impacts on tourism are significant in that the species people want to see can be eliminated from protected areas. In Cuc Phuong National Park in Vietnam mammals that are highly sought after, such as Delacour's langur, have been either lost or greatly reduced and for bird watchers many species have low visibility due to hunting shyness and low population levels. Primack and Corlett (2006) maintain that the increase in human population, use of guns, vehicles and spotlights at night and the demand for bush meat from urban centres as well as for traditional medicines have led to a substantial depletion of larger wildlife in many tropical forests. Where there is effective protection, the abundance of wildlife is high and it can be readily seen because the animals are not so shy of humans. According to Corlett (2009), there is not likely to be a forested area anywhere in tropical East Asia where an intact fauna comprising large mammals occurs at natural levels; such forests, even those ostensibly protected, are mostly devoid of large wildlife.

Currently there is a soaring world demand, especially from China, for animal and plant products and products from Asia and Africa are sought after for traditional medicines and beliefs. One example is the poaching of rhino in Africa for horn. There is a large population of unemployed people living along the western border of the Kruger National Park in South Africa and poaching rhino horn has become a serious problem in recent years, with 448 rhinos killed in the park in 2011. As part of a wider management response to poaching incursions along the borders of the park, a 150 km electric fence will be constructed along the eastern border in order to deter poaching by people from Mozambique. An additional 150 rangers are also to be employed in an effort to reduce the loss of rhino to poaching (BBC News, 2012).

Conclusion

The purpose of this chapter has been to illustrate the possible consequences of recreation and tourism in a wide range of natural environments. Although it appears to paint a negative picture of natural area tourism, the objective has been to highlight potential problems so that they can be anticipated and managed. Never before have so many people been interested in the natural world. Moreover, our social and political systems would not allow nature to be 'locked away' and left only for scientists and film-makers. The future lies in making natural landscapes, flora and fauna available for people to experience and enjoy. Local communities and economies also need to benefit from natural area tourism. Indeed, this is usually the major justification for the protection of nature in both the developed and developing world. Impacts brought about by tourism, however, can spoil the resource and diminish visitor experience. This, over time, can lead to a decline or change in tourist interest in an area, resulting in social conflict, economic impacts and environmental degradation.

The need for ecological understanding remains clear and being able to see how a natural area fits into the bigger landscape context is also critical in reducing impacts and protecting the resource. This exemplified by the situation in the Kruger National Park, which is being affected by largely external factors.

Although there is an extensive literature on the impacts of tourism there are many issues and specific cases that remain to be understood. The data, for example, on the impacts of tourism on wildlife are just not available for many countries in the developing world and even for places like Australia, which is a major destination for nature-based tourism. Furthermore, there is a new type of user in natural areas, arising from activity-based recreation and organised sporting events, and there are no direct data on the biophysical or socio-cultural effects of such new uses. Tourism in protected areas cannot be planned, managed and monitored without data on resource condition and environmental impacts. This chapter has provided an overview of a range of potential impacts and sets the scene for the planning, management and monitoring of various activities that fall under the umbrella of natural area tourism. We have also considered impacts in perspective, where, compared with many uses of the landscape (agriculture, mining, forestry, extraction of forest products), ecotourism and wildlife tourism are passive and constructive. It is vital that managed tourism is valued as a means to conserve nature and foster the public's interest in conservation.

Further reading

Over the last decade, there has been an exponential rise in the number of papers published documenting the environmental impacts of recreation and tourism. However, previous syntheses of data and reviews by Hendee *et al.* (1990a), Kuss *et al.* (1990), Liddle (1997) and Hammitt and Cole (1998) are still relevant. These publications contain a large amount of information that spans many years of work. New work includes Buckley (2004a), Newsome *et al.* (2005), Smith and Newsome (2006), Cater *et al.* (2008), Davies and Newsome (2009), Pickering (2010), Pickering and Mount (2010) and Pickering *et al.* (2010a, 2010b). Disturbance to wildlife is a major issue in natural area tourism and has received much attention. For example, Knight and Gutzwiller (1995), Shackley (1996), Liddle (1997), Hammitt and Cole (1998), Higginbottom (2004), Newsome *et al.* (2005) and Higham and Liick (2008) have produced important accounts, case histories and discussions on recreational and tourism disturbance to wild animals in various contexts and settings.

Information on the positive social and economic impacts of tourism, especially from a community perspective, may be found in books by Scheyvens (2002) and Singh *et al.* (2003). Other readings which take a similar approach include examples from Australia (Beeton, 2006), Africa (Jones, 2005), the Amazon (Stronza & Gordillo, 2008) and Costa Rica (Koens *et al.*, 2009). A sound reading on the economic impacts of ecotourism is provided by Lindberg (2001) and an analysis of the complexities surrounding the social and economic impacts of tourism can be found in Hall and Lew (2009).

4 Visitor Planning

Introduction

If natural area tourism and its potential impacts are to be managed in effective and cost-efficient ways, then planning is essential. This chapter focuses on recreation/tourism planning frameworks as a means of planning for visitor use of natural areas. The subjective nature of planning and the need to engage stakeholders throughout planning processes are emphasised. The concepts of carrying capacity, 'acceptable' change and the spectrum of recreation opportunities are described because an understanding of them is essential before discussing the frameworks themselves. Details of six visitor planning frameworks follow, plus suggestions as to how to choose between them. Benefits-Based Management is also included as another approach to planning.

A variety of planning processes can be used to manage natural area tourism. Planning for and with the tourism industry is of central importance. Visitor and tourism industry management may also be part of a broader suite of issues considered in natural area management (also referred to, with slight variations in meaning, as protected-area, heritage and environmental management). Planning for visitor use of natural areas desperately requires detailed attention to the rapid increases in visitation over recent years (Chapter 1) and the associated potential for increased impacts (Chapter 3). As such, it is the subject of this chapter.

Confusion continues to exist regarding the interface and overlaps between visitor planning frameworks and management plans for natural areas. Most often, these frameworks contribute to sections in these plans on visitor management. Management plans are usually much more broad ranging than visitor planning frameworks, addressing management of ecological communities and rare species, wildlife, fire, introduced weeds and pests, adjacent land use and water bodies, in addition to visitors and the tourism industry. For reasons explored in this chapter, visitor planning frameworks have not been widely used in management planning for natural areas. Their omission from such processes continues to compromise the quality of natural area management.

Much of the material here is drawn from the wealth of research and practice in recreation planning and management for national parks and wilderness areas worldwide. Planning activities in the USA, Canada and Australia were a particularly rich source of ideas. The frameworks have been applied to individual, often large, natural areas and more broadly across a region or group of natural areas.

Definition

Planning is a process of setting goals and then developing the actions needed to achieve them. For natural area tourism, it allows managers to define what experiences visitors will have, the experiences they want to produce, the visitors they want to attract and the limits to environmental modification deemed acceptable. This type of planning focuses on managing to achieve desired outcomes. Planning helps achieve these outcomes in the face of changing internal conditions, such as funding and staff changes within management agencies. It also helps managers to cope with external changes, such as swings in public opinion and changing demographics. Planning is essential for dealing with complexity and uncertainty (McCool *et al.*, 2007) and so is especially important given the uncertainties associated with the world's climate, energy supplies and political and economic stability. All potentially affect natural area tourism and its management.

Planning has a number of distinguishing characteristics. For natural area tourism, these coalesce around the idea that natural areas used by tourists are products. Thus, the basic task of planning is to visualise the area, that is, the product, as visitors and managers wish it to be in the future. Planning is a process continuing over time, sometimes resulting in a written plan, but not always. It generally includes establishing goals and objectives, determining strategies and actions, and guiding implementation and review (Figure 4.1). Goals are general statements of desired future conditions, whereas objectives are specific, measurable and attainable. Strategies are management directions, for example providing interpretive opportunities for visitors, and actions are specific details on what will be done, for example providing a visitor centre in a specified place. Until implementation has occurred, planning is not completed. Hall (2000a) described this process as strategic planning and emphasised its iterative nature. The process must be flexible, adaptable, iterative and ongoing,

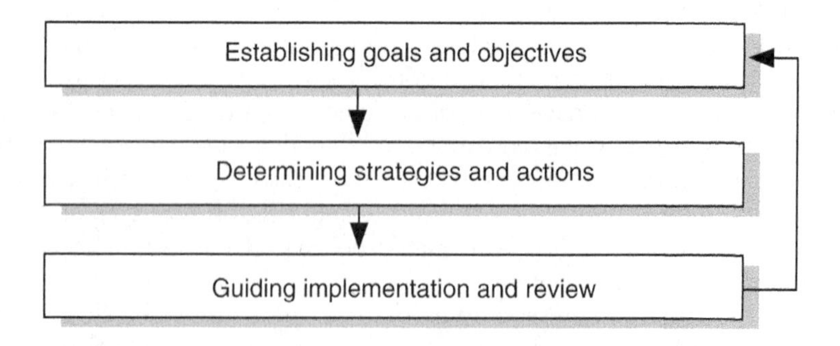

Figure 4.1 A generic planning process

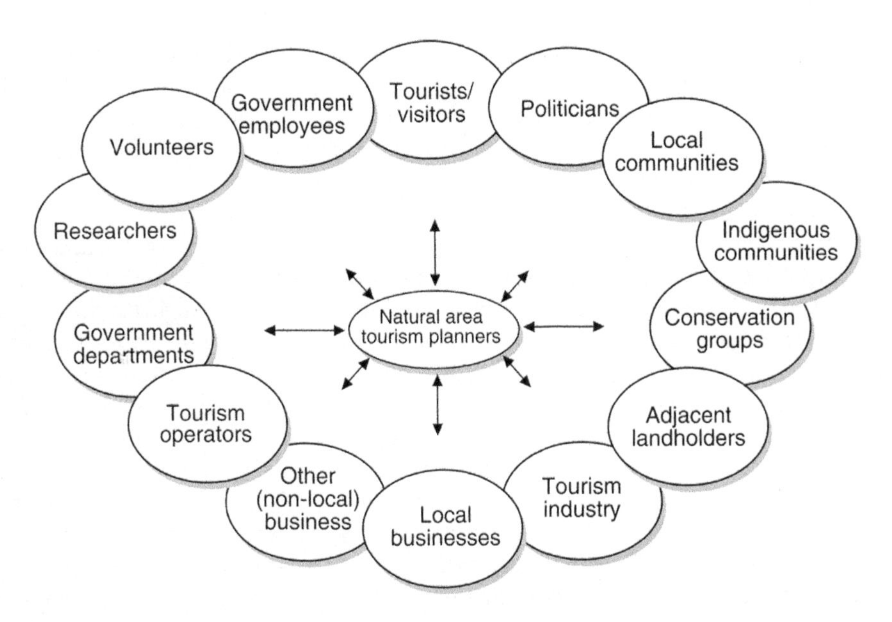

Figure 4.2 Possible stakeholders of natural area tourism (Derived from Hall & McArthur, 1998; Sautter & Leisen, 1999)

allowing objectives and strategies to be adjusted while still providing a means for consistent management.

The last distinguishing feature of planning for visitor use of natural areas is its participatory nature. All the planning frameworks described in this chapter have provision for stakeholder involvement. Where and how involvement occurs depend on the framework. Stakeholders include those directly affected, such as visitors themselves, plus those indirectly affected (e.g. local communities), as well as those managing or providing tourism opportunities, such as land managers and tour operators. Those further afield, while not actively involved in or influenced by tourism use, may watch such activities with interest. They are also stakeholders. Potential stakeholders are tourists/visitors, local communities resident in or near natural areas, indigenous peoples, conservation groups, local and other businesses, politicians, governments and their employees, and competitors (Hall & McArthur, 1998; Sautter & Leisen, 1999) (Figure 4.2).

Reasons for visitor planning

Planning for visitor use of natural areas is a relatively recent phenomenon, emerging only in the final decades of last century. Such planning has been a response to the dramatically increasing use of natural areas worldwide. This increase is due to several factors, including changing mobility. Motor transport places many natural areas within easy reach of major population centres, while air travel means remote corners of the world are only a day away. Technology, and particularly lightweight camping and hiking equipment, has made extended stays in remote areas possible for increasing numbers of visitors. Also, at least in the 1960s and 1970s, people had increased leisure time to access and enjoy natural areas, and to become concerned about their management.

Other reasons for increased use of natural areas are related to changes in education levels, lifestyles and spirituality. Over the last few decades, people have become more educated, mainly through mass access to tertiary education. An increased appreciation of the natural environment is an outcome. Such an appreciation is also a product of the stresses of urban life, with many urban dwellers relying on natural environments for relaxation and regeneration. Nature shows on television are also a contributing factor. Additionally, some people's spiritual practices depend on natural areas. Beliefs such as Gaia focus on the Earth and the interconnectedness of all associated living and non-living matter. These relationships are perceived as most harmonious in natural areas. Other societal groups, such as the men's movement, rely on natural settings for ceremonies and rights of passage.

Several problems that make planning imperative have arisen from this increasing use. The very things that attracted visitors in the first place, whether spectacular landscapes, unusual plants or animals or high biodiversity, can be degraded by human use. Conflict between users is also a possibility. For example, hikers often come into conflict with others enjoying the same area on horseback or in motor vehicles. Planning helps to avoid or at least minimise such conflicts. It can also help avoid problems created by successive minor decisions. Often, one decision leads to another and before long undesirable and even irreversible actions have been taken. This tyranny of small decisions can result in the values that drew visitors to an area being inadvertently lost.

For many natural area tourism destinations, such as protected areas in the USA, Canada, Australia, New Zealand, China, Taiwan, Indonesia, Thailand, India, Malaysia, Austria and Finland, managers have devoted their energies to acquiring and/or reserving land. In the face of limited staff and financial resources, acquisition has been the highest priority. The next steps are planning and management. Limited resources for these activities have been accompanied by a lack of systematic and widely available planning approaches. The frameworks presented in this chapter fill this gap and are crucial in managing protected areas and regional tourism for sustainability (Wearing & Neil, 1999; Ahn *et al.*, 2002).

The last reason for visitor planning is a legislative one: planning for protected areas is mandated by legislation in many countries. For example, in countries such as the USA, where management of protected areas is the responsibility of the federal government, planning is mandated by the Wilderness Act of 1964, the National Forest Management Act of 1976 and the National Environmental Policy Act of 1969. In countries such as Australia, where states/provinces are responsible for managing protected areas, legislation such as the Conservation and Land Management Act 1984 (Western Australia) similarly requires planning. Such plans usually include visitor management.

Planning as a value-laden activity

Above, planning was defined as a process of setting goals and then developing the actions needed to achieve them. This is very much a rational, objective view of planning. Do value judgements have a place in such a process? The answer is a resounding yes, for two reasons. First, planning is about determining what should be, as well as what is (Lipscombe, 1987). Making decisions regarding 'shoulds' always involves value judgements. Second, planning includes planning for visitors as well as recognising the concerns and interests of managers. As such, the value judgements of visitors, managers and other stakeholders must be considered (McCool *et al.*, 2007).

Stakeholder Involvement in Visitor Planning

Stakeholders in the developed world generally expect to have a choice regarding whether or not they become involved in visitor planning for natural areas. In contrast, in the developing world citizens may not expect to participate, although this is changing. Planning may be perceived by local residents as the responsibility of government, which then informs people of its decisions. An Indonesian government tourism planner interviewed as part of a study of participatory planning commented that decisions should be made by government for the good of society (Timothy, 1999). Another interviewee commented that Indonesian's customary approach to authority based on respect and subservience may inhibit grassroots involvement. Additionally, involvement may be curtailed by it being considered a luxury, people being more concerned with basic survival than long-term planning, and citizens' beliefs that they know too little to get involved. Not all stakeholders are excluded. Elites, such as influential businesspeople and political figures, have long had access to governmental decision making (Timothy, 1999).

A range of very different intentions and possibilities can underpin the inclusion of stakeholders in visitor planning (Figure 4.3). Arnstein's (1969) 'ladder' provides a useful summary of these possibilities, ranging from no opportunities for involvement (i.e. non-participation) through to stakeholders having complete control of planning and management of an area (i.e. citizen control). In the developed world (e.g. in the USA, Canada and Australia) stakeholder involvement in planning for protected areas has relied on consultation and in some cases partnerships. Movement beyond partnerships is unlikely, given that public agencies are legally responsible for managing most protected areas in these countries. As such, devolution of responsibility for management through delegated power or citizen control is highly unlikely. Buchy and Ross (2000) noted the major influence of tenure – that is, who has management responsibility for, or owns an area – on the level of stakeholder involvement. In the developing world, involvement is more likely to be on the lower rungs of the ladder.

Benefits and costs

Involving stakeholders in visitor planning incurs both benefits and costs. Benefits may include better decisions, increased accountability, stakeholder acceptance, local

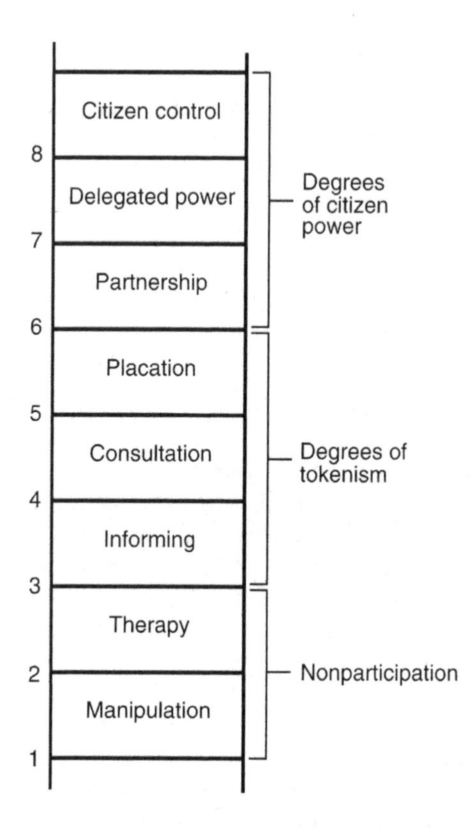

Figure 4.3 Ladder of citizen participation (Derived from Arnstein, 1969)

community empowerment and clarifying visitor preferences (Bramwell & Lane, 2000). Better decisions result from stakeholders providing and assessing collectively held information. Accountability of land managers is increased through stakeholder involvement, as planning and associated actions are subject to scrutiny. Stakeholder and especially local involvement may help communities understand and then accept planners' proposals. Involvement may also empower local communities, through control over local resources and decisions becoming possible. And, most importantly, benefits from these resources may be accrued. Lastly, stakeholder involvement helps make explicit the values, norms and preferences of visitors. The success of a number of the visitor planning frameworks discussed in this chapter relies on these being made explicit. Stakeholder involvement in such processes also helps clarify the multiple goals and preferred futures associated with most natural areas (McCool *et al.*, 2007).

Involving stakeholders also has disadvantages and costs. It requires more time and more staff. Not only are more resources needed to undertake consultation, but such consultation also has indirect resource effects. For example, through consultation, communities are often able to exert pressure to have services extended beyond those originally planned, leading to increases in implementation costs. Other costs can flow from 'losing control' of a planning process. Land management agencies may lose control as communities struggle, either with them or other groups. Groups or individuals may seek control of the planning process or the process may inadvertently become part of broader community disputes.

Techniques

The objectives of stakeholder involvement in visitor planning range from providing information (the lower rungs of the ladder in Figure 4.3), through information receiving and sharing, to participatory decision making (the higher rungs) (Table 4.1). The planning frameworks discussed in this chapter work best with participatory decision making. Other techniques, as listed in Table 4.1, can be used separately or

Table 4.1 Stakeholder involvement techniques in visitor planning for natural areas

Technique	*Objective*			
	Information giving	*Information receiving*	*Information sharing*	*Participatory decision making*
Information sheets	✓			
Displays	✓			
Media campaigns	✓			
Draft plans				
Review of plans	✓	✓		
Discussion papers	✓	✓		
Telephone hotlines	✓	✓		
Stakeholder interviews	✓	✓		
Phone polling/surveys		✓		
Focus group	✓	✓	✓	
Public meetings	✓	✓	✓	
Stakeholder meetings	✓	✓	✓	✓
Joint field trips	✓	✓	✓	✓
Advisory committees	✓	✓	✓	✓
Task forces	✓	✓	✓	✓
Workshop	✓	✓	✓	✓

Derived from Hall and McArthur (1998)

in conjunction with participatory approaches. Information sheets are widely used to provide information, while reviewing plans is a means of disseminating and collecting information. Stakeholder meetings, information days and field trips are used to share information. Often, different techniques are used at different stages of planning because each stage has different information requirements. Early on, information giving may dominate, as planners inform stakeholders about the planning initiative. As planning progresses, this may change to information sharing, as stakeholders describe existing visitor use and associated conditions. Later, participatory decision making is likely, as desired future conditions are discussed and determined.

The internet and social media provide a wealth of opportunities for stakeholder involvement. It may be information provision through websites or social media such as Facebook or Twitter. Blogging and social media also create opportunities for information exchange, while wiki approaches can enable stakeholders to contribute content to plans and planning processes. For example, Parks Victoria (PV, responsible for managing parks and reserves in Victoria, one of Australia's six states) used a wiki approach to develop online its management plan for the Greater Australian Alps National Parks (Figure 7.1, p. 329). The public could log on and contribute to the plan content (PV, 2012). Such an approach can also be used to develop indicators and standards in an interactive way. Web-based conferencing (e.g. Skype) can bring together stakeholders to make decisions about park planning in an interactive way.

The choice of techniques depends on several factors, including the ability of land managers to share power with stakeholders and the desire of stakeholders to do so. If power sharing is possible and desired, then a task force or advisory committee can be used. Choice is also influenced by the complexity of the resource management problems, the levels of stakeholder interest, and the types of knowledge and expertise stakeholders bring to planning (Smith & Moore, 1990; Tritter & McCallum, 2006). If the resource problems are simple or localised in a low-use area, then minimal stakeholder involvement such as information giving (e.g. information sheets) and an opportunity for information sharing (e.g. plan review) may be sufficient. For a large natural area with many stakeholders, a range of techniques will be essential. If there are divergent values and preferences between these stakeholders, participatory processes such as workshops will be necessary to explore and if possible resolve some of these differences.

Planning Concepts

An understanding of the concepts of carrying capacity, 'acceptable' change and spectrum of recreation opportunities is essential before progressing to the planning frameworks themselves. Over the years, countless managers and researchers have attempted to determine a numerical carrying capacity for a natural area, generally without success. Yet the search continues. At least one of the following frameworks (e.g. Limits of Acceptable Change) was established to provide an alternative approach to the vexed issue of determining when the conditions of an area had become unacceptable. The concept of 'acceptable' change underpins this alternative approach. Providing a spectrum of recreation opportunities forms the basis of another framework (the Recreation Opportunity Spectrum) as well as underpinning most of the other approaches.

Carrying capacity

Carrying capacity is a fundamental concept in natural resource management. It is the maximum level of use an area can sustain, as determined by natural factors such as food, shelter and water (e.g. three sheep per hectare). Beyond this limit, no major increases in the dependent population can occur (Stankey *et al.*, 1990; Manning, 2011). The term has been widely applied in rangeland management worldwide and wildlife management in the USA. If the balance between animals and the range's capacity is upset, either by an increase in animals or by a decline in the resource conditions, then problems will occur. Fewer animals can be supported and, in the worst case, irreversible environmental damage occurs.

In the early 1960s the concept of carrying capacity was carried across to recreation, and especially wilderness management, as wilderness conditions deteriorated in the face of rapidly escalating levels of use. Managers hoped to be able to determine a visitor carrying capacity, below which the natural environment could be sustained. Wagar (1964) broadened 'capacity' to include social as well as ecological factors. Thus, recreation and tourism carrying capacity has two main components: an ecological capacity – the impact on the biological and physical resources (i.e. soils and vegetation) – and a social capacity – that is, the impact on the visitor experience (Morin *et al.*, 1997).

Continuing and growing impacts in natural areas fuelled research on carrying capacity as a way of helping make decisions about controlling impacts. By the early 1980s, more than 2000 papers had been published on the topic (Drogin *et al.*, 1986, in Stankey *et al.*, 1990). Why so many and where are we now regarding the application and use of the concept of carrying capacity? In reality, the concept has failed to generate practical limits on visitor use. There are four main reasons (Stankey *et al.*, 1990; McCool & Patterson, 2000; McCool *et al.*, 2007):

(1) *Different recreation/tourism experiences have different carrying capacities*. Natural areas are used by many different people seeking many different experiences. Some want solitude, some want companionship. What are regarded as reasonable encounter levels by some are regarded as overcrowded or too isolated by others. Every person and form of use seems to have a different experiential carrying capacity.

(2) *A strong cause-and-effect relationship between amount of use and impacts does not exist*. Numerous studies have failed to link amount of use and impact. Much of the biophysical impact observed occurs at very low levels of use. There may then be a period of time when no impacts are observable, until levels of use become such that impacts become evident again. This relationship is anything but simple and linear, plus a number of variables affect it. Type of activity is usually a better predictor of impact than intensity of use. For example, low levels of horse riding may have greater impacts on trail condition than large numbers or frequent use

of the same trails by hikers. The season of use may also be more important in explaining impacts than amount. Hiking in wet, winter or monsoonal conditions, for example, potentially has far greater impacts on trail condition than increases in use during the dry season.

(3) *Carrying capacity is a product of value judgements and is not purely a product of the natural resource base and therefore determinable through careful observation and research.* The idea of carrying capacity as a product of the natural resource base was taken directly from range management, where carrying capacity was a direct product of natural factors such as soils and rainfall. It was seen as a scientific idea the identification of a specific carrying capacity would be constrained only by the level of effort and ingenuity exerted by managers in measuring biophysical impacts (Stankey *et al.*, 1990). However, it became increasingly evident that carrying capacities are as much the product of value judgements as they are of science. These are the values of visitors and managers. Visitors' values influence the experience they are seeking and their perceptions regarding the acceptability or otherwise of impacts. This broadens carrying capacity from a solely scientific assessment into the political arena of stakeholder involvement.

(4) *Carrying capacity does not help determine the balance between protecting the pristine qualities of a natural area and allowing visitor use.* Managing visitor use of natural areas is inherently complex and must be based on recognising that allowing use leads to some degradation. Managing for protection and visitor use requires that protection is ultimately constraining but can be initially compromised (Cole & Stankey, 1998). Initially, protection and pristine conditions are compromised as visitor use impacts on the environment. Such use continues, accepting some level of environmental impact, until further change becomes socially unacceptable. Then, visitor use is managed to prevent further impacts. Protection becomes the constraining goal.

McCool and Patterson (2000: 116) noted that research and planning have now advanced to the point where carrying capacity is recognised as 'a reductionistic, naïve and inappropriate paradigm upon which to base actions that protect recreational settings or tourism dependent communities'. They suggested instead focusing on understanding what conditions are desired, what impacts are acceptable and unacceptable, and what actions will lead to accepted goals. Such a re-focusing is clarified by rephrasing the question from 'how much use is too much?' to 'how much change is acceptable?' or 'what are the desired conditions?' (Lindberg *et al.*, 1997; McCool *et al.*, 2007).

'Acceptable' change

For many stakeholders associated with the natural environment, no change is acceptable. However, managers and other stakeholders are increasingly realising that changes inevitably accompany visitor use (Lindberg *et al.*, 1997). Thus, natural areas need to be managed to limit change to levels 'acceptable' to stakeholders. The value judgements made about acceptable levels of change reflect philosophical, emotional, spiritual, experience-based and economic responses. As such, few people will have identical responses and therefore few will make identical value judgements. The task for managers is resolving fundamental differences between stakeholders to determine desired conditions and how to achieve them.

Spectrum of recreation opportunities

Not everyone wants the same experience or to be involved in the same activities when they visit a natural area. Also, not all activities can occur at the same site at the same time, or conflict inevitably results. The opportunities for recreation sought by people range from easily accessible, highly developed areas with modern conveniences to undeveloped, primitive areas in remote locations and all the opportunities in between. The assumption that quality is best assured by providing a diverse array of opportunities underpins the application of the recreation opportunity spectrum (Manning, 2011). The concept can be applied within a single natural area, such as a national park, or to a group of natural areas.

Recreation/Tourism Planning Frameworks

Over the last two decades, a number of frameworks have been developed to plan and manage visitor use of natural areas. All aim to protect the natural environment while providing desirable opportunities for visitors (Cole & Stankey, 1998). The most widely discussed and applied, most often to one or a small number of natural areas, are the Recreation Opportunity Spectrum, Limits of Acceptable Change and Visitor Impact Management frameworks. Recently, the Tourism Optimisation Management Model has been developed specifically for tourism. It has been applied to both individual and groups of natural areas.

The following recreation/tourism planning frameworks are not mutually exclusive. As noted by Boyd and Butler (1996), this field has been evolutionary rather than revolutionary (Figure 4.4). Thus, a number of frameworks have common

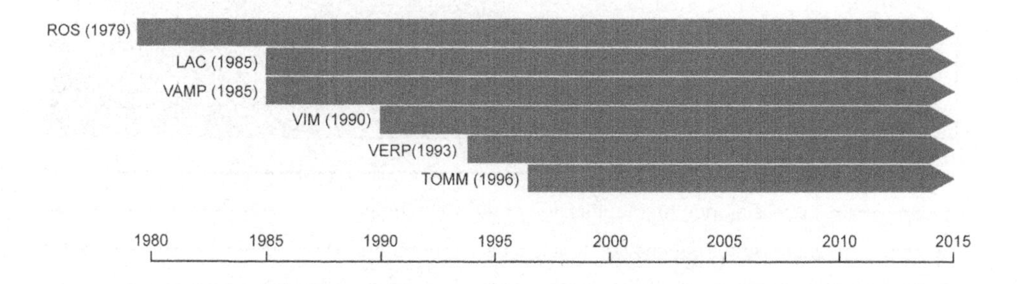

Figure 4.4 Chronological relationship between the recreation/tourism planning frameworks (Derived from Nilsen & Tayler, 1998; acronyms are defined throughout the chapter)

features and on first glance the differences between them may not be apparent. Also, features of one framework may be subsumed within another. For example, the Limits of Acceptable Change framework includes most of the elements of the Recreation Opportunity Spectrum. To help tease out these similarities and differences, the following sections include flow diagrams and examples; a section at the end focuses on choosing between them. Benefits-Based Management is also included as part of this overview of frameworks because it provides another way of determining how a range of visitor experiences can be ensured.

Recreation Opportunity Spectrum

The Recreation Opportunity Spectrum (ROS) was developed in the 1970s by researchers associated with the US Forest Service (Driver & Brown, 1978; Clark & Stankey, 1979). The 1960s and 1970s were a time of unprecedented growth in recreation use of natural areas and managers were very concerned regarding levels of use and maintaining the experience that drew users in the first place. ROS was offered as a means of identifying and determining the diversity of recreation opportunities for a natural area, based on the idea that the quality of visitors' experiences is best assured by providing diversity and helping visitors to find the settings providing the experiences they are seeking (Clark & Stankey, 1979).

Those who crafted ROS assumed that by providing diversity the adverse effects of increasing levels of use both on the natural environment and on visitors' experiences would be mitigated. These effects would be reduced in large part by allocating high-impact activities to more resilient sites and low-impact activities to less resilient locations. The diversity recognised by ROS is usually categorised as a number of opportunity classes, ranging from primitive to developed (Figure 4.5, Table 4.2). Today, the term 'zone' is often used rather than opportunity class.

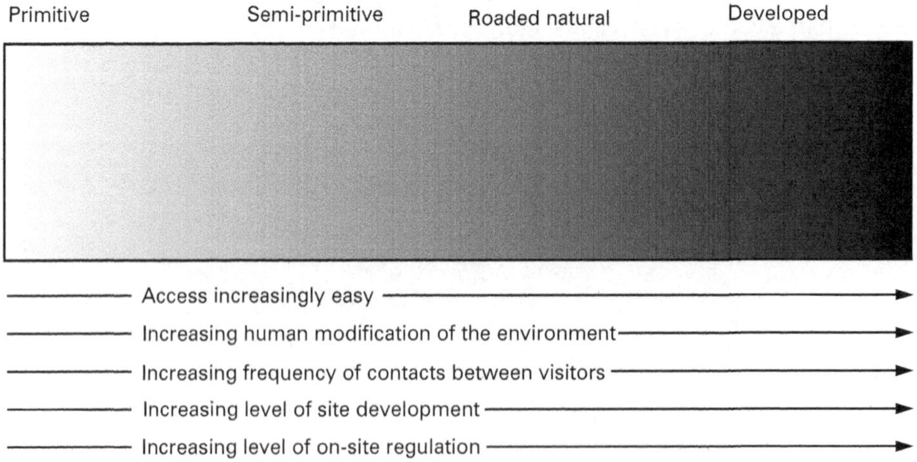

Primitive Semi-primitive Roaded natural Developed

Access increasingly easy
Increasing human modification of the environment
Increasing frequency of contacts between visitors
Increasing level of site development
Increasing level of on-site regulation

Figure 4.5 The Recreation Opportunity Spectrum (Derived from Clark & Stankey, 1979)

ROS spans places accessible only on foot with no facilities through to freeways and facilities with many comforts, such as resorts and lodges. As such, it is as applicable to natural area tourism as it is to the recreational use of wilderness areas, the focus of its initial development. Butler and Waldbrook (1991) used a similar approach in their tourism opportunity spectrum for ecotourism planning, although they paid more attention to the developed end of the spectrum.

Table 4.2 Recreation opportunity classes

Management factors	*Classes*			
	Primitive	*Semi-primitive*	*Roaded natural*	*Developed*
Physical				
Access	No motorised use	No motorised use	Motorised use and parking	High levels of motorised use and parking
Remoteness/ naturalness	Remote and completely natural	Completely natural	Appears predominantly natural	Natural background, site dominated by modification
Size	Large	Moderate	No size criteria	No size criteria
Social				
Contacts with other visitors	Few contacts	Low to moderate	Moderate along roads and tracks	High to very high along roads and tracks and at developed sites
Acceptability of visitor impacts	Not acceptable	Minor impacts accepted	Moderate impact accepted in specific areas, such as campsites	Substantial impacts evident and accepted
Managerial				
Level of site development	No site development, no structures	Natural-appearing setting, structures rare and isolated	Roads, site facilities for comfort and security	Roads and site facilities for intensive use including resorts
Regulation	No on-site regulation, reliant on self-policing	On-site regulation subtle, if present at all	Moderate regimentation/ regulation via site design and signs	Controls obvious and numerous via design, signs and staffing
Example				
Natural area tourism site	'Wild' campsite in a wilderness area	Designated site for hikers in a national park	Campsite/picnic area in most national parks	Built accommodation/ interpretation centre/ resort village in or next to a natural area

Derived from Clark and Stankey (1979), Leonard and Holmes (1987), McArthur (2000a)

ROS uses physical, social and managerial characteristics to describe and compare opportunity classes (Clark & Stankey, 1979). Physical characteristics include access, remoteness, naturalness (degree of human modification of the natural environment) and size. Social characteristics are contacts with other visitors and acceptability of visitor impacts. Managerial characteristics are the level of facility development and the amount of on-site regulation (e.g. site hardening, fencing, signs). One or more of these characteristics can be manipulated to provide a chosen recreation opportunity, ranging from primitive to developed (Table 4.2).

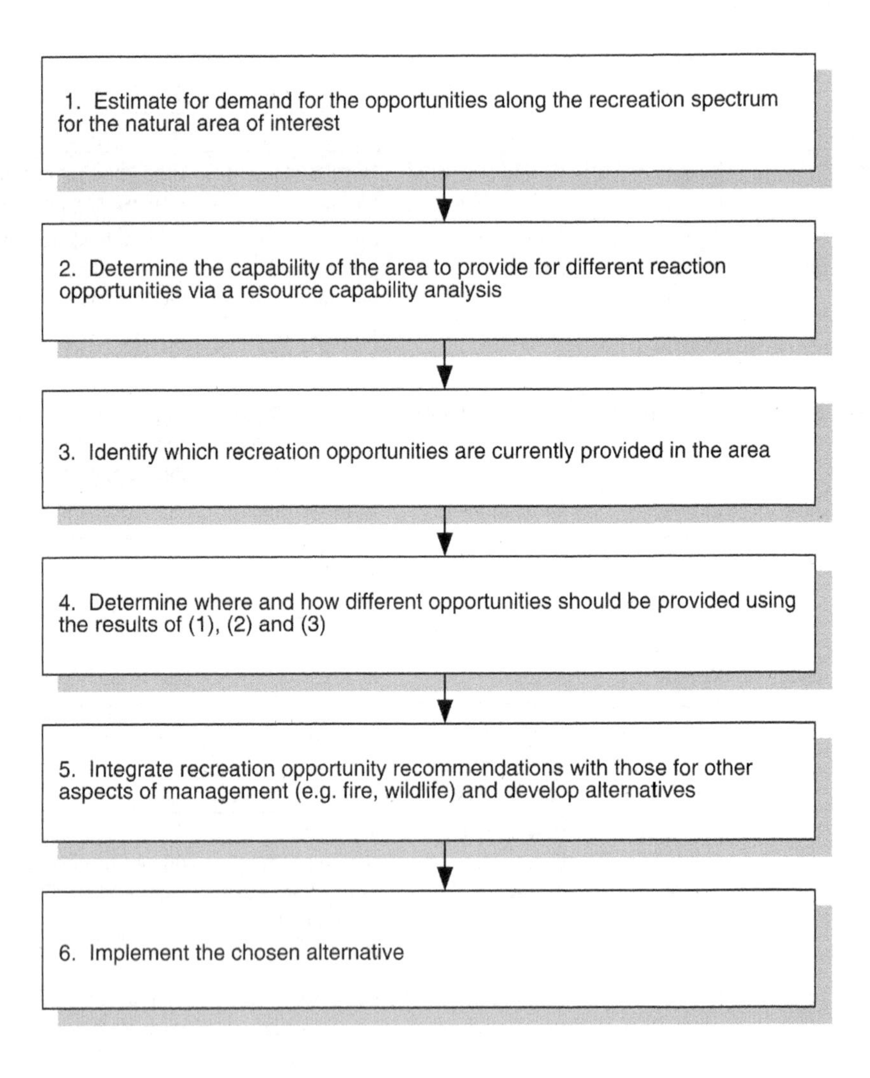

Figure 4.6 Process for applying the Recreation Opportunity Spectrum planning framework (Derived from Stankey & Brown, 1981)

Steps in the ROS framework

This framework has been used in various ways to plan visitor use of natural areas. The most widely used approach is given in Figure 4.6 (Stankey & Brown, 1981). It was developed for flexibility, not as a prescriptive set of steps. As such, every application is slightly different. Most importantly, ROS is a process, dependent on collecting and analysing biophysical and social information, for making management decisions.

The first and most difficult step in ROS is determining the demand for recreation/tourism opportunities (step 1). These opportunities are a product of the settings and experiences and to a lesser extent the activities sought. Existing demand is sometimes known from visitor surveys; however, often there are only informal records from field staff and in some cases nothing. Future demand can usually be only weakly predicted from current levels of use. Supply, the capability of the area to support various visitor uses and the opportunities currently provided, is easier to

Box 4.1 Applying ROS to Mount Cole Forest, Victoria, Australia

Because levels of recreational use and associated conflicts in the Mount Cole Forest (Figure 4.7), with an area of 12,150 ha, had increased, ROS was applied (Leonard & Holmes, 1987). Recreation uses included camping, day use, hiking, trail biking, off-road driving and hang-gliding. The area supports rare fauna of great public interest, including koalas, platypus and echidnas.

The Mount Cole project began by assessing the demand for various recreation opportunities (step 1 in Figure 4.6) and the recreation opportunities that could be offered, that is, supply (steps 2 and 3). Demand was determined from information provided by field staff, booking systems and assessment of the physical effects of use. Most use was occurring in the middle part of the spectrum, in the semi-primitive and roaded natural parts rather than the primitive and developed ends. Supply evaluation included mapping and describing the biophysical environment, existing recreation features and opportunity classes, and reviewing the recreation opportunities that could be offered. Maps of landscape features such as streams and lakes, vista points and old sawmill sites likely to influence patterns of use were one output. Another was existing opportunity classes and associated site features, such as walk trails and campsites.

The supply and demand information was used to determine where different opportunities classes should be provided (step 4). Opportunity classes were allocated to provide choice and overcome conflicts between users. The next step was integrating recreation planning with other management concerns, such as fire protection, timber production, water catchment values and fauna conservation (step 5). Overlays of the various uses were used to 'adjust' the opportunity classes to avoid conflict and provide an equitable balance between uses. Implementation, the final step in Figure 4.6, was sought via a three- to five-year plan. The plan detailed the developments required, such as campgrounds and walk trails, ways of shifting a site from one class to another on the spectrum (e.g. from roaded natural to semi-primitive one) and implementation costs.

Box 4.2 Using ROS to classify the recreation opportunities offered by Thailand's national parks

Tanakanjana (2008) used ROS to classify Thai national parks. Thailand has 103 national parks covering 10% of the country's area. The opportunity classes offered by these parks were analysed using a sub-set of sites, including waterfalls, rivers and lakes, caves, hot springs, geomorphological sites, scenic areas, nature trails, islands and beaches. At each site the management factors of access, remoteness, naturalness, opportunity for social encounters, evidence of human impact, site and facility management, and visitor management were determined using aerial photos, on-site measurements and a visitor survey. These management factors were based on the study by Clark and Stankey (1979) and are similar to those given in Table 4.2.

Using the results for these factors and sites, five opportunity classes were identified: primitive (5% of sites); semi-primitive non-motorised (34%); semi-primitive motorised (35%); modified natural or rural area (19%); and urban area (7%). Statistical analysis showed the strong influence of naturalness, remoteness and social encounters on opportunity class. The concentration of sites in the middle of the spectrum showed that recreation diversity had not been achieved in this particular national park system (Tanakanjana, 2008).

Although ROS was not fully implemented in this case study – it was used only to identify current recreation opportunities (step 3) – having this information makes it easier to engage with the other steps in the ROS planning process and comprehensively plan for visitor use of this park system.

determine (steps 2 and 3). Capability is based on factors such as an area's resilience to visitor use, remoteness, size, naturalness and landscape features appealing to visitors. Current opportunities can be mapped, with particular attention given to facilities such as campgrounds, roads and walk trails.

Determining the 'best' mix of recreation opportunities and allocation of land uses for a given area is not easy (steps 4 and 5). Selection of opportunity classes and balancing recreation and other land-use allocations draw on the preceding steps, considered within the constraints of budgets, statutory and non-statutory policy requirements, and other potential uses of the resource, such as fire and wildlife management (Stankey & Brown, 1981). A number of options are usually considered, with maps of alternatives and costs and benefits compared. The last step is implementation, with management objectives given for each class (step 6). Example applications of ROS to planning in Australia and to the classification of recreation opportunities in Thailand are given in Boxes 4.1 and 4.2.

Application, strengths and weaknesses

ROS has been widely applied to recreation planning in North America, Australia and New Zealand. It remains a mainstay of planning in North American parks and wilderness areas (Brown *et al.*, 2006). In Australia, ROS has been applied to protected areas in five of the country's six states and two territories. Its longest-running and most effective application has been in zoning natural areas (McArthur, 2000a). The most usual place to find the decisions resulting from ROS is in a zoning plan for recreation, usually included in a management plan, where the need to provide a spectrum of opportunities is recognised. It is unusual to find a document such as a management plan or strategy based solely on ROS.

The greatest strength of ROS is that it ensures that a range of recreation opportunities is considered in planning at local and regional levels. Also, it allows visitor management to be integrated with other forms of planning (Nilsen & Tayler, 1998). Its weakness, as with other frameworks, is that if agreement is lacking regarding opportunity classes and their characteristics, then decisions and implementation cannot follow.

Limits of Acceptable Change

The Limits of Acceptable Change (LAC) planning framework builds on and goes beyond ROS to set measurable standards for managing recreation in natural areas (Cole & Stankey, 1998). ROS is a process for recognising and designating opportunity classes with different levels of use. However, it does not provide guidance on the setting of standards and their subsequent use in managing visitors and their impacts. LAC provides a process for deciding what environmental and social conditions are acceptable and helps identify management actions to achieve those conditions.

It represents a major alternative approach to the concept of carrying capacity. Instead of asking 'how much use is too much?' and trying to link the number of visitors to environmental changes and failing, the LAC approach rephrases the question by asking 'how much change is acceptable?' (Prosser, 1986). Acceptability is a social phenomenon, and thus stakeholder involvement in the LAC process is essential. Stakeholders can provide judgements regarding the acceptability of impacts and in some instances can monitor to see if management is working. They can also provide a substantial amount of expertise on areas that are impacted and what management actions are likely to work.

Like ROS, LAC was developed by US Forest Service researchers (Stankey *et al.*, 1985), in close collaboration with management staff, to address concerns regarding increasing levels of recreational use of wilderness areas and associated environmental consequences. Application to a large wilderness area in the north-west of the USA, the Bob Marshall Wilderness Complex (Figure 4.7), guided its development (Stokes, 1990). The description of LAC in this book has been broadened to include more developed environments, such as tourism resorts and other built, intensively used facilities.

Steps in the LAC framework

This framework, similarly to ROS, is a flexible process with managers expected to adapt and modify it using their and stakeholder's experience and knowledge of a given area. It is a process for making management decisions, not a prescriptive series of steps to be unthinkingly followed (Stankey *et al.*, 1985). These steps can, however, be generically described (Figure 4.8).

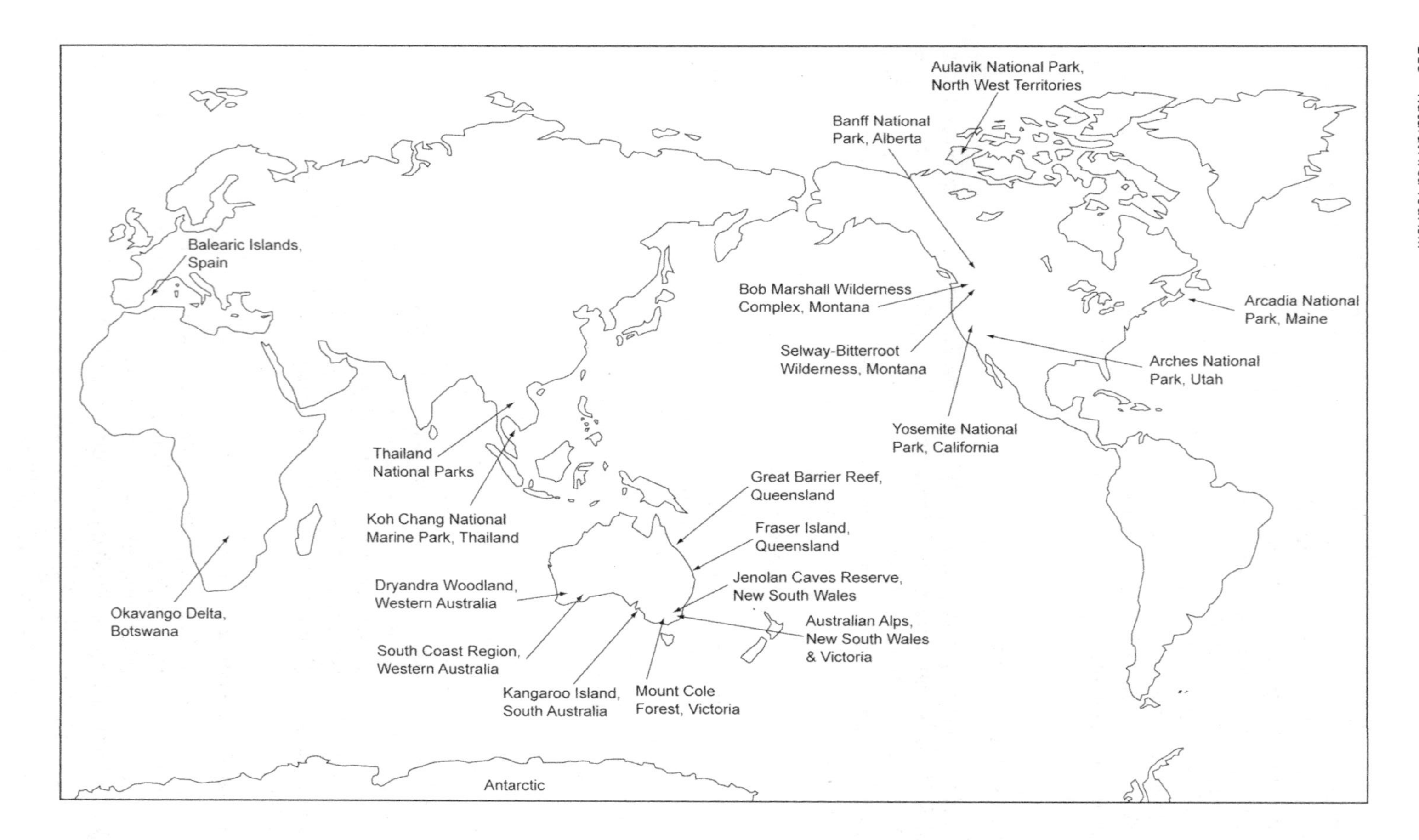

Figure 4.7 Location map of important nature-based tourism destinations referred to in this chapter

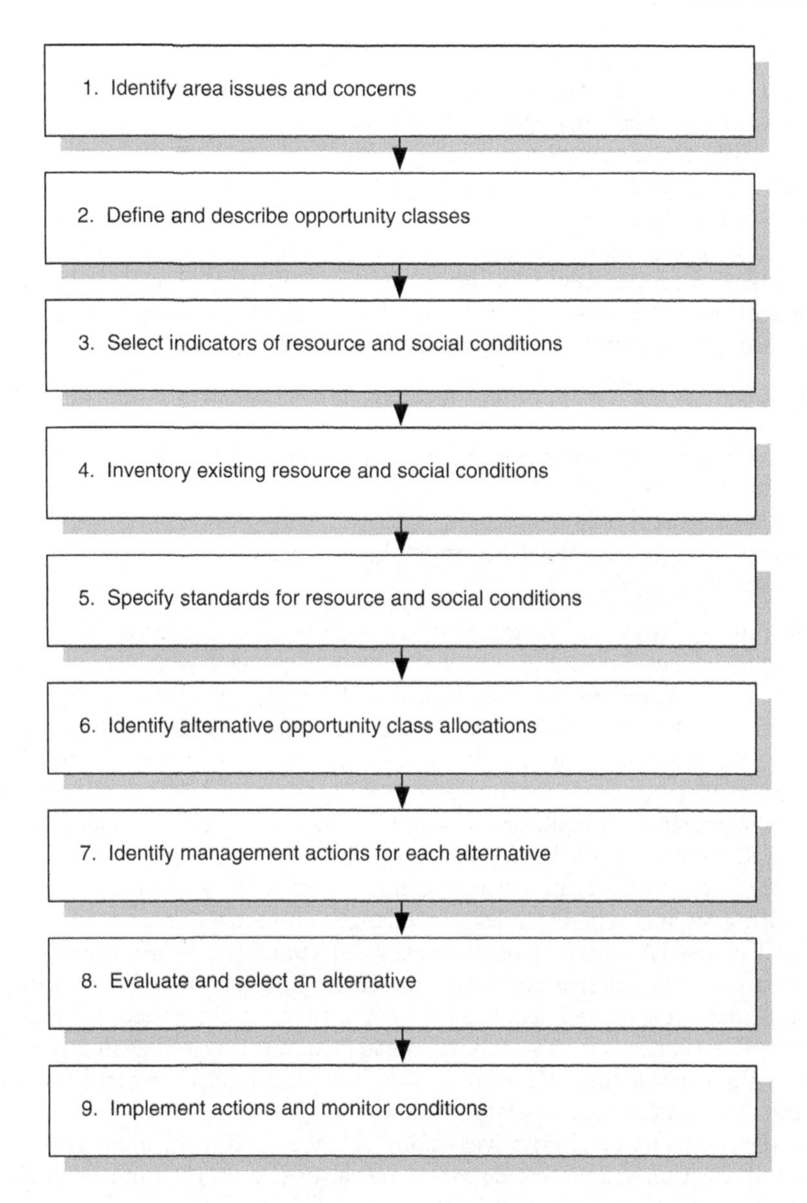

Figure 4.8 Process for applying the Limits of Acceptable Change planning framework (Derived from Stankey *et al.*, 1985)

The first two steps describe the management issues for an area and the recreation opportunity classes (i.e. zones), providing the context for the remainder of the planning process. The opportunity classes are those managers wish to provide, not necessarily those currently available. Next, indicators are selected to measure existing resource and social conditions and acceptable standards are determined (steps 3, 4 and 5). Possible physical indicators are water quality, soil compaction and erosion, and air pollution. Biological indicators include vegetation cover and fauna. The numbers of people seen or heard are indicators of social conditions. Chapter 7 (on monitoring) has further details on possible indicators and associated standards.

Indicators must have the following attributes:

- they must be capable of being measured in cost-effective ways, at acceptable levels of accuracy;
- the condition of the indicator should reflect some relationship to the amount/type of use occurring;
- social indicators should be related to user concerns;
- the condition of the indicator must be responsive to management control (Stankey *et al.*, 1985).

A number of indicators may be required to cover the state of desired conditions.

Standards, a level beyond which further change is unacceptable, are selected for each indicator. A standard for an indicator will usually vary between opportunity classes (i.e. zones). In the most pristine zone the highest standard will be set, while a lower standard will be set in the most developed zone. For example, in the Selway-Bitterroot Wilderness in Montana (Figure 4.7), the standard for the maximum number of campsites per square mile was 1 for the most pristine opportunity class (class 1) and 4 for the most developed class (class 4) (Ritter, 1997).

Standards may reflect the existing condition of an area or provide targets for rehabilitating an area where the level of change is no longer acceptable. Standards are crucial in the LAC process because they determine the future character of an area. There is usually uncertainty about the accuracy of standards and in most places where LAC has been applied people have been hesitant to set standards in case they are wrong. This concern is unnecessary. Because monitoring and evaluation are the central component of the LAC system, standards and indicators can be revised as better information becomes available.

The development of alternative zoning schemes and associated management actions and selection of a preferred scheme are the next steps (6, 7 and 8). Implementation accompanied by monitoring is the last step (9). Evaluating the effectiveness of management actions is also part of this last step (Morin *et al.*, 1997). An example of the application of the process to the Bob Marshall Wilderness Complex in Montana is presented in Box 4.3. Although this is an old example, it is valuable because here the LAC framework was developed in tandem with the realities of planning for visitor use, plus it continues to be used by managers. Another example, with an emphasis on the early steps, concerns its application to snorkelling tourism in Thailand (Roman *et al.*, 2007) (Box 4.4).

Box 4.3 Applying LAC to the Bob Marshall Wilderness Complex, Montana, USA

The most often used example of LAC is recreation planning for the Bob Marshall–Great Bear–Scapegoat Wildernesses initiated in the 1980s and continuing today. This wilderness complex encompasses 682,000 ha (Figure 4.7, Figure 4.9, overleaf). The area provides opportunities for extended hiking and horseback trips and is renowned for big-game hunting and river-rafting (McCool, 1986). It covers complete ecosystems, from river bottom to ridge top, on both sides of the North American continental divide (USDA FS, 1985). By 1980 there was increasing conflict among user groups and between some user groups and the managers, the US Forest Service. There was also a growing perception among users that the wilderness was deteriorating (Stokes, 1987, 1990).

A task force with representatives from local and national stakeholder groups and including land managers from the US Forest Service was brought together in 1982 and worked together to prepare an action plan for recreation management (Stokes, 1987, 1990). Each of the following steps was undertaken consultatively by the US Forest Service and the task force. The first step was identifying area issues and concerns (Figure 4.8, step 1). These included lake and range management, trail conditions, visitor encounters, level of regulation, wild and scenic river management, campsite numbers and condition, and management structures. Other issues raised but beyond the scope of the plan were use levels by commercial horseback tour operators ('outfitters') and wildlife management. Four existing opportunity classes were described, from unmodified natural environment with no facilities and very infrequent encounters with other visitors through to predominantly unmodified with facilities for resource protection and visitor safety and moderate to high levels of encounters with others (step 2).

Once the opportunity classes were defined, indicators were selected (step 3). Both biophysical and social indicators were selected because both were of concern to visitors and managers. Area of bare soil and number of damaged trees were the chosen environmental indicators of campsite conditions. Number of trail encounters and others camped within site and sound were the social indicators. To determine standards and set a baseline for on-going monitoring, indicators were surveyed by stakeholders and the US Forest Service (step 4). Standards for a given indicator varied between opportunity classes. For example, the standard for area of bare soil at a campsite ranged from 100 square feet in the most pristine class through to 2000 square feet in the most developed class. This was step 5.

The task force then developed a number of alternative opportunity class allocations (i.e. zoning schemes) and selected a preferred alternative (steps 6 and 8). One alternative reflected current conditions on the ground. Another provided additional recreational opportunities. Four alternatives were also developed by user groups. All were overlaid and a composite produced and debated. Some modifications to the composite were made and through consensus this modified version was adopted. The preferred alternative leaned toward pristine conditions, except along heavily used trail corridors.

Once the opportunity class allocations were made, the third-to-last step was determining management actions (step 7). The task force agreed actions were needed where standards were currently exceeded or would be exceeded when the area was moved to another opportunity class. In many places the proposed standards were violated; for example, there were too many campsites in a given area or numbers of trail and campsite encounters were excessive. Information and education were popular proposed actions for all opportunity classes. The last (ongoing) step is implementation and monitoring (step 9).

Figure 4.9 Bob Marshall Wilderness Complex, USA (Photo: Steve McCool)

Box 4.4 Applying LAC to Koh Chang National Marine Park, Thailand

The LAC approach has been used by Roman *et al.* (2007) to identify resource (i.e. biophysical) and social indicators and standards for managing snorkelling tourism in Koh Chang National Marine Park, Thailand (Figure 4.7). This park, with an area of 650 km^2, located in the Gulf of Thailand, encompasses Koh Chang Island and over 40 smaller, surrounding islands. Concerns have been expressed about the impacts of tourists on its coral reefs. This concern provided a starting point for the LAC process (step 1). An estimated 30,000 people per year participate in snorkelling tours, with snorkelling the main tourism activity on these reefs. Step 2 – defining opportunity classes – was not undertaken.

Steps 3 and 4 of LAC (Figure 4.8) involve identifying and inventorying resource and social conditions. The indicators identified and used by Roman *et al.* (2007) to measure resource conditions included coral mortality, diversity and vulnerability to trampling, with data recorded from linear transect surveys. The social conditions measured were perceptions of coral mortality and diversity, and numbers of other snorkellers at a site (i.e. crowding). This information was obtained via visitor questionnaires.

A standard of 35–50% coral mortality was chosen based on the questionnaire results and supported by the reef transect data. A generic crowding standard of 30–35 snorkellers was proposed, with this adjusted downwards for less intensively visited zones and upwards for high-use parts of the park (step 5). Zones and management actions (steps 6 and 7) were suggested based on high diversity and vulnerability. Roman *et al.* (2007) recommend different levels of tourist access, with limited access to the most vulnerable and highly valued sites.

Application, strengths and weaknesses

LAC has been undertaken predominantly in the USA, with the next highest level of use in Australia. The Australian applications to national parks, rivers and a World Heritage site have not, however, been fully implemented (McArthur, 2000a). Three incomplete treatments have been documented in New Zealand (McKay, 2006). Part of the framework has been applied to tourism in the Okavango Delta, Botswana (Mbaiwa *et al.*, 2002), to the Antarctic (Davis, 1999), to marine tourism activities including snorkelling on the Great Barrier Reef, Australia (Shafer & Inglis, 2000), to Koh Chang National Marine Park, Thailand (Roman *et al.*, 2007) (Box 4.4) and to recreational boating in the Balearic Islands, Spain (Diedrich *et al.*, 2011).

Where LAC has been used, most attention has been paid to the first few steps (Figure 4.8, steps 1–5), which centre on identifying indicators and standards and using them to describe existing conditions. Box 4.4 provides an example of this focus. It is much less common to find progression to consideration of alternative opportunity classes and evaluation of alternative management actions (steps 6–8). Additionally, it is unusual to find a document, such as a management plan, with LAC as its central focus. If the LAC process has been used, it generally manifests itself in the recreation section of a management plan, as indicators and standards. The Bob Marshall Wilderness Complex Action Plan (USDA FS, 1985) was an exception (Box 4.3) both in encompassing all the steps and presenting LAC in a stand-alone document.

The greatest strength of LAC is determining when 'enough' change has occurred. The two weaknesses of LAC have been selecting standards and gaining stakeholder support (McArthur, 2000a). Little information may be available to help choose a standard. Additionally, environmental changes rather than visitor activities may lead to fluctuations around a standard, with these two very different causes often impossible to separate. Gaining stakeholder support has long been a concern in natural area planning. LAC – indeed, planning frameworks in general – is doomed to fail if stakeholders cannot agree. In the case of LAC, agreement on indicators, standards and allocation of zones is crucial, as planning is very much a political process in a politicised setting (McCool & Cole, 1998).

Visitor Impact Management

The Visitor Impact Management (VIM) planning framework was developed for national parks by researchers working for the US National Parks and Conservation Association (Graefe *et al.*, 1990). Similarly to LAC, VIM was developed as an alternative to carrying capacity; however, it was intended to be simpler, narrowing the focus to visitor impacts rather than broader concerns with opportunity classes. Its purpose is developing strategies to keep visitor impacts within acceptable levels. Recognising that effective management is part science, part subjective judgement is fundamental to its application (Graefe *et al.*, 1990). Also fundamental is recognising that limits to use are only one possible way of managing unacceptable impacts. Other management strategies, such as education and site design, may be more effective. As research on carrying capacity has shown, the relationships between overall use levels and impacts are weak. Controlling use alone may not reduce impacts.

Steps in VIM framework

VIM is described by eight steps (Graefe *et al.*, 1990) (Figure 4.10). Together, they lead the manager from reviewing existing data and management objectives, through selecting indicators and standards and using these to identify unacceptable impacts, to identifying causes and suitable management strategies. The first five steps are the problem identification phase of VIM. Step 5 requires observation of existing conditions for the indicators selected in step 3. If the selected standards are exceeded then the causes need to be determined (step 6). The challenge of step 6 is

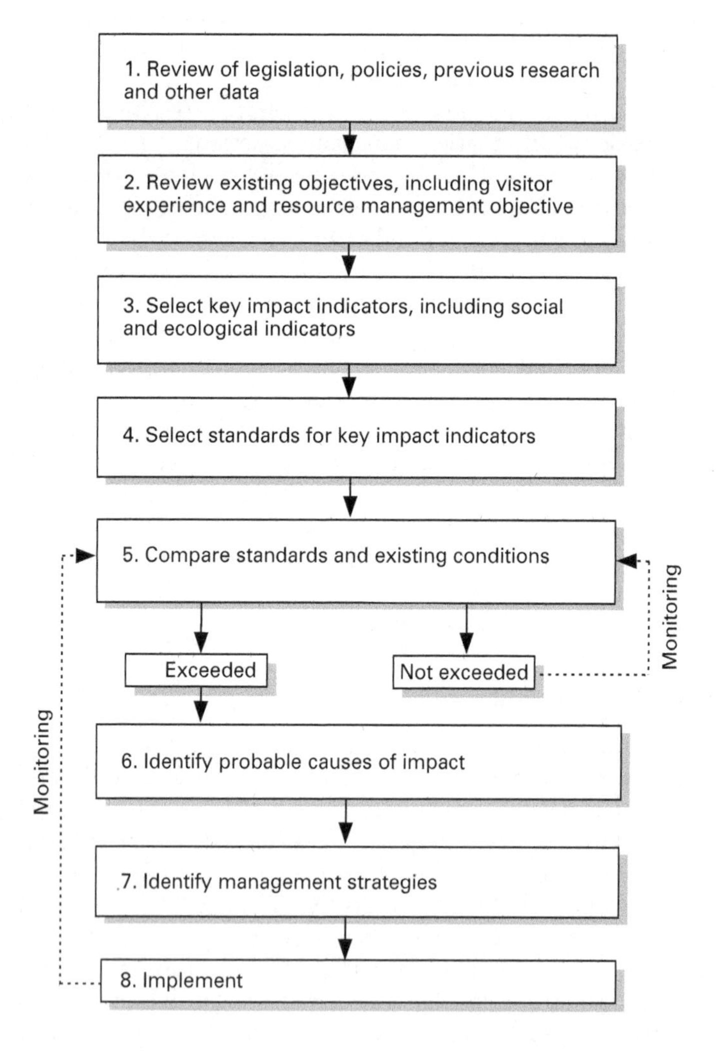

Figure 4.10 Process for applying the Visitor Impact Management planning framework (Derived from Graefe *et al.*, 1990)

isolating the most significant cause of the impact. This step may require studies of the relationships between key impact indicators and visitor use patterns.

Having followed these steps, there is unfortunately no one 'right' management strategy. Graefe *et al.* (1990) recommended using a matrix for evaluating alternatives (step 7). One side of the matrix lists possible management strategies and the other side criteria such as consistency with management objectives, difficulty in implementing, probability of achieving desired outcome, effects on visitor freedom and effects on other impact indicators. Achieving a balance among criteria is the basis for selecting a particular management technique (Graefe *et al.*, 1990). For example, a strategy such as enforcement may have good odds of reducing the impact but may cause even more problems through lack of visitor acceptance. The matrix helps make explicit these trade-offs. Similarly to the frameworks already described, VIM is intended as a guide rather than a prescriptive set of steps. The flexible use of VIM is evident from the example in Box 4.5, where several steps were adapted or combined with others.

Box 4.5 Applying VIM to the Jenolan Caves Reserve, New South Wales, Australia

The Jenolan Caves in eastern Australia is one of the country's best-known cave systems (Figure 4.7). In the 1990s, visitor numbers were increasing by 5–6% per year and resource degradation was an ongoing concern (Manidis Roberts Consultants, 1995). The system encompasses 45 km of known passageways divided into about 350 caves, with 16 of these for public use. Infrastructure includes car parks, accommodation for guests and staff, and other amenities and service facilities.

A modified version of VIM was applied using a three-day workshop of management staff and physical and social scientists. Steps 1 and 2 (Figure 4.10) were subsumed within an issue identification stage. Issues included the need for clear objectives for the area and a better understanding of visitor needs. Also of concern were problems with traffic. The next steps were selecting indicators and standards for four resource management units (=zones) (steps 3 and 4). Carbon dioxide and water quality were two of the indicators for the caves management unit. Infrastructure capacity was an indicator for the above-ground management unit. Visitor satisfaction was selected for all four units. Quantitative standards were selected for each indicator. For example, the desired condition or standard for visitor satisfaction was 90% of visitors rating their satisfaction as high to very high (Manidis Roberts Consultants, 1995). After the workshop, a monitoring programme was initiated, with an annual 'State of the Environment' report produced (Mackay, 1995). Collection of these data allows comparison of current and desired conditions (step 5).

The workshop culminated with the determination of 'key limiting conditions' or issues for management attention. These were near-capacity vehicle parking, vehicle–pedestrian conflicts, overcrowding above and below ground, and hydrological disturbance to the caves from above-ground developments. These limiting conditions were derived from probable causes of impacts (step 6) as well as earlier steps in the VIM process (Figure 4.10).

The Jenolan Caves Reserve Trust has, in addition to refining indicators and desired conditions, developed methods for and undertaken monitoring of indicators, and suggested causes of impacts and possible management responses (steps 6 and 7) (Mackay, 1995).

Application, strengths and weaknesses

VIM has been applied to at least 10 national parks/reserves/wildlife refuges in the USA, in three states of Australia and in Canada, Argentina, Mexico and The Netherlands (McArthur, 2000a). In general, VIM has not appeared as a separate planning document; rather, it has been used to develop strategies for localised problems, either directly for managers or as part of a larger management plan. A useful exception is the report funded by the Australian Department of Tourism applying VIM to the Jenolan Caves (Box 4.5) (Figure 4.7).

The main strength of this framework is its reliance on both science and subjective judgement to guide visitor management. It is particularly suited to smaller sites as there is no recognition within the framework of different opportunity classes, making it most useful where only one opportunity class exists (McCool *et al.*, 2007). As such, one of its weaknesses is not making use of ROS. Another is addressing current rather than potential impacts (Nilsen & Tayler, 1998). Successful visitor management relies on dealing with visitor impacts before or as they occur, not afterwards.

Tourism Optimisation Management Model

The Tourism Optimisation Management Model (TOMM) was developed in the 1990s by the Sydney-based consulting firm Manidis Roberts, through application to Kangaroo Island (Figure 4.7), off the coast of southern Australia (McArthur, 1996; Brown, 2006). Unlike the frameworks discussed so far, it was developed specifically for tourism planning in natural areas. An important early step in this framework is describing the political, socio-cultural and economic context within which planning is occurring. Although it builds on the emphasis in LAC on monitoring, it differs in its broader regional application and its coverage of a number of land tenures, both public and private. Involving a diversity of stakeholders throughout the planning and implementation process is the final essential feature of this framework. The name – TOMM – was selected to take the emphasis away from limits, which had led the tourism industry to equate LAC with anti-growth and anti-business sentiments (McArthur, 2000a).

Steps in the TOMM framework

TOMM has three major parts – context description, monitoring programme and implementation (McArthur, 2000a) (Figure 4.11). Similarly to VIM and LAC, this framework leads the manager from describing the planning context, through selecting indicators and standards and using these to identify optimal conditions, to identifying causes and suitable management strategies. The description of the planning context includes current policies and plans, community values, product characteristics, growth patterns, market trends and opportunities, positioning and branding, and alternative scenarios for tourism in the region (Manidis Roberts Consultants 1997) (Figure 4.11, steps 1 and 2). Box 4.6 describes the Kangaroo Island application of TOMM.

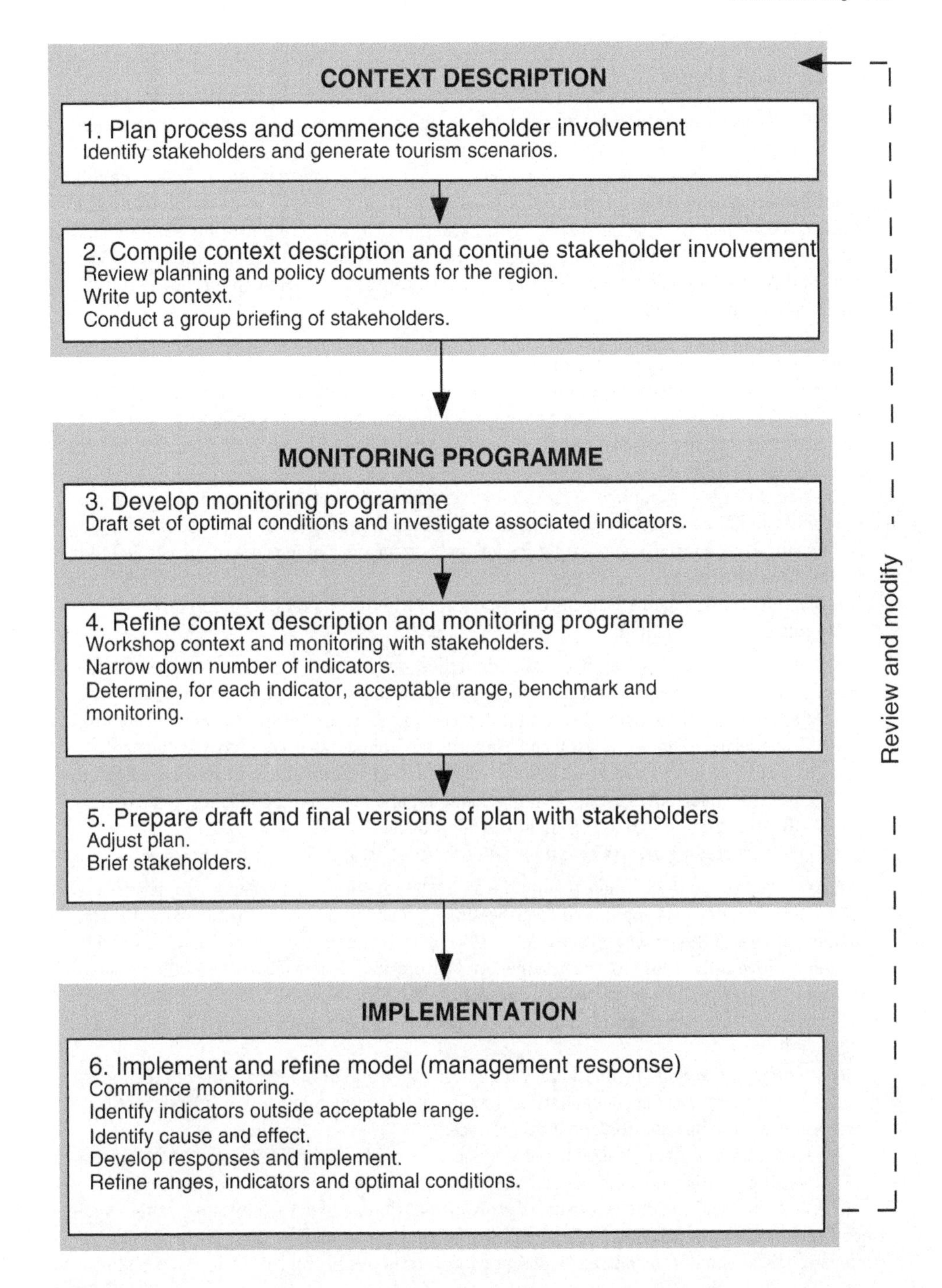

Figure 4.11 Process for applying the Tourism Optimisation Management Model (Derived from Manadis Roberts Consultants, 1997; McArthur, 2000a)

Box 4.6 Applying TOMM to Kangaroo Island, South Australia, Australia

In 1996 TOMM was initiated for Kangaroo Island, lying 21 km off the mainland of southern Australia (Figure 4.7) and with an area of 4500 km², to facilitate tourism development and management. Most of the island is private farms and residential properties, although about 24% is conservation reserve. The population of 4000 depends on agriculture and tourism. TOMM had three major parts – contextual identification, monitoring program and management response. These equate directly to the three major parts in Figure 4.11. The following description is from Manidis Roberts Consultants (1997), McArthur (2000a, 2000b) and Colmar Brunton (2010).

Part I. Context identification (steps 1 and 2)
This part involved identifying and describing the political and social context of planning. Also included was generating alternative management scenarios. A day-long briefing was held with the steering committee established for the project. The following aspects of the context were described.

Existing policies and political issues. Relevant policies included government ecotourism plans and the Kangaroo Island Sustainable Development Strategy. Political issues included dealing with the negative and positive aspects of tourism and making sure that TOMM helped the tourism industry.

Community values included the natural landscape, wildlife and relaxed lifestyle.

The tourist product included:

- *Natural assets* such as a spectacular coastline and natural bushland; unique natural attractions (e.g. Seal Bay and Remarkable Rocks); wildlife (e.g. seals); cultural assets (e.g. European maritime history); island industries (e.g. freshwater lobster and eucalyptus oil); parks (e.g. Flinders Chase); and a healthy environment with limited development.
- *Activities*, including nature-based activities such as sightseeing and viewing wildlife; recreational activities such as fishing, camping and swimming; and cultural activities such as farm-stays and experiencing local produce.
- *Themes*, including diverse coastline, wildlife, rural lifestyle and natural produce.

Growth trends. In 1995, when planning was initiated, the island received an estimated 150,000 visits. The most popular site was Seal Bay, receiving 72–86% of total visits. Visitor numbers grew at 8.8% per year from 1992 to 1997, with the majority coming from the nearby mainland (67%). By 2009/10 the number of visitors had increased to 185,000 (Colmar Brunton, 2010), with 26% of these from overseas in 2007/08 (Access Economics, 2009).

Market opportunities. Suitable markets are people who are environmentally aware and enjoy wildlife.

Positioning and branding. The island will be positioned and branded to achieve excellence in nature-based tourism accommodation, low-impact development, visitor infrastructure, interpretation and information and local produce.

Ten scenarios were generated and examined, including significant increases and decreases (15% a year) in tourism demand, a decrease in overnight stays and an increase in day visits. The benefits and costs of each were listed. The information needed to determine if there had been a benefit or cost was also identified. Generating the scenarios helped to make TOMM relevant to stakeholders as well as identifying information needs for future decision making.

Box continues opposite

Box 4.6 continued

Part II. Monitoring programme (steps 3, 4 and 5)

Optimal conditions. These were desirable yet achievable conditions generated from planning documents and reworked at a stakeholder workshop. They were developed for economic, marketing, experiential and socio-cultural as well as environmental conditions. 'Major wildlife populations attracting visitors are maintained and/or enhanced in areas where tourism activity occurs' is an example of an optimal environmental condition.

Indicators. Indicators were selected so that conditions could be measured. They were identified through two workshops and further discussions with stakeholders. For example, an indicator of the optimal condition of a wildlife population was the number of seals at designated sites.

Acceptable range. Acceptable ranges for each indicator were developed using information from previous research, observations and estimations from those with experience. For the seal population, a 0–5% decrease in numbers sighted per annum was the acceptable range. This step was really the most sensitive and difficult part of planning as decisions were made regarding whether an indicator was outside its acceptable range.

Monitoring and benchmarking. A monitoring programme was developed to collect information on the indicators and especially how close each was to its acceptable range. A benchmark was set for each indicator based on the best information available in 1996. For wildlife, a benchmark was the number of seals at designated sites.

Part III. TOMM management response (step 6)

The management response involved identification, exploration and action (Figure 4.12).

Identification of poorly performing indicators. This involves annual measurement of indicators and then identifying those outside their acceptable range. For example, if the number of seals was an indicator and the acceptable range was a 0–5% decrease and there was a sudden decrease of 10%, then this indicator has performed poorly.

Exploration of causes. For a poorly performing indicator, the next step is to work out if tourism was responsible. In the case of the seals, poor performance may be tourism or non-tourism related. A tourism-related cause might be increased numbers of visitors transgressing the boundaries of a designated viewing area, leading to a decline in the number of seals hauling out on to the beach.

Action. The third step is deciding on the action needed. For declining seal numbers at Seal Bay, if tourism was responsible, solutions might include closing the beach to tourists, advising tourists they will see fewer seals, and/or developing an alternative wildlife viewing opportunity. Brainstorming can be used as part of TOMM to determine the effects of proposed actions. For example, if beach access is closed, what will be the effects on other indicators, like native vegetation cover and the proportion of visitors experiencing wildlife in the wild?

Monitoring of indicators for optimal conditions against their acceptable ranges continues today. A review of indicators was completed in 2009. The monitoring associated with TOMM is overseen by the KI Management Committee, who are supported by and represent the community, industry and government agencies (Colmar Brunton, 2010).

Developing a monitoring program is the heart of this planning process (Figure 4.11, steps 3, 4, and 5). Included are identifying optimal conditions, indicators, acceptable ranges and benchmarks (Box 4.6). An optimal condition is a desirable yet realistic future. Indicators are then selected to measure these conditions. Indicators must be relevant to visitors and the associated data must be cost-effective to collect, available and accurate. The next step is developing an acceptable range and benchmark for each indicator. In the Kangaroo Island application, the value of an indicator at the start of monitoring was taken as its benchmark. As data are collected from monitoring, optimal conditions, indicators and acceptable ranges can be refined.

The last major part is implementation, including the response by managers when an indicator falls outside its acceptable range (Figure 4.11, step 6). Before acting, the manager needs to determine whether the result of concern is part of a longer-term trend or a one-off, and also its cause, which may or may not be related to tourism (Figure 4.12). The essential question is whether or not an indicator exceeding its acceptable range was caused by tourism (McArthur, 2000b). Many indicators are subject to effects from other sources, such as ecological processes, the actions of local residents, initiatives by other industries, technological innovation, and national and global influences (Manidis Roberts Consultants, 1997). The last element of the implementation part of TOMM is taking action. A management response can be selected by brainstorming – thinking how the preferred response will affect other indicators and the likely results for the indicator of interest.

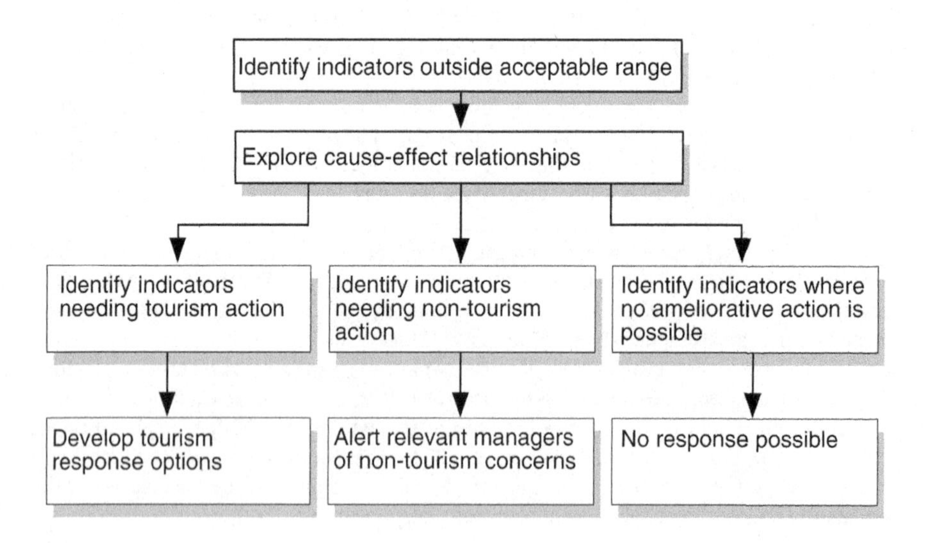

Figure 4.12 Using the Tourism Optimisation Management Model to take management action (Derived from Manidis Roberts Consultants, 1997)

Application, strengths and weaknesses

TOMM has not been widely applied, with the exception of Kangaroo Island, Australia (McArthur, 1996; McCool *et al*., 2007). It has been partially implemented in Canada at Lake Louise in Banff National Park and for Aulavik National Park, and for Dryandra Woodland in Western Australia (Moncrieff, 1997; McArthur, 2000a) (Figure 4.7). In the case of planning for Kangaroo Island, the TOMM analysis is the planning document. For Dryandra, TOMM was applied after the management plan had been completed and published.

The strength of TOMM rests on its explicit inclusion of the political and economic environments in which use of natural areas occurs and of stakeholders throughout planning. Its main limitation is the amount of information needed, given it can cover a breadth of tenures across a region as well as including market, economic and socio-cultural as well as biophysical information. This limitation also means that data management and manipulation require a significant level of resources. Locating and working with stakeholders across large areas and a complexity of issues is also resource-intensive.

Other planning frameworks

Two other planning frameworks developed for natural areas, plus Benefits-Based Management, also warrant mention. The Visitor Activity Management Process was created by Parks Canada in the 1980s to guide national park planning and management (Nilsen & Tayler, 1998). The Visitor Experience and Resource Protection process was created by the US National Park Service in the 1990s for a similar reason (McArthur, 2000a).

Visitor Activity Management Process

The Visitor Activity Management Process (VAMP) is one element of a broad, integrated planning and management process. It is part of a whole-of-park approach to management. Other parts include a national parks management planning process and a natural resource management process (Figure 4.13). VAMP employs an overt marketing approach to integrate visitors' requirements with the resource opportunities provided by a given area (Nilsen & Tayler, 1998). It also provides a flexible framework for integrating social and natural science data (Lipscombe, 1993).

VAMP begins by establishing terms of reference and management objectives for an area (Figure 4.14, steps 1 and 2). Next is creating a database of park ecosystems and settings, visitor activities and opportunities, and the regional context (step 3). Analysis to produce alternative visitor activity concepts follows (steps 4 and 5). The last two steps are creating a park management plan and implementation (Nilsen & Tayler, 1998).

VAMP has had limited application in its birthplace, Canada, and virtually none elsewhere. In Canada it has been used to establish a national park, assess the impacts of cross-country skiing, and for interpretation planning. Typical outputs include information on visitor activities and associated development options for assessment,

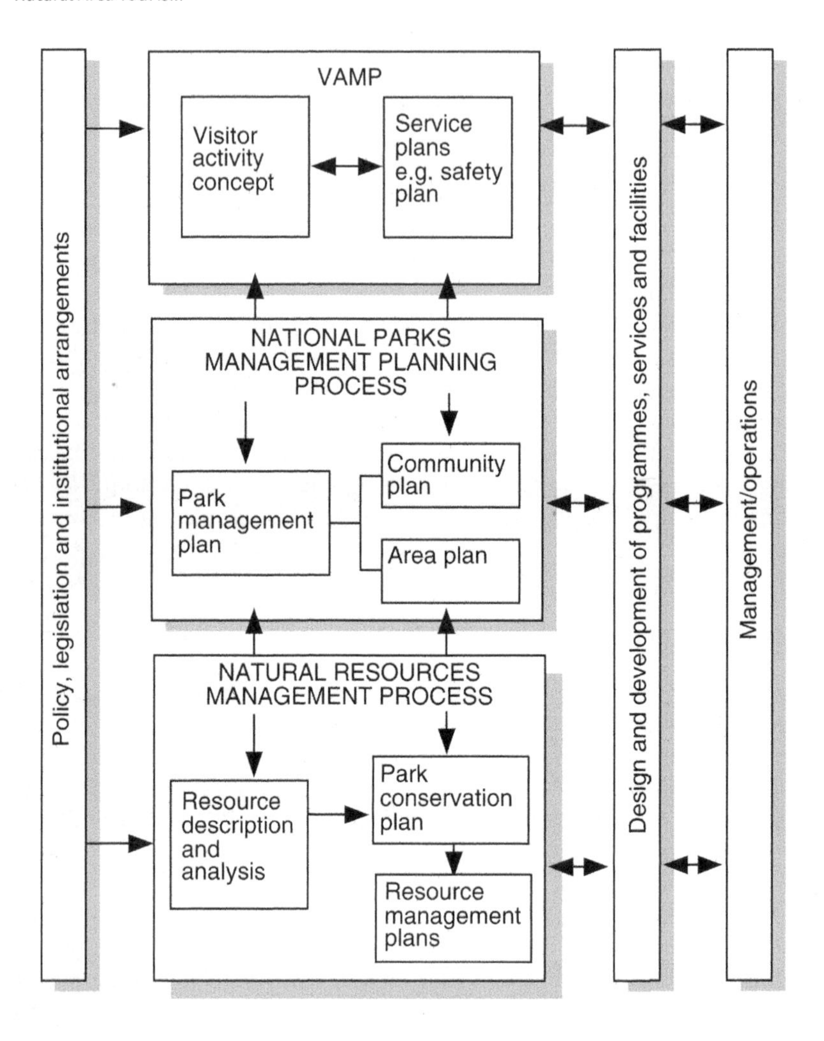

Figure 4.13 The Visitor Activity Management Process as part of national park management (Derived from McArthur, 2000a)

or operational plans for visitor services or interpretation. As such, it can contribute to a management plan for a natural area or may appear as an operational plan for one aspect of visitor management, such as interpretation or visitor safety. By the late 1990s it had been abandoned in favour of ROS (McArthur, 2000a).

Its strength is recognising the demand as well as supply side of natural area management. This is also a weakness, as it may be very difficult to shift managers from a product-centred to a market-centred approach (McArthur, 2000a). Another weakness is VAMP's failure to develop limits or acceptable ranges for impacts, although ROS similarly does not include this level of detail.

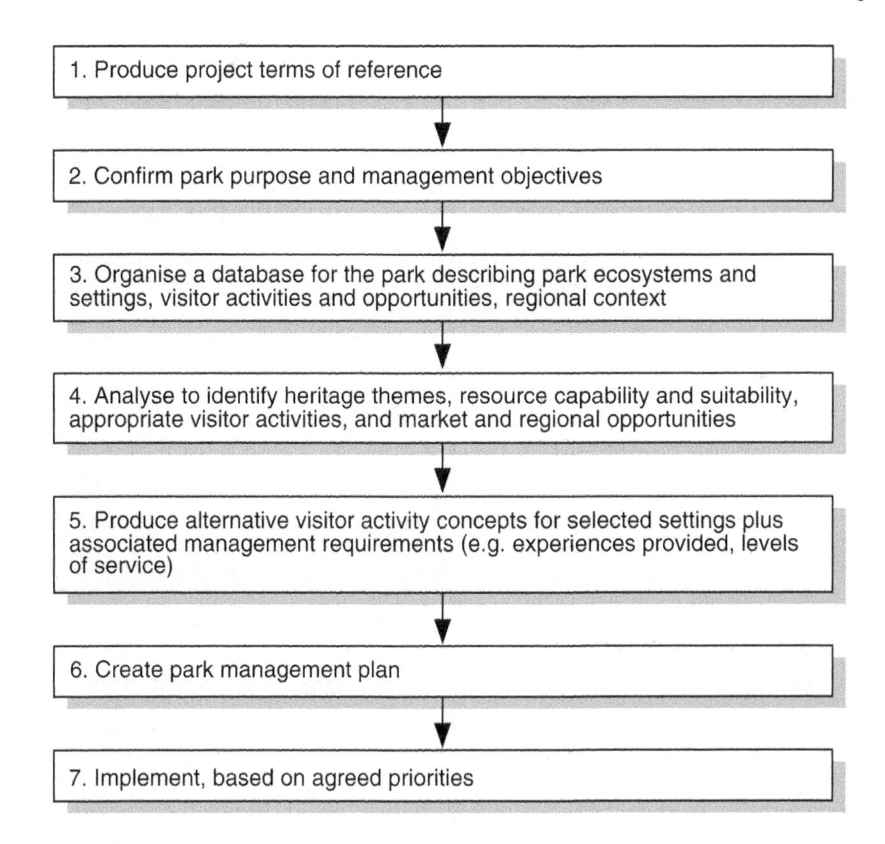

Figure 4.14 Process for applying the Visitor Activity Management Process (Derived from Nilsen & Tayler, 1998)

Visitor Experience and Resource Protection

The Visitor Experience and Resource Protection (VERP) planning framework was developed by the US National Park Service as a means of addressing concerns about the carrying capacity of their national parks (Hof & Lime, 1998; Manning, 2009). Integral to this process, which is very similar to LAC, is determining the appropriate range of visitor experiences for a chosen area. Thus, zoning is a focus (Nilsen & Tayler, 1998). VERP underpins management plans by producing a series of zones for inclusion. An important premise underlying VERP is that zoning should be resource related and not determined by the location of existing facilities (McArthur, 2000a).

The steps in the process include describing the context, analysing existing resources and visitor use, determining a potential range of visitor experiences and resource conditions and allocating zones, then selecting indicators and standards and monitoring. Specifying carrying capacities is avoided by providing desired ecological and social conditions. The last step is taking management action (Figure 4.15).

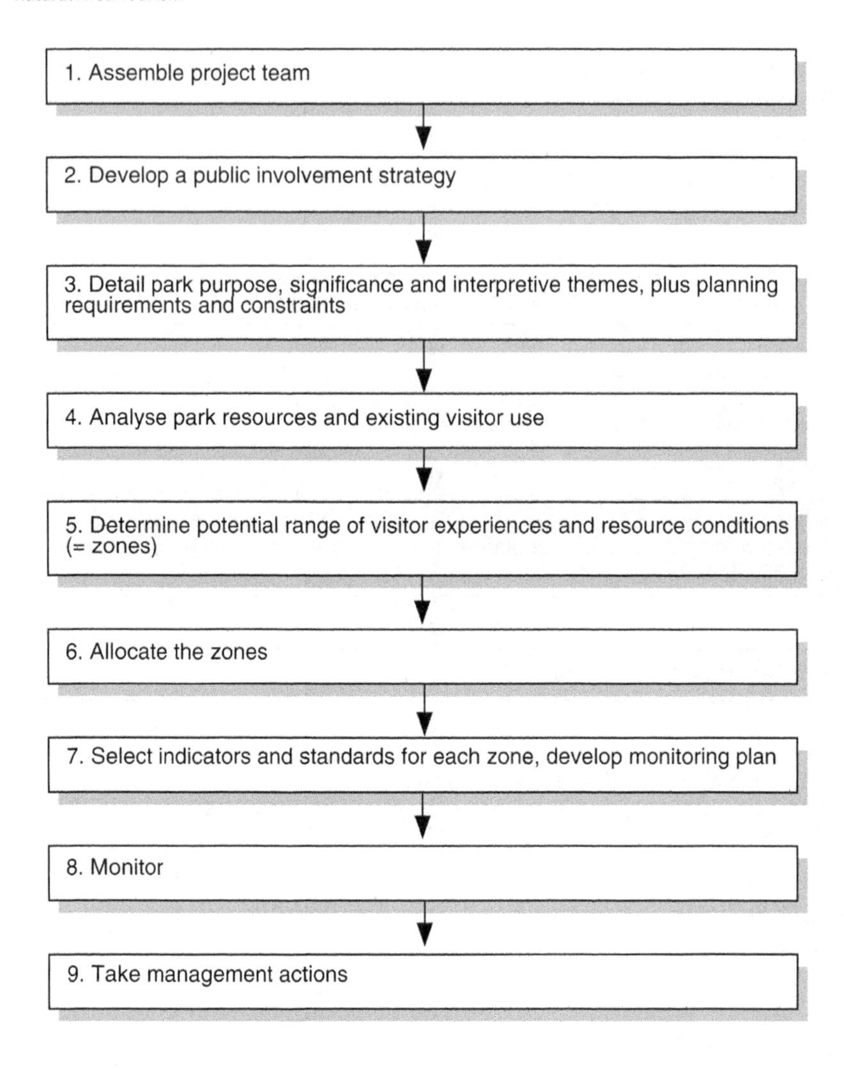

Figure 4.15 Process for applying the Visitor Experience and Resource Protection planning framework (Derived from Nilsen & Tayler, 1998)

VERP has been applied in full in only a few places (Hof & Lime, 1997). The most notable application examples are Arches National Park in south-eastern Utah, carriage roads in Arcadia National Park in Maine, and Merced River in Yosemite National Park, California (McCool *et al.*, 2007; Manning, 2009, 2011) (Figure 4.7). In these three cases, where the VERP process has been conducted separate from management planning, it has been completed through to identifying indicators and standards. Where it has been subsumed in management planning, identification of indicators and the associated monitoring have been deferred to the implementation plan and then not done (McCool *et al.*, 2007). Two partial applications integrated with other frameworks are known from Australia. For Fraser Island off the eastern coast (Figure 4.7), VERP was combined with LAC. For the Australian Alps, elements of ROS, LAC, VIMM and VERP were used (McArthur, 2000a).

The strengths of VERP are its usefulness as a management planning framework, explicit inclusion of a public participation strategy and ready inclusion in management plans. Its weakness is shared with many of these frameworks – without implementation and monitoring the acceptability or otherwise of impacts cannot be determined and neither can the effectiveness of management.

Benefits-Based Management

Benefits-Based Management (BBM) was developed to focus decision-making and management on desired outcomes (McCool *et al.*, 2007; Driver, 2008; Manning, 2011). This approach, also known as the Benefits Approach to Leisure and the Net Benefits Approach, was developed by Driver and colleagues (e.g. Driver *et al.*, 1991; Driver, 1996, 2008). Recent interest has been a result of growing public concerns about the accountability of land managers regarding their expenditure of public monies. BBM can potentially help by determining the benefits the public want from these lands and then report on the extent to which these benefits have been achieved.

Although BBM has been extensively addressed in research papers, there is no single, definitive document describing the processes or steps for planners to follow (McCool *et al.*, 2007). The following phases are derived from Driver (1996).

- *Phase 1* is identifying a study area and then describing the known and potential benefits to visitors and the local community. Identified benefits are assigned a relative importance to the agency's mission, a relative social importance and a cost of provision. Information sources include a visitor questionnaire and interviews, expert judgement and public involvement. Benefits may be psychological, social/cultural, economic or environmental (Driver, 2008).
- *Phase 2* centres on planning and involves evaluating the changes needed in the agency to provide the desired benefits mix. Comparison of alternative mixes and selection of the preferred alternative follows. Development of objectives, prescriptions and standards related to the preferred benefits conclude this phase.
- *Phase 3* involves a marketing plan and its dissemination, implementation of actions to provide the benefits, and monitoring benefit achievement. BBM shares its interest in alternatives, prescriptions and monitoring with the other planning frameworks.

BBM has been widely adopted in Canada and the USA and to a lesser extent in Australia (Brown *et al.*, 2006). The lack of broadscale implementation has been attributed to the lack of definitive steps that are easy to follow. Other impediments include the complexity of information required and a lack of commitment to social research. BBM provides a way of thinking about recreation and tourism use of natural areas rather than a practical planning framework (McCool *et al.*, 2007).

Choosing a planning framework

All the planning frameworks have the following features:

- They provide a framework for adaptive management.
- They focus on and manage human-induced change.
- They rely on the natural and social sciences.
- They depend on clearly articulated management objectives.
- They recognise and use opportunity settings that are a combination of physical, social and managerial conditions.
- They base planning on a spectrum of recreation opportunities.
- They require monitoring and evaluation (Nilsen & Tayler, 1998; McCool *et al.*, 2007; Manning, 2011).

In addition, some (but not all) of the frameworks require identification of indicators and associated standards.

How, then, do we choose between these frameworks? Managers and planners continue to struggle to identify which they should use (Nilsen & Tayler, 1998). This question can be answered by considering seven choices (Table 4.3).

First, are managers planning regionally or locally? If managers want to establish management directions for a region or group of natural areas, ROS, TOMM, VAMP and VERP are the most useful. All four provide a range of recreation opportunities in optimal locations. ROS is particularly suited to regional planning and has been used to allocate national parks to different places on the recreation opportunity spectrum, for example in the south coast region of Western Australia (Watson, 1997). TOMM can be applied across multiple land tenures, although it does not explicitly cater for opportunity classes. Both LAC and VIM focus on site-level impacts, so their application has tended towards individual protected areas. VIM is most often applied to a single site.

Second, is planning being applied to a largely undeveloped natural area or to areas encompassing developments and/or a variety of environments? TOMM has proven its usefulness in planning for different land tenures, spanning terrestrial and marine environments. ROS and LAC are also likely to be suitable but to a lesser extent, given they have rarely been applied, within the one planning exercise, to a diversity of tenures. Most applications have also been to natural areas rather than highly developed tourism facilities, so the indicators and standards used to date are best suited to wilderness and back-country scenarios. VIM, with its lack of zoning, is really only suitable for a single-site use, irrespective of whether the site is highly developed or not. It is of limited use where there is a diversity of land tenures, settings and activities. VAMP and VERP are designed specifically for national parks, meaning they are of limited use when consideration of a diversity of tenures within one plan is required.

Third, what are stakeholders' expectations regarding their involvement? LAC was explicitly developed to include stakeholders throughout the planning process, although, as mentioned above, gaining stakeholder support and then agreement on indicators

Table 4.3 Choosing the 'best' recreation/tourism planning framework

Planning framework	*ROS*	*LAC*	*VIM*	*TOMM*	*VAMP*	*VERP*
1. Suitable for regional planning (i.e. for more than a single natural area)	•••	•	–	•••	•••	•••
2. Applicable to many different settings (e.g. marine, terrestrial, pristine, highly developed)	••	••	•	•••	–	–
3. Explicit provision for including stakeholders in planning	–	•••	–	•••	–	••
4. Integrates readily with other forms of planning (e.g. management or tourism plans)	•••	•	•	••	••	••
5. Provides information on impacts of visitor use needed for management action	–	••	•••	•••	–	••
6. Requires identification of indicators and standards and whether they measure progress against objectives	–	•••	•••	•	•	•
7. Requires management actions and monitoring to be taken as part of framework	•	•••	•••	•••	•	•••
Means	• (1.1)	•• (2.4)	•• (1.6)	•• (2.4)	• (0.9)	•• (1.9)

ROS, Recreation Opportunity Spectrum; VIM, Visitor Impact Management; TOMM, Tourism Optimisation Management Model; VAMP, Visitor Activity Management Process; VERP, Visitor Experience and Resource Protection

••• matches criterion well, •• matches criterion, • poorly matches criterion, – does not match criterion, (x.x) mean score where each '•' was given a value of 1

Derived from Moore *et al.* (2003)

and standards continues to be problematic. This problem is not unique to LAC; it plagues all the planning frameworks (McCool *et al.*, 2007). TOMM, similarly to LAC, was designed to make stakeholder involvement an integral part of the planning process. The TOMM process for Kangaroo Island has a management committee with representatives from the community, industry and government agencies. VERP has a single step directing the development of an involvement strategy. ROS, VIM and VAMP do not explicitly consider stakeholders.

Fourth, how readily can the framework be integrated with other planning processes and documents? For those responsible for preparing management plans for natural areas or tourism plans for an area or region, great appeal lies in using a recreation/ tourism planning framework for the visitor management component of such a plan. Management plans for natural areas address visitor management as well as facets of the natural environment such as fire management and the protection of rare species. Tourism plans encompass managing visitors as well as the tourism industry. None of the frameworks, though, is exemplary in its ability to be explicitly integrated with other forms of planning. The principle behind ROS, of providing a range of recreation opportunities, appears in many management plans. Few of these plans, however, explicitly describe ROS and its application. TOMM has been explicitly considered and included in tourism plans (McArthur, personal communication, 2000). To date, however, it has not been integrated into an area management plan. VERP and VAMP have both been designed to integrate with management planning processes but their value is uncertain given their very low levels of application. Both LAC and VIM provide guidance to planners in preparing the visitor management part of plans but mention is rarely made of them in planning documents.

Fifth, do managers need information on the impacts of visitor use to implement management actions? If so, any of the planning frameworks relying on indicators and standards – LAC, VIM, TOMM or VERP – would be suitable. Two of these provide additional information on the likely causes of these impacts. VIM helps to identify causes of impacts and select suitable management strategies. TOMM also investigates the causes of impacts, seeking to differentiate tourism-related ones from those not related to tourism.

Sixth, does the framework explicitly rely on indicators and standards that are clearly linked to management objectives? The frameworks provide a range of choices. LAC and VIM have both objectives and indicators able to report on performance against these objectives. TOMM, VAMP and VERP are more mixed – the first two require indicators but no objectives, while the third prescribes objectives but no indicators. ROS has neither.

Seventh, does the framework explicitly require management actions and monitoring? All the frameworks require management, and all with the exception of ROS and VAMP require monitoring.

LAC and TOMM appear to be the 'best' frameworks, in terms of the items used to construct Table 4.3. However, they are so only if those applying them want to satisfy the criteria listed in the left-hand column of the table. For example, if regional planning (criterion 1), including stakeholders (criterion 3) and taking into account impacts (criterion 5) is required then TOMM is a good choice. On the other hand, if a planning process directed towards regional allocation of zones and integration with other forms of planning (criteria 1 and 4) is needed then ROS is a good choice. Table 4.3 makes clear these trade-offs and choices. There is no one 'best' framework.

Reasons for lack of implementation

Given all the work on frameworks it seems odd that implementation has not been more widespread. There are a number of explanations (Lipscombe, 1993; McCool *et al.*, 2007). In the USA, Australia and Canada, park managers have been highly focused on preparing management plans for their parks. These are usually general documents prepared to meet legal requirements and covering a plethora of resource issues. The recreation/tourism frameworks outlined in this chapter are much more specific and managers have struggled to incorporate this level of specificity in management plans and associated planning processes. Another reason for low levels of adoption is the confusion regarding the purpose of each framework and deciding which to choose. This is not helped by imprecise and vague language, and confusion regarding the terms 'opportunity class' and 'zone' (Nilsen & Tayler, 1998). The extensive use of acronyms such as LAC and VAMP also tends to alienate those unfamiliar with the frameworks.

Lack of resources for managing natural areas is a problem worldwide. In developing countries, this problem is particularly evident. All of the frameworks have been initiated and applied in developed countries, primarily the USA, Canada and Australia. Few applications in developing countries exist, the exceptions being applications to marine areas in the eastern Carribean (Suba) and Belize (McCool, personal communication, 2000). Many natural areas in developing countries have only recently been protected, meaning management efforts are directed towards park or reserve establishment. Reservation and management, not planning, are seen as the priorities.

Lack of resources is also a problem for these frameworks, with their reliance on biophysical and social data. The data needed range from information on the recreational opportunities offered by an area (for ROS) to measures of indicators such as vegetation loss or size of visitor groups (LAC, VIM, TOMM). Many land management agencies have apparently more pressing concerns than data collection; to many it seems an unaffordable luxury. Additionally, many have histories of biophysical data collection but little expertise or practice in the social sciences (Lipscombe, 1993). Also, managers may be concerned about how to select the 'right' indicator, a problem that can thwart applying LAC, VIM and TOMM (McArthur, 2000a).

Lack of resources is particularly a problem for monitoring, that is, data collection over time. Buckley *et al.* (2008) refer to chronic budget shortages as contributing to monitoring being a low priority for management agencies responsible for protected areas. Senior bureaucrats may be concerned about the costs of ongoing monitoring or collecting data that may indicate their agency is performing poorly and then not having the budget to address performance problems. Institutional commitment, as evidenced by formalised processes and structures within an agency, is essential for framework development and the accompanying monitoring.

Conclusion

Planning is crucial if natural area tourism is to be sustainable. Such planning allows impacts to be recognised and managed. This chapter has described six visitor planning frameworks, all offering viable alternatives to carrying capacity, a previously popular concept but increasingly acknowledged as unsuitable for sustainable tourism. All these frameworks focus on determining how much change is acceptable rather than trying to determine how much use is too much.

They all follow the stages of rational planning: setting objectives; data collection, collation and analysis; the development of alternatives and a final plan; and implementation as a crucial last step. And all can facilitate adaptive management by establishing management objectives and associated actions, and requiring monitoring, feedback and subsequent changes to management. These are all features central to adaptive management (McLain & Lee, 1996). For most of the frameworks, stakeholder involvement is an integral part, even if involvement is not explicitly detailed in the framework itself. There is no one 'right' framework: the choice depends on a number of factors, such as requirements for regional planning, for indicators and standards, and the extent to which visitor planning is to be integrated with other forms of planning.

This chapter plays a crucial, bridging role in this book. It provides, through the planning frameworks, a systematic means of linking the impacts detailed in Chapter 3 with possible management strategies, especially interpretation in Chapters 5 and 6, and monitoring in Chapter 7. Several of the planning frameworks, such as VIM and TOMM, explicitly require planners to develop and assess a range of management strategies as part of the planning process. Both these frameworks also require analysis of the cause of impacts before management strategies are considered. Cause can be determined only if the ecology of the natural area is understood (Chapter 2). Most of the frameworks, for example LAC, VIM and TOMM, also require the identification of indicators and standards for monitoring conditions. Chapter 7 describes monitoring of the impacts of visitors on the natural environment and on each other.

Further reading

It is well worth reading the various reports in which each planning framework is first described or applied. For ROS, refer to Clark and Stankey (1979). For LAC Stankey *et al.* (1985) describe the framework and USDA FS (1985) its application to the Bob Marshall Wilderness Complex in the Northern Rocky Mountains. VIM is described by Graefe *et al.* (1990) and Manidis Roberts Consultants (1995) provide a good example of its application to the Jenolan Caves in eastern Australia. The development of TOMM, and its application to Kangaroo Island off the southern coast of Australia, is covered by Manidis Roberts Consultants (1997).

Excellent contemporary reviews of planning frameworks include McCool *et al.* (2007) and Brown *et al.* (2006). Cole and McCool's (1998) compilation of papers on Limits of Acceptable Change gives extensive insights to the promises, pitfalls and progress with planning frameworks. The paper by Nilsen and Tayler (1998) in this compilation, comparing the various frameworks, helpfully summarises a vast amount of information into a few pages. Simon McArthur's doctoral dissertation (McArthur, 2000a) has a wealth of detailed information on the frameworks and their implementation. The journal article by Moore *et al.* (2003) places planning frameworks within the broader activities of natural area managers and the tourism industry.

Including stakeholders is integral to the success of planning. Hall and McArthur's (1998) book has several useful and easy-to-read chapters on stakeholder involvement. The IAP2 Public Participation Summary (IAPP, 2004) organises and clearly describes the different types of public participation according to level of meaningful engagement. It provides a recent interpretation of Arnstein's (1969) ladder of participation described in this chapter.

5 Management Strategies and Actions

Introduction

A wealth of strategies and actions are available for managing tourism in natural areas. Very often the difficulty is knowing which can be employed and then choosing among them. This chapter seeks to address this difficulty initially by describing the suite of possible management strategies, in the order they are likely to be implemented, and then by outlining the plethora of site and visitor management actions. Here, *strategies* are defined as the mechanisms and processes by which objectives are achieved. In this chapter an example of a strategy is reserving and/or zoning an area as a protected area (e.g. as a national park). They are general approaches to management. *Actions* are more specific: they are what must be done (Hall & McArthur, 1998). Examples of actions are providing educational materials and closing campsites for restoration.

For the majority of natural areas, being reserved as a protected area is the crucial first management strategy. Zoning generally follows. Managers then have many possible actions to choose from. In this chapter these actions are discussed as either site or visitor management. Site management actions rely on manipulating infrastructure and the natural environment to influence where visitors go and what they do. Campsite and trail design and management are the most well known of these actions. Visitor management, on the other hand, relies on managing visitors themselves through regulating numbers, group size and length of stay, providing information and education, and enforcing regulations. A number of factors influence the actions chosen by managers, including the cause, location and extent of the impact of concern, the cost and ease of implementation of actions and their effectiveness, and the preferences of visitors and managers. A brief comment on and an example of how actions are usually implemented in combination conclude this part of the chapter.

The chapter then goes on to describe current strategies being used to manage the tourism industry in natural areas. Voluntary strategies include codes of conduct, certification, environmental management systems and best environmental practice. Government organisations can choose to use regulatory approaches such as licensing and leases.

Reasons for managing natural areas

Worldwide, government and non-government organisations and private interests continue to acquire and seek to protect natural areas for their biodiversity and a range of other environmental, social and economic values. In almost all circumstances, this protection requires management. It is insufficient to reserve an area for conservation and then hope that the values that led to its reservation will be retained. Climate change and the threat it poses to protected areas further suggest the need for management.

Natural areas are a focal point for tourism. Without proper management, tourism in these areas will result in environmental degradation. For developing countries, such tourism is viewed as a means of generating much-needed foreign currency without degrading the environments on which the industry depends. This form of tourism is also favoured by environmental organisations,which believe it will increase support and the value attributed to natural areas, particularly in developing countries. All these outcomes rely on management of these areas.

Creating Protected Areas

Protected areas as a recent phenomenon

Designation as a park or reserve or some other formal acknowledgement of special status (e.g. designation of Indigenous Protected Areas in Australia) is usually the first, critical step in managing a protected area. A protected area is a 'clearly defined geographical space, recognised, dedicated and managed, through legal or other effective means, to achieve the long term conservation of nature with associated ecosystem services and cultural values' (IUCN, 2012).

When countries such as the USA, Canada and Australia were initially colonised, Europeans occupied very small areas, with the remainder of the country occupied by the original inhabitants – indigenous people, plants and animals. By the late 19th century it was becoming apparent that most of the land would eventually be occupied by Europeans and used for resource production, such as agriculture and timber, mining, fishing and urban settlement. This concern led to national parks being created. The first, Yellowstone in the USA (Figure 5.1), was set aside for its recreation and landscape values in 1872. Banff Hot Springs in Canada followed in 1885 and Yosemite in the USA in 1890. The purpose of Banff was 'public park and pleasure ground for the benefit, advantage and enjoyment of the people of Canada'. All three included recreation as a major focus.

For many parts of the world, however, establishing protected areas is a relatively recent phenomenon. In North Africa and the Middle East, China, South and South-East Asia, the former republics of the Soviet Union, Central America and the Caribbean, more than 50% of the area protected has been established since 1982 (McNeely *et al.*, 1994). Interestingly, sub-Saharan Africa, currently the focus of extraordinarily high levels of wildlife tourism, had almost half of its current protected area system in place by 1962. This early adoption has been attributed to concerns centred on watershed protection and erosion control (Burnett & Butler Harrington, 1994).

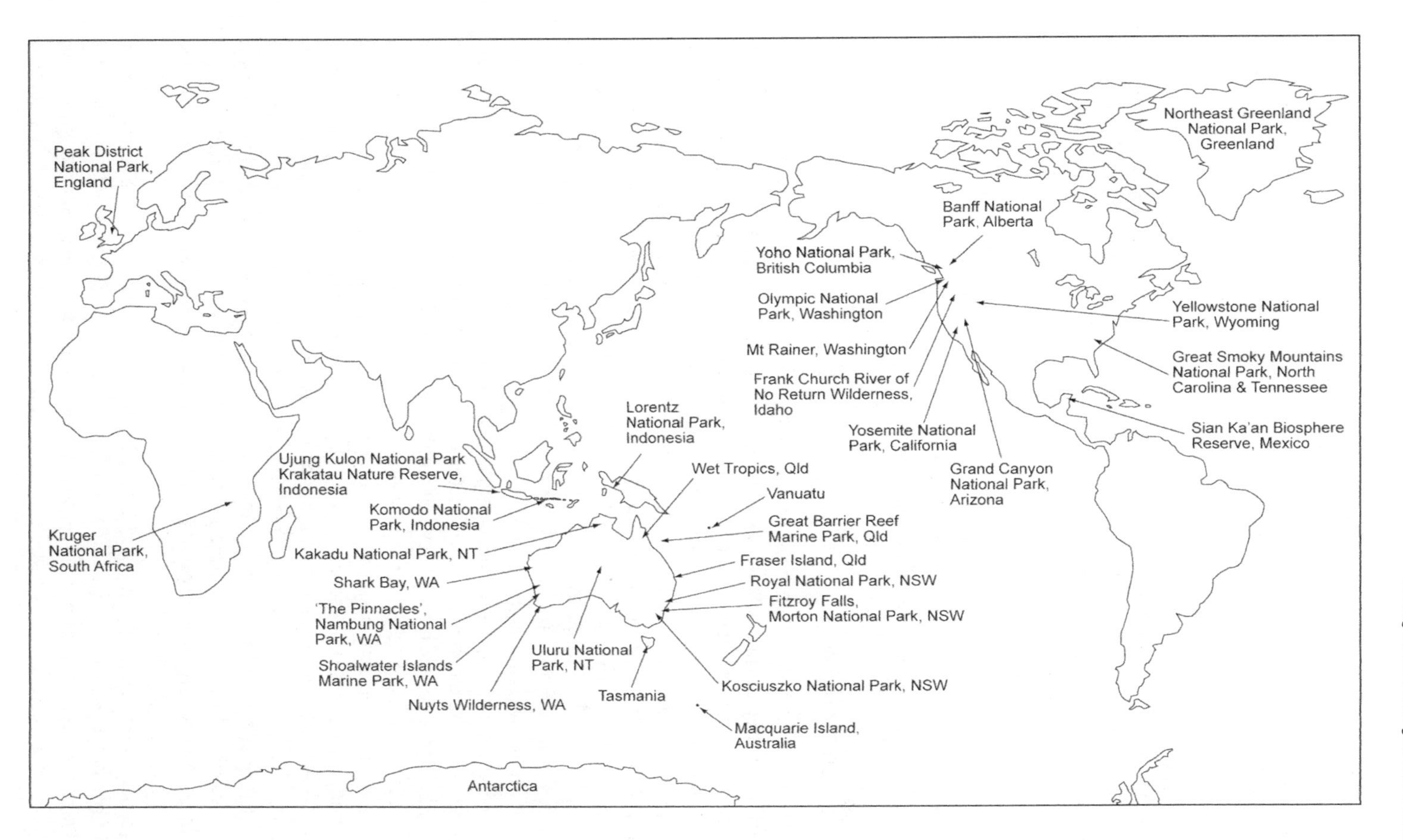

Figure 5.1 Location map of important nature-based tourism destinations referred to in this chapter

Over the last three decades, protecting biological diversity has become important to many people. Given that most species, and therefore biological diversity, exist in the wild, the best way to protect them is by protecting habitats in the wild. One of the most widely accepted ways is by establishing protected areas. Agardy (1993) noted that ecotourists are more inclined to visit well managed protected areas where landscapes look natural, water quality is good and biodiversity is high, rather than degraded, poorly kept areas.

Reservation and associated management by a government agency, such as a national parks service, and purchase and management by non-government conservation organisations (e.g. Australian Wildlife Conservancy) are the two most common approaches. Reservation by governments is usually achieved by legislation. The objective behind reservation is protecting biodiversity and other values of an area from exploitation by a few, for enjoyment by broader society. This is the reason for government ownership in many cases. It is generally recognised that governments are responsible for helping to meet the needs of society at large today and in the future, rather than a few individuals gaining personally.

Designing protected areas

Protected areas have often been created where lands or waters are not regarded as useful for other purposes, such as agriculture, forestry or fisheries. Additions to protected area systems have been opportunistically made when lands or waters become available. In the face of this haphazard approach, considerable knowledge now exists regarding how such areas should be selected and protected area systems designed. Many of these ideas are drawn from MacArthur and Wilson's theory of island biogeography (MacArthur & Wilson, 1967, in Primack, 1998). More recent work emphasises the importance of the dynamic interdependencies among sites and species' populations and how these need to be taken into account in designing protected areas (Williams *et al.*, 2005).

One of the most important design questions concerns the size a protected area should be. Is one single large reserve better than several small ones? This known as the SLOSS debate (single large or several small). A single large area has the advantage of providing for large, wide-ranging, low-density species such as large carnivores. It also has lower extinction rates than smaller reserves and lesser edge effects. On the other hand, creating more reserves, even if they are small, reduces the possibility of a single catastrophic event, such as a fire or disease, destroying a species because all individuals are located in one large reserve. Also, several small reserves are likely to contain more species than a single large one because they are expected to contain more types of habitat. And, a large number of small reserves is more likely than a single large one to preserve diversity among smaller species such as invertebrates, fungi and bacteria (Jordon, 1995).

Another way of thinking about the optimal size for a protected area is to consider the size of the population(s) the area is required to support. Protected areas should be sufficiently large to conserve large populations of important species, such as rare and endangered ones, and keystone and economically important species. Generally, populations of at least several hundred reproductive individuals are needed to ensure the long-term viability of a vertebrate species, and several thousand individuals are preferable (Primack, 1998).

The other important elements of design, in addition to size, are shape and connectivity. Protected areas should be shaped so as to minimise both the distances over which species have to disperse and harmful edge effects, including weed invasion and fertiliser drift into the area (Caughley & Gunn, 1996). Protected areas should be round shaped to minimise the ratio of circumference (edge) to area, thereby minimising external influences. Fragmentation of protected areas, by roads, farming and other human activities, should be avoided, as it can separate populations, reduce their ability to disperse and create undesirable edge effects.

Connectivity, achieved by creating or maintaining vegetated corridors between protected areas, has been a popular notion for decades, but the empirical evidence supporting its value is still limited. Corridors may be valuable in facilitating the movement of wide-ranging species, such as the mountain lion and wolf in the USA. They may also assist in providing habitat for migratory species, for example those using riparian corridors. Some ecologists suggest that corridors to higher elevations and latitudes will prove useful for species migrating in the face of climate change. Drawbacks associated with corridors include their role in facilitating the movement of pests and diseases and their attractiveness to predators, including humans, because of the wildlife they contain (Primack, 1998).

Landscape ecology provides information pertinent to reserve design at a regional scale. A landscape with large patches of protected habitat and minimal edge effects is favoured by many ecologists because such a pattern minimises habitat disturbance and protects species that rely on long-undisturbed (i.e. old growth) or 'interior' habitat. Take, for example, two landscapes, each with an area of 100 ha, of which half is forested and the other half cleared. One landscape has an alternating checkerboard of 1 ha patches of fields and forest. The other landscape has four patches, each 25 ha in area, with two of the four forested. The latter landscape would be favoured by such ecologists as it provides large patches and minimises the length of edge.

Three objectives – comprehensiveness, adequacy and representativeness (CAR) – are widely espoused to guide the design of reserve systems. Comprehensiveness means including the full range of ecosystems across a region, country or series of countries, at an appropriate scale, in the reserve system. Adequacy requires that sufficient areas are reserved to ensure the ecological viability and integrity of populations, species and communities. Representativeness refers to selecting areas for inclusion to reflect the biotic diversity of all the ecosystems across the areas being considered (ANZECC TFMPA, 1998).

Systematic conservation planning helps to prioritise the acquisition and management of lands and waters for biodiversity conservation. Ideally, it can be used to design a reserve system, but has more often been used to suggest additions to existing systems. This form of planning recognises the contribution to conservation goals made by existing reserves and then uses simple methods for locating and designing new reserves to complement existing ones (Margules & Pressey, 2000). Explicit goals or targets guide these selection processes. For example, a common target is to include

10% of each community/habitat/ecosystem in the reserve system (Stewart *et al.*, 2007). A number of mathematical approaches are used in systematic conservation planning, with MARXAN being the most widely known and applied (University of Queensland, 2011). MARXAN uses spatial data to identify the 'best' areas for addition to a reserve system.

Extent and types of protected areas

As of 2010, 12.2% of the Earth's land area was in nationally designated protected areas (WDPA, 2011a). This figure excludes the now considerable natural areas protected by private individuals and organisations. The largest protected area, Northeast Greenland National Park, Greenland (Figure 5.1), covers 972,000 km^2. Marine protected areas currently cover 6.4% of the Earth's seas (WDPA, 2011a).

Almost half of the world's biomes (43%) remain under-represented, based on the target of 10% of all ecosystems set by the International Union for Conservation of Nature and Natural Resources (IUCN) (Jenkins & Joppa, 2009). Least well represented are temperate grasslands, savannas and shrublands, with less than 4% in protected areas. Tropical and sub-tropical dry broadleaf and coniferous forests, boreal forests, deserts and Mediterranean ecosystems also fall short of the 10% target. In

Table 5.1 IUCN protected area categories

Category	*Description*
Ia	*Strict nature reserve* strictly managed to control and limit human visitation, use and impacts; area containing regionally, nationally or globally outstanding ecosystems, species and/or geodiversity features
Ib	*Wilderness area* managed to protect its long-term ecological integrity, where natural forces and processes predominate; large unmodified area without permanent or significant human habitation
II	*National park* managed for ecosystem protection and recreation/visitor enjoyment; large natural or near-natural area with cultural and recreational opportunities
III	*Natural monument or feature* managed for conservation of specific natural features (e.g. sea mount, submarine cavern); area containing outstanding natural features and associated biodiversity
IV	*Habitat/species management area* managed for conservation through intervention
V	*Protected landscape/seascape* managed for conservation and recreation; area where traditional interaction between people and the land/sea needs safeguarding to maintain the area's distinct character
VI	*Protected area with sustainable use of natural resources* managed to protect natural ecosystems and use natural resources sustainably; large area, mostly in a natural condition with part under sustainable natural resource management

Derived from IUCN (2012)

Box 5.1 Government-managed protected areas in Australia

In Australia, state governments are responsible for managing the majority of protected areas. This is in contrast to countries such as the USA, where these responsibilities rest mainly with the federal government. National parks occupy 5.2% of Australia's land area. They are managed to provide recreation and tourism opportunities while ensuring that natural ecosystems are protected. Including all types of terrestrial protected areas – indigenous protected areas, nature reserves, wilderness and state recreation areas as well as national parks – places 12.8% of Australia's land mass in protected areas (DSEWPC, 2011). State forest is not included in this figure.

Marine protected areas, most often reserved as multiple-use marine parks, are a more recent phenomenon. They occupy 27.9% of Australian waters (WDPA, 2011b). The Great Barrier Reef Marine Park (Figure 5.1), with an area of 345,400 km^2, is one of the largest protected areas in the world. Marine parks in Australia include commercial fishing, although mining and oil and gas production are usually excluded. They include general use and sanctuary zones, with these zones generally managed as IUCN category VI and II reserves, respectively (Table 5.1).

contrast, tropical and sub-tropical moist broadleaf forests are well represented, with most of the land area added annually to the protected area system being to this biome in the Amazon, Brazil (Jenkins & Joppa, 2009). In October 2010 this 10% target was raised to 17%, as part of the Aichi Biodiversity Targets (UNEP, 2012).

Once land or water comes under government control, it is then a matter of determining how it will be managed. A key concern is managing human disturbance, including tourism. How an area is managed depends on how much disturbance and use have occurred in the past, the biological values at stake and their vulnerability, and visitor numbers and activities. The IUCN has developed a system of classification for protected areas that ranges from minimal to intensive human use (Table 5.1) (IUCN, 2012). National parks (IUCN category II, Table 5.1), the oldest and best-known form of protection for natural ecosystems, account for 20% of the world's protected areas. 'Protected area with sustainable use of natural resources' (IUCN category VI, Table 5.1) adds another 20% (Jenkins & Joppa, 2009). Included in category VI are production forests and marine parks with commercial fishing. How these categories are implemented and match current practice varies from country to country. Box 5.1 describes the protected area systems in Australia.

Other forms of protection

There are a number of other ways in which areas are protected, in addition to reservation. The most significant are international designations, such as World Heritage sites, Ramsar wetlands and biosphere reserves.

World Heritage Convention

UNESCO's World Heritage Convention aims to identify, protect and preserve natural and cultural heritage sites worldwide. It does this via World Heritage sites – natural areas and culturally significant structures, settlements and places with outstanding universal value (Anon., 1997). The Convention was adopted by UNESCO in 1972. Sites may be listed as natural, cultural or mixed properties (having both natural and cultural values). There are currently 936 listed properties, comprising 183 natural, 725 cultural and 28 mixed (UNESCO, 2012a).

Natural properties should:

- be outstanding examples representing major stages of the Earth's history, significant ongoing geological processes influencing landform development, or significant geomorphic or physiographic features; or
- be outstanding examples representing significant ongoing ecological and biological processes in the evolution and development of terrestrial, freshwater, coastal and marine ecosystems and communities of plants and animals; or
- contain superlative natural phenomena or areas of exceptional natural beauty and aesthetic importance; or
- contain the most important and significant natural habitats for *in situ* conservation of biological diversity, including those containing threatened species with outstanding universal value for science or conservation (Anon., 1997).

The intention of listing is to protect outstanding areas by preventing them from falling into disrepair, as well as influencing development projects so they do not jeopardise a site's heritage values. If a site is in danger, the World Heritage Committee can place the site on the List of World Heritage in Danger. If a country does not fulfil its obligations under the Convention, it risks having its sites removed from the World Heritage list (Anon., 1997). Recreation, tourism and interpretation are usually provided for at World Heritage sites; however, some may be of such significance and fragility that such uses are prohibited.

To give an indication of the types of natural properties on the World Heritage list, a brief description of these properties in Australia and Indonesia follows. Australia has 16 World Heritage natural and mixed sites, ranging from the largest monolith in the world in inland Australia (Ayers Rock at Uluru) to the wet tropics of Queensland, Tasmanian wilderness, offshore islands such as Macquarie Island, to Shark Bay in the west, with its outstanding marine and terrestrial ecosystems. Australia's first World Heritage site, the Great Barrier Reef, was listed in 1981. Indonesia has three World Heritage natural and mixed sites, the most recent, Lorentz National Park, listed in 1999, is the largest protected area in South-East Asia. The other two are Ujung Kulon National Park, on the extreme south-western tip of Java and the last viable natural refuge for Javan rhinoceros, and Komodo National Park, inhabited by around 5700 Komodo dragons (UNESCO, 2012a) (Figure 5.1).

Ramsar Convention on Wetlands

The Convention on Wetlands, signed in Ramsar, Iran, in 1971, is an international treaty providing for the conservation and wise use of wetlands. It relies on goodwill and has no means of penalising countries that fail to conserve protected sites (Hollis & Bedding, 1994). A total of 160 countries are signatories, with 1997 wetland sites designated for the Ramsar List of Wetlands of International Importance. The number of wetlands designated by individual countries is highly variable. For example, the UK has 168 designated sites, Australia has 65 and Rwanda 1 (Ramsar, 2012).

Two criteria are used to select Ramsar wetlands: the site must contain representative, rare or unique wetland types and/or be of international importance for conserving biological diversity. Wetlands are defined by the Convention as including rivers, lakes, swamps and marshes, wet grasslands and peatlands, estuaries, deltas and tidal flats, near-shore marine areas, mangroves and coral reefs, oases, and human-made sites such as fish ponds, rice paddies, reservoirs and salt pans (Ramsar, 2012).

Biosphere reserves

The Man and the Biosphere (MAB) programme was initiated by UNESCO in Paris in 1968 (Batisse, 1982). Biosphere reserves are intended as places where conservation and development can occur together. Involving local communities in management is integral. They are connected as a worldwide network of representative natural areas where monitoring, research, management, training and education are undertaken (Batisse, 1982).

Each biosphere reserve is intended to fulfil three complementary and mutually reinforcing functions (UNESCO, 2012b):

(1) a conservation function, contributing to the conservation of landscapes, ecosystems, species and genetic variation;
(2) a development function, fostering economic and human development which is environmentally and socially sustainable and culturally appropriate;
(3) a logistic function, providing support for research, monitoring, education and training related to local, national and global issues of conservation and development.

Biosphere reserves ideally consist of protected core areas surrounded by buffer zones and a broader transition area (Figure 5.2). The buffer zones are used for cooperative activities that complement sound ecological practices. Environmental education, recreation, ecotourism and research are suitable activities. Sustainable development is the focus of the transition area, where local residents, economic interests, scientists, management agencies and non-government organisations work together linking conservation and economic development (UNESCO, 2012b).

There are 580 sites in 114 countries, with many of these reserves providing for tourism. For example, Sian Ka'an Biosphere Reserve in Mexico, regarded as one of the flagships of the UNESCO MAB programme, accommodates ecotourism (Box 5.2). To give some indication of numbers of these reserves in various countries, Australia has 15, Canada 16, Indonesia 7, the UK 8 and the USA 47 (UNESCO, 2012b).

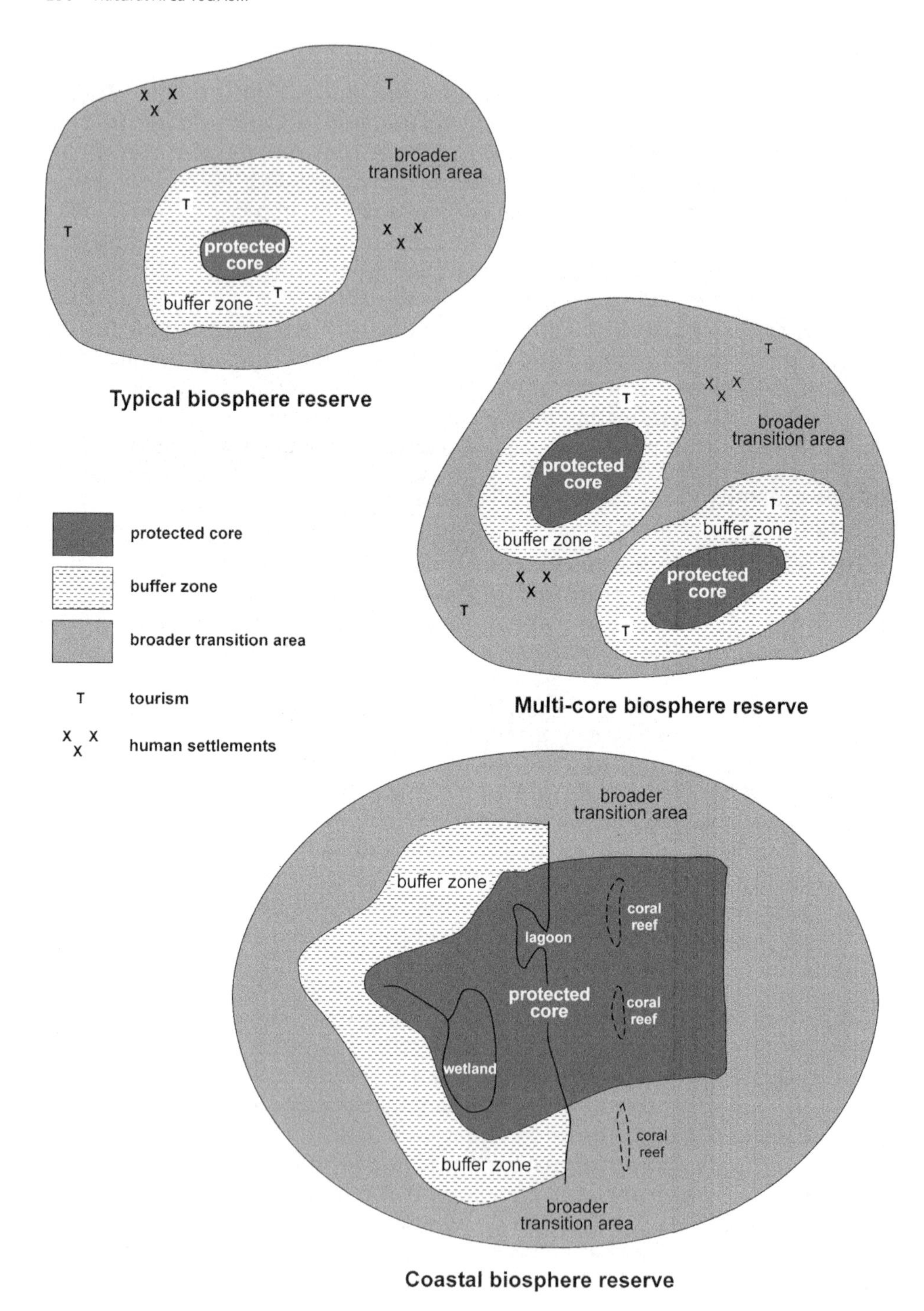

Figure 5.2 Zoning of international biosphere reserves (Derived from Batisse, 1982)

Box 5.2 Sian Ka'an Biosphere Reserve and natural area tourism

Sian Ki'an on the eastern shore of the Yucatan Peninsula in Mexico (Figure 5.1) was officially dedicated as a Biosphere Reserve in 1985. It encompasses coastal dry forest, extensive shallow-water lagoons, part of the second largest barrier reef in the world, Mayan temples and other ruins, and numerous limestone sinkholes with associated biota (Agardy, 1993). Within the boundaries live about 800 Mayan people. The reserve has a core area where extractive uses of the forest and coastal areas are banned while buffers accommodate the needs of fishers, snorkellers and sightseers.

Tourism in the state of Quintana Roo in which Sian Ka'an is located is booming. Sian Ka'an has facilities for tourists, with Mayan residents building several visitor centres and running nature tours. Mayans have been trained to build and maintain park infrastructure, to guide visitors and market their crafts. They are also involved in sea turtle conservation, through guarding nesting turtles and their young from natural predators and human poachers (Agardy, 1993).

Governance and Joint Management

Protected areas can contribute to conservation and tourism only if they are managed effectively. Governance strongly influences effectiveness, by providing the structures within which management can take place. Governance is the regulatory processes, mechanisms and organisations through which people ('actors') influence actions and outcomes. Governance and government are different. Governance includes the actions of government plus those of communities, businesses and non-government organisations (Lemos & Agrawal, 2006).

Most protected areas are still managed by governments; however, this is changing as powers and responsibilities are shared with local communities and non-government organisations. Bramwell (2010) explains that the top-down model of governance, with governments in charge, has been augmented and sometimes replaced by a diversity of approaches, including collaborative management, public–private partnerships, delegated authority and community management.

Eight different forms or models of governance have been identified for protected areas (Eagles, 2008, 2009): national park, parastatal, non-profit organisation, ecolodge, public and for-profit combination, public and non-profit combination, Aboriginal and government, and traditional community (Table 5.2). The national park model, where the government owns and manages the park and is funded by societal taxes, has been the traditional approach in countries such as the USA, Canada and Australia but is increasingly being replaced by models that include new management arrangements and income sources, such as the public and for-profit combination (Eagles, 2009). These forms are differentiated by who owns the land or water (e.g. government-owned), the sources of income (e.g. user fees) and the type of management body (e.g. government agency, non-profit corporation).

Table 5.2 Governance models for protected areas

Model	*Resource ownership*	*Income sources*	*Management body*	*Examples (from Eagles, 2008)*
1. National park	Government	Taxes	Government agency	US National Park Service pre-1950s
2. Parastatal	Government	User fees	Government-owned corporation	SANParks in South Africa
3. Non-profit organisation	Non-profit corporation	Donations	Non-profit corporation	Royal Society for the Protection of Birds, UK
4. Ecolodge	For-profit corporation	User fees	For-profit corporation	Ecolodges on the western edge of Kruger National Park, South Africa
5. Public and for-profit combination	Government	Taxes and user fees	Government and for-profit corporation	Management of provincial parks of British Columbia, Canada
6. Public and non-profit combination	Government	Taxes and user fees	Government and non-profit corporation	'Park friends' groups (often provide interpretation)
7. Aboriginal and government	Aboriginal/ government	Taxes and user fees	Government/ Aboriginal	Uluru-Kata Tjuta, Kakadu National Parks, Australia
8. Traditional community	Aboriginal	Taxes and user fees	Aboriginal	Australian Indigenous Protected Areas

Derived from Eagles (2008, 2009)

Buteau-Duitschaever *et al.* (2010) compared a public and for-profit combination approach to the governance of British Columbia provincial parks with the parastatal approach of Ontario provincial parks. British Columbia administers all parks visitor services, including fee collection, through private contractors. It obtains only 20% of its funding through user fees, with the rest coming from the provincial government. In contrast, the parastatal approach in Ontario involves in-house staff providing most visitor services, with just a few specialised services (such as rental equipment and garbage collection) administered through private contractors. A total of 80% of its funding is from tourism-generated income, including user fees.

These researchers surveyed 367 visitors to parks in the two provinces. Respondents were asked how their park system performed relative to criteria for 'good' governance, with these criteria including accountability, transparency, efficiency, effectiveness, responsiveness and equity. Visitors to Ontario parks ranked their park system higher against all the governance criteria than did visitors to the British Columbia parks. These results suggest that the parastatal model in Ontario provides for 'better' governance, as perceived by visitors, than the public and for-profit combination in British Columbia. Although the latter approach may be politically popular and seen as cost-shedding or sharing, it may result in poorer governance, as perceived by visitors, than retaining a greater involvement in the delivery of services, as in the parastatal model.

Joint management – most commonly referring to a sharing of responsibility by indigenous people and a government agency – equates with the Aboriginal and government governance model (Eagles, 2009; Buteau-Duitschaever *et al.*, 2010). The indigenous people may derive their living from the protected area or have long-standing spiritual links with it. Eagles (2009) scored this model the lowest in terms of good governance because of perceived weaknesses in public participation, consensus, responsiveness, efficiency, accountability and transparency. These weaknesses help explain current concerns regarding joint management, and suggest where governance might be improved, and hence joint management.

Joint management is nonetheless an increasingly popular management strategy as, importantly, it provides formal recognition by government agencies of the relationships between indigenous people and protected areas. It is relying more and more on a formal agreement between the involved parties. For example, joint management of Kakadu and Uluru National Parks in northern Australia (Figure 5.1) is based on the park management agency formally leasing the land from the traditional Aboriginal owners. Joint arrangements also include a board of management with majority membership from the Aboriginal owners and a jointly developed statutory management plan. Cooperation is sought in both longer-term planning and day-to-day management (De Lacy, 1994).

Where lands within a national park are privately owned, as is the case in England and Wales, joint management is also evident. Here, national parks have been designated in landscapes used for agriculture and forestry, with long histories of human occupancy and private ownership. Joint management involves fostering partnerships between national agencies responsible for conservation, individual national park authorities, national agricultural interests (e.g. Ministry of Agriculture), and individual farmers and landowners. The intent is to manage changing agricultural practices in order to maintain the 'wildness' of these parks and thus their appeal to visitors (Swinnerton, 1995).

Possibilities also exist for involvement of indigenous people in governance arrangements different to the joint management – Aboriginal and government model – detailed above. Traditional community ownership, where these communities may be indigenous, and land management for conservation is a rapidly growing approach. Eagles (2009) notes such approaches in Ghana and Africa. Indigenous Protected Areas (IPAs) in Australia are owned by Aboriginal people and managed for nature conservation. They are declared voluntarily by the 'traditional owners' and managed with support from a range of government and non-government sources (Ross *et al.*, 2009). As of 2012, there were over 40 IPAs across Australia, totalling over 26 million hectares (DSEWPC, 2012). Indigenous people may also be involved in a number of other models, for example running tours within a government-managed park, which is the parastatal model.

New thinking about the governance of protected area suggests moving away from a somewhat restricted focus on the 'traditional' national park model and narrowly defined joint management, to think and act creatively with up to eight different governance models. Research by Eagles (2008, 2009) and colleagues suggests that the choice of model(s) should be based on (and performance subsequently evaluated using) criteria such as efficiency, effectiveness, accountability and transparency.

Zoning

It is generally insufficient to reserve an area of land and then hope its values will automatically be protected. The next crucial stage is management. Management is differentiated from governance, with the former concerning plans, actions and resources and the latter ideally laying a foundation for management through the allocation of the authorities and responsibilities of participating organisations (Lockwood, 2010). Many parks in developing countries are at the point where they have been reserved but are still vulnerable to exploitation. One of the key strategies for managing protected areas is zoning. This involves recognising smaller units or zones within the area, each with prescribed levels of environmental protection and certain levels and types of public use. Most of the planning frameworks described in Chapter 4 include identifying and managing zones. Zoning for protected areas generally has two purposes – protecting the natural environment and providing a range of recreation/tourism opportunities.

Providing a choice of experiences for visitors through zoning is fundamental to the Recreation Opportunity Spectrum planning framework and subsequent approaches, such as the Limits of Acceptable Change (Chapter 4). Zones range from primitive, with few to no facilities and little likelihood of encountering others, through to developed areas, with extensive facilities, such as resorts, and numerous interactions with others (Box 5.3). Zoning also clarifies future intentions. The management actions explored in the following sections are employed to maintain existing levels and types of use, or to manipulate use to return a zone to a more pristine state or conversely develop it further.

Another reason for zoning, in addition to providing choice for visitors, is separating incompatible visitor uses in space and time (i.e. spatial and temporal zoning). Spatial zoning is often used to separate hikers from motor vehicles, and motorised and non-motorised watercraft. At a finer scale, cliffs may be zoned to separate abseilers and climbers. Zoning is also used to protect the natural environment from damaging use by visitors. It can be used to exclude people or strictly regulate them at fragile geological formations, such as the Burgess Shales in Yoho National Park, Canada (Figure 5.1). Or, as happens more frequently, types of use may be regulated. For example, spear fishing may be prohibited from areas with vulnerable species or ones of special interest.

Zoning may also be applied temporally. For example, fairy terns (*Sterna nereis*), a seabird breeding on beaches, are highly susceptible to disturbance. Colonies can be zoned to exclude people during the bird's breeding season. Seasonal closures have also been used in south-western Australia to prevent the spread of the water mould *Phytophthora cinnamomi*, a soil-borne pathogen affecting many native plant species. Wet conditions enable the pathogen to be transported in mud attached to vehicles and

Box 5.3 National park and marine park zoning in Western Australia

National park zoning (terrestrial)

Five zones have been applied to national park planning in Western Australia – four with very little development and one with an emphasis on facilities and services. This approach was developed in the 1980s, based on the approach taken by Parks Canada (CALM, undated). *Special conservation* zones contain the most intact examples of ecosystems in the park that would be threatened by uncontrolled access. Access and use are strictly controlled and may be prohibited. No motorised access or built facilities are permitted. Areas zoned *wilderness* must be extensive, minimally disturbed by humans and with no motorised access. *Natural environment* zones can support low-density outdoor activities with minimum facilities. Non-motorised access is preferred, although in many parks access by four-wheel-drive vehicles will continue. *Recreation* zones can accommodate a broad range of recreation activities while protecting the natural environment. Motorised access is permitted. *Park services* zones may include visitor centres, park headquarters and/or towns. Not all zones are used in every park.

Marine park zoning (marine)

Marine park zoning in Western Australian state waters is a more recent development, a product of the late 1980s and 1990s and influenced by zoning of the Australian Great Barrier Reef Marine Park (Figure 5.1). The purpose, similarly to zoning terrestrial parks, is to protect sensitive habitats and provide a range of recreational opportunities. Equity and minimising conflict are emphasised. Four zones are delineated. The first, *general use*, provides for commercial activities such as sustainable commercial and recreational fishing, aquaculture, pearling and petroleum exploration and production, plus recreational uses, so long as they do not compromise the ecological values of the park. In the *recreation* zone, commercial fishing is not permitted; however, commercial activities associated with recreational uses are allowed, for example commercial viewing of whale sharks and game fishing. *Sanctuary* zones provide for total protection – fishing and removing other organisms are not permitted. This type of zone is usually specified to cover areas with vulnerable or special-interest biota requiring the highest level of protection; it also protects representative areas of the park's ecosystems from human disturbance. Non-extractive tourism activities are usually permitted. *Special purpose* zones are specified if the other zones are not appropriate. Uses may include a combination of commercial and recreational uses. Each such zone has a stated purpose that is legally recognised. For example, in Shark Bay Marine Park, on the north-west coast of Australia (Figure 5.1), Cape Peron special purpose zone has the stated purpose of 'wildlife viewing and protection'. The area has abundant wildlife, including dugongs, dolphins, humpback whales, manta rays and many fish species. Commercial and recreational fishing, diving, and non-motorised water sports are allowed; spear fishing, aquaculture and motorised water sports are not.

shoes and introduced into previously disease-free areas (see Box 2.2, p. 66). Zoning is used to close areas in wet conditions so the pathogen is not introduced or spread.

Site Management Actions

Rationale for approach taken

The impacts of natural area tourism can be managed through either *site* or *visitor* management (Hammitt & Cole, 1998) (Figure 5.3). Site management seeks to control visitors through actions at the sites where the use occurs. Sites include linear transport corridors, such as roads and walk trails, and terminus points, such as resorts and lodges, campgrounds and individual campsites, picnic areas, water bodies and coastlines. Site management relies on locating use in the more durable parts of

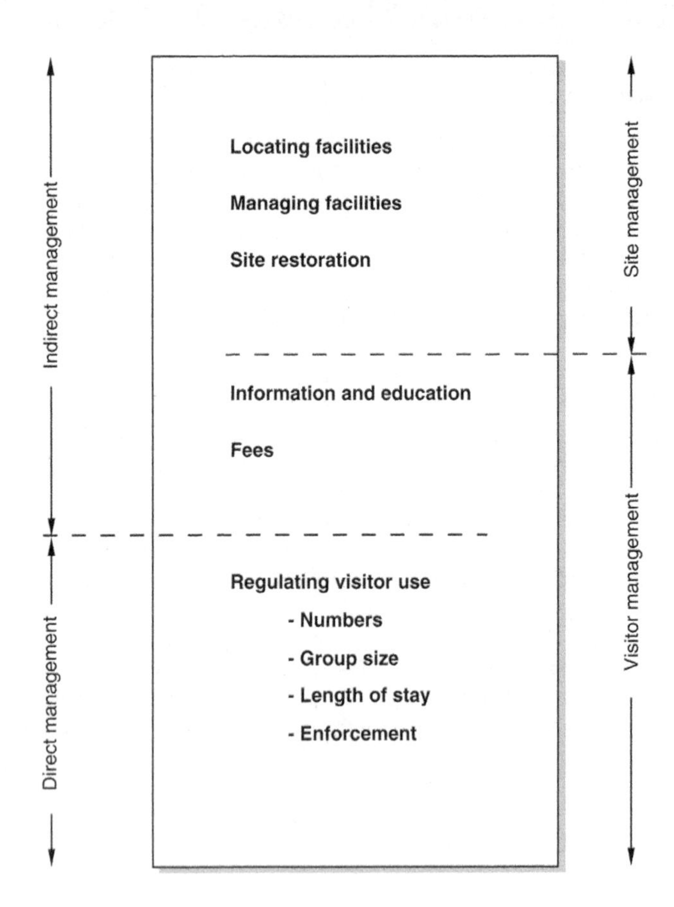

Figure 5.3 Common ways of classifying approaches to managing visitors to natural areas (Derived from Lucas, 1990a; Hammitt & Cole, 1998)

Table 5.3 Ways of classifying approaches to managing visitors to natural areas

Classification	*Purpose of classification*	*Description of approach*
1. Management actions grouped as site or visitor management (Hammitt & Cole, 1998)	Actions grouped to make them accessible to managers and according to what is to be managed	Site management seeks to influence visitor activities by manipulating the natural environment and infrastructure, while visitor management relies on regulating visitors and information and education. This is the classification system adopted in this book
2. Management actions grouped as direct and indirect (Lucas, 1990a)	Actions grouped to make them accessible to managers	Direct techniques, such as limiting visitor numbers, restrict individual choice, while indirect techniques, such as education, seek to influence visitors, leaving them free to choose. This is the most widely used classification system
3. Management actions grouped as containment or dispersal approaches (Cole, 1981a)	Actions grouped to make them accessible to managers	Containment limits where activities occur, while dispersal reduces the frequency of use at each site so that permanent impacts are avoided
4. Management actions organised into five groups (Anderson *et al.*, 1998)	Actions grouped to make them accessible to managers	The five groups are: site management, rationing and allocation, regulations, deterrence and enforcement, and visitor education. Developed for US National Park Service managers
5. Management actions organised according to eight strategies (Cole *et al.*, 1987)	Actions grouped (within eight strategies) to make them accessible to managers	The eight strategies are: reduce use of entire wilderness, reduce use of problem areas, modify location of use within problem areas, modify timing of use, modify type of use and visitor behaviour, modify visitor expectations, increase resistance of the resource, and maintain/ rehabilitate the resource
6. Management actions listed in general groupings (Hall & McArthur, 1998)	Actions listed to make them accessible to managers	Management actions loosely grouped as regulatory, fees based, site modification, research, marketing, interpretation and education, profile management, and assisting alternative providers
7. Grouping of actions according to impact location and source (Cole, 1990b)	Actions grouped according to the location and source of visitor impacts	Management actions organised as responses to campsite, trail, and pack and saddle stock impacts
8. Strategies based on supply and durability of resources used by visitors (Manning, 1979)	Strategies described and organised	The four strategies are increasing supply, reducing the impact of use, increasing resource durability, and limiting use
9. Typology of spatial strategies (Leung & Marion, 1999a)	Strategies described and organised	The four strategies are spatial segregation, spatial containment, spatial dispersal and spatial configuration. The classification is spatially based

Derived from Manning (1979), Cole *et al.* (1987), Cole (1990), Lucas (1990a), Anderson *et al.* (1998), Hall & McArthur (1998), Hammitt & Cole (1998), Leung & Marion (1999a)

the landscape and designing and managing sites and associated facilities to minimise visitor impacts. Wherever possible, site management should aim to enhance visitor understanding of the area's values and issues (Eagles *et al.*, 2002). Visitor management, covered in the next major section, focuses on managing visitors themselves through regulating use, communicating with them and providing education.

Three broadly different purposes in organising and describing management practices are apparent (Table 5.3). The most common is detailing and grouping actions so they are readily accessible to managers. The related classifications, numbered 1–6 in Table 5.3, typify this grouping and are described by Manning *et al.* (1996a) as focusing on tactics. Possible actions are described and grouped to make the wealth of information digestible and accessible to managers. Within this group, the most widely used classification has been the direct/indirect approach. Direct actions restrict individual choice (e.g. by regulating visitor numbers) while indirect actions seek to influence visitors (e.g. by education), leaving them greater freedom to choose. Here, we adopt Hammitt and Cole's (1998) site/visitor management classification, rather than the more widely known indirect/direct classification, because it makes clear at the outset *what* is to be managed – whether it is a site or visitors themselves.

The two other ways of grouping management practices are either according to the location and source of visitor impacts (classification 7 in Table 5.3) or to describe and categorise management strategies (classifications 8 and 9). Neither of these purposes matches the needs of this book as well as the site/visitor management approach. Organising actions based on the location of impacts, for example grouping actions for locations such as campgrounds and trails, is potentially useful to managers. However, there are significant areas of overlap and repetition because an action such as managing visitor numbers applies to all locations. The last grouping focuses more broadly on strategies (classifications 8 and 9) so is not directly comparable or relevant to the suites of actions covered by the other classifications.

Roads and trails

Locating roads and trails

Natural area tourism relies on major highways, such as the Interstate 1 traversing Banff National Park in Alberta, Canada, four-wheel-drive tracks in the desert nature reserves of central Australia, paved walk trails such as those along the edges of the Grand Canyon, and unmarked routes in remote wilderness areas such as the Frank Church–River of No Return in Idaho, USA (Figure 5.1). Although it is difficult to generalise across this diversity, some guidance is possible regarding locating these linear features to minimise their adverse impacts. Through careful siting, construction and maintenance they can support substantial visitor traffic while at the same time protecting off-road and trail areas (Leung & Marion, 1999a; Marion & Leung, 2004a).

Constructing roads and trails results in environmental impacts. Minimising the area cleared is the simplest management measure. Also, minimising changes to natural drainage patterns is important. This is best achieved by selecting a well drained location, assisted by engineering if necessary. Using materials that blend with the colours of the existing landscape is another way of reducing the potentially intrusive appearance of such developments.

One of the most common ways in which soil pathogens such as dieback are spread is through road and trail construction and maintenance. Dieback spreads through the movement of infected soil as well as in water and by root contact with infected plants. Infected soil may be spread either attached to earth-moving equipment or in soil brought on-site for construction purposes. The introduction of infected material can be minimised by washing earth-moving equipment and making sure infected soil is not moved to uninfected sites.

The key influences in selecting a durable route are slope and soil characteristics such as soil moisture and erodibility (Hammitt & Cole, 1998; Marion & Leung, 2004a). The degree of slope of the linear development itself and the extent to which it intercepts run-off are critical. Ideally, trails and roads should have gentle grades so that they do not erode but have some slope to prevent water pooling on them. Steep grades can be avoided by using switchbacks. Where this is not possible, engineering works are necessary.

In many places, especially in mountainous and tropical settings, muddy roads and trails caused by excessive soil moisture are a concern. Very often users skirt around muddy parts, leading to trail widening and braiding. Locating these developments where they will not collect water from up-slope or intercept high water tables is a useful preventive measure. High water tables can usually be identified by characteristic vegetation types. Trails and roads that have eroded below the surrounding ground level can become *de facto* streamlines.

Erodibility is also a key soil characteristic in deciding where to locate linear developments. Trails and vehicle tracks that are not hardened or engineered into the landscape are best located on well drained soils such as loams with a substantial organic matter component (Hammitt & Cole, 1998). Clays tend to pond water while sands rapidly disperse when the vegetation is removed. Linear developments over clays and sands therefore generally require engineering attention.

Stream banks with their steep slopes and high moisture content are best avoided, as are coastal dunes with their sandy, poorly consolidated soils subject to wind erosion. Cole (1990b) drew on studies reporting that trail erosion and trampling led to increased iron and phosphorus levels in heavily used lakes. As such, trails should be set back from the edges of water bodies and run-off filtered by fringing vegetation. Erosion of stream banks can be minimised by locating trails where the banks are low and stable. The same applies to coastal dunes, with the added proviso that, where possible, trails should be aligned at right angles to prevailing winds. Trails used by stock are best located on rocky substrates, such as ridges, rock scree and bedrock, to minimise soil erosion.

Multiple or 'braided' trails are the result of conditions where it is difficult for visitors to walk or drive. Walking and driving are difficult in muddy, sandy, very rocky or slippery conditions. The best solution is relocation where possible. Short cutting switchbacks, another problem created by visitors leaving the designated trail, can be effectively managed by minimising the number used, locating them out of sight of each other, building barriers between them, and using wide turns. Informal use, that is visitors using non-designated areas, resulting in environmental damage, is difficult to control. The best approach is to provide designated trails to features of interest (Cole, 1990b; Hammitt & Cole, 1998).

The safety of visitors and agency staff using roads and trails is a fundamental concern in managing natural areas. Safety can be enhanced by attention to road and trail design, construction and maintenance. Good design includes managing user speeds, sight distances, and road and trail surfaces and widths. Safety may be compromised by conflicts between users. Conflict within a single user group may occur, for example when passing areas on four-wheel-drive tracks are not provided. The different requirements of mountain bikers, hikers and horse riders, for example, may compromise user safety on multiple-use trails (North Carolina State University & Department of Parks, Recreation and Tourism Management, 1994). Providing crossings for wildlife, such as underpasses, protects them from vehicles as well as reducing accidents from drivers swerving to avoid them.

Where roads and trails are located (and not located) in natural areas is a powerful management tool. Road and trail location is an acceptable, unobtrusive way to influence visitor use. Areas deliberately kept free of roads and trails will remain little used. Trails can be built or rebuilt to be gentle or steep, short or long, navigationally easy or difficult. Each permutation will support different numbers and types of visitors. Roads and trails can take visitors past vistas or through monotonous landscapes, with them lingering at the former and passing rapidly through the latter. Trails can be routed through varied vegetation, to water points and into areas where there are increased chances of seeing wildlife. Trail design allows managers to programme the experience that visitors have (Lucas, 1990a).

Despite the usefulness of trail and road design as a management tool, its application for this purpose has been limited. Most roads and trails follow routes pushed through the landscape for fishing or hunting access, for the extraction of resources such as minerals or timber, or to reach other destinations. Many walk trails follow vehicle tracks constructed for fire management or the earliest routes followed by hikers, usually traversing the highest points in the landscape. Limited money is allocated by management agencies to trail construction and maintenance, with most going to relocating poorly located short sections of existing trails. Roads are relocated to reduce maintenance costs by moving them to a more durable part of the landscape, improve the available vistas and ensure the road can be safely travelled at designated speeds.

Visitor use of areas accessible by non-motorised means only (usually by foot) can also be strongly influenced by road locations. Use can be changed by closing, shortening, lengthening and upgrading access roads. Closing or shortening roads makes non-motorised access deep into the area less likely and may decrease day use. Reducing use can increase the opportunities for solitude. Improving or extending roads has the opposite effect. Upgrading and constructing new roads provides greater access and levels of use and makes day use more likely. Improving access should be undertaken with great care, as new problems due to increased numbers and types of visitors may be created (Lucas, 1990a).

In addition to the management of roads and trails, the facilities provided at access points can be used to influence how visitors use an area. For example, horse facilities such as loading ramps and corrals encourage horse riders. Where there are water bodies, launching ramps encourage use by larger trailer-based boats. Where visitors have to carry their boats from their motor vehicles to the water, canoe-sized vessels are more likely. The size of parking areas influences the number of users at one time, although people will park somewhere even if the area is full. However, parking areas should not be expanded where increased use is not wanted.

Managing roads and trails

Engineering is often an essential tool for managing roads and trails. In many instances it may not be possible to locate or relocate a linear development to improve its durability or the levels of use may be such that engineering actions become essential. Cole (1990b) listed erosion and muddiness as the common trail problems amenable to engineering solutions. Additional problems with roads in natural areas include safety concerns, such as poor visibility, tight corners and deterioration in surface conditions. Management becomes necessary when the safety of users or staff is jeopardised, the linear feature becomes difficult to use, maintenance becomes expensive, or the natural environment is being obviously damaged. Cole (1990b) advocated, where naturalness is the primary goal, keeping engineering to a minimum because of the intrusiveness of human structures.

To prevent erosion, walk trails can be aligned on the contour and built to slope slightly away from hillsides. They can have a rolling grade of dips and rises to prevent water building up speed and eroding the trail (Cole, 1990). Water bars are a common tool, comprising rocks, logs, boards or mounded soil, angled across the trail, usually at 20–40°, and securely anchored to prevent them being washed away. Steps, oriented perpendicular to the slope, can also be used to slow down water and soil movement. To prevent water getting onto trails and roads in the first place, earth or rock ditches, culverts of concrete, metal or other materials, or drains parallel to the linear feature are all engineering options. All this work must be carefully planned and maintained, otherwise it can be intrusive and/or ineffectual (Cole, 1990b; Hammitt & Cole, 1998).

Muddiness can be addressed by 'hardening' the site using decking/board walks, forming up using earth and gravel, or sealing using asphalt or bitumen. Board walks, slightly elevated across the landscape, are common in valley bottoms and in alpine areas subject to prolonged waterlogging (Figure 5.4). Raising the trail or road using earth and gravel is usually accompanied by some form of drainage to ensure the formed trail or road is not washed away. For high-use roads and even trails, forming up and permanent sealing with asphalt/bitumen, although expensive, may be the best way to provide a durable, low-maintenance surface. All of these approaches are intrusive and show clear evidence of humans in the landscape. Box 5.4 gives examples of trail management from England and Australia.

Geosynthetics can be employed to improve drainage, and hence reduce muddiness, and also to reduce the amount of fill material needed. These are usually sheet-like, synthetic materials that are laid across the ground, often with cells or gaps where soil

Figure 5.4 Walking track in Tasmanian Wilderness World Heritage site, Australia (Photo: Parks and Wildlife Service, Tasmania)

can accumulate. Such materials provide the soil with structure and support as well as facilitating drainage. Another stabilising approach is adding chemical binders to soils to increase their density, moisture resistance and stability (Marion & Leung, 2004a).

Another engineering solution for muddiness is bridge construction. Bridges can prevent bank erosion and also improve the safety of river crossings (Hammitt & Cole, 1998). The bridge may be nothing more than a series of stepping stones, or it may be a raised gravel walkway with culverts to shed the water downstream. A more elaborate approach is laying boards across a timber base. For major roads, steel and concrete fabrication will be required. In all cases, the structure must be sufficiently well anchored to prevent it washing away.

Signs are another built element associated with trails and roads. Directional, warning and interpretive signs are expected along major roads. Warning signs may be the only ones found on four-wheel-drive tracks. Visitors to wilderness areas in the USA and Canada preferred a few simple signs, whereas in an Australian study visitors were more tolerant of such structures (Morin *et al.*, 1997). Most managers

Box 5.4 Management to improve trail conditions: Examples from England and Australia

The Pennine Way, Peak District National Park, England

The Peak District National Park (Figure 5.1) through which the Pennine Way passes receives more than 10 million visitors a year (PDNPA, 2012). The Way is intensively used by long-distance walkers and day trippers. Prior to management intervention, the bare areas spanning the Way extended to 7.8 m, with trample widths approaching 70 m in places. To improve visitors' experiences as well as reducing trampling and disturbance by keeping people on the trail, part of the Way was resurfaced using flagstones. This management action reduced the number of walkers straying off the trail from 30% before resurfacing to 3.8%. This concentration of use on the paved Way means large areas of the adjacent moorland are relatively undisturbed (Pearce-Higgins & Yalden, 1997).

Walking track management, Tasmania, Australia

The Tasmanian Parks and Wildlife Service is responsible for managing over 1700 km of walking tracks in national parks and reserves in the state of Tasmania (Figure 5.1). Landscapes include alpine vegetation, especially low-growing buttongrass plains, temperate rainforest and eucalyptus forests. The alpine areas are popular for walking, while at the same time being vulnerable to damage because of water ponding and low-growing vegetation making it easy for visitors to stray off paths and create new ones. A dramatic increase in visitor numbers over the last three decades has led to escalating biophysical and social impacts. Issues of concern are the deterioration of existing tracks and the development of new ones in formerly trackless areas. The spread of the soil pathogen *Phytophthora cinnamomi* is also of concern.

A comprehensive assessment of the walking track system was completed in 2011, with the Limits of Acceptable Change system used as a framework for managing and monitoring track conditions. A Walking Track Management Strategy for Tasmania's National Parks and Reserves has been published (Dixon & Hawes, 2011). Strategies in it include stabilising walking tracks through installing new infrastructure (e.g. duckboarding, gravel surfacing, drainage works), realignment, encouraging walkers to disperse or 'fan out' and closures. Figure 5.4 illustrates the use and environmental benefits of duckboarding. Doing nothing is also included as an option. Wash-down stations and infrastructure isolating walkers from the ground, through elevated boardwalks and gravel-surfaced paths, are proposed to reduce the introduction and spread of *P. cinnamomi*.

The Strategy also relies on walker education to reduce impacts. Such education emphasises the traditional dos and don'ts of minimal-impact bush-walking: not lighting campfires, not blazing trees, carrying a fuel stove and disposing of faecal wastes hygienically. Track rangers will be advocates for these 'leave no trace' messages and enforce restrictions such as no fires. Integral to the Strategy is having a database with information on all the walking tracks and a regular monitoring programme to report on the status of the tracks, assign priorities for management and evaluate the success or otherwise of management strategies.

of wilderness areas try to provide only limited, simple directional signs at confusing intersections (Lucas, 1990a). Generally, the less developed an area, the fewer signs should be provided.

Managing informal trails

Informal trails will have an impact on the environment and their proliferation is a problem in many natural areas (Randall & Newsome, 2008; Wimpey & Marion, 2011). The strategies given in the previous sections apply to formal trails: those designed and managed for visitor use. However, they apply equally to informal trails, where visitors move 'off-trail' to access features of interest and take shortcuts, especially to access facilities such as toilets and look-outs.

With informal trails an important first choice is whether or not to proceed to formal trail development and management. Such a decision must be based on an assessment of supply (other trails and opportunities in the immediate vicinity and further afield) and current and future demand, plus the capability of the site to support likely future uses. If, after this assessment, the choice is made to turn an informal trail into a formal trail, careful consideration of its location (and possible relocation) must follow, along with adherence to the design strategies detailed above. Closing informal trails and associated restoration activities is also a management choice. Details regarding restoration, for both trails and facilities, are provided below.

Built accommodation, campgrounds and other facilities

Locating facilities

Built accommodation, campgrounds, campsites and other facilities are ideally located in durable parts of the landscape, in a similar manner to trails and roads. Durability is influenced by the soil's erodibility, drainage and depth, and the site's dominant vegetation type (Hammitt & Cole, 1998). Soils should be well drained loams. Deep soils are preferred because many human waste disposal systems require on-site disposal. Because regeneration of vegetation around built accommodation and heavily used campsites is unlikely, such sites should be located in stands of relatively young, long-lived trees or tall shrubs (Hammitt & Cole, 1998). Species known to shed limbs should be avoided because of the safety risks.

Another essential consideration in locating facilities is visual impact. Such facilities should blend with the landscape and, where possible, should not be visible from vista points, roads and trails. Building sizes and shapes should be in scale with, and borrow from, natural landforms, while materials should blend with the colours and textures of the natural environment. The area cleared during construction should also be kept to a minimum, as should changes to natural drainage patterns.

Leung and Marion (1999c) noted the importance of design in reducing visitor impacts. For example, campgrounds can be designed to prevent shortcuts between campsites and between sites and facilities such as toilets. They can also be designed to accommodate a variety of group sizes, so that large groups do not need to damage vegetation to fit in. All facilities should be designed to allow access by wheelchairs and for people with limited mobility where this is possible, given the topography of the site and the extent of development regarded as acceptable.

Managing facilities

Site management concentrates use, by providing facilities wanted by visitors. Such facilities include day-use areas, campgrounds and campsites, shelters, huts, horse camps, interpretive centres, lodges and resorts. Although concentrating use significantly impacts on the site itself, surrounding the natural areas receive greater protection (Marion & Leung, 2004b). A plethora of accommodation types is now offered in natural areas, ranging from resorts and lodges through to unmarked campsites in wilderness areas. For example, hotels are provided in Grand Canyon National Park, USA, rest-camps such as Olifants and Lower Sabie in Kruger National Park, South Africa, wilderness cabins in Cradle Mountain–Lake St Clare National Park, Tasmania, tent platforms in Great Smoky Mountains National Park, USA, and unmarked campsites in Kakadu National Park, Australia (Figure 5.1). Suggestions for managing built accommodation, such as resorts, are given later (see Box 5.11).

The most common form of site management in non-wilderness areas is hardening using gravel, paving or asphalt/bitumen and channelling use into these hardened areas. Surfacing facility areas and associated trails minimises muddy areas and soil compaction. It can also be used to improve site durability around built accommodation, interpretive facilities and across campgrounds.

Managing vegetation to prevent site deterioration from trampling, erosion and muddiness, to maintain the site's visual attraction, to provide a visual and sound buffer between different activities and protection from the weather, and for educational purposes is also common (Van Riet & Cooks, 1990a). Vegetation management at sites other than the most pristine can include introducing hardy vegetation, over-storey thinning, watering and fertilising (Hammitt & Cole, 1998). The most common example of introduced hardy vegetation is lawns. Many picnic areas use lawn to increase durability and it often provides the surrounds for built accommodation. In many parts of the world it also provides a food source for grazing animals, such as deer in the USA and kangaroos in Australia. These animals appeal to natural area tourists. Trees may be planted to provide shade and windbreaks for built accommodation, campgrounds and day-use areas.

Thinning over-storey trees encourages more vigorous growth of ground covers, especially grasses (Hammitt & Cole, 1998). Grasses are more resistant to impacts than other plant types and hence their presence increases the robustness of a site. In the USA, a study of campgrounds in the southern Appalachians found that reducing canopy cover from 90% to 60% doubled grass cover and reducing the canopy cover to 30% more than tripled grass cover (Cordell *et al.*, 1974).

Watering and fertilising can be important if lawns or shade/screening vegetation are planted in dry climates. Either flood or sprinkler-based systems can be used. Flood irrigation has been used to maintain trees and shrubs in the developed campgrounds at the bottom of the Grand Canyon (Hammitt & Cole, 1998). A permanent irrigation system is required to ensure the endemic plants survive at Berg-en-Dal, a rest-camp in Kruger National Park (Van Riet & Cooks, 1990a) (Figure 5.1).

Box 5.5 Deciding when to install toilets in natural areas

Wilderness areas are often appreciated for their absence of facilities, including having no toilets. In these circumstances visitors are encouraged to practise minimal-impact camping, which requires them to walk at least 100 m away from campsites and water sources to defaecate. Material should then be buried in a hole 15 cm deep and covered with soil.

At the Tasmanian World Heritage site, recent observations suggest that visitors are not complying with these minimal-impact guidelines, with faecal deposits recorded within 100 m of huts and the Overland Track. Bridle *et al.* (2007) linked poorly buried deposits to increased faecal bacteria being detected in a nearby alpine lake. They suggest a 'green–yellow–red light' management system, based on the Limits of Acceptable Change. For green, no action is needed, as there are fewer than two deposits within a 20 m radius of a hut or campsite. A yellow light is a warning to instigate a monitoring programme because there are two to five deposits within the monitoring area. A red light signifies the need for immediate action, because there are more than five deposits. Usually the result of a red light assessment is installation of a toilet or a significant upgrade of an existing one. A yellow light assessment is most likely to be followed by education rather than construction of a new or upgraded facility. This assessment approach and strategy provides a socially acceptable way of dealing with toilet waste across different zones and parks (Bridle *et al.*, 2007).

The value of fertilising depends on the soil conditions. Soil testing is essential to determine what nutrients are limiting. It is critical not to over-apply phosphorus or nitrogen, as these can leach out of the soil and lead to nutrient enrichment of nearby waters. Although watering and fertilising individually lead to increases in ground cover, when both are applied the results may prove twice as effective as either one by itself. Both approaches are effective only when combined with careful site design and hardening of higher-use trails and facilities (Hammitt & Cole, 1998).

Other elements of sites that can be managed to protect the natural environment are toilets, fireplaces and rubbish disposal. All three concentrate use where the facilities are located, with the level of facilities provided depending on visitor numbers and the experience visitors are seeking. Research several decades ago in wilderness areas in the USA found that visitors prefer not to see or have built toilets, as they are regarded as human intrusions (Lucas, 1990a). Toilets are standard, however, in developed natural areas and increasingly common in heavily used wilderness areas. As such, it is important to be able to determine when toilets should be installed (see Box 5.5).

For built accommodation such as resorts and lodges, a permanent, enclosed human waste disposal system is essential. In areas where water is scarce or leach fields and ponds are environmentally unacceptable, waterless systems are a possibility. For example, the Fitzroy Falls Visitor Centre in New South Wales, Australia (Figure 5.1) has a large conventional composting toilet facility that caters for about 5000 visitors per week; the toilets are pumped out annually. Waterless septic systems can also be used where water is scarce. Water-flushed septic systems are more common, however,

Table 5.4 Methods for managing human wastes in natural areas

Method	*Level of visitor use*	*Visitor acceptability*
No facilities - individual's responsibility		
Buried in a shallow hole (American 'cat hole') dug by the visitor, at least 60 m from water	Very low; dispersed recreation areas with light use and soil cover	Good with educated users
Carried out in bag or container	Very low; river-rafting and climbing settings	Compliance can be difficult
Direct ocean disposal	Very low; infrequently visited areas	Fair with educated users
Pit toilet (Australian 'long-drop dunny')		
Hole at least several metres deep, covered by a toilet seat and structure. Pit is covered with dirt and the structure moved when the hole is full	Low; available sites may be limited by water tables close to the surface and shallow soils; can fill rapidly	Moderate, although flies and smell may be a concern to users
Drum toilet		
Toilet seat and structure placed over a transportable, replaceable drum	Low; as use levels increase the frequency of drum replacement increases; suitable where soils, water tables or risks of water contamination preclude using pit toilets	Moderate, although use of vehicles for maintenance may not be accepted by users in wilderness areas
Composting toilet		
Waste decomposes in digester tank to which a carbon source such as paper, straw or wood chips must be regularly added; waste can be reduced in volume by as much as 80%	Moderate; requires frequent visits (at least 1/week) and regular maintenance, so not suited to remote locations	High
Waterless septic system		
Waste drops into tank and moves through one or more tanks	Moderate to high; requires regular maintenance; may require electricity to pump waste through the treatment system	High, where leach fields and/or transpiration beds are acceptable
Flushing septic system		
Material is flushed into septic system which is pumped regularly into a leach field/pond	High; requires maintenance and regular pumping of septic system and hence access to power source	High, where plumbing systems and leach fields/ ponds are acceptable
Deep sewerage/on-site treatment plant		
Material is flushed into sewerage system and is then treated on- or off-site	High; requires significant investment in infrastructure	High

Derived from Land (1995, in Hammitt & Cole, 1998), Cilimburg *et al.* (2000), Western Australian Department of Environment and Conservation (personal communication, 2012)

given many visitors' preference for a flushing system. Larger resorts and other very high concentrations of visitors require on-site treatment plants or access to deep sewerage. Table 5.4 lists alternative methods of managing this form of human waste, associated levels of use and visitor acceptability.

Enjoying a campfire in natural areas is a contested practice. Campfires are integral to many people's overnight experience. There are safety concerns, however, regarding fires escaping and becoming wildfires, as well as regarding the ecological and aesthetic effects of fuel collection and burning (Smith *et al.*, 2012). Solutions include not allowing campfires in some areas, using loose rock rings that are pulled apart following use, providing wood for established fireplaces, and gas-fired or electric designated fireplaces/cookers. As with other facilities, the choice depends on levels of use, maintenance requirements, visitor preferences and environmental consequences. For example, gas or electric cookers are most likely to be found where there are high levels of use, regular maintenance is possible and visitors are comfortable with built structures at the site. At the more primitive end of the recreation spectrum, a 'no campfires' approach seems more likely where signs of human intrusion are unwanted and managers aim to minimise the ecological effects of human use.

The disposal of solid rubbish and littering are both concerns. In wilderness areas, visitors are expected to pack up their rubbish and dispose of it elsewhere. Many national parks now have bin-free sites, instead providing rubbish disposal facilities at a central location. This approach reduces the amount of time staff devote to rubbish collection, freeing them for other activities. For built accommodation and visitor centres with large numbers of visitors, often over extended periods, waste management must be multi-pronged. Purchasing goods with little packaging and recycling are two waste-minimisation strategies. Ideally, solid waste can be disposed of outside the protected area. If this is not possible, the site should be located to avoid contaminating surface or ground waters and managed to prevent scavenging by wildlife.

Riverbanks, lakes and coastlines

Water bodies such as lakes, rivers and the sea hold a deep attraction for humans. Visitors enjoy vistas that include water and prefer to picnic and camp near water. Water bodies also have functional uses for visitors, as sources of drinking water and for washing and swimming. However, 30% of all wilderness areas in the USA prohibit camping close to streams and lakes (Washburne & Cole, 1983): setbacks in wilderness and back-country range from 2 m to 1 km, the most common being 30 m (Washburne & Cole, 1983; Marion *et al.*, 1993; Leung & Marion, 2004b). Setbacks reduce the risk of pollutants moving from the site into the water, protect fringing vegetation from trampling and other damage, and prevent erosion of steep banks and edges. There are also social reasons for setbacks, especially in relation to camping. Camping on shorelines effectively means that those who do not get one of these premier sites may not be able to access the water because to do so entails walking through someone else's camp. Also, campsites on shorelines are highly visible, decreasing perceptions of naturalness for other visitors. Because the social reasons are more compelling

Box 5.6 Locating and managing facilities in coastal settings: Fraser Island World Heritage site, Australia

Fraser Island, located off the coast of south-east Queensland (Figure 5.1), is the world's largest sand barrier island. Connected to the mainland by ferry, it provides some of the best coastal four-wheel-driving and camping opportunities in Australia. More than 90,000 people camp on the island each year, many in the dunes behind the beaches (Thompson & Schlacher, 2008). Camping 'zones' encompass 23% of the island's dunes. Thompson and Schlacher (2008) found that vehicle tracks have destroyed 20% of the dune front in these zones. Shoreline retreat and scalloping of the shoreline are also associated with these tracks.

Management actions that have improved the island's coastal zone are improved signage, closure and rehabilitation of sites, and public education. Rationalisation and rehabilitation of trails have also been recommended (Hockings & Twyford, 1997; Thompson & Schlacher, 2008). Confining and concentrating camping in fore-dunes and swales are also potentially part of the solution (Hockings & Twyford, 1997).

than the ecological ones, setbacks are more appropriate in high- rather than low-use natural areas (Hammitt & Cole, 1998). Setbacks stop visitors camping or visiting where they most want to. As such, they should be undertaken only when necessary for ecological and/or social reasons. Ecological outcomes can often be accomplished through education about not damaging shorelines or polluting waters.

Several other strategies can also be employed to reduce the movement of pollutants into water bodies. Maintaining or planting fringing vegetation between facilities and water bodies can filter out nutrients and pesticides. Locating toilets away from water bodies or ensuring that a closed system is installed, where warranted by visitor numbers, also minimises the flow of nutrients and pathogens into the water body. Hughes *et al.* (2008), in their review of the effects of recreation on the quality of drinking water, note that contamination of natural surface water bodies is most likely to be linked to the presence of animals, residential areas, or agricultural, mining and industrial activities, with recreation posing a very minor risk relative to these other factors.

Given the mobility of dunes and active erosive processes of coastlines, facilities should be located where they are not susceptible to wave or wind erosion or sand inundation (Oma *et al.*, 1992). Location should be where some protection from the wind is achieved – in low areas such as dune swales and deflation areas and next to vegetation if available (Box 5.6). Planting shrubs or trees, preferably those native to the area, is also a way of achieving some protection at sites. Plantings are not appropriate at primitive or semi-primitive sites where signs of humans are undesirable. Raised walkways are often used in coastal areas to enable visitors to access beaches (Hammitt & Cole, 1998). Plants can still grow underneath; however, on the downside, structures are often damaged or removed by storms.

Site restoration

Some sites do not have the capability to support visitors, while others need temporary closure to allow regeneration. Sites may be permanently closed to ensure setbacks from water bodies, because they are poorly located or to reduce the number of sites. Washburne and Cole (1993) and Marion *et al.* (1993) noted that a total of 37% of wilderness areas in the USA and the same percentage of national parks with back-country have closed campsites, with 16% and 27% respectively having active revegetation programmes. Another reason for site closure is to protect rare or vulnerable plants or animals. Hammitt and Cole (1998) use the example of the endangered flowering plant the sentry milk vetch (*Astragalus cremnophylax cremnophylax*) at Grand Canyon National Park (Figure 5.1). Managers re-routed walk trails and placed a fence around the population because of concerns regarding its viability.

Sites may be temporarily closed to allow them to recover. Temporary campsite closures have been in both developed and primitive parts of natural areas. For this approach to work, managers need to know the relationship between the time it takes for impacts to occur, the threshold beyond which they become 'unacceptable' and the required recovery period. Also crucial is realising that most impact occurs in the first few years a site is used. Impacts at wilderness campsites, canoe-accessed campsites and car camping sites increase dramatically for the first year or two and then level out (Hammitt & Cole, 1998). As such, a two-year use period followed by a recovery period of uncertain length has been suggested. The length of the recovery period depends on moisture availability and growing season length, with active revegetation helping shorten this period. Rotation of recovery areas is not generally recommended but, if used, works best in resilient environments where active revegetation is feasible.

Several general comments can be made about restoration actions, but site-specific approaches are usually required (Cole, 1990b). The first crucial step is effectively closing the site to all use. Barriers to access and information often work. All evidence of human use should be removed – fire rings, site furniture and litter. The site may then be left to regenerate naturally. Alternatively, active restoration through scarifying and/or direct seeding, planting seedlings, compost amendments or transplanting from adjacent areas can be pursued (Marion & Leung, 2004b). Locally collected and propagated material should be used. At Olympic National Park in the USA, for example, Scott (1998) referred to approximately 25,000 plants having been propagated annually for restoration projects. Rocks were dug in like 'icebergs' and dead wood placed vertically in closed areas to make the site unappealing and uncomfortable-looking.

Visitor Management Actions

Visitor management seeks to influence the amount, type, timing and distribution of use as well as visitor behaviour. Actions include regulating visitor numbers, group size and length of stay, using deterrence and enforcement, communicating with visitors and providing education (Figure 5.3). Charging fees as a means of regulating numbers is also discussed here.

Regulating visitors

For many years, managers have been encouraged to use light-handed management approaches, such as communicating with visitors and education programmes, rather than restricting visitor numbers or activities (Shindler & Shelby, 1993). These approaches were assumed to work and to be preferred by visitors. Minimal regulation was regarded as essential to satisfactory experiences by visitors to natural areas, especially more primitive places (Hendee *et al.*, 1990a). Cole (1990b) has suggested that effectiveness should be a primary consideration in selecting actions. He noted that education and restoration efforts at campsites have been ineffective in many places. Regulation of numbers, regarded by managers as a last resort, may be more effective and is better initiated earlier rather than later – taking regulatory action only after significant physical damage has occurred is like relying on bandages until the situation becomes so bad that surgery is required (Cole, 1993).

Researchers have questioned the assumption that visitors prefer education and communication efforts and will be opposed to more restrictive approaches, such as limiting numbers or activities within natural areas. Increases in ecological damage and visitor encounters at popular sites have increased visitor support for more direct approaches, such as use limits. Shindler and Shelby's (1993) study of frequent visitors to three wilderness areas in Oregon found that setting limits on the number of users was generally supported. Studies of visitors to a Nuyts wilderness area in Western Australia (Morin *et al.*, 1997) and Bako National Park, Borneo (Chin *et al.*, 2000), identified support for limiting overall numbers of visitors, the number of people per group and length of stay. Manning (2011) notes that ultimately the choice of management strategy, whether it is regulation or education or both, depends on the management problem and its context.

Visitor numbers

Although there is evidence that regulating access is supported by visitors, such an approach conflicts with one of the central objectives of management for most natural areas – that of providing opportunities for visitors. Limiting access to protected areas is one of the most controversial aspects of management (Eagles & McCool, 2002). As such, other options should be considered first. Also, because the relationship between amount of use and impact is not linear, reducing use may not necessarily reduce impacts. In many situations, a little use causes considerable impact and further increases in use levels have less and less additional effect on the natural environment (Hammitt & Cole, 1998) (Chapter 3).

Most of the controversy associated with use limits is centred on determining how and when they should be implemented. Much of the debate concerns whether empirical data can be directly translated into use limits. Use limits are actually subjective judgements made by managers and should be based on two factors: stakeholders' perceptions of impacts and scientists' understanding of the ecological impacts (Cole *et al.*, 1997). The planning frameworks outlined in Chapter 4 provide the best way of determining the levels of impact acceptable to stakeholders. Cole and Landres (1996) suggested that ecological impacts are a function of the impact's intensity, its areal extent and the rarity or irreplaceability of the attributes being affected.

Scientists can then assess the relationships between amount of use and impacts, and the maximum levels of use that can be supported without exceeding the acceptable level of impacts can be determined. Once maximum levels have been determined, simulation models and computer programs can be used to set entry limits for individual trail-heads. More sophisticated programs allow travel routes to be determined, linking the availability of a number of sites with a visitor's route preference (Hammitt & Cole, 1998).

Other approaches have been summarised by Hammitt and Cole (1998). In Yosemite National Park (Figure 5.1) use limits are based on the number of acres in a zone, the miles of trails it contains and its ecological fragility, based on ecological rarity, vulnerability, recuperability and repairability (van Wagtendonk, 1986). In nearby Sequoia and Kings Canyon National Parks, existing campsites have been assessed to determine if they are acceptable or unacceptable. Unacceptable ones were those within 8 m of water or 30 m of another heavily impacted site. The total number of acceptable sites was used to determine the maximum number of groups to be permitted at any one time (Parsons, 1986).

Hammitt and Cole (1998) and Cole (1990b) suggested there is little point in restricting visitor numbers in high-use areas unless it is accompanied by confining use to certain sites. Otherwise, the reduced numbers of visitors will continue using all available sites, none of which will recover. Reduced numbers will improve the quality of the experience, by being less crowded, but there will be no ecological benefits, as all sites will still be impacted. Cole *et al.* (1997) cautioned that reducing use levels will deny access to many people as well as increase visitor impacts in nearby natural areas. At higher levels of use, only large changes in visitor numbers have an effect on impact levels.

Table 5.5 Ways of allocating visitor access to natural areas

Allocation system	*Equity outcome*	*Visitor acceptability*
Advance reservation	Benefits those able to plan ahead	Generally high
Queuing/first-come first-served	Favours those with lots of time and who live nearby	Low to moderate
Lottery	No group obviously benefited or disadvantaged	Low
Fees	Favours those able to pay	Low to moderate
Eligibility requirements	Favours those with time (and money) to meet requirements	Not known

Derived from Stankey and Baden (1977, cited in Hammitt & Cole, 1998), Manning (2004, 2011)

In lightly used areas, given that at low use levels differences in amount of use can have significant effects on the amount of impact, use limits can contribute substantially to keeping impact levels low. Use needs to be kept low at all sites, with visitors avoiding fragile sites and not undertaking destructive behaviour. To encourage these types of suitable behaviour, use limits in such areas need to be accompanied by communication and education programmes on low-impact use.

Having decided to limit visitor numbers, the issue then becomes one of equity and allocation. With restrictions, some people get to visit the area and others do not. Those permitted to enter can enjoy greater solitude. Those who are excluded do not get to enjoy the natural area (Lucas, 1990a). Table 5.5 summarises ways of allocating access, who benefits and acceptability to visitors. Access is usually allocated via a permit. A mixture of approaches may be used to manage access to a protected area. For example, as described by Lucas (1990a), for the seven allowed float trips per day down the Middle Fork of the Salmon River, Idaho (Frank Church–River of No Return Wilderness), three permits are allocated to commercial guides and four are available to private parties through a lottery. Support for a particular allocation system appears to be related to what visitors are most familiar with and which one they think will give them the best chance of gaining access (Manning, 2011).

Fees are generally used to raise revenue, although they are also a means of rationing use. Many protected areas have entrance fees. Ideally, fees would encourage those

Table 5.6 Nature and extent of restrictions on visitor use in natural areas

Restriction method	*Description*
Limit entry to an area	May apply to day-use or overnight visitors, more often the latter
Whole area	Number of visitors to the whole area is regulated. Applications include the number of parties floating along a wild river, number of visitors entering all trail-heads, number of groups/individuals camping overnight
Entry points – all or specified ones	Use managed through individual trail-head quotas, with visitors free to travel and camp where they want once they have entered
Limit activities once in the area:	Most likely to apply to overnight visitors
Campsites/zones specified	Visitors must indicate where they intend to camp each night – either a site or within a specified area (e.g. travel zone). There may be restrictions on how long they can stay at one site/within one zone
Travel routes specified	Permits may be issued for itineraries linking campsites, rather than for individual campsites in isolation. This allows itineraries to be adjusted and alternative routes selected if space is not available at sites on the preferred route

Derived from Lucas (1990a), Hammitt and Cole (1998)

who place a low value on protected areas to go elsewhere. Unfortunately, however, those who value these areas but have low incomes may also be discouraged. Variable fees have been widely discussed, with higher fees for heavily used areas and lower or no fees for seldom-used sites. The hope with this approach is that higher fees would discourage use and vice versa.

Park visitors, including those visiting wilderness areas, generally support fees. Visitors are most supportive when the collected revenues are reinvested in recreation facilities and services (Manning, 2004). Among wilderness users, support is strongest when the fees are used to restore human-damaged sites, remove litter and provide information (Vogt & Williams, 1999). These users preferred their fees to go to maintenance of wilderness conditions rather than to developing new facilities and services.

Visitor numbers can be managed in other ways in addition to simply regulating the numbers entering an area (Table 5.6). In many places, overnight use is limited but day use is not. Marion *et al.* (1993) noted that in two-thirds of US national parks that limit back-country use, limits apply only to overnight users. Also, overnight limits may be set in different ways. The number of visitors entering an area may be restricted, but once in they are free to travel and stay where they want. Alternatively,

Box 5.7 Regulating visitor use: Michaelmas Cay and Reef, Great Barrier Reef Marine Park, Australia

The Cay, with an area of 1.8 ha, provides a breeding site for up to nine species of bird and is one of the most important seabird breeding sites for the Great Barrier Reef (Figure 5.1). It is also a popular and accessible destination for tourist boats, both commercial and private, operating from Cairns, a major tourism centre in north Queensland. Six tour operators with combined visitor and crew numbers of 586 per day are permitted to use the Cay, although actual daily visitor numbers rarely exceed half this permitted figure. Total visitor numbers per annum are 90,000.

Management strategies include:

- limits to the number of moorings (two moorings on the reef and 19 associated with the Cay);
- installed moorings to be used where available (if not, anchor in sand or mud rather than coral);
- limit of one vessel booking per company per day;
- vessels over 35 metres not permitted to anchor in the locality (i.e. within one nautical mile radius of the Cay);
- vessel speeds restricted to six knots in the locality and four knots in the access channel to the Cay;
- no aircraft in the locality and no use of horns or sirens;
- access onto the Cay allowed only within a roped-off area between 9.30 am and 3.00 pm;
- limit of 50 people on the Cay at one time.

Derived from AG GBRPA (2008), QG (2012), Anon. (undated), Queensland Department of Environment and Resource Management (personal communication, 2012) where they go within an area may be regulated by permit (e.g. to camp at certain sites on specified nights) (Hammitt & Cole, 1998). Stewart (1989) argued that trail-head entry quotas may be as efficient, simpler to implement and found by visitors to be easier to apply for and comply with than travel route and campsite quotas.

Restrictions can be developed and applied in various, apparently endless combinations. Tables 5.5 and 5.6 summarise different allocation systems and their varying nature and extent. Other elements that can be manipulated include the time of year when an area is available or rationing is in place, group size (see below), cost of an access permit, length of stay at one site or within one travel zone (see below), use of campfires, and number of access permits issued per person per season (Lucas, 1990a). Box 5.7 provides an example of combined restrictions regulating visitor numbers and use.

Completely excluding or spatially separating different types of visitor use are other ways managers seek to control impacts. Horses are excluded from 14% of wilderness areas in the USA (McClaran & Cole, 1993) and many Australian national parks. Spatial separation can be achieved by designating trails for different kinds of use. For example, at Yellowstone National Park (Figure 5.1), snowmobilers are restricted to roadways while cross-country skiers may go off-road. In the USA, 12% of national parks designate different trail uses. Managers have also segregated campsites by type of use, with separate sites for general visitors, groups, stock users, commercial outfitters and even llama users (Leung & Marion, 1999c).

Visitor group size

Common sense and some research suggest that limiting the size of groups visiting wilderness areas has ecological and social benefits (Lucas, 1990a; Hammitt & Cole, 1998). Larger groups occupy and impact larger areas than a small group, especially in less developed sites or where sites are designed for small groups. Most visitors to wilderness areas in the USA support limits on group size (Cole, 1990b). Maximum specified group size, regulated in almost half the wilderness areas in the USA, has ranged from 5 to 60, with a median size of 15 and most common limit of 25 (Washburne & Cole, 1983). Lucas (1990a) suggested a group size of 6–12 is reasonable, while Hammitt and Cole (1998) noted that visitors preferred groups of 10 or less. An Australian study in Nuyts wilderness in Western Australia (Morin *et al.*, 1997) showed that visitors preferred groups of six people or less.

Limits on group size are most effective in lightly used natural areas, such as wilderness, where use levels are low and camping is dispersed (Cole, 1990). In more developed natural areas, catering for larger groups may be possible and desirable. Picnic areas and beach access points, for example, need to provide for larger groups. With careful site planning some geological features, such as the Pinnacles in south-western Australia (Figure 5.1), may be able to cater for hundreds of visitors per day. Wildlife viewing opportunities, such as whale watching from built platforms or headlands, may also be able to cater for larger groups so long as the viewing structures are designed with larger numbers in mind.

In many natural areas, limits have been placed on the number of horses, and other pack stock such as mules and llamas, in a group. Horses are used in many wilderness areas in the USA and in some protected areas in other parts of the world. Research is confirming that pack stock have greater impacts on natural areas than hikers (De Luca *et al.*, 1998) and they cause considerable damage to natural areas (Newsome *et al.*, 2008). Stock limits for US wilderness areas have ranged from 5 to 50 head (Washburne & Cole, 1983; McClaran & Cole, 1993), with a median of 10 (Marion *et al.*, 1993). Hammitt and Cole (1998) suggested that one approach could be to set composite limits in the form of the total number of 'bodies' (i.e. humans and animals combined), say to 15 (e.g. 6 humans and 9 horses).

Visitor length of stay

Length of stay generally does not contribute significantly to overuse (Lucas, 1990a). Additionally, in heavily used popular areas limiting length of stay is unlikely to reduce impacts (Hammitt & Cole, 1998). However, such limits at popular sites may allow more people to use them, while maintaining the existing levels of use and associated impacts (Cole, 1990b). In lightly used areas, length-of-stay limits can reduce ecological impacts. Hammitt and Cole (1998) suggested visitors should stay no more than a night or two at an individual site in remote areas. Parks in a number of countries have length-of-stay limits, most often 14 days (e.g. Canadian provincial parks, US national parks and national wilderness areas). For many areas (e.g. US national parks) stay limits vary through out the year; for example, for Yosemite National Park there is a 30-day limit in a calendar year, but this reduces to 14 days over summer (NPS, 2012).

Deterrence and enforcement

Field staff, including protected area rangers, have a multitude of roles, one of which is deterring inappropriate behaviour and if necessary enforcing the law. A study of visitor behaviour at Mount Rainier in the USA (Figure 5.1) showed that the main deterrent to visitors wandering off trails was the presence of a uniformed employee. Most people knew what they were supposed to do, but chose to do otherwise unless regulated (Swearingen & Johnson, 1995). Visitor surveys indicate rangers are well accepted by wilderness users (Lucas, 1990a).

Visitor communication and education

Education is regarded as crucial to reducing impacts by visitors to natural areas, for all sites, from primitive through to the most developed (Cole, 1990b; Lucas, 1990a; Hammitt & Cole, 1998; Marion & Reid, 2007). It is particularly important in addressing illegal, careless, unskilled and uninformed actions (Marion & Reid, 2007). Education is a favoured and widely accepted management approach because it does not overtly regulate or seek to control visitors directly (Eagles & McCool, 2002). Visitors retain the freedom to choose, plus receive information that potentially makes their experience more rewarding. More than half of US wilderness areas have an educational programme (Washburne & Cole, 1983), while Marion *et al.* (1993) noted that 91% of back-country areas in the national park system educated visitors about 'pack it in, pack it out'. Given the importance of education, communication and the associated process of interpretation, Chapter 6 is devoted to the latter and addresses in detail how education and interpretation enrich the visitor experience, in addition to reducing impacts.

Visitors to natural areas provide a good audience for communication/education programmes (Lucas, 1990a). Most studies agree that wilderness users generally have high education levels (Morin *et al.*, 1997). Lucas (1990b) noted that 60–85% of visitors to US wilderness areas have some form of tertiary education, while Morin *et al.* (1997) noted the same for 70% of those visiting an Australian wilderness (Nuyts Wilderness in south-western Australia).

The effectiveness of providing education and information compared with other actions is poorly known, although Cole (1995) noted they have been preferentially favoured because of their palatability to visitors. He noted that managers believe they should try education and communication before restricting access, even though the comparative efficacy is unknown. Also, research has tended to focus on attitudinal rather than behavioural change, which makes it difficult to draw conclusions about the effectiveness of interpretation in promoting more sustainable visitor behaviour (Munro *et al.*, 2008). Education efforts at campsites to address deteriorating conditions have been found ineffective in many places. Also, research shows that regulatory actions, such as site closures and restricting visitor numbers, are generally supported by visitors (Shindler & Shelby, 1993; Watson & Niccolucci, 1995; Morin *et al.*, 1997; Chin *et al.*, 2000).

The values of communication and education in impact management are fourfold: they support other, more direct actions such as restricting access; they can be applied from the most primitive to most developed settings; they enable managers to start being proactive rather than reactive; and visitors have the opportunity to make informed choices. Education should not be expected to solve problems in the short term (Cole, 1995). Rather, specific problems may require immediate, direct responses while education is used as a longer-term, complementary strategy.

Communication and education can be used to reduce impacts by redistributing use (Lucas, 1990a). Redistribution may be within or to natural areas outside the area of concern. If managers can provide descriptive materials on a range of sites then they can make sure that particular sites are not overused. Visitors can also select the sites most closely matching their needs. Use redistribution is most likely to occur when visitors have access to pre-visit trip information (Roggenbuck & Lucas, 1987).

The other main use of communication and education in reducing impacts is through encouraging minimum-impact use of natural areas. In the USA, most wilderness areas have minimum-impact education programmes. These are provided through schools and colleges, at wilderness access points and on wilderness trails and campsites. Educational materials include brochures, staff in agency offices, agency-run community education programmes, maps, signs, field staff such as rangers in the back-country and trail-head displays. Most of these materials focus on resource impacts but some relate to effects on other visitors' experiences (Lucas, 1990a).

The international Leave No Trace (LNT) programme aims to help visitors to natural areas reduce their impacts through seven guiding principles (Table 5.7). The

Table 5.7 Principles of the Leave No Trace low-impact education programme

Principle	*Description*
(1) Plan ahead and prepare	Know the regulations and special concerns for the area being visited; travel in small groups, with appropriate equipment
(2) Travel and camp on durable surfaces	In popular areas, concentrate use on existing trails and campsites; in pristine areas, disperse use and impacts; avoid places where impacts are beginning to be evident
(3) Dispose of waste properly	'Pack it in, pack it out'; dispose of human waste in toilets or by burial; wash yourself and dishes at least 60 m away from water bodies
(4) Leave what you find	Preserve the past: examine but do not touch; avoid tree damage, moving soil or plants, building structures or digging trenches
(5) Minimise campfire impacts	Cook on stoves, with minimise use of, or do not use, campfires
(6) Respect wildlife	Avoid scaring or harassing wildlife
(7) Be considerate of other visitors	Respect solitude experiences being sought by others

Derived from LNT (2012)

principles encourage forward planning to reduce impacts, behaviours such as 'pack it in, pack it out' for waste and being considerate of other users. It is an educational programme rather than a prescriptive set of rules (LNT, 2012). Tread Lightly (TL) is another international programme with an interest in reducing the impacts of recreational use of natural areas. Its focus is the use of motorised and mechanised vehicles. Principles include travelling responsibly, respecting the rights of others, educating yourself, avoiding sensitive areas and doing your part (i.e. protecting and restoring the natural environment) (TL, 2012).

Choosing Management Actions

Making the choice

A number of factors influence the selection of management actions by managers (Figure 5.5). If an unacceptable impact is identified, it is essential to determine whether visitors were the cause. Not all impacts in natural areas are due to visitors. Fluctuations in wildlife populations, for example in an elk population, may be due to variations in food availability, predation patterns or disease. Impacts can also result from outside influences (Buckley, 2004b). Poor water quality, for example, may be due to activities upstream of an area rather than within it. Although a complete understanding of the underlying causes may not be possible, if a relationship exists between visitor use and impacts it can usually be identified. Having determined

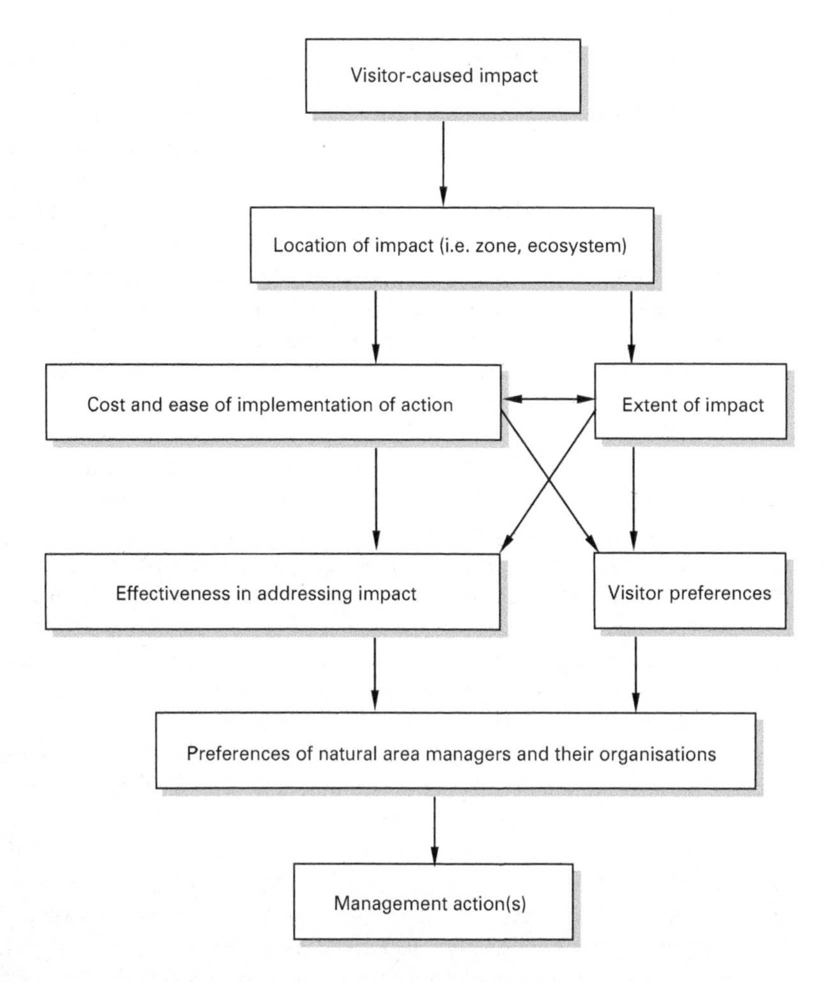

Figure 5.5 Factors influencing the choice of management actions by managers

visitors as the source of the impact, a wide range of actions is available, as outlined in the preceding sections.

The location and extent of impacts influence the choice of management actions. At the 'primitive' end of the spectrum – in wilderness – engineering, extensive environmental modification and even restrictions on where visitors can go are undesirable. The wilderness experience is based on freedom, solitude and little evidence of humans. The available actions are then limited to restricting the number of visitors, and communication and education. Towards the 'developed' end of the spectrum, site engineering, buildings and extensive environmental modifications are more appropriate. Here, communication and education are still useful. For example, managers of a lodge may rely on extensive engineering and site management as well as communicating with visitors about protecting wildlife and waste minimisation.

The extent of impacts also influences the actions chosen. Cole (1995) suggested that where impact problems are widespread at particular locations or require rapid attention and amelioration, then direct actions are needed, such as regulating visitor numbers and where they go. For example, a campsite denuded of vegetation, increasing in size and showing signs of erosion requires direct intervention as well as visitor education. Interventions could include site closure, redesign or hardening, and limiting visitor numbers. Cost and ease of implementation are crucial considerations. Natural area managers worldwide face declining budgets and increasing visitor numbers. Management actions must be cost-effective, in terms of both the initial action taken and the associated maintenance required. Engineering solutions, such as bridges and board walks, are expensive to construct and maintain. Also, work requiring the movement of materials into or out of remote locations is expensive and time-consuming. In the USA, helicopters are routinely used to service cabins and toilets in wilderness areas, as well as moving trail-building materials. The associated cost would be prohibitive for natural area managers in many other countries.

The ease of implementation is influenced by both the initial and the ongoing time commitment managers must make. Introducing a permit system, for example, necessitates time spent informing visitors it is in place and for its administration. Site engineering requires design skills as well as staff to do the construction work. Communication and education require an initial commitment of time; however, once materials are prepared little ongoing input from managers is necessary.

There is little point in undertaking management unless it is effective in ameliorating the impact. Surveys of wilderness area managers in the USA by Marion *et al.* (1993) and Manning *et al.* (1996a) showed that site design, management and facilities provision (e.g. designating campsites, using a formal trail system and plan, providing toilets), as well as regulating use (e.g. limiting group sizes, implementing quotas), are regarded by managers as highly effective practices. Moore and Walker (2008), through their survey of visitors and managers at two national parks in Western Australia, identified site management as effective in reducing visitor impacts, whereas the effect of providing information was less certain. Hughes and Morrison-Saunders (2005), in another study of visitors to national parks in Western Australia, noted that site design can have more influence on visitor attitudes than on-site interpretive materials. In marine environments, the location of access roads and boat-launching facilities can be just as influential in guiding human use as regulation, communication and education programmes (Smallwood *et al.*, 2012).

Generally, restricting and rationing use is regarded as more effective than information and education (Cole *et al.*, 1995). Regulating use, if enforced, should change the behaviour of most visitors, whereas providing information only increases the likelihood of people behaving as desired (Buckley, 2011). However, managers and visitors alike, when asked to give their relative preferences for management actions, favour providing information over closing sites, regulating visitor numbers and charging fees (Vistad, 2003). In Vistad's (2003) study of visitors to the Norwegian landscape, although there were broadly similar pattern of preferences expressed by managers and visitors, significant differences existed, with managers being more supportive of information provision and regulations than visitors.

Managers' interpretations of visitor preferences have been highly influential in selecting management actions. Managers and researchers have generally believed that visitors prefer information and education to direct actions such as restricting access and rationing use (Cole, 1995) or using site infrastructure such as rails to guide where visitors can go. Research suggests this is not always the case. Shindler and Shelby (1993), in their survey of visitors to three wilderness areas in Oregon, found that regulating visitors through restricting access and site closures was generally supported. A survey of wilderness users in Western Australia showed that although educating visitors was the most strongly supported management action, more than two-thirds of visitors supported limiting use, length of stay during peak times and number of people in a group, and temporarily closing areas (Morin *et al.*, 1997). A similar survey in Borneo by Chin *et al.* (2000) showed substantial support for regulatory actions such as limiting forest use and the number of people, as well as for indirect actions such as education.

These preferences can vary depending on the level of site development. Bullock and Lawson (2008) concluded from their study of the iconic and heavily visited Cadillac Mountain (Arcadia National Park, Figure 4.7) in Maine, USA, that visitors support management structures such as signs and rock borders as a means of keeping people on walk trails while being opposed to limits to public access. This contrasts with studies of wilderness use, where visitors prefer few facilities and are more amenable to use limits (Manning, 2011). Cahill *et al.* (2008) similarly found visitors to a popular, highly developed site were more willing to accept higher levels of walk trail development (e.g. use of gravel and wooden planks) than visitors to a more remote, undeveloped site.

An influential, but little discussed, element is how the preferences of protected area managers and their organisations guide the choice of management actions (Figure 5.5). Managers who see their role as a policing one are likely to prefer regulation. Those who see their role as educational are likely to prefer communicative and educative approaches. Others with training and experiences in site design and engineering are likely to prefer these approaches. Similarly, past preferences and practices of organisations will influence current preferences. For example, managers in organisations with few resources and limited political support may favour softer, inoffensive approaches such as education.

A combined approach

Tourism management in most natural areas relies on a combination of various management strategies and actions. Where there are few resources for management, the main management thrusts are reservation followed by zoning. Such is the case for many protected areas in developing countries (Wallace, 1993). For protected areas in developed countries, reservation, zoning and often joint management are accompanied by a suite of management actions, including site design, regulating visitors and education. Box 5.8 describes the management of Kakadu National Park in northern Australia, where many of the strategies and actions outlined in this chapter have been applied.

Box 5.8 A combined approach to management: Kakadu National Park, northern Australia

This case study illustrates the use of a range of strategies to manage an iconic natural and cultural area for tourism and other values. These strategies begin with international agreements, here the World Heritage and Ramsar Conventions, and culminate in site and visitor management. Joint management by the Aboriginal traditional owners and the Australian Commonwealth government is a key feature.

Kakadu National Park is one of only 28 World Heritage sites in the world listed for both its natural and cultural values. Located in northern Australia (Figure 5.1), with an area of almost 20,000 km^2, it contains both ancient and modern landforms. The ancient bedrock that contains the uranium ore mined near the park and the Arnhem land plateau that contains numerous spectacular cliffs, waterfalls and caves are examples of the former; the park's extensive floodplains that include spectacular wetlands, listed under the Ramsar Convention, are examples of the latter.

Kakadu is regarded as an Aboriginal living cultural landscape, with a number of clan groups having associations to the area (KNPBM, 2007). Additionally, Kakadu has one of the most important rock art collections in the world, dating back 20,000 years (AG DSEWPC, 2012). Many tourists are attracted to Kakadu, with 227,000 visitors reported for 2007–08 (Haynes, 2010). Major reasons for visiting include appreciating the scenery, viewing wildlife and rock art, and learning about the area's ecological and cultural heritage (Wellings, 1995) (Figure 5.6).

Kakadu National Park has been identified as a successful example of joint management (De Lacy, 1994; Press & Hill, 1994) involving the Aboriginal traditional owners and the Australian Commonwealth government. This is an example of the Aboriginal and government governance model (Eagles, 2008, 2009). Concerns have been expressed, however, about unequal power arrangements, cultural differences and appropriation of Aboriginal ideas and technologies by the Park Service, resulting in park management in ways that are alien to traditional owners, all making joint management activities extremely difficult and complex (Palmer, 2007; Haynes, 2010). A fluctuating population of 500 Aboriginal people live in the park (Palmer, 2007). The Kakadu Board of Management, established in 1989 and with 10 of its 15 members nominated by the traditional owners, determines policy and is responsible, along with the director, for preparing management plans for the park.

Joint management is achieved through enabling legislation, lease agreements with the Aboriginal traditional owners, and general management arrangements between the traditional owners and park staff. The legislation allows traditional owners to claim land and lease it to the director of National Parks as well as providing for boards of management for parks on Aboriginal lands. The lease agreements provide for the rights and protect the interests of traditional owners, employment and training for Aboriginal people in park management, public education and information services, lease payments and a share in park revenue (Kakadu Board of Management & Parks Australia, 1998).

General management arrangements include the Board of Management, the plan of management and day-to-day liaison. Day-to-day liaison is an informal but nevertheless crucial element and includes local meetings, employing senior traditional owners as cultural advisers, day-to-day working contact with traditional owners, and employing young Aboriginal people (Press & Hill, 1994; Press & Lawrence, 1995).

The fourth management plan (the one preceding the 2007–14 plan) precluded non-Aboriginals or companies without Aboriginal involvement from developing new commercial tourism activities in the park (Kakadu Board of Management & Parks Australia, 1998). The 2007 plan is much less prescriptive, noting instead that Aboriginal people will determine how and when they will be involved in tourism, and that they will determine the pace and level of tourism development in Kakadu. Haynes (2010) comments that Aboriginal people are still not fully benefiting from the tourism opportunities provided by Kakadu, although there have been recent improvements, with a

50% increase (albeit off a very low base) since 2006 in the income generated from tourism accruing to Aboriginal people.

Site management is an important, integral part of managing natural areas, such as Kakadu National Park, for tourism. In Kakadu, site management encompasses roads, tracks and walk trails, visitor and cultural centres, campgrounds, safari camps and other visitor infrastructure. A range of camping opportunities is provided, from highly developed and maintained sites to bush camping outside designated areas (Commonwealth of Australia, 1991). Some developed campsites have been landscaped, the ablution blocks have hot water and lighting, and picnic tables and benches are provided. Other sites, accessible on unsealed rather than sealed roads, have toilets only. The least developed option is bush camping, with no facilities, primarily for hikers. The 2009 Tourism Master Plan identified opportunities to develop safari camps and a nature lodge (AG KNP, 2009). Accommodation facilities associated with the park, but managed by other parties, include three hotels, two caravan parks and a backpacker hostel.

Managing visitors, for example through fees and limiting visitor numbers, is the other important part of managing tourism use of natural areas. At Kakadu, visitors are charged a fee to enter the park and to use the major developed camping areas. Revenue from fees contributes to running the park, with traditional owners also receiving a share. Depending on the season, 40–60% of park visitors rely on commercial package tours to see Kakadu (KNPBM, 2007). Tours include boat cruises at Yellow Water (South Alligator wetlands) and the East Alligator River, scenic flights, visits to art sites, four-wheel-drive excursions, bird watching and general sightseeing. Limits to visitor numbers to sensitive sites have been achieved by limiting the number of permits available for certain types of commercial tour operations and for camping in some areas. For example, the number of permits for safari camps has been limited to nine, and for boat tours to three, each for a different location.

Figure 5.6 Aboriginal shelter and interpretation, Nourlangie, Kakadu National Park (Photo: Sue Moore)

Managing the Tourism Industry

Managing natural area tourism relies on three parties – land managers, visitors and the tourism industry. The industry includes those operating tours through natural areas as well as the owners/managers of built accommodation and other facilities, such as interpretation centres. The previous two sections have been devoted to management strategies and actions available primarily to land managers. This section explores strategies focusing on the tourism industry. These may be voluntary, such as codes of conduct, or they may be regulatory, such as licences and associated conditions administered by management agencies.

Voluntary strategies

A range of voluntary means are available for assisting tour operators and those who own or manage tourism facilities, such as resorts or interpretive centres, to conduct their business in ways that minimise its environmental consequences. Available strategies include codes of conduct, certification, environmental management systems and best environmental practice. There is lots of overlap between these categories. For example, a tourism destination might be eco-certified and nested within that certification is a requirement for the company to have an environmental management system. Research in the hotel sector suggests that of all these voluntary strategies, environmental management systems provide the greatest guarantee of improving a tourist operator's environmental and sustainability performance. Part of their success can be attributed to them including one or more other voluntary strategies, such as codes of conduct and best environmental practice (Ayuso, 2007).

Codes of conduct and guidelines

A code is a set of expectations, behaviours or rules written by industry members, government or non-government organisations (Holden, 2000). Its principal aim is to influence the attitudes and behaviour of tourists or the tourism industry. A code may be informal and adopted by a group, or more formal and instituted for industry members and/or tourists. The former are often referred to as codes of ethics and tend to be philosophical and value-based, whereas the latter are usually known as codes of practice or conduct and are more applicable and specific to actual practice in local situations. Guidelines are also used to direct how tourism activities are undertaken and, similarly to codes, may be written by industry members, government or non-government organisations.

A code of ethics provides a standard of acceptable performance, often in written form, that assists in establishing and maintaining professionalism (Jafari, 2000). A Global Code of Ethics for Tourism has been developed by the UNWTO (2012b). The code includes nine principles, covering tolerance of and respect for others, safeguarding the natural environment, respecting heritage and culture, providing benefits for local people, honesty and transparency in business, participation and freedom of movement rights for all, and protection of workers in the tourism industry. A tenth principle focuses on implementation. Although the code is not a legally binding document and is reliant on voluntary implementation, perceived breaches can be referred to the World Committee on Tourism Ethics.

Codes of conduct and guidelines also exist for tourism in natural areas. Some have been developed for specific places, such as Antarctica (Box 5.9), and others for activities such as river rafting or kayaking. Destination-specific codes recommend how tour operators and/or visitors should behave at a destination. Codes or guidelines for specific activities are intended to help tour operators and visitors improve their environmental management and minimise their impacts. Guidelines for white-water rafting and kayaking note that environmental management is important because most rafting rivers have limited campsites and minimal-impact techniques are essential for client satisfaction as well as environmental protection. Such guidelines suggest that tourists plan ahead, tread lightly, camp with care, carry out their litter and continue learning about the environment.

Certification

Certification is another means of assisting industry members to act responsibly. It is often discussed as part of eco-labelling. Buckley (2002) included voluntary codes, awards, accreditation and certification under the umbrella term 'eco-label'. There is enormous overlap in terms and concepts within the area of eco-labelling. In general, such a label implies a certain level of environmental performance. Certification within the tourism industry specifically involves testing a facility, product, service or management system using specified standards (Haaland & Aas, 2010). The associated assessment is ideally, but not always, achieved through external review.

Certification aims to improve industry performance (in terms of environmental protection and improving sustainability) and influence markets (Buckley, 2002; Font & Harris, 2004). Environmental impacts can be reduced and the efficiency of natural resource use increased by certification and the associated improvements in practice. Certified tourism companies can gain a market advantage over other market segments. A certification system allows tourists to identify those companies operating to achieve sustainable tourism. It is also a way of exposing those who purport to be providing sustainable tourism but are not.

Governments and other parties continue to confuse certification and accreditation, which is inevitable given the potential overlap in intention between them. Certification involves testing individuals to determine their mastery of a specific body of knowledge. Accreditation is based on an agency or organisation evaluating and recognising a programme of study or institution as meeting certain predetermined standards or qualifications. Thus, certification applies to professionals such as tour operators while accreditation concerns programmes and institutions (Morrison *et al.*, 1992). Increasingly, tourism has moved towards more accurately using the term 'certification' (rather than incorrectly using 'accreditation') to refer to mastery of a certain level of performance and associated knowledge acquisition.

The Eco Certification Program, administered by Ecotourism Australia, is a widely cited example of an implemented certification scheme. It was established in 2001, although it was preceded by earlier programmes that also focused on setting performance standards for the tourism industry. Three levels of certification are

Box 5.9 Antarctica: Guidelines for the conduct of tourism and tourists

Antarctica, the last continent to be 'discovered' and the most isolated, occupies almost 10% of the world's land surface (Splettstoesser, 1999) (Figure 5.1). Nearly all of the continent is ice-bound (98%), although its extreme aridity means some areas of coast are permanently free of ice. In contrast to the limited extent and biological diversity of its land, the Antarctic waters are rich in wildlife, including whales, seabirds, seals, squid and fish (Dingwall, 1998; Mason & Legg, 1999). Unlike the countries of the Arctic, the Antarctic has no indigenous inhabitants.

Antarctica has become a tourist destination relatively recently, with the first tourist aircraft in December 1956 and the first tourist ship in 1958. Over 45,000 tourists now visit the Antarctic each year (Haase *et al.*, 2009). Ship-based visitors continue to account for more than 95% of all tourists (Stonehouse, 1994; Haase *et al.*, 2009). Their activities centre on the Antarctic Peninsula region over the austral summer of November to March. Attractions include the remoteness, extreme climate, beauty of the physical setting and abundant wildlife. The remains of explorers' huts as well as the scientific bases are also of great interest.

Traditional ship-based expedition cruises, with an educational component, are the main form of sea-borne tourism, although cruise-only tourism is growing in popularity (Haase *et al.*, 2009). Visiting ships vary in size from 13 to more than 500 passengers, with most carrying between 13 and 199 people (BAS, 2012). Cruise length also varies, although most last 12–15 days, with 4–5 days spent landing at different sites using inflatable boats. A typical itinerary includes penguin rookeries, scientific bases, historic sites and trips in inflatable boats to scenic areas and to see seals on icebergs. Most ships provide educational briefings on history, geology, wildlife and scientific research (Enzenbacher, 1992; Haase *et al.*, 2009).

Management responsibility for Antarctica is shared by more than 40 countries and is guided by the international Antarctic Treaty, which came into force in 1961. The Treaty freezes sovereignty claims (seven nations have territorial claims), demilitarises the area, guarantees free access and gives pre-eminence to scientific research. Industry guidelines, along with the Treaty, play an important role in managing tourism in the Antarctic.

The most widely known guidelines are those produced by the International Association of Antarctic Tour Operators (IAATO), an organisation founded in 1991. The IAATO now has more than 102 member companies (BAS, 2012). It has two sets of guidelines, one for those organising and conducting tourism in the Antarctic and the other for visitors. The tourism operator guidelines have a strong emphasis on safety, but they also recommend an assessment of potential environmental impacts and the monitoring of these during the visit, the employment of a sufficient number of trained, experienced guides, and the removal of most wastes from the area. The number of passengers ashore at one time and place is limited to 100, with a typical ratio of one member of staff for every 10–20 passengers when ashore (BAS, 2012).

The guidelines for visitors emphasise wildlife and environmental protection, the value of historic huts, the vulnerability of protected areas and research activities, and ensuring safety by staying in groups (Enzenbacher, 1995; IAATO, 2012). Both sets of guidelines are voluntary and seek to influence behaviour. There is no recourse to punitive action if the guidelines are not followed, unlike the case with regulations and associated penalties under the law of a country (Mason & Legg, 1999).

provided: nature tourism (focus on minimal impact), ecotourism (learning about the environment and contributing to sustainability) and advanced ecotourism (innovative ecotourism plus the features of the preceding categories) (Ecotourism Australia, 2012). The programme distinguishes *bona fide* ecotourism products on the basis of 10 principles: ethical business management; customer satisfaction; responsible marketing; natural area focus; environmental management; climate change action; interpretation and education; contribution to conservation; working with local communities; and cultural respect and sensitivity. Operators complete a self-assessment workbook, which is then submitted to Ecotourism Australia for independent assessment.

A total of 570 operators are currently certified under the categories of tours, accommodation, attractions and 'other' (e.g. restaurants, car hire) (Ecotourism Australia, 2012). Eco Certified operators receive marketing benefits from Ecotourism Australia, including representation at trade shows and industry conventions, and preferential marketing support from state tourism bodies and from the Australian government via the T-QUAL 'Tick' programme (which formally recognises sustainable, capable tourism enterprises). A number of Australian protected area management agencies offer extended permits to Eco Certified operators. For example, the Great Barrier Reef Marine Park Authority offers 15-year permits to operators with products certified at 'ecotourism' or 'advanced ecotourism' levels.

In early 1994, the World Travel and Tourism Council initiated the Green Globe certification programme for the travel, tourism and hospitality industries. This sets international standards for good environmental performance in these sectors. The main areas covered are sustainable management: social economic, cultural heritage and environmental. Included are attractions, businesses, meeting venues, cruise ships, golf courses, hotels and resorts, meetings/events, organisations, restaurants, health centres, transportation (e.g. bus company) and the travel industry (e.g. tour operators) (Green Globe, 2012). Of most relevance to natural area management is the certification of tour operators, visitor attractions and hotels if they are located within or adjacent to natural areas.

Environmental management systems

Environmental management systems are tools to manage and continually improve the environmental performance of a company using a planned strategy (Ayuso, 2007). An associated ideal is having every individual in the company accepting responsibility for environmental improvements. Tourism certification programmes such as Green Globe require the preparation of an environmental management system. Box 5.10 describes the environmental management system developed for ski resorts in the Australian Alps. The analysis in Box 5.10 is provided to assist in understanding the intent and benefits of such systems.

Environmental management systems developed from concerns in the 1990s regarding the implementation of environmental policy. Many companies in the mining and manufacturing sectors had developed environmental policies and reviewed their practices but were unable to implement the necessary changes. Quality systems such as BS 5750, ISO 9000 and ISO 14000 were developed as

Box 5.10 Environmental management systems for ski resorts in the Australian Alps

The Australian Alps encompasses a well watered, snow-clad mountainous area with 1.2 million ha of protected areas in an otherwise predominantly arid continent. These protected areas are located in the states of New South Wales and Victoria, and the Australian Capital Territory. The Alps attracts hikers, skiers and a variety of other outdoor enthusiasts. A total of 10 ski resorts provide facilities for snow-based activities in the winter months and for an increasing number of visitors who come to enjoy these alpine areas in the snow-free summer months. Many of these recreational facilities are located in protected areas, a number of which contain plant and animal species restricted to the Alps. Of particular importance is the mountain pygmy possum, an endemic marsupial that survives the winter beneath the snow cover. Montane bogs, containing a number of threatened species, are also of concern and potentially affected by the increased salt in the environment, sourced from de-icing activities on the roads and car parks within the resorts. Invasive weeds and increased storm-water as a result of modified ground cover and snowplough activities also pose a threat to the unique alpine environment (OEH NSW, 2011).

All four of the alpine resorts within Kosciuszko National Park (in the New South Wales part of the Australian Alps) (Figure 5.1) have environmental management systems (EMSs), with the first one produced in 2002; all have been completed and final approval was given by the park managers, the New South Wales National Parks and Wildlife Service (NSW NPWS), in 2009. These EMSs use the ISO 14000 series of international standards to describe system elements, including commitment to the environment, identifying environmental risks, complying with legislation and other regulations, reducing impacts, setting objectives, developing procedures, and monitoring and reporting on performance (OEH NSW, 2011).

These EMSs are seen as a tool to help protect the values of Kosciuszko National Park while providing for recreation opportunities. Through being publicly available, they also provide the community with confidence that the resorts have a focus on and are required to manage and minimise their environmental impacts. For example, all four resorts have plans in place for high-risk activities such as sewage processing, water use, and hydrocarbon storage and management. Highlights of environmental reporting for the 2010/11 period include: a reduction in water consumption per visitor-night by 7.7% over the last three years by the Perisher lodges; 25% reduction in fuel consumption by Charlotte Pass for resort transportation through improved technology; 30% reduction in electricity consumption by Mt Selwyn Resort through the installation of power factor correction units on snow-making facilities; and achievement of 29.1% recycling of total waste generated, including aluminium cans, glass bottles and jars, plastic bottles, steel cans, paper and cardboard by Thredbo (OEH NSW, 2011).

The main focus is on continual improvement, with the adaptive principles of 'plan, do, check, act' underpinning the development and application of the EMSs in all four resorts. Risk assessment is also part of the broader context for these EMSs. Continual improvement and pollution prevention have been the central interest, rather than documenting each element of the system as prescribed in the ISO 14001 guidelines. Current activities by the NSW NPWS, in working with resort managers to continue to enhance environmental performance, include aligning future environmental reporting more closely with each resort's environmental objectives and defining common key performance indicators and targets across the four resorts. These changes will enable management effectiveness to be more readily assessed over time (NSW NPWS, personal communication, 2012).

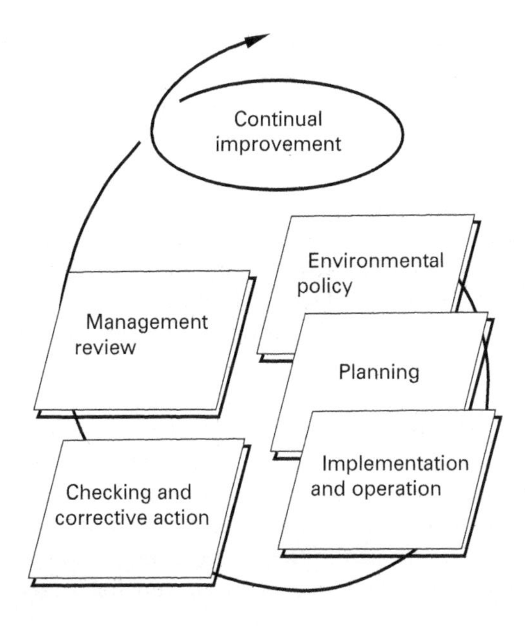

Figure 5.7 Environmental management system model (Derived from Standards Australia, 1997)

possible solutions. Such systems assisted organisations to conform to specified quality standards. The philosophy behind quality systems, of continual improvement in performance, underpins environmental management systems (Figure 5.7).

Today, the most widely recognised environmental management system is the International Standard ISO 14001, published in 1995. This standard describes five principles for a complying organisation – commitment and policy, planning, implementation, measurement and evaluation, and review and improvement (Standards Australia & Standards New Zealand, 1996). These principles are an integral part of the continuous improvement process, as illustrated in Figure 5.7. The success of such a system depends heavily on involving staff in developing procedures, record keeping and conducting audits. Being able to collaborate with external organisations on waste management and green purchasing is also important (Ayuso, 2007).

Best environmental practice

Best environmental practice, similarly to codes of conduct, certification and environmental management systems, is another means of encouraging responsible, self-motivated behaviour by members of the tourism industry. It aims to improve the environmental performance of companies (Ayuso, 2007) and may be embedded in other voluntary approaches, such as certification. Best practice involves minimising environmental impacts, particularly through careful use of resources and their disposal. Andereck (2009), in her survey of visitors to tourism centres in Arizona, found the greatest support for the following best environmental practices: landscaping with native plants, energy efficiency, and conservation and recycling. Such approaches may often reduce costs, although not always. They may also add to a company's market advantage if it is apparent that best practice approaches have been adopted.

Environmental impacts can be minimised directly – by changing or modifying the facilities and services provided and how they are provided – or indirectly – by work methods, processes or initiatives that improve organisational effectiveness, service delivery and employee satisfaction (Department of Tourism, 1995). The latter is a focus on best practice in organisational activities rather than on-the-ground actions for minimising environmental impacts. Box 5.11 explains the full range of best environmental practices, based on a toolkit prepared for the Canadian tourism industry. Much of the tourism in this country takes place in protected areas.

In the general tourism industry, such approaches are referred to as greening programmes. The most common activities for hotels are saving energy and water, sorting waste and waste management, correct handling and storage of hazardous substances, and control of atmospheric emissions, water effluents and noise (Ayuso, 2007). Becoming carbon neutral is an aim for many hotels, for example

Box 5.11 Environmental best practice: A toolkit prepared for the Canadian tourism industry

The toolkit was prepared to progress sustainable tourism (Marr Consulting Services, 2008). It provides a sector-by-sector guide to best environmental practice, followed by organisational and business management advice and concluding with the more traditional focus on direct management of waste, energy, transportation and carbon emissions. Food services, accommodation, tour operators, travel agents, hunting and fishing outfitters, and attractions and venues are the sectors addressed. Business management advice ranges from product development and marketing to the internal processes of administration, purchasing and operations, and to 'giving back' to the community. Suggestions for giving back include donations to local charities, promoting local suppliers and industries (e.g. local arts and crafts) and volunteering. Direct management has a strong focus on reducing energy use, resulting in significant cost savings and reductions in greenhouse gas emissions.

Waste, water use, transportation, indoor air quality, being socially responsible and becoming carbon neutral are the best environmental practices given particular attention. For carbon emissions management, companies are encouraged to determine their carbon footprint and to purchase carbon offsets for the carbon they produce as part of their business (e.g. through travel and temperature control in buildings). Becoming carbon neutral must also include strenuous efforts to reduce the consumption of manufactured products that have contributed to the planet's carbon load. Reductions in energy use and a shift to alternative energy sources (such as solar) are also critical given that the burning of fossil fuels is the greatest contributor to the production of greenhouse gases.

the Mercure group. The airline industry has been active in greening programmes through addressing noise and emissions reductions, fuel efficiency and encouraging passengers to purchase carbon offsets for their flights. Restaurant programmes have focused on the reduction of solid wastes and energy consumption as well as broader community conservation issues (Todd & Williams, 1996). For example, the five Starwood restaurants in Abu Dhabi have joined the World Wide Fund for Nature 'Choose Wisely' campaign to serve only fish species from stocks known to be sustainable (WWF, 2012).

Regulatory strategies

Several regulatory strategies are widely used by government agencies managing tourism in protected areas. Two key ones considered here licences and leases. Licences with associated conditions are issued to tour operators, while leases are issued to tourism businesses occupying fixed premises for longer periods. Both provide legally based guidance, usually enshrined in legislation (i.e. laws), to tourism companies as to how they must conduct their business within protected areas. Buckley (2011) noted that laws and regulations for protected areas, as well as for planning developments and controlling pollution, are the most effective means for improving environmental management by tourism.

Licences

A licence is a certificate or document giving official permission to undertake an activity. The possession of a licence is often mandatory for tourism on government lands and waters and they can therefore be used to ensure the natural environment is conserved and managed. Licences allow the governing agency to control and monitor access and use of the areas under its control and to ensure that conservation values are maintained. It is self-evident that by protecting these values, tour operators will be able assure themselves of continued use of these natural locations over the longer term. Licence holders agree to abide by a set of rules and regulations with regard to the natural areas in which they operate. The rules are both general – for example protecting plants and animals, obeying road rules, prohibiting firearms, and removing rubbish – and specific – with regard to camping and specialised activities such as abseiling and caving, for example.

Licences usually but not always apply to non-exclusive activities, where the activities of one operator do not preclude or exclude the activities of another. For example, wildflower tours and bird watching can support a number of operators. Where the resource is potentially susceptible to damage from overuse and/or a significant level of capital investment is required, some agencies, such as the Department of Environment and Conservation in Western Australia, have the capacity to issue an exclusive licence to one operator only. An example is the licence for tours and a ferry service provided in Shoalwater Islands Marine Park off the Western Australian coast (Figure 5.1). The operator has significant capital invested in boats and the market is too small to support more than one operator. Also, the associated marine and coastal environment is not robust enough to support more than one operator.

Leases

Leases are generally issued where operators require exclusive rights to land or waters. Such rights are likely when the operator intends to construct or manage a substantial facility such as a lodge, restaurant or visitor centre. Such leases usually involve major capital investment over an extended period of time.

Conclusion

This chapter has described the wealth of approaches available to manage natural area tourism. Management generally begins with some form of protection, whether it is designation as a protected area such as a national park, or through an international convention such as the Ramsar Convention. Such designation, although a crucial starting point, is generally insufficient for sustainable management of a natural area as a tourist destination. Development of shared governance arrangements, for example between local people and national park agencies, and/or zoning often follow.

This chapter then described the plethora of actions from which a manager can choose. Site management relies on designing and then managing linear features such as roads, tracks and trails and terminus points such as campgrounds and parking areas to keep environmental impacts to acceptable levels. Visitor management focuses on visitors, with management through either direct regulation or communication and education. Education is emphasised in both research and practice as an essential element in managing natural area tourism. Chapter 6 explores in detail the importance of education and interpretation for the sustainable management of natural area tourism.

The tourism industry, as well as the destinations themselves, can be managed to protect the environment and the experiences of visitors. Such management can be through voluntary strategies, including certification and environmental management systems, and regulatory approaches, for example licences and leases.

In all instances, effective management relies on managing with some objective in mind. A fundamental management objective for many natural areas is offering experiences that satisfy visitor needs while at the same time protecting and maintaining natural systems and processes. Acceptable conditions provide a measure of how managers are performing against this objective. They are best determined using a planning framework, such as the Limits of Acceptable Change or Visitor Impact Management, as described in Chapter 4. The success or otherwise of management is then determined by monitoring and evaluation, the subject of Chapter 7. Good management relies on adequate planning, knowledge, implementation and monitoring (Cole, 1993). Where management strategies are employed without planning and monitoring, they may be inefficient and ineffective.

Further reading

Wildland Recreation: Ecology and Management by Hammitt and Cole (1998) and *Wilderness Management* by Hendee *et al.* (1990a) are still two of the best reference books on strategies and actions for managing the recreational use of wilderness. Both are directly relevant to managing natural area tourism. Eagles and McCool's (2004) *Tourism in National Parks and Protected Areas: Planning and Management* has a useful chapter on management strategies plus information elsewhere in the book on planning and policies. The edited book *The Environmental Impacts of Ecotourism* (Buckley, 2004a) has very good summary chapters on trail and campsite management, and rationing and allocation of visitor use.

Manning's (2011) *Studies in Outdoor Recreation: Search and Research for Satisfaction* provides important insights into the efficacy of various visitor management strategies and most importantly how these strategies are perceived and supported (or otherwise) by visitors and managers. Another edited book by Buckley (2001), *Tourism Ecolabelling: Certification and Promotion of Sustainable Management* gives the reader ready access to voluntary means for behavioural changes within the tourism industry, in particular certification and eco-labelling.

An important aspect of managing natural area tourism not considered in this chapter is wildlife management. Very often, it is wildlife that draws visitors to natural areas and brings them back again and again. There are a number of books and articles addressing wildlife tourism and its management. The most useful are the classic *Wildlife Tourism* by Shackley (1996) and the more recent *Wildlife Tourism* (Newsome *et al.*, 2005).

6 Interpretation for Nature Tourism

Introduction

Interpretation is an integral part of 'best practice' ecotourism. Its importance lies in communicating ideas, the provision of minimal-impact messages and enriching visitor experiences. McArthur (1998a, 1998b) stated that much of the interpretation practised by the tourism industry was of poor quality and he recommend two things to remedy the situation: the first was to value the significance of interpretation and the second was that the deliverers of ecotourism needed to understand how 'best practice' interpretation is achieved. Notwithstanding this observation, the theory and application of interpretation have come a long way since the late 1990s, as evidenced by its practice in the field, the publication of books on the subject, the outputs from journals devoted to the subject (e.g. *Journal of Interpretation Research*), the inclusion of many papers on the subject in the tourism and related literature and workshops and conferences devoted to furthering interpretation as a communication strategy in natural area tourism. Furthermore, there are many tour-guiding associations – such as the Field Guides Association of South Africa and the Interpretation Australia Association – that offer training courses, promote good practice and offer accreditation and the exchange of ideas. Furthermore, Greig (2011) writes that the World Federation of Tourist Guides has 65 member countries (and 12,000 individual members), with the majority of these having compulsory licensing and training of guides.

The objectives of this chapter are therefore to provide some background on the theory and application of interpretation and offer some discussion and examples of the issues surrounding and what comprises 'best practice' interpretation. Wearing and Neil (1999) described the various situations in which interpretation can be applied. These include promotional, value-adding, educational, economic and ecological contexts. This chapter is principally concerned with the last of these, that is, the role of interpretation in the management and conservation of the tourism resource. This is in keeping with the major focus of the book and completes the suite of planning and management strategies that are available for sustainable tourism.

Principles

Interpretation can be defined in a number of ways but is generally described as being an educational activity that brings out meaning and enriches visitor experience (Figure 6.1). In a wildflower guided walk, for example, it is not simply informing people about plants by systematically providing a list of names. It is a 'hands on' involvement where visitors learn and self-discover the answers to questions like:

- How do plants survive?
- Why does it look like this?
- How does it relate to the animal life around it?
-

Tilden's (1957) definition (Box 6.1) has been widely used but various other definitions have emerged reflecting the particular objectives of the organisations involved (e.g. McArthur, 1998a). These definitions, however, always embrace the fundamental principles presented by Tilden's original work. Tilden (1957) emphasised that interpretation is an art and techniques have to be designed to accommodate visitor needs, attitudes and expectations. For example, there need to be different approaches for adults and children. Moreover, visitors need to relate to what is being conveyed, with the observation that it is much more effective when it is directly relevant to the individual experiencing the interpretation.

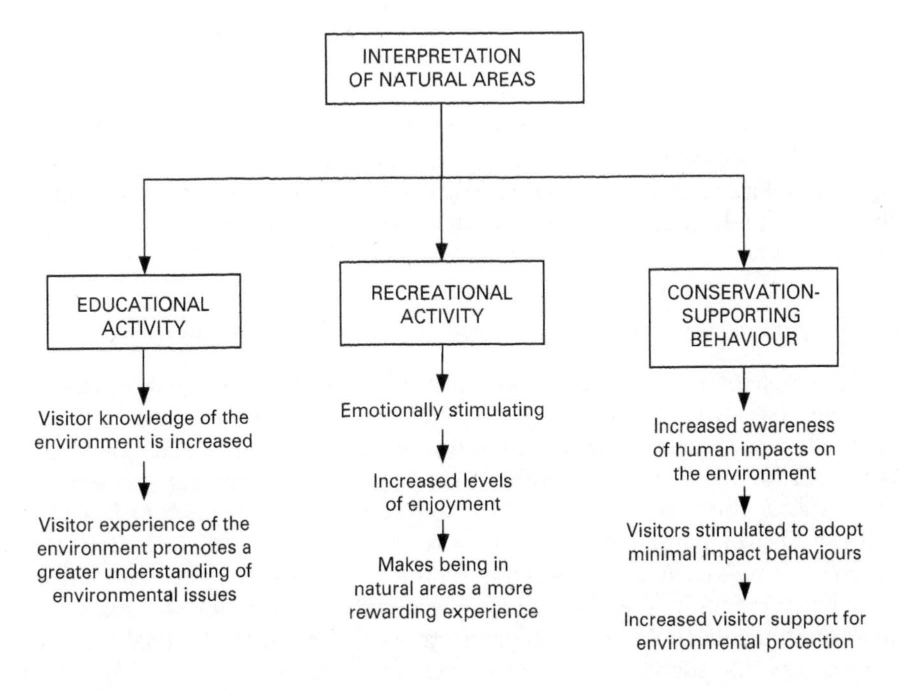

Figure 6.1 The education–knowledge–awareness relationship embodied within interpretation

Box 6.1 Definitions of interpretation

Tilden (1957)
'An educational activity which aims to reveal meaning and relationships through the use of original objects, by first hand experience, and by illustrative media, rather than simply to communicate factual information.' Meaning is achieved through stimulation and revelation. Whereas education per se is generally a more formal provision of facts, interpretation is more concerned with concepts and messages.

McArthur (1998a)
'Interpretation is a coordinated, creative and inspiring form of learning. It provides a means of discovering the many complexities of the world and our role within it. It leaves people moved, their assumptions challenged and their interest in learning stimulated.'

Moscardo (2000)
'Interpretation is any activity which seeks to explain to people the significance of an object, a culture or a place. Its three core functions are to enhance visitor experiences, to improve visitor knowledge or understanding, and to assist in the protection or conservation of places or cultures.'

Weiler and Ham (2001)
'Services which provide meaning and understanding for tourists about what they are visiting and experiencing.'

Following on from the concept outlined by Tilden (1957), several basic principles have emerged as the mainstay of interpretation. These concepts have been outlined by McArthur (1998a, 1998b), Wearing and Neil (1999) and Weiler and Ham (2001). The fundamental principles of interpretation are summarised under separate headings below.

Interpretation should centre on a theme and associated messages

Writers on interpretation frequently state that it is important to develop a theme that contains concepts and messages. Crabtree (2000) justifies the use of themes in that it allows interpretive ideas and information to be organised and easy to follow. Ham (1992) stated that themes embrace entire ideas and all-encompassing messages that a visitor can reflect on after the interpretive experience. McArthur (1998a) used the idea of natural disturbance in Australian forests to show how theme, concept and message relate to one another: the theme was that forests are subject to natural disturbance by fire; the concept of naturally changing forests which contain fire-adapted plants could be developed from this theme; and the final interpretive message was that the structure and composition of such forests are shaped by fire. Various techniques can then be applied in delivering such a message (see p. 303).

Interpretation entails active involvement and first-hand experiences

Getting actively involved and 'doing', rather than just passively listening to straightforward instruction, make the interpretive activity easier to appreciate and a more enjoyable experience (see for example Figure 6.2). In continuing the theme of naturally changing forests, participants could be asked to look for evidence of disturbance. Such active involvement is more likely to engage the audience in a sense of discovery in the field.

Interpretation facilitates maximum use of the senses

Encouraging use of the senses is an approach that is likely to bring the interpretive experience 'alive' and make it more enjoyable and satisfying. By analogy, simply looking at food is not as satisfying as smelling it, feeling its texture in the mouth and then tasting it! Similarly, the smells (oils and resins in leaves), textures (bark and spiny vegetation) and even the taste (various fruits) of a forest will deepen the visitor experience by forging a greater connection with the forest.

Figure 6.2 Learning about skulls and teeth is best achieved if visual aids and specimens are available as a first-hand experience and active involvement (Photo: David Newsome)

Interpretation seeks to foster self-discovered insights

Such insight comes from active involvement and maximum use of the senses. A guided walk or elephant trek in an Asian rainforest clearly leaves more scope for insight than a vehicle excursion. This is because of the opportunity to use all of one's senses in discovering the forest. Walking in the humid environment also brings the visitor closer to the specific environmental conditions that characterise rainforests. Good trail design maximises the opportunities to see specific features in a self-discovered fashion. Wild animals are difficult to see in the rainforest and good interpretation will build on this and develop a thrill of anticipation. Active participation in searching for wildlife and learning about forest ecology then provides the scope for self-discovered insight.

Interpretation is of relevance to the visitor and clients find the imparted knowledge and insights useful

Crabtree (2000) recommends asking the audience about their interests and motivations and suggests that this can be achieved by talking to a group for a few minutes before the activity commences. Making the whole visitor experience relevant to the

Box 6.2 Scottish Seabird Centre

The Scottish Seabird Centre is situated close to the Bass Rock gannet (*Sula bassana*) colony at North Berwick, Scotland (Figure 6.3). The centre contains educational and interpretive features (e.g. photographs, diagrams, specimens and interactive displays) that are common to visitor and education centres around the world. The materials are arranged in a series of themes that are designed to capture the attention of all age groups and account for different levels of interest. Displays titled 'What is a seabird?', 'The shore', 'Survival', 'Built for the job' and 'High-rise living' provide information and interpretation on the biology and ecology of seabirds. There is also a multimedia show that focuses on the gannets themselves and a viewing deck with telescopes that can be trained on to Bass Rock and the island of Fidra, where puffins (*Fratercula arctica*) are visible, particularly during April and May. Furthermore, interactive technology is used to provide visitors with a unique view of birds on seabird breeding islands with otherwise restricted access. From inside the centre the 'seabirds live' exhibit video-screens provide people with the opportunity to see birds going about their breeding activities on the two islands. This is achieved by the use of remote cameras positioned on the islands with zoom, pan and rotate capabilities that allow visitors in the centre to scan the islands for birds and focus in on nests and chicks during the breeding season. Trained volunteers are also present to assist and answer questions. A theatre is also present, in which films about seabirds are screened. In addition, there is a website as well as the traditional boat trip that circumnavigates Bass Rock. In recent years additional video technology has been employed in the form of webcams that broadcast live pictures from various sites in the vicinity of the centre. A migration flyway tunnel has also been added, where visitors can learn about bird migration. The tunnel leads to an exhibition area, which, for example, focuses on climate change and renewable energy.

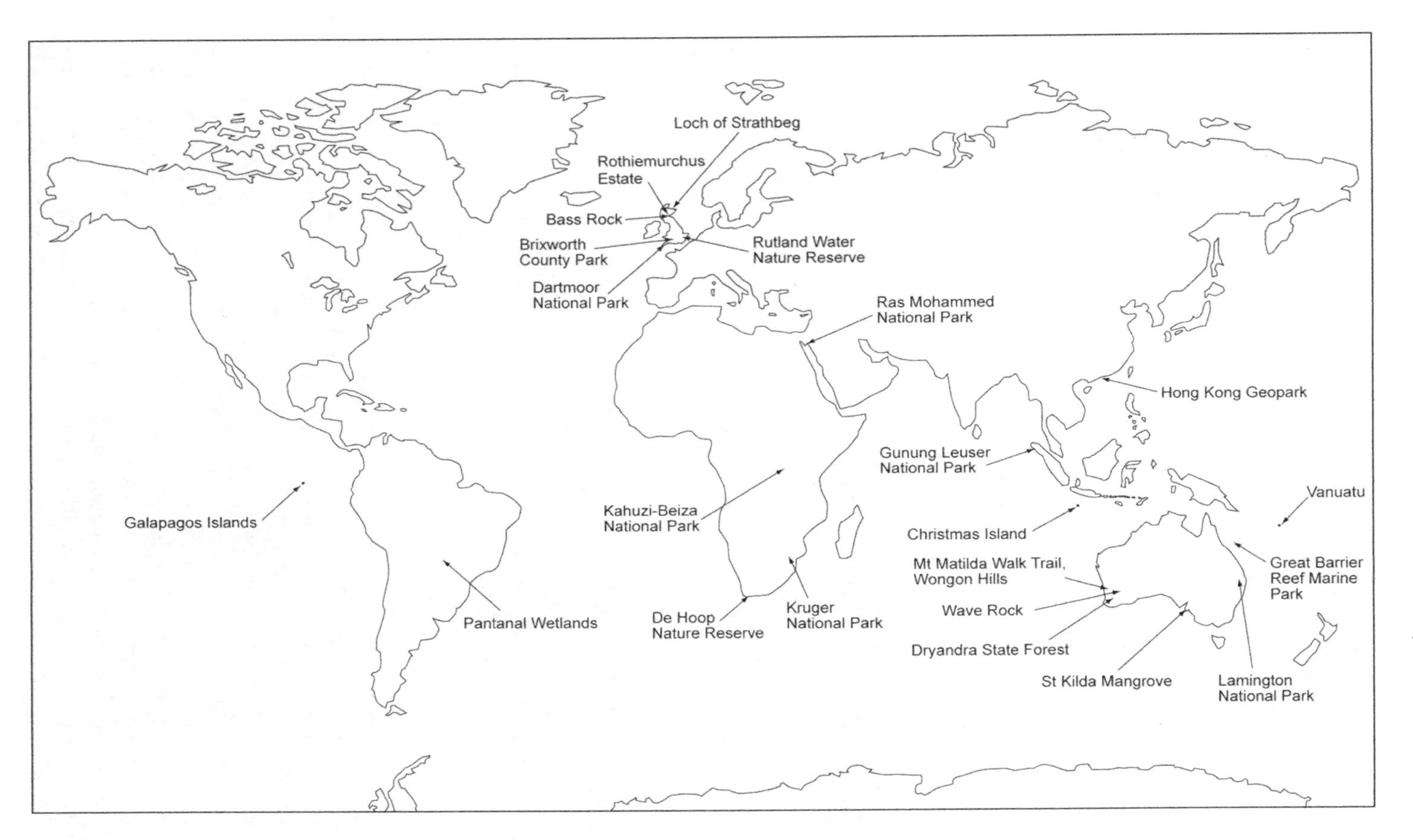

Figure 6.3 Location map of important nature-based tourism destinations referred to in this chapter

chosen site or particular activity through self-discovered insight is also important. In this way people can appreciate the importance of a particular theme and link the interpretive experience to the situation and site in which it takes place. Most people are seeking accurate information in relation to the interpretive experiences they are having and this is best achieved through personal, face-to-face interpretation.

Crabtree (2000) also maintains that the experience can be made unique by layering what is delivered, to accommodate both adults and children with different backgrounds. In this way everyone can become involved, with technical information supplied to some people while children are involved in some other activity. Such layering is the approach taken in the design and operation of many visitor centres. Features listed by Crabtree (2000) can be seen in the Scottish Seabird Centre (see Box 6.2). The centre contains material for children (simple bird identification, colouring and drawing), opportunities for people to use the centre at their own pace, static displays, interpretive displays, photographs, a touch table and even the provision of information on research methodologies and associated data. All of this provides for variety and the different levels of visitor interest.

Visitors generally like to learn about a particular area or wildlife situation and develop some understanding of what it means to them. Realising why they have to keep to footpaths and understanding why animals should not be fed makes management relevant to them. The public is also interested in learning about why natural areas are important and what efforts are being made to preserve what remains. The major goal of interpretation is to make people aware of the natural world, enhance their sense of wonder and capacity to enjoy nature and stimulate an interest in environmental protection.

Stages of the Interpretive Experience

From a psychological standpoint it is recognised that visitor experience can exist in several cognitive states during a particular interpretive experience. Cognition relates to the processes of perception and reasoning in acquiring knowledge. Particular cognitive states, as they apply to a nature-based tourism experience, have been explained in the Forestall and Kaufman model as pre-contact, contact and post-contact phases (see Orams, 1995b).

The first cognitive state – the pre-contact phase – is where the visitor lacks knowledge of a particular tourism experience, and this gap in knowledge can potentially be filled by finding the answers to various questions that can be generated. Fennell (1999) offered the example of whale watching, where participants need to learn about and then carefully observe whales. The dearth of information at the start of the interpretive process can start to be satisfied long before a participant gets on a boat or travels to a designated viewing spot. It is usually achieved through pamphlets and information boards, which frequently include wider information about marine ecosystems. Such information can also be obtained from natural history books, conversations with friends and from websites.

The idea is that most visitors will then have been stimulated through this initial phase to seek further information. The intention is that people are left with a desire for more knowledge. This then provides the appropriate conditions for the contact phase, which is the real essence of interpretation. It is here that the fundamental principles of interpretation, as already described, can be applied in answering questions. The particular approach is clearly dependent on the situation. In the case of whale watching, participants may want to know how to identify species of whales, require information on whale behaviour and have their own knowledge gain verified (Fennell, 1999). Encouraging a questioning approach, allowing visitors to talk more and posing return questions create the conditions for interaction with guides and results in greater visitor satisfaction.

Table 6.1 Role of interpretation in fostering appropriate and sustainable tourism at natural attractions using a de-marketing approach

De-marketing strategies	*Application in Mulu National Park, Malaysia*
Educating potential visitors	Promotion of rainforest nature experiences. Website attended to on a daily basis
Educating media e.g. newspapers and television	Role of park in conservation and promotion of natural experiences
Encourage desirable markets	Promotion of rainforest nature experiences (e.g. direction of visitors to interpretive plant trail)
Discourage undesirable markets	Management objections to use of all-terrain vehicles on the borders of the park. Promotion of a rainforest experience as a tourism product
Notification of prohibited activities	No organised adventure-based activities, such as mountain biking events, permitted
Publicise alternative sites	Redirect to facilities available at nearby resort. Use alternative sites if visitors wish to engage in sporting activities
Limiting activities permitted	Park available for ecotourism as defined by nature appreciation, hiking and visiting caves
Warning visitors of environmental circumstances under which activities may be curtailed	Visitor rules for park users available at reception, via staff and visitor centre
Permitting certain activities under supervision	Long-distance walks and certain caves open to 'adventure' caving. Information available at reception, via staff and visitor centre
Re-imaging destination to attract a certain demographic	Park not available for organised sporting events and activities such as Via Ferrata
Make access to fragile areas more difficult and promote less fragile options	Provision of boardwalks for access to main show cave. Interpretive signage in place

Adapted from Beeton (2003)

The post-contact phase is where the imbalance between the initial pre-contact phase and the new state of awareness is realised and can be followed through and built upon. A successful interpretive experience will have engaged people's emotions. Moreover, a successful emotional involvement can leave the visitor with a long-lasting experience and a desire to take a deeper interest. Accordingly, the motivation to return and relive the experience is fostered. Furthermore, it is also here that the opportunity for participants to contribute and to become involved in conservation could be realised.

Orams (1995a) nevertheless pointed out that there will be many situations where interpretation cannot take place within the framework of pre- and post-contact activity. This can be especially so in the case of pre-contact information. Despite this potential deficiency, interpretive staff need to be aware of, and sensitive to, the different stages of the interpretive experience so that interpretation can be tailored accordingly (Fennell, 1999).

Within the context of recognised stages of the interpretive experience, some workers have investigated how interpretation can be used to reduce the number of visitors to an area and/or discourage certain types of visitors (e.g. Beeton, 2003). Given the recognition that some destinations may not be able to sustain increased visitation or that a site may be at such capacity that there is the risk of environmental damage, 'de-marketing' can be used to educate potential visitors, discourage certain types of visitors and enforce required codes of behaviour (Beeton, 2003; Wearing *et al.*, 2007; Buckley, 2011). Table 6.1 provides an example of how pre-contact information can be used as a de-marketing tool in promoting ecotourism at Mulu National Park in Malaysia.

Application of Interpretation

The immense scope of interpretation is indicated by the set of objectives developed by the US National Park Service. These objectives include: providing information and orientation of visitors; educating visitors in relation to park resources and the national park system; bringing about understanding and appreciation of the natural world; fostering environmental protection; and opening up dialogue between the public and park management (Sharpe, 1982). Moreover, there are a number of situations where interpretation is critical in ensuring sustainable tourism and fostering visitor satisfaction. These include the case of fragile environments such as caves and the situation with sensitive species, as in whale shark and gorilla tourism, and situations where members of the public feed animals. In cases such as these ignorance and lack of awareness can be a major problem and can lead to accidental and unwitting negative impacts.

Various approaches have been prescribed in planning interpretation (e.g. Bradley, 1982; McArthur & Hall, 1996). McArthur (1998a) identified three important components: defining the target audience; content and structure of the materials; and the selection of a technique that suits the audience. He stressed that it is important to define the target audience because of the values and interests that people of contrasting ages, origins and various levels of education have. For example, the requirements of a local school group will be quite different from those of a group of foreign adults.

This initial stage should be followed by a consideration of interpretive structure and content, based on target audience characteristics and using the theme and message approach. The development of a theme focuses attention onto a particular aspect of the area being visited. This allows an interpretive activity to be completed in a particular time slot, and provides scope for including aspects of management and the interests of the interpreter.

The final step in planning interpretation is the selection of an appropriate technique to deliver the interpretive experience. McArthur (1998a) maintained that many interpreters make the mistake of selecting techniques before understanding the target audience or developing themes. Techniques can be personal, as in guided walks, or non-personal, where the reliance is on a visitor centre or signage, for example. The effectiveness of any particular technique will depend on group size, the age and interests of participants and their level of education. Different techniques can be utilised to cater for these different visitor characteristics.

A general approach that contains five main elements in the application of interpretation is discussed by Orams (1995a). This consists of establishing objectives, developing a specific theme, selecting appropriate techniques, engaging aspects of the psychological theory behind interpretation and then evaluating its effectiveness at the end. Establishing objectives consists of determining whether the focus is to be a combination of visitor orientation, enhancing awareness and increased understanding of management issues or a specific topic such as bird watching on a guided walk. As discussed by McArthur (1998a), themes and messages need to be developed. Orams (1995a) also highlighted the importance of utilising the psychological theory behind the Forestall and Kaufman model in providing answers to visitor questions and providing opportunities for motivated individuals to become involved in a particular environmental issue. Finally, interpretation plans and activities need to be evaluated and the cost and effectiveness of interpretation require monitoring for evidence of success or failure, so that modifications can be made (see p. 315 and p. 321).

Techniques Used in the Delivery of Interpretation

Overview

Various approaches and techniques can be applied in delivering interpretation. The major techniques include interpretive panels, brochures, guides and booklets, electronic educational resources (e.g. iPod and touchpad technologies, dedicated phone applications), visitor centres and displays, lecture programmes on cruise ships and tour boats, self-guided trails and guided tours undertaken as part of a walk, night drive or bus trip. Interpretation can be delivered to individuals, small groups (4–8 people) and larger groups (10–20 people, sometimes more depending on the situation). It can last for only several minutes or up to an hour or be part of a much longer tourism experience (perhaps for up to two weeks). The most widely used and important techniques (Table 6.2) are now considered in turn.

Table 6.2 Summary of major interpretation techniques

Technique	*Application*	*Strengths/advantages*	*Weaknesses/ disadvantages*
Publications and websites	Supply of pre-contact information; visitor orientation and trip planning; support for visitor centres and self-guided trails; information on landscape, fauna and flora	Cost-effective and portable information; many possible distribution/access points with wide dissemination	There is no active visitor involvement; does not necessarily cater for different visitor needs; can be expensive if subject to frequent updates and alterations
Electronic educational resources	Support for visitor centres and self-guided trails; information on geology, landscape, fauna and flora	Portable information and visitors able to explore trails at their own pace; high appeal to young people who are used to operating electronic devices; opportunities for frequent updating	There is no active visitor involvement; does not necessarily cater for different visitor needs; can be expensive if subject to frequent updates and alterations
Visitor centres	Information on landscape, fauna, flora and management; opportunity for face-to-face contacts with staff; located at the entrance gates to national parks and within popular nature based recreation areas	Recognisable sites where visitors can obtain information; scope for the application of a wide range of techniques (e.g. audiovisual, verbal interpretation, interactive displays and original objects)	Can be expensive to set up; may not be designed to cater for different audiences (e.g. focus may be entirely on school groups)
Self-guided trails	Focus of attention for visitors in various natural settings; opportunities to provide messages through signage	Always available and visitors can explore trails at their own pace	Signs and displays subject to vandalism; signage may contain too much information; generally not suitable for children
Guided touring	Wide application in all environments; especially important in forests, geological and wildflower tourism and during wildlife observation; time frames can be only 1 hour or up to 2 weeks	Very powerful and highly effective if applied properly; interpreter can respond to client needs and deal with various levels of complexity; information can be constantly updated; interpreter can facilitate active involvement	Requires the availability of well trained, certified and effective interpreters; training courses likely to be expensive; requirement for audience attention and commitment to being involved

Publications and websites

Publications and websites have the important role of orientating visitors to a natural area. They also frequently contain messages on visitor impact minimisation and wider environmental conservation; they can be used for both the marketing and de-marketing (see Table 6.1) of a nature-based tourism attraction. Many contain site maps on which footpaths are marked, which is the first stage in advising visitors where they can go to discover an area for themselves.

In most cases published material takes the form of brochures, pamphlets and information sheets that provide information on access, major site characteristics and wildlife. A typical example would be the pamphlet produced by the Royal Society for the Protection of Birds (RSPB) reserve at Loch of Strathbeg Nature Reserve in Scotland (Figure 6.3). It describes the importance of the reserve, the facilities that are available, the location of observation hides, its wildlife, elements of management and the seasonal occurrence of birds; it also presents environmental conservation messages.

While many pamphlets are essentially just two pages of information, others take more of a booklet form. The Rothiemurchus Estate (Figure 6.3) Visitor Guide, for example, contains some 20 pages of information about the area. Besides the usual visitor orientation and map, all of the recreational options for the area are described. These include guided walks, tours in off-road vehicles, fishing activities, bird watching, cycling and mountain biking. The specific management objectives and activities of the estate are explained, in addition to information on educational activities for schools. The location and nature of tourism facilities such as the Rothiemurchus Visitor Centre, camping and accommodation sites are also included.

Published material can also take the form of more substantial guides and books, as are for sale at visitor centres in many North American, European and Australian national parks and wildlife refuges. Well established tourism destinations such as the Kruger National Park in South Africa (Figure 6.3) have maps and 'where to find' guides that contain ecological information. The 'where to find' guides are designed to assist with self-discovery about the park and its wildlife. They are important resources that visitors can use for the largely self-drive wildlife viewing that the Kruger National Park provides for. Specific field guides to wildlife that occur in the park through to larger, more expensive books that contain numerous colour photographs also support the 'where to find' guides.

Websites are becoming increasingly important as providers of information and as a means of orientating potential visitors to various national park systems around the world. Information may include facilities available, activities undertaken and details of the key attractions. Many of them provide maps and wildlife checklists that can be downloaded and that contain suggested trip itineraries. Where remote video-cameras have been installed, some sites feature online streaming of a natural attraction.

Box 6.3 Case study: Electronic interpretation

Mobile tourist guides are becoming increasingly available in various forms and they have great application for use in natural areas where telecommunication is available. The use of mobile phones for navigational systems supported by destination information has increased dramatically in recent years (Kenteris *et al.*, 2011). Visitors to many natural areas can now access all kinds of multimedia information on GPS-supported digital tours, utilising smart phones and tablet computers.

Tailor-made geographical and thematic maps and satellite navigation can lead visitors to a destination's sites of interest and provide environmentally appropriate guidance on the location. The digital character of the system enables easy and real-time tour arrangements; thus, tours can be changed at any time and visitors can reach points of interest via different routes. Once a visitor reaches a point of interest, this is usually indicated with an acoustic message. At the point of interest the application employs location-based services with a variety of media, including text, verbal information, video and pictures.

Multimedia is the most advanced and up-to-date way of providing travel information on specific locations. The smartphone acts as a 'multimedia signboard' and provides a number of advantages over traditional methods of interpretation. Multimedia may be used to replace expensive interpretive signage in the natural environment. The electronic system also facilitates new ways of providing tourist and educational information. Such applications enhance both the efficiency of the learning experience and people's enthusiasm for understanding their surroundings, as they are both fun and instructive.

In addition to the mobile application, the system can be implemented in a stationary format as an information resource in tourist information centres (Figure 6.4). Tablet computers (e.g. iPads) can be integrated into the furnishing of tourist facilities to serve as a modern and technologically advanced tourist terminal, enhancing the professional capability, visibility and image. Barrier-free, multi-touch experiences, live information, social media and multimedia data are all integrated aspects of the applications, whether mobile or stationary. Updates and maintenance are routinely carried out using a content management system (CMS). The CMS can also serve as the web presence of the destination, providing pre-trip planning services to tourists.

The concept is suitable for all tourist destinations and is particularly useful for natural areas (e.g. national parks, geoparks, biosphere reserves and World Heritage sites). Furthermore, it is also appropriate and useful as a visitor resource in museums, zoos, stage-managed destinations and on recreational and adventure trails.

The service has been applied for geological trails in the Sultanate of Oman, where 30 geological attractions have been developed for geotourism (see Figure 8.3, p. 386). Another example is the GPS Adventure Park in Germany, where 15 GPS nature trails cover all relevant aspects of the natural and cultural heritage of the Teutoburger Forest Nature Park. Both projects accommodate applications for Android and Apple smartphones and tablets, and incorporate web portals, printed maps and brochures. The service has also been recognised by UNESCO for making an official contribution to the UN Decade of Education for Sustainable Development (2005–14).

Contributed by Henning Schwarze, INTEWO World Habitat Society

Electronic educational resources

Such resources now take the form of iPod and touchpad technologies, dedicated phone applications, interactive maps and audio trails (see Box 6.3 and Figure 6.4). For example, the USA National Park Service has developed a phone app that is free to download. It provides links to all national park websites in the USA. Visitors can also use the app to add their own description and photographs of a visit. Staff at the Hong Kong Geopark developed a 'talking pen' (Figure 6.5), which visitors to the park

Figure 6.4 Stationary terminal in a visitor centre for a network of protected areas in the cultural landscape of the Hoexter district in Germany. At such terminals visitors can access all the multimedia information on iPads. In addition, they can download the application for their personal device (smartphone or iPad) for free, to take the information into nature (Photo: Henning Schwarze)

Figure 6.5 The talking pen and interactive map developed by staff at the Hong Kong Geopark in order to facilitate self-guided touring (Photo: David Newsome)

can use when on a self-guided visit. The pen is placed on an interactive map that is carried by the visitor, who touches points of interest and is then able to listen to a commentary about the geology in their selected language.

Moncrieff and Lent (1996) describe an audio drive trail in the Dryandra Woodland in Western Australia (Figure 6.3) to facilitate self-guided touring. Six radio concealed transmitters have been placed along the 25 km road trail, powered by 60-watt solar panels (although batteries provide backup). Commentary points are indicated with a drive trail logo and a brochure provides additional information. A radio broadcast band is indicated at which to access the commentary (recorded on microchip) about conservation and various aspects of forest management at Dryandra. Moncrieff and Lent (1996) speculated on various possibilities associated with audio trails, such as visitors actually having to complete an activity, like crawling through a cave, in order to activate the transmitter, thus encouraging active participation.

The value of audio electronic media has been investigated by Kang and Gretzel (2012), who report on the importance of the human voice in engaging visitors and conveying information via podcast tours. Their study indicates that the human voice creates a constructive social context that facilitates positive tourist experiences and attitudes, with podcasts having the potential to increase the effectiveness of interpretation.

Visitor centres

Visitor centres (Boxes 6.2 and 6.4) provide a focal point for the tourist to obtain information, find out about walks and commence a sense of discovery about a particular area. They contain site maps, static and interactive displays, brochures and sometimes live exhibits. They also reinforce the interpretive panels that are provided on self-guided trails and can add to information that has been delivered on guided walks. Visitor centres can present material in different ways, such as aerial photographs or models of the landscape (e.g. Figure 6.6) and provide opportunities to understand content that has not been delivered via a pamphlet or fixed panel. With a visitor/interpretive centre there is the opportunity to deliver integrated content in the form of botanical, geological, faunal and cultural information. Electronic content, video technology and moving displays require funding and maintenance, otherwise, if non-functional, their presence can give the impression of neglect and a lack of professionalism.

The visitor centre at Brixworth Country Park in England (Figure 6.3) has a live tank exhibit (of pond life) that provides an example of the sorts of organisms that can be found in the ponds within the park. This display is then supplemented by the opportunity to borrow a magnifying glass, tray and net. Visitors can then go to a nearby artificially created pond and sweep for aquatic organisms. Identification charts and books are also available and in this way visitors are able to identify and find out about pond life for themselves.

Box 6.4 Bird watching at Rutland Water Nature Reserve, England

Rutland Water (Figure 6.3) has become one of the most important bird-watching sites in the British Isles. Originally developed as a water supply reservoir, the site now supports significant populations of birds and was designated a Ramsar site in 1991. The conservation area consists of two reserves, comprising woodlands, ponds and extensive lagoons juxtaposed with open grassy areas and adjacent agricultural land. The two reserves are actively managed to create a range of habitats in order to maximise the number and diversity of winter-visiting and breeding birds. Staff and observers have recorded 250 species of bird. During winter the area supports up to 20,000 waterfowl and comprises an important over-wintering site for various species of duck, goose and swan.

Facilities include two visitor centres, 19 bird-watching hides, walkways and nature trails. The main visitor centre sells field guides and educational material and contains an indoor gallery from which woodland birds and waterfowl can be observed. There is an environmental interpretation section containing displays about birds and information on how and why the reserve is managed as it is. There is also an interpretation centre on the south side of Rutland Water, at Lyndon. The reserve has its own website containing information on bird sightings, meetings and educational programmes. The weather and wildlife exhibit at the Lyndon centre highlights how weather affects wildlife and humans. Facilities such as these play a significant role in public education, especially with regard to climate change.

Figure 6.6 Interactive landscape model in a geo-interpretive centre, Grand Canyon National Park, USA (Photo: David Newsome)

Self-guided trails

Self-guided trails are many and varied but usually involve the visitor following a designated route along which various items of interest are indicated by signs. For example, the Mt Matilda Walk Trail, Wongan Hills, Western Australia (Figure 6.3) comprises a 7 km loop with a 4.8 km option plus lookout points. Design of such a trail requires an introductory sign (orientation/maps), an overall theme and a focus on observable features (Figure 6.7). This is the case for the Mt Matilda Wildflower Trail, which has one type of panel that brings attention to, and identifies specific, plants and, situated in different locations, larger panels where the focus is on local ecology, geology, landscape and wildlife. Such trails are often supported by pamphlets that explain, in further detail, features such as specific plants, geological features and plant–animal interactions. In the case of the Mt Matilda Walk Trail, a guidebook pertains to wildflowers that occur in the area.

There are many types of interpretive panel but there are several fundamental rules relating to accuracy, use of colour, the amount of information (not too much), prose style (text composed of short sentences), explanation via diagrams and photographs, a portrayal of things that can be readily seen at the site and the fostering

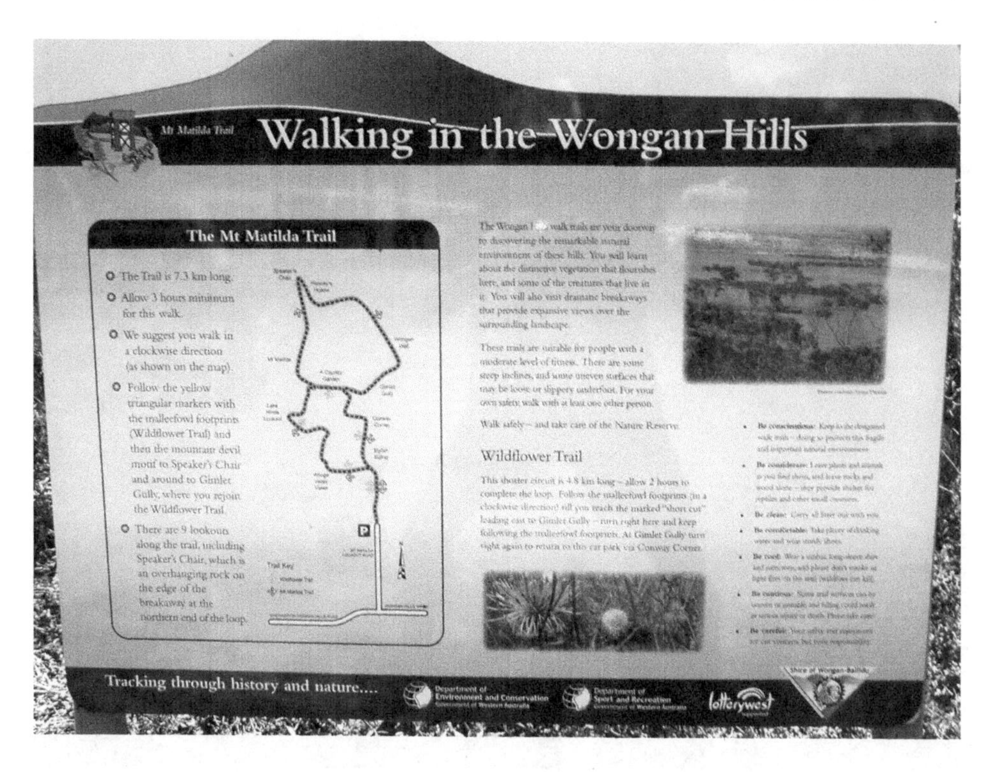

Figure 6.7 Information panel indicating the site, walk trails and main features of the Mount Matilda Walk Trail, Western Australia (Photo: David Newsome)

of active engagement (e.g. by posing questions, requesting the reader to examine something or by making an eye-catching statement) (e.g. Figure 6.8).

Besides walking, self-guided tours can also be completed via bicycle, on horseback or in a vehicle. Such self-guided trails have been developed in a range of settings, such as underwater (coral reef snorkelling trails), along riverbanks, in caves and in many forested environments. At Lamington National Park, Australia (Figure 6.3), there is a senses trail for the blind (and for sighted people, blind-folded). At Wave Rock in Western Australia a self-guided trail as been marked out by numbers that correspond to numbers in an illustrated field guide. In this way large interpretive panels (highly visible on exposed rock) are not required and instead the visitor searches for the corresponding information in the book once a less obtrusive numbered marker on the trail has been located.

The St Kilda Mangrove trail near Adelaide, Australia (Figure 6.3) is a 1.7km boardwalk that has been constructed within a mangrove system with the aim of developing a wider understanding of mangroves. Although guided walking does take place, there is a self-guiding visitor map for those who wish to undertake a

Figure 6.8 Interpretive panel located at a lookout at the Shark Bay World Heritage site, Western Australia. The panel presents the main elements of the region's ecology, illustrated with diagrams (Photo: David Newsome)

walk without a guide. The map contains information on flora and fauna and arrows highlight various aspects of the mangrove ecosystem along the boardwalk.

The Christmas Island (Figure 6.3) Nature Trail is 1 km long and serves to interpret the Christmas Island rainforest. There is a supporting booklet that describes 22 stops and visitors are urged to walk slowly and look carefully at and into the forest. Illustrated descriptions are used to assist in observation and help describe aspects of forest ecology and the nature of individual plants and wildlife. In this way the visitor can, for example, search the forest for buttress roots and locate them and then read the corresponding text by referring to the appropriate illustration.

Guided touring

A guided tour requires the services of a guide and, although usually undertaken on foot, can also involve the use of pack animals, boats, off-road vehicles and even coaches. They take place in all kinds of settings and involve a range of activities such as bird watching, wildflower appreciation, whale watching, tracking mammals, spotlighting and snorkelling. Sometimes the guide will be located at a fixed point, as is the case for the commentary that is provided at the Monkey Mia dolphin-feeding site in Western Australia. In some cases it is very difficult to experience the natural environment and wildlife without employing the services of a tour guide. Wearing and Neil (1999), Weiler and Ham (2001), Ham and Weiler (2002) and Black and Weiler (2005) maintain that guided touring is a powerful interpretation technique, with a guide being able to influence the experience of tourists in a natural area (Box 6.5). Tour guiding will be most effective when group sizes are small and when they are under the supervision of trained and certified guides (e.g. Black & Weiler, 2005).

An example of appropriate guided touring from the perspective of interpreting the natural environment for the visitor is provided by tour guides at the Buchu Bushcamp (Box 6.6), which is situated close to the De Hoop Nature Reserve on the southern coastline of South Africa (Figure 6.3). De Hoop offers opportunities to observe rare mammals and birds and to undertake land-based whale watching, as well as being an important flora reserve. These activities can be achieved on foot, by self-drive touring or through mountain biking.

Buchu Bushcamp Tours cater for small groups (1–5) and up to 60 people at a time. Large coach tour groups are given pre-visit information as to what books and equipment (hand lens, binoculars) to bring. On arrival they are split into smaller groups and leaders are assigned to each group. On a coastal tour visitors are asked to state what they can see and questions are posed as to why an animal 'is here and not there'. The sensitivity of coastal environments is discussed and explanations given as to why it is important to remain on footpaths. People are asked to feel and smell various plants and animals. There is an overall philosophy of active involvement and visitors are encouraged to contribute what they think about what has been covered in the tour.

Box 6.5 Role of the tour guide in natural area tourism

- Orientation of visitors
- Supervision of access to protected areas
- Safety of visitors
- Provision of information
- Mediator
- Role model
- Interpreter
- Visitor engagement
- Minimal-impact messages
- Monitor and manage tourists in reducing environmental impacts
- Demonstrate ethical practices
- Encourage tourists to make a contribution towards ecological sustainability

According to Weiler and Ham (2001); Ham and Weiler (2002) and Black and Weiler (2005)

Box 6.6 Buchu Bushcamp: Interpreting the South African fynbos vegetation

Fynbos means fine-leaved bush and is characterised by plants belonging to the plant families Proteaceae, Ericaceae and Restionaceae. These particular plant associations are very diverse and unique to South Africa. The reduced leaf size has evolved to reduce water loss under dry, windy conditions. Fynbos is also adapted to cope with periodic fire, which helps to maintain species richness. These aspects of fire adaptation and diversity provide ideal themes for interpretive guiding.

The approach taken by Buchu Bushcamp Tours is considered within the framework discussed by Wearing and Neil (1999), namely that of active involvement, use of the senses, self-discovered insights and usefulness of knowledge. Active involvement consists of walking through the vegetation, feeling its texture, noticing colours, smelling flowers and leaves and interacting with the tour guide. Use of all senses is encouraged. Visitors are asked to look for detail in the vegetation. The leaves of the plant family Rutaceae are crushed so that the different odours can be appreciated. The reason why the vegetation smells the way it does is discussed. The question of plant chemical defence is raised and participants are given the opportunity to taste various crushed leaves. Other leaves and plant structures are then handled for their textural characteristics. Aspects of herbivory and the grazing preferences of various animals follow on from this. The interpretation naturally moves on to plant–animal interactions and visitors are asked to listen for birdcalls. The different birdcalls are discussed and any observations noted. An exploration of fynbos diversity provides for self-discovered insights. Participants are asked to take a straight-line drop point from where they are standing down to soil level. They are asked to look at the detail and consider what is in the fynbos. People are asked to compare their observations. How many different plants did they see? The guide then discusses the different drop points and the number of plants that they did not see is revealed to them. Questions as to why the fynbos is so diverse and has so many endemic plants are raised.

All of these approaches provide for first-hand experiences of the fynbos. Examining pollination mechanisms and considering why plants have been given their scientific names support this further. The handling of *Heteropogon contortus* provides a striking example of the latter because the contorted nature of the grass can be readily appreciated in the hand. The importance of conserving fynbos and the role that diverse ecosystems have to play in our lives provides for the 'usefulness of knowledge' component discussed by Wearing and Neil (1999). Visitors are shown what constant trampling can do to fynbos vegetation and the importance of minimising impact is emphasised.

Allardice (2000) further maintained that the public-education benefits derived from walking through a designated area, crushing leaves and picking flowers, so that their structure can be fully appreciated, far outweigh the damage to the vegetation that may be a consequence of tourism conducted along the trail.

As already noted, the tour leader is the interface between the tourist and product/experience and has a role in orientation (knowing the route), locating species, providing information, identifying species, managing visitor expectations, giving explanations and facilitating sightings (e.g. provision of audio equipment to attract birds and managing a feeding interaction). They also perform a role in raising awareness of wider issues and managing visitor behaviour/adherence to codes of conduct. Accordingly, tour guiding is a critical component of sustainable tourism (Curtin, 2010).

Randall and Rollins (2009) explored the role of tour guides in shaping visitor experiences and particularly in relation to environmentally responsible and conservation-supporting behaviour. Using the example of tour-guided kayaking they found that clients highly rated the role of the guide as a communicator on environmental matters and motivator of environmentally responsible behaviour.

Ham *et al.* (2009) discuss approaches to communication as part of their guide for managers on how to influence visitor behaviour. The essence of their message is that interpreters need to provoke effortful thought, motivate visitors to engage and achieve motivation via relevance. Such approaches are especially important where visitors are engaging in ill-informed or inappropriate behaviours. For example, in the case of hikers who do not want to take everything out of an area that they carried in (non-compliers) it is important to understand not only their beliefs but also the beliefs of hikers who would carry everything out (compliers). A comparison of the beliefs of the two groups enables a targeted approach to communication. The manager/tour guide can then focus on the beliefs that differ most and target the beliefs with relevant messages that support the bring-in bring-out target behaviour. Following exposure to the intended message, visitors would have the opportunity to engage in the targeted behaviour with the intention that it leads to increased compliance.

Black and Weiler (2005) posit that the key mechanisms for improving the quality of guiding are codes of conduct, training, awards for excellence, professional associations and professional licensing. In relation to this Black and Ham (2005) discuss a model for tour guiding certification based upon the Australian EcoGuide Program. The main elements of the tour guide certification model they propose are: a sponsor, tangible benefits, appropriate fees, assessment options, qualified and experienced assessors, a code of ethics, implementation plans and a distinctive and marketable logo.

The Role and Effectiveness of Interpretation

Context

The purpose of this section is to highlight the importance of interpretation in visitor impact management and in improving environmental awareness among tourists. As already discussed in different contexts, interpretation is a vital component in managing tourism in sensitive environments and is particularly important in the case of cave tourism (p. 174), at many geological sites (p. 187) and in wildlife tourism (p. 178).

Tubb (2003) reiterated the work of Sharpe (1982), O'Riordan *et al.* (1989), Orams (1995b) and others in relation to the importance of interpretation as a tourism management strategy and noted that the fundamental goals of interpretation are knowledge gain and increasing awareness of visitors, fostering attitude change and behaviour modification, all of which can lead to better environmental outcomes. In a British study Tubb (2003) went on to explore knowledge gain, attitude change and behaviour modification as key indicators in assessing the effectiveness of interpretation delivered at a visitor centre in Dartmoor National Park in England (Figure 6.3). Using pre- and post-visit questionnaires, Tubb (2003) confirmed that interpretation added to visitor knowledge, which in turn resulted in greater awareness and thus provided the scope for behaviour modification and attitude change. It was noted, however, that the design of the interpretation material was important and that interactivity was key. Tubb's study and the work of others (e.g. Ham & Weiler, 2006; Weiler & Smith, 2009) provide an increasing body of evidence that highlights the importance of interpretation as a fundamental strategy in delivering visitor information and positive engagement. What follows are two case studies of the pivotal role of tour guiding as an essential component of best-practice ecotourism.

The importance of quality tour guiding: Orang-utan viewing at Bukit Lawang, Gunung Leuser National Park, Sumatra, Indonesia

Bukit Lawang lies at the southern edge of Gunung Leuser National Park (Figure 6.3) and is a focal point of tourism activities and local tourism development. The main attraction are the semi-wild orang-utans that are at various stages of rehabilitation following displacement from areas undergoing logging or that have been confiscated as captive animals.

As of 2009 the centre had 18 orang-utans, with some kept in cages as they were more difficult to rehabilitate. No more orang-utans were being accepted and instead new animals were to go to an orang-utan conservation project near Medan, to be subsequently released into Jambi National Park.

The future of ecotourism at Bukit Lawang is considered to lie in forest trekking and seeing orang-utans in the wild. Currently, the main attraction is a viewing platform where orang-utans are fed each day. Tourists hire a guide and take a guided tour that involves a river crossing and a short walk that terminates at a viewing area close to the feeding station. However, other activities at Bukit Lawang include short walks and one- to three-day treks, visits to hot springs, waterfalls, a bat cave and a tour to study medicinal plants.

Groups of 10 or more people are taken to the viewing platform by a guide. There is a fundamental reminder about the need for being informed about the ecological requirements of an animal, as visitors have to pass an old viewing platform en route to the newer platform located further into the forest. The old feeding platform was built too close to its viewing platform and was situated in an area of bamboo that was not very high and unlikely to support the weight of an adult orang-utan, so the only way animals could access that feeding station was from the ground. The orang-utan is arboreal and prefers to view the feeding station from trees to check for safe conditions. In addition, if rehabilitated orang-utans associate food with the ground, it is likely that they will be more vulnerable to predation by tigers and leopards. The currently used second feeding platform allows for good space between the platform and the ground and the orang-utans are able to descend to it from the safety of the trees. Orang-utans are then called to the site by a ranger banging on a tree. Visitors observe and photograph the orang-utans from a viewing area (Figure 6.9).

Figure 6.9 Orang-utan feeding platform at Bukit Lawang, Sumatra, Indonesia. Orang-utans frequently leave the platform and move among the tourists. Such interactions necessitate ranger presence to manage visitors and deliver interpretation (Photo: David Newsome)

A brochure in support of the activity is available in local guest-houses but this is not always apparent or provided when paying for park access. A visitor centre is located 1 km away, in the village, and a lack of interpretive panels on site may relate to the panels being destroyed by the orang-utans. Accordingly, there is likely to be no interpretation prior to or at the feeding experience except that provided by the guides. No contact with orang-utans is sanctioned and a 7 m approach distance is encouraged and actual contact time with the animals is set at a maximum of one hour. The information provided by guides is of variable quality and quantity. It is possible for visitors to leave the site without learning anything about orang-utans. Interpretation, when provided, centres around what the orang-utans are fed and how much. Orang-utans are known to disperse 200 species of plants and a deliberate monotony of diet (bananas and milk) at the platform encourages the orang-utans to seek their own food. Other information provided includes threats, predators, conservation status, the role of rehabilitation, diet and breeding behaviour.

An aspect of the experience, which is in particular need of guiding and interpretation, is that the orang-utans move around and come into close contact with visitors. Two aspects are problematic: firstly, tourists can lose items such as cameras and rucksacks; and secondly, orang-utans are susceptible to parasites, diarrhoea and respiratory tract infections that can be transmitted from humans and orang-utans at the centre may not have been vaccinated. Additionally, there is no tour guide training at present and thus no accreditation as a tour guide. This results in differences in quality, attitude and knowledge among the guides. Given that there are around 170 guides operating at Bukit Lawang at present, the lack of training is a major deficiency in delivering much-needed interpretation for visitor awareness. Furthermore, there is no coordinated ranger programme, no training in park management and rangers receive very low wages and supplement their income from guiding. Currently Bukit Lawang has 37 rangers, with 10 servicing the feeding station; moreover, the government budget changes from year to year.

The Pantanal: A tour operator's perspective

The account provided here serves to provide an historical example of poorly organised and executed tourism. Although the situation will have improved since 1990 there is the risk that tour operators and guides may engage in bad practices. However, these days, with a better informed public and information being readily available on the web (e.g. via TripAdvisor), potential tourists have the opportunity to study reviews and tourists readily provide comments on tourism practices, especially for high-profile nature tourism destinations such as the Pantanal.

The Pantanal (Figure 6.3) is a large wetland and is a major habitat for migratory birds, endemic species and species threatened with extinction. Spectacular and charismatic species include the hyacinth macaw (*Anodorhynchus hyacinthinus*), giant anteater (*Myrmecophaga tridactyla*), jaguar (*Panthera onca palustris*), maned wolf (*Chrysocon brachyurus*) and yellow anaconda (*Eunectes notaeus*). Trent (1991) considered the effects of uncontrolled visitation, ignorance and a lack of sensitivity towards wildlife. Littering was a widespread problem and its effects on wildlife was reflected in observations of birds that had swallowed plastic netting and cigarette butts. The lack of interpretation and widespread ignorance about Pantanal ecology and potential impacts of tourism were lamented. A tour operator himself, Trent observed visitors attempting to drive over snakes, approaching colonies of breeding birds too closely and deliberately putting birds to flight. In addition to these problems he noted the illegal collection of wildlife for the pet trade and hunting carried out by so-called tourists. Besides the obvious impacts on wildlife, many other visitors at the time would have found such behaviour offensive as well as reducing the effectiveness and quality of their tourism experience. Trent (1991) clearly highlighted the problems associated with ignorance. He maintained that many of the problems stemmed from a lack of education about the Pantanal as an ecosystem and the natural history of its wildlife.

The tours conducted by Trent aimed to provide such information. Furthermore, the company developed policies on waste disposal, recorded the presence of species, conducted population counts, reported on illegal activities and fostered environmental education. Clients were educated on the value of tropical habitats, wildlife and rainforests. A percentage of company profit was donated to conservation projects. The example provided by Trent (1991) clearly demonstrated the positive contribution that private tour operators can make in enhancing visitor experience and in fostering sustainable tourism.

In relation to some of the problems highlighted by Trent (1991), Randall and Rollins (2009) maintain that unsatisfactory guide performance is likely to be associated with an absence of monitoring and enforcement by managing agencies. They state that this can be resolved by the application of stricter licensing requirements for commercial tour guides, especially where they are expected to assist in resource protection (see discussion on certification in Chapter 5, p. 285).

Enhancing and Valuing the Role of the Tour Guide: Some Important Issues

Cochrane (1996), in a discussion on the sustainability of tourism in Indonesia, spelt out some of the problems associated with poorly planned and organised ecotourism development at Bukit Lawang, Gunung Leuser National Park, discussed above. A major problem was the attitude of poorly paid and poorly educated park wardens at the orang-utan rehabilitation station. At the time, park wardens ignored regulations and allowed visitors to feed and hold orang-utans. Cochrane (1996) maintained that there was sufficient evidence that animals constantly subjected to close contact with humans become or remain habituated, hindering successful rehabilitation. Cochrane (1996) cautioned that all of the work on orang-utan rehabilitation indicates that it should not be carried out as part of a tourist attraction. This has now been recognised and the focus has shifted to self-discovered insights. These include guiding in a variety of situations, such as tree-top walkways, trails with views, visits to hides and salt licks, and trails that offer walks of different distances. Informed guided forest walking leads to greater visitor satisfaction than do overcrowded, unsatisfying knowledge-deficient situations. There needs to be a sense of anticipation that compensates for the elusive nature of many rainforest species. Listening for calls, checking spoors and searching for footprints will add a sense of adventure and excitement. This is the real contribution that interpretation can make in eliciting visitor interest.

In the case of reducing impacts on rare and endangered species, ecotourism features as a motivating factor for the conservation of habitats that are under pressure from expanding and economically poor human populations. For example, the viewing of wild chimpanzees and gorillas has been developed in a number of central and east African countries. The endangered mountain gorilla (*Gorilla gorilla berengei*) continues to remain under threat from habitat loss and poaching. Continued clearing for agriculture and grazing has forced the animals into higher and less favourable habitat (Barnes, 1994). Tourism, which is promoted as an argument for conserving the gorillas, has been successfully developed in Rwanda, Uganda and in the Democratic Republic of Congo. In these countries various gorilla groups have been deliberately habituated so that tourists can gain close access. Some of the early tourism focusing on the eastern lowland gorilla in Kahuzi-Beiga National Park in the Democratic Republic of Congo (Figure 6.3) started with no controls over tourist group size and behaviour. It was reported that some groups numbered up to 40 people and that guides provoked displays in order to impress and entertain tourists (Butynski & Kalina, 1998). When park staff and guides are poorly paid they can be pressured to ignore rules and allow visitors to approach the animals closely. This, coupled with extended and twice-daily visits, disrupts social activity and increases the risk of physical contact between tourists and gorillas (Butynski & Kalina, 1998).

The importance of informing visitors and reducing contact with the gorillas is emphasised by Butynski & Kalina (1998), because of the animals' susceptibility to human disease. In 1988 six gorillas died as a result of respiratory disease. In addition to this 27 individuals were successfully treated with antibiotics. Moreover, these infections took place in three out of four tourist-habituated groups. Because of the threat and susceptibility of gorillas to measles, a vaccination programme was implemented and 65 gorillas were vaccinated. Vaccination and treatment with antibiotics are, however, not always successful. This was evident during the 1990 bronchopneumonia outbreak in a group of 35 gorillas visited by tourists. The disease affected 26 animals, four of which were given antibiotics, but two gorillas still died. The impact of disease is a significant issue where populations of a rare animal are isolated. In terms of bringing tourists into contact with gorillas, sustainable tourism therefore depends on adequately paid and trained interpretive guiding. Moreover, and as mentioned previously, interpretation needs to start at the pre-contact phase (p. 300) so that tourists are fully aware of the disease risk posed by humans before they visit the park.

The Tour Operator as a Role Model

In some cases tour operators may think that their practices are sustainable but participant observation has indicated otherwise. Wiener *et al.* (2009) assert that just because interpretation is in place we cannot assume that it is working or that it is being delivered properly. They investigated the interpretive programmes of commercial tour operators conducting dive, snorkelling, swim-with-dolphin and whale-watching tours around Hawaii. Friedlander *et al.* (2005) report that Hawaii receives large numbers of tourists (7 million a year), 80% of whom engage in marine tourism activities, serviced by 1000 ocean tourism companies; there is, thus, scope for overcrowding and the risk of negative impacts. Wiener *et al.* (2009), utilising participant observation on a sample of 29 tour boats, found that up to 50% of these operators were dumping food scraps, feeding fish, harassing wildlife and trampling coral. The interpretation being delivered was focused on personal safety and equipment use rather than on environmental issues. In many cases there was no interpretation relating to degradation of the marine environment or any encouragement of client involvement in conservation. Wiener *et al.* (2009) concluded that as well as the importance of tour guide training, guides and tour operators need to demonstrate knowledge, passion and concern for the environment and be good role models for their clients.

An example of what a passionate and interested tour operator can do is provided by Powell and Ham (2008), who report on the effectiveness of interpretation programmes conducted by Lindblad Expeditions on the Galapagos Islands. Powell and Ham (2008) in particular set out to investigate whether targeted interpretation programmes can lead to measurable outcomes in terms of conservation-supporting behaviour. Lindblad Expeditions have a comprehensive interpretation programme, part of which involves tourists directly in the conservation needs of the islands. Interpretation is provided on a range of subjects but there is also a presentation by a member of the Charles Darwin Research Station on threats to the Galapagos National Park. Opportunities are subsequently provided for tourists to contribute to the Galapagos Conservation Fund set up by Lindblad Expeditions (for further details see Powell & Ham, 2008).

Powell and Ham (2008) showed that there was almost 100% satisfaction and that 70% of the tourist sample indicated a moderate to strong intention to donate money to the Galapagos Conservation Fund. Lindblad Expeditions is a company passionately concerned with conservation and environmental protection of the Galapagos Islands. The Galapagos Conservation Fund that it set up has raised US$3 million to support conservation works on the Galapagos Islands and is a case where tourism development (with interpretation as an integral component) can benefit conservation in important natural areas.

Views on the Effectiveness of Interpretation

Interpretation can increase visitor knowledge, encourage behaviour modification, enhance environmental awareness and foster conservation-supporting behaviour (e.g. Madin & Fenton, 2004; Andersen & Miller, 2006; Christensen *et al.*, 2007; Dearden *et al.*, 2007). However, in the same way as pointed out by Wiener *et al.* (2009) with tour operators, Howard (1999) reported on a survey conducted in Vanuatu in the South Pacific, where divers were asked whether their activities had any impact on the coral reef. It was found that 90% of respondents indicated that there was no impact on the coral reef. Howard (1999) nevertheless found that when divers were observed in the water they frequently made contact with the coral while trying to counter buoyancy problems or while examining a particular coral feature. It was also observed that the number of contacts depended on the approach taken by the instructor leading the dive. In a similar study Medio *et al.* (1997), in the Ras Mohammed National Park system along the Egyptian Red Sea coastline (Figure 6.3), set out to show how briefings about coral sensitivity and awareness of diver impacts on the reef reduced damage. Diver behaviour and activity were observed and the number of contacts divers made with coral was recorded over eight weeks. During this period environmental briefings were delivered to divers with the aim of testing whether they had any effect on diver behaviour. The briefings consisted of the same points raised by Howard (1999), covering coral biology and impacts as well as ideas about the role of protected areas in conserving coral.

Recordings of diver behaviour after the briefings showed that the rate of contact with coral substrate decreased from 1.4 to 0.4 per diver for every 7-minute observation period. While there was increased and allowable contact with non-living coral, contact with living coral declined from 0.9 to 0.15 instances per diver per 7-minute period. Medio *et al.* (1997) calculated that there is a large potential impact from the many dives that take place. Moreover, such potential impact can be significantly reduced by the use of environmental briefings and interpretation delivered by instructors and guides.

Although interpretation is a key component of visitor management and satisfaction, there will always be some tourists who do not want to engage with it, especially in the case of reading static displays such as interpretation panels. The role interpretation plays in modifying visitor attitudes may therefore be overstated (McNamara & Prideaux, 2010). Munro *et al.* (2008) also caution that assessing visitor knowledge and attitudes remains separate from quantifying actual behaviour change resulting from an interpretive experience. The views of Munro *et al.* (2008) and McNamara and Prideaux (2010) therefore indicate that there is scope for further investigations into the long-term effectiveness of interpretation. The work of Tubb (2003), Madin and Fenton (2004) and Powell and Ham (2008), however, provides us with evidence that well designed and engaging interpretation, particularly when delivered by passionate and interested guides, achieves the goals of high visitor satisfaction, environmental protection and conservation-supporting behaviour.

Conclusion

The value of interpretation in enhancing the visitor experience and reducing impacts is not in doubt. Although some writers express caution as to the effectiveness of interpretation in managing natural area tourism and fostering public interest in the protection of nature, the available evidence indicates that it can be effective in achieving these aims. Nevertheless, we must be mindful as to how interpretation is delivered and constantly strive to improve performance in order to achieve desired objectives. This can be achieved via codes of conduct, training (Australian EcoGuide Program), awards for excellence, professional associations (e.g. Interpretation Australia, Ecotourism Australia), professional licensing, and monitoring and evaluation.

There is a trend for tourists from Western countries to seek natural experiences, undisturbed conditions, ecological integrity and authenticity. Such people are also increasingly focused on the need for information and inspiration from nature. This trend, coupled with the need for park agencies to manage increasing visits to national parks and other natural areas, provides for much scope in the wider use and application of interpretation.

Many people can be unaware of and disconnected from nature, and interpretation therefore plays a vital role in bringing understanding to the wider public so that people are able to make informed choices about the impacts of climate change, habitat loss, pollution, weeds and feral animals. It is hoped that in the future various approaches to interpretation can be used in crowded and mass tourism situations to help alleviate associated problems of graffiti, lack of respect for wildlife, off-trail impacts, litter, lack of interest and cheap amusement and noise. Engagement through interpretation is relevant in the case of many Asian national parks that are crowded with domestic visitors during weekends and public holidays (e.g. Mt Pangerango-Gede in Java) and in the case of mass tourism at Halong Bay in Vietnam. Significant numbers of visitors to Cuc Phuong National Park, Vietnam (100,000 year) show a lack of awareness about the local ecosystem and domestic visitors have articulated that quiet is boring and that it is fun to be noisy. These visitors are also of the impression that the forest is scary and it is necessary to create noise to scare away snakes.

The importance of interpretation therefore lies in educating, engaging, informing and inspiring people about nature. A major objective is to increase understanding of the natural world and provide a view as to how we see and value nature in the hope that we can offset and turn around exploitative attitudes and indifference. Interpretation can enhance our understanding of the need for sustainable practices and the preservation of nature and biodiversity.

Further reading and sources of information

There are now many associations and sources of information regarding the art of interpretation and further details can be obtained, for example, from the web links of Interpretation Canada, Interpret Scotland, Interpret-Europe, the National Association for Interpretation (US), Interpretation Australia, the Association for Heritage Interpretation (UK), Interpretation Network New Zealand (NZ) and the ICOMOS International Committee on Interpretation and Presentation.

Sharpe (1982) gives a comprehensive account of interpretive techniques, while a particularly useful coverage of the design of interpretive signage is given by Moscardo *et al.* (2007). McArthur and Hall (1996) and McArthur (1998a) provide overviews and discussion on planning for interpretation. Ham (1992) provides excellent coverage of the principles and practice of interpretation and focuses on the practicalities of conducting activities and self-guided tours. The important practice of tour guiding is discussed further by Weiler and Ham (2001). Additional readings include Beck and Cable (1998) and Pastorelli (2002).

Ballantyne *et al.* (2000) developed a workbook and DVD that are designed to enable trainees to develop guiding skills and to stimulate discussion of the application of interpretive skills in various natural settings.

Ham and Weiler provide strong leadership in the area of environmental interpretation and education in the tourism context. Additional recommended readings include Ham and Krumpe (1996), Ham *et al.* (2008), Weiler and Ham (2010) and Curtis *et al.* (2010).

7 Monitoring

Introduction

Monitoring has long been a neglected element of natural area management. Today, however, monitoring is absolutely essential for managers, who are increasingly being required to report on the outcomes of their activities. This chapter describes why and how monitoring must be a part of managing tourism in natural areas. It begins by defining monitoring and explaining the reasons for doing it before detailing guiding principles.

Most of the chapter is devoted to describing ways of monitoring the impacts of visitors on the natural environment and visitors' attributes, activities, needs and perceptions. Built facilities and campsites, roads and trails, water bodies (taken to include marine environments) and soundscapes are covered. Approaches to monitoring visitors are then described. Given the importance of setting standards against which changes can be assessed, a section is devoted to this. Details on monitoring visitor satisfaction and their future behavioural intentions regarding recreating in natural areas follow. Information on interactive techniques, including the internet, concludes the section on visitor monitoring. The last part of this chapter covers system-wide monitoring, especially how to evaluate the effectiveness of management actions using a framework developed and promulgated by the World Commission on Protected Areas. The conclusion links this chapter to earlier ones, especially those on impacts and planning.

Definition

Monitoring is the systematic gathering and analysis of data *over time*. Griffin *et al*. (2010: 11) go further, commenting that monitoring for the effective management of a protected area 'requires the systematic gathering, analysis and integration into management systems of data relating to both the natural environment and visitors over time'. For natural area tourism, a comprehensive monitoring programme will collect data on the natural environment and its visitors. Information on the natural environment could include vegetation cover, damage to vegetation, weed invasion, soil properties (especially erosion), water quality and wildlife populations (e.g. changes in breeding success, distribution). Such data can be used to identify and quantify site-specific impacts (Monz, 2000).

Four areas of focus for visitor monitoring can be identified (Moore *et al.*, 2009; Griffin *et al.*, 2010):

(1) visit/visitor counts – total number of visits to a site, park or park system;
(2) visitor characteristics – demographic and socio-economic attributes of individuals (usually collected as quantitative information) as well as their psycho-social characteristics (collected as quantitative and qualitative information, such as reasons for visiting, attitudes, motivations, expectations and preferences);
(3) visit characteristics – mostly quantitative information on use patterns, group size, length of stay, frequency of visits and activities undertaken;
(4) visitor outcomes – quantitative and qualitative information on satisfaction, experiences and behavioural intentions.

Ideally, visitor preferences (point 2) help inform the selection of indicators and standards as part of the planning frameworks described in Chapter 4. These preferences can include biophysical conditions, for example the acceptable level of erosion on a walk trail, as well as social conditions, for example the acceptable number of parties encountered per night while camping.

Reasons for Monitoring

The lack of adequate monitoring of visitors to protected areas and their impacts continues to be lamented. Despite increasing visitor numbers and impacts, especially to iconic sites such as waterfalls and mountain tops, and increased reporting requirements being placed on the managers of protected areas, monitoring appears to be inadequate or non-existent (Buckley, 2004a; Leung & Monz, 2006; Hadwen *et al.*, 2007). Additionally, where visitor monitoring has been undertaken, it is often done so in an *ad hoc* manner and/or the results are not incorporated in management decision making (Griffin *et al.*, 2010).

Numerous, sound reasons for monitoring the impacts of natural area tourism and of tourists themselves exist.

(1) Management of natural areas – monitoring provides the information needed to ameliorate impacts and assess management effectiveness

Managing the impacts of visitor use is essential if the natural environment is to be protected and visitor experiences maintained. Monitoring provides information, not only on when management intervention is required; it can also improve managers' understanding of the cause–effect relationship between levels and types of visitor use and the resultant impacts (Cole, 1989; Marion, 1995; Monz *et al.*, 2010a). Such understanding is essential if use is to be managed to prevent further impacts or, in many cases, to reduce the existing levels of impact. Additionally, if an impact can be detected early, it is probably cheaper and easier to remedy before it reaches a threshold of irreversible change (Buckley, 1999b).

Monitoring also allows managers to assess the effectiveness of management strategies once in place. The outcomes of different strategies, assessed by measuring changes in resource conditions and/or visitors' perceptions, can be compared. Data from one point in time are insufficient for this assessment, hence the need for monitoring rather than a single inventorying exercise.

(2) Planning – monitoring provides the information needed for management plans, recreation/tourism planning and site design activities

Essential data for planning include visitor numbers and characteristics, activities, resource impacts, patterns of use, satisfaction and expectations. These data can be provided by a one-off inventorying exercise or preferably from an ongoing monitoring programme. Monitoring also allows the success of these plans to be determined, as well as indicating when revision is needed and assisting in the revision process. The planning frameworks outlined in Chapter 4 rely on monitoring for measurements of initial baseline conditions and then subsequent repeat measures of selected indicators to determine the amount of change.

(3) Resource allocation – monitoring provides managers with a systematic basis for allocating funds and resources

Managers need a systematic basis for allocating funds and resources, such as staff, within and between natural areas. Without reliable data on environmental impacts, visitation levels and patterns of use, allocation is based either on a manager's intuitions or on external pressures such as finance, staff constraints and political directives (Marion, 1991; Pitts & Smith, 1993; Griffin *et al.*, 2010). Although managers' intuitions will often stand them in good stead, frequent staff turnover mean that changes in the natural environment and visitors often go unnoticed. Using reliable data enables resources to be allocated where they are most needed, for example to a site where damage to vegetation or overcrowding needs management attention, rather than to a less affected site that a manager intuitively believes needs attention. Such strategic allocation of resources improves the economic efficiency of natural area management.

Environmental impact and visitor monitoring data also have a crucial role to play at the corporate level in helping land management organisations seek funding from government and other sources (Pitts & Smith, 1993; Wardell & Moore, 2004). Senior agency staff may use total visitor numbers, visitor satisfaction or total areas impacted (e.g. kilometres of trail impacted by horses) to justify requests for financial support. Also at the corporate level, these cumulative data, such as total visitor numbers to a state's national parks, may be used to develop community awareness by demonstrating the extent of use and support for such areas.

(4) Public accountability – monitoring provides information to the corporate levels of land management agencies to assist with accountability and transparency

Increasingly, the activities of natural area managers and their organisations are being subject to public scrutiny. Performance reporting, where monitoring data are made publicly available, is one way of meeting public requests for accountability. In

Australia, organisations such as the Western Australian Department of Environment and Conservation (WA DEC) report annually to the state government on their performance. Examples of tourism performance measures include number of visits to DEC-managed sites and visitor satisfaction (WA DEC, 2011).

'State of the Park' reporting, where an agency systematically evaluates its management across a protected area system over time, is a relatively recent phenomenon (Hockings *et al.*, 2009). Cope *et al.* (2000) provided an early review in describing how the North York Moors National Park Authority used State of the Park reports to develop, record and review sustainability indicators and trend indices. Landres *et al.* (1994) emphasised the increasing importance of park managers being able to assess and report on the status and trends over time of the natural area they manage.

At a basic level, a description of the system is required: the number of natural areas and their sizes. Objective data documenting the values and threats to this system are also useful, as are trends in conditions and visitor demand, reported via a few core indicators. Lastly, understanding how society values natural areas and the benefits of such areas to society is essential for political forums.

In Australia, the New South Wales National Parks and Wildlife Service has been through two iterations of State of the Parks reporting for its statewide system of national parks and nature reserves. At the time of the first review, in 2004, the New South Wales protected area system included 639 parks with a combined area of 6 million ha. The first iteration provided basic-level reporting, essentially an inventory of park attributes. The second iteration was more comprehensive, describing values and threats to the park system and evaluating key issues, such as visitor impacts, and influences on visitor experiences, such as the adequacy of visitor facilities (Growcock *et al.*, 2009; Hockings *et al.*, 2009).

(5) Enhancing visitors' experiences – monitoring can provide information on what experiences visitors expect and whether these expectations have been realised

Many protected areas have a dual mandate of protecting the natural environment and providing nature-based recreational experiences for visitors. There is a growing interest in ensuring these experiences are high quality, as part of best practice, but also to foster loyal visitors who will behave well in protected areas, visit again, recommend such areas to others and be advocates for these areas in the wider world.

Monitoring can provide information on the types of experiences visitors are seeking and obtaining, their levels of satisfaction with their visit, and what contributes to or detracts from quality experiences for them. It can also provide information on their 'loyalty', for example whether they would recommend the area to others. The resulting information can help managers to create new experiences or modify existing ones and manage the contributing factors.

(6) Marketing and interpretation – monitoring provides the information needed to successfully market and interpret natural areas

Understanding who visitors are and what they want is fundamental to successfully marketing a natural area. And increasingly, managers are trying to meet the needs of specific user groups, each with their own particular information and interpretation requirements. Such an approach requires information on the motivations and characteristics of different segments of the user population (Beh & Bruyere, 2007; Konu & Kajala, 2012).

(7) Legislative and legal requirements – monitoring may be legally required

For many land managers, the legislation guiding their operations (for example the US National Park Service is guided by the National Park Service Organic Act of 1916) requires that visitor experiences are to be provided, but only to the extent that the natural environment is unimpaired. Monitoring shows when 'impairment' has occurred and is therefore needed to meet this legislative mandate. Management policies and guidelines may also explicitly require visitor impact monitoring. Marion (1995) noted that in several places in the US National Park Service management policies, monitoring of resource impacts is prescribed.

Monitoring may also be legally prescribed as part of the approval of tourism developments subject to environmental impact assessment (EIA). A study of tourism developments in Australia by Warnken and Buckley (2000) noted that of the 175 developments subject to EIA from 1980 to 1993, only 13 had formal resource monitoring. Obviously, such a requirement is not widespread. Most of the projects requiring monitoring were coastal resorts and day-visit pontoons on the Great Barrier Reef (Figure 7.1). Parameters included reef fish and coral communities, mangrove and seagrass communities and water quality.

(8) Obtaining information to include in simulations to assist and improve decision making

Visitor management requires information on the spatial and temporal movements of visitors. It is often prohibitively expensive to collect these data from on-the-ground observations. Simulations, where data are strategically collected and then used in models to simulate visitor behaviour more broadly in an area, can provide a cost-effective means of describing where visitors go and what they do. Simulations can be used: (1) to describe where visitors go in large natural areas (by recording their movements at the periphery and extrapolating these to the interior); (2) to model the number of encounters on trails over time; (3) to test the consequences of proposed management changes, such as building a new visitor centre, on visitor numbers and movements; and (4) to provide visual representations of different management options so that visitors' preferences can be determined (Lawson, 2006; Lawson *et al.*, 2008).

Most of these simulations require information on visitor numbers and their patterns of use. To test proposed management changes, such as constructing new campgrounds or new or extended access roads, simulations using autonomous agents (also known as individual-based models) can be undertaken. RBSim (Recreation Behaviour Simulator) (Itami *et al.*, 2003) is an example of this type of simulation. The additional information required for these simulations, usually collected with questionnaires, includes when and where visitors arrive, their travel movements, and how long they stay at particular places within the destination.

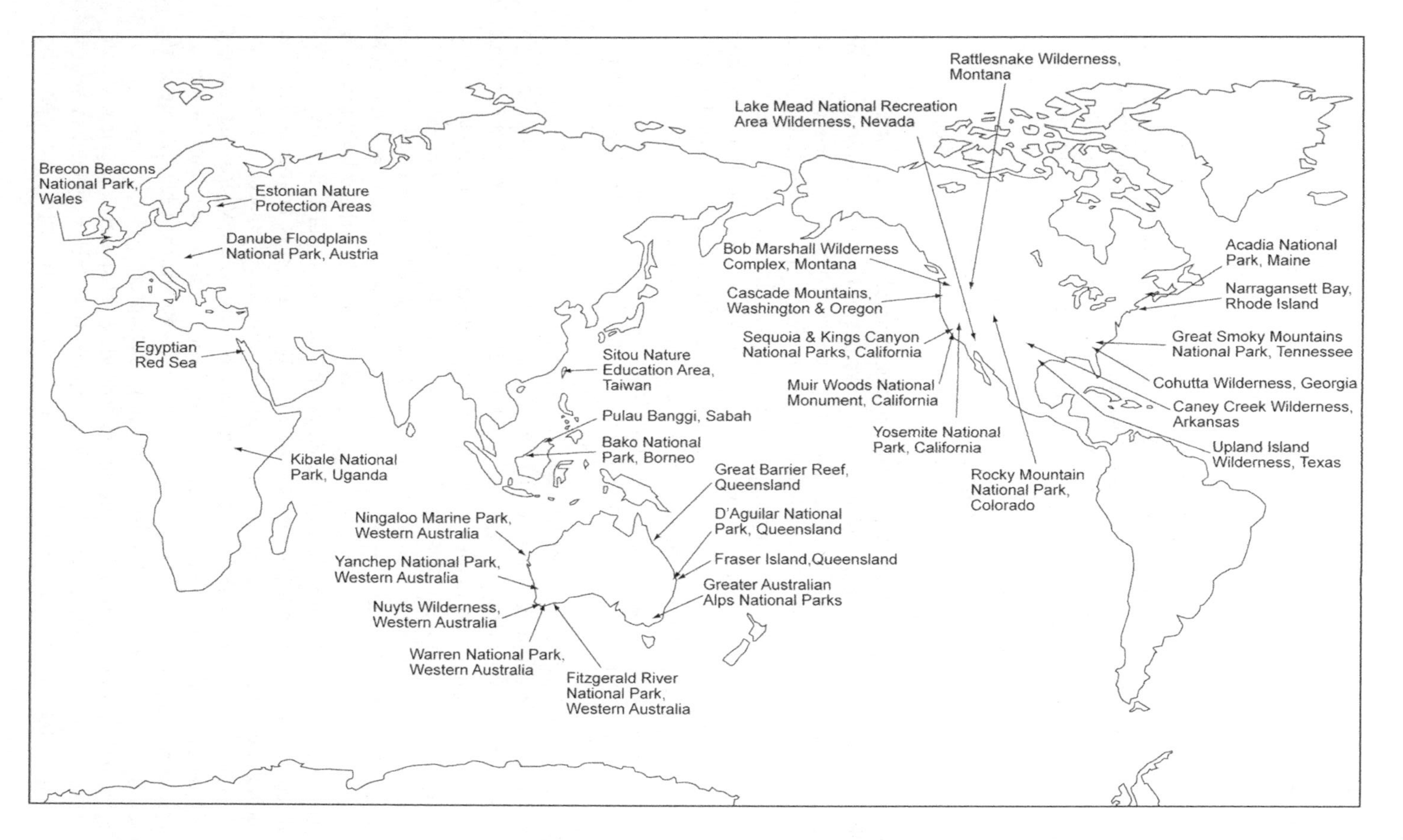

Figure 7.1 Location map of important nature-based tourism destinations referred to in this chapter

Principles of Monitoring

The following principles should guide the development of a monitoring programme for natural area tourism.

(1) Clear objectives are integral to the success of a programme

The objectives will depend on the reasons for monitoring. Monitoring should never be an end in itself. Reasons range from managing the impacts of visitor use through to legal requirements accompanying approval of tourism developments. Objectives must be clearly identified before commencing activities, otherwise important information may not be collected and time and money could be wasted on collecting unnecessary data (Marion, 1991; Eagles *et al.*, 2002). Cope *et al.* (2000), from their study of visitor monitoring by resource management organisations in the UK, concluded that the value of data is greatest when monitoring objectives are explicit and clear at the outset and remain the focus throughout data collection.

(2) A well planned and managed information storage and retrieval system is essential

Monitoring requires comparison of repeat measures of the same parameters. As such, data need to be stored where they are safe but accessible so that new data can be compared with the old. It is also desirable that the form of storage allows simple manipulations and analysis so that comparisons between data collected at different times can be readily made. Such information systems need not be grandiose and expensive so long as they are based on a systematic, standardised process for data collection and entry, a structured computer-based storage system, and clear instructions and simple procedures for entering, manipulating and retrieving information (Wardell & Moore, 2004; Griffin *et al.*, 2010). Where possible, data should be geo-referenced.

(3) A sampling strategy providing cost-effective, robust data is crucial

Careful consideration of where and how much data are collected is crucial to the success of a monitoring programme. This must take place during the design phase. Environmental impacts may be monitored by census, with all sites surveyed, or by sampling only a subset of sites (Marion, 1991). With sampling a subset, the sites must be selected in such a way that the results are generalisable to all sites and measurements can be repeated over time. For trails, for example, if a description of the condition of the whole trail is the objective, then either the whole trail can be censused or a subset of randomly located points can be sampled and the results generalised to the whole trail (Leung & Marion, 2000; Marion & Leung, 2001). The first approach will be much more expensive and accurate than the second. The choice depends on the objectives of monitoring and available resources. Visitor monitoring generally relies on sampling, and inferences about the whole visitor population are drawn from this subset of the larger population (Pitts & Smith, 1993; Moore *et al.*, 2009). Sampling strategies are discussed later in this chapter.

Ideally, environmental monitoring associated with natural area tourism would involve measurements before and after a development takes place, as well as records from the impacted site and a similar, nearby, undisturbed control site. This approach –

before, after, control, impact, paired (BACIP) sampling – is widely applied in ecology (Bernstein & Zalinski, 1983). Unfortunately, there are few opportunities for its application in natural area tourism because most of the sites requiring monitoring have been in use for many years. However, this does not preclude managers using control sites where possible and when major developments are proposed ensuring that pre-development monitoring is done.

(4) Quality assurance is critical to the success of a programme

Quality assurance is a key element of any monitoring programme. It is achieved by training staff involved in data collection, entry and analysis, calibration and regular checking of counting devices (e.g. traffic counters), regular feedback to staff and regular reviews of performance (Marion, 1991). Performance reviews could include regular checks of how surveys are conducted, completed data sheets or visitor questionnaires, overall system performance, and the information needs, objectives and priorities of the whole monitoring programme (Pitts & Smith, 1993).

Quality is also improved by providing standardised means for conducting monitoring and recording the data obtained. Field data forms can be used to collect information on environmental parameters such as vegetation (e.g. Marion *et al.*, 2011) and visitor information such as activities and numbers at a particular site (e.g. Moore *et al.*, 2009). Questionnaires are a common means of collecting information from visitors themselves. Computer databases are the best way of providing a standardised data storage, manipulation and retrieval system. Manuals to guide monitoring, such as the ones produced by Marion (1991), Hornback and Eagles (1999) and Horneman *et al.* (2002), also assist in quality control.

(5) Skilled managers must lead and take responsibility for designing and implementing monitoring programmes for their natural area

The responsibility for monitoring generally rests with managers, as part of their suite of management responsibilities. Marion (1991), in his manual on monitoring visitor impacts in US national parks, noted that the usefulness of monitoring programmes is entirely dependent on the managers who initiate and manage them and that programmes developed in isolation from other resource protection decision making will be short-lived. To help managers cope with limited resources, seasonal staff or students may be contracted, so long as they are adequately trained and supervised. Roles for scientists could include assistance in designing a monitoring programme, including the use of new technologies and analytical techniques, and periodic reviews to ensure that the quality of data is being maintained and programmes are scientifically defensible. Scientists may also have ongoing research interests that could form part of such a programme.

Any new monitoring programme should build on what already exists. For visitors, existing data may include direct counts from traffic counters or entry tickets or indirect counts from permits, trail registration or records of attendance at an interpretive programme. Building on and modifying such systems recognises years of practical experience by field staff and should help them adopt the 'new' programme (Pitts & Smith, 1993).

Developing a Monitoring Programme

Many of us would like to get straight out into the field and start measuring environmental impacts or interviewing visitors. However, to make sure resources are used efficiently and effectively and that the measurements taken are useful and repeatable, planning is necessary before any fieldwork is undertaken. This thinking is best guided through developing a monitoring programme based on the principles above and the steps outlined in Table 7.1. Key steps include evaluating the need for the programme, establishing explicit objectives and documenting the data collection methods. Such documentation is essential if data are to be consistently and reliably collected, given that repeat measures over time will be made, potentially by different people.

In the past, monitoring programmes were regarded as the responsibility of managers and researchers. Today, contributions to monitoring programmes by tour operators and community members are widespread. For example, a worldwide monitoring programme for coral reefs (Reef Check) draws on community contributions (Hodgson, 2001). Reef Check relies on training, guidelines and protocols for surveying and data collection to ensure that accurate, useful data are collected (Reef Check, 2012a).

Table 7.1 Steps in a monitoring programme for resource impacts and visitors in natural areas

Step	*Description*
(1) Evaluate need for the programe and determine objectives	• Review of legislative, legal and policy reasons for monitoring; often critical in enlisting organisational support • Determine agreed objectives
(2) Review existing approaches	• Examine what has been done elsewhere and previously in the area to be monitored
(3) Develop monitoring procedures	• Select techniques (e.g. photo points, visitor surveys) based on considerations of accuracy, precision, sensitivity and cost • Test techniques, including pilot testing of any visitor survey forms, and modify as needed
(4) Document monitoring protocols and provide training	• Produce monitoring manual, field data forms, survey forms for visitors, computer database • Provide staff training
(5) Conduct monitoring fieldwork	• Plan fieldwork. For resource impacts, this is best conducted by a small number of evaluators working full time for a short period towards the middle or end of the recreation season. For visitor monitoring decide where and when to survey (surveys based on personal contact give the best response rate)
(6) Develop analysis and reporting procedures	• Use computer, if possible, to store, analyse and retrieve data
(7) Apply monitoring data to management	• Establish priorities based on data • Undertake actions

Derived from Cole (1989), Marion (1991), Wardell and Moore (2004)

Monitoring Visitor Impacts on Natural Areas

A wealth of research on monitoring visitor impacts on the natural environment in the back-country areas of national parks and wilderness areas in the USA exists (Leung & Marion, 2000). Over the last two decades, this research has extended to include the more developed front-country of national parks in the USA, Australia, UK, Africa and China (Obua & Harding, 1997; Deng *et al.*, 2003; Mende & Newsome, 2006; Randall & Newsome, 2008; Olive & Marion, 2009; Kincey & Challis, 2010; Kim & Daigle, 2011). Most of this work is directly applicable to natural area tourism, with some minor modifications and additions to cover the monitoring of facilities at the highly developed end of the spectrum, such as lodges and resorts.

Techniques are grouped in this chapter according to whether they are used for terminus points, such as resorts, campgrounds and campsites, or linear features, such as roads and walk trails. Techniques for monitoring visitor impacts on water bodies (including marine environments) are also briefly reviewed. This section concludes with information regarding monitoring sound. For each grouping, details are provided on the monitoring techniques, indicators and assessment procedures, and associated sampling strategies.

Built facilities, campgrounds and campsites

Campsites in wilderness and back-country in protected areas in the USA have been the focus of significant monitoring efforts. More recently, more developed campgrounds in national parks have been incorporated in these monitoring efforts, in countries like Australia (Smith & Newsome, 2002) and Uganda (Obua & Harding, 1997). Such efforts are highly relevant to natural area tourism, both in wilderness areas and in more developed natural areas, including large campgrounds with built facilities. These techniques can also be applied to monitoring the environmental impacts of facilities such as resorts; however, there is little evidence that these approaches have been applied at this more developed end of the recreation spectrum.

Four general approaches to monitoring campsites exist. All aim to provide information for managers so they can take action before impacts become unacceptable. These approaches and their advantages and disadvantages are summarised in Table 7.2. As these approaches were developed for back-country sites in North America, some of the indicators will not be suitable or will need modifying before they can be applied elsewhere. For example, for front-country in many national parks, campsites will be delineated by a gravel pad or cleared area. As such, bare ground will not be a useful indicator of impact in these situations.

Most of these approaches include campsite area as an indicator. This can prove time-consuming to measure but is essential if changes over time are to be detected. The variable radial transect method (Marion, 1991) is widely advocated as the most accurate (Figure 7.2), with the geometric figure method providing a quicker but less accurate alternative (Figure 7.3). The former is based on flagging as many points on the campsite boundary as are necessary to define a polygon approximating the area. The distance from a marked, fixed centre-point to each point, plus the

Table 7.2 Summary of campsite monitoring techniques

Monitoring technique	*Advantages*	*Disadvantages*	*Application*
(1) Photographs of the site: (a) from a fixed point; (b) through 360° to create a panorama; (c) of vegetation quadrats; and/or (d) using overhead digital imaging followed by software analysis, plus GPS location data See Newman *et al.* (2006), Monz and D'Luhosch (2010) for information on (d)	Quick, relatively low cost; visually documents extent and location of impacts; may document campsite size	(a)–(c) may not provide accurate quantitative measures of changes in indicators; comparisons between photos are often impossible due to differences in cameras and lens; (d) practically limited to sites of 25 m^2 or less	Best used as a supplement to other data collected in the field
(2) Condition class rating – a single rating (1–5) is given to each site based on the degree of vegetation loss, tree damage, bare mineral soil and erosion See Frissell (1978) for original rating system and Marion (1995) for system based on ground cover conditions only, as many sites lack trees	Quick, relatively low cost; provides a rapid survey when large numbers of campsites are spread over large areas; rapidly provides data on which campsites are most seriously impacted	Does not provide accurate quantitative measures of changes in indicators; the single rating does not provide information on the severity of each impact type (e.g. is vegetation loss or erosion more serious?) and thus selection of an appropriate management response is difficult; rating classes are so broad that major changes may occur before allocation to another class occurs	Best used for assessing large natural areas where there are numerous sites and the resources are not available to spend more than a few minutes at each site; may be used in combination with multiple indicator approaches
(3) Multiple indicator ratings – ratings for individual indicators recorded and then summed to give a summary impact score* for each site See Parsons and MacLeod (1980), Cole (1983a)	Quick – a campsite can be evaluated in 5–10 minutes – as well as providing a lot of information at relatively low cost; accuracy is sufficient to detect changes over time in campsites as well as categorising the status of existing sites; can identify the most serious impacts	Information on individual indicators is not very precise (large variations in ratings for an indicator made by different evaluators) although the overall score does not vary greatly between different evaluators	Particularly useful for rapidly assessing a large number of sites
(4) Multiple indicator measurements – measures taken of multiple indicators See Cole (1989), Marion (1991), Cole *et al.* (2008)	Provides a large amount of precise information; amenable to statistical analysis (e.g. multivariate techniques)	Time-consuming – 30 minutes to 2 hours per site; vegetation quadrat and soil sampling skills may be required	Useful for checking if rapid assessment measures are accurate; feasible only for small numbers of campsites

*Such summing is regarded as a statistically improper procedure because of the wide variation in assessment units between indicators (e.g. number of trails versus m^2 of site area)

Derived from Frissell (1978), Parsons and MacLeod (1980), Cole (1983a, 1989), Marion (1991), Hammitt and Cole (1998), Leung and Marion (1999c, 2000), Monz (2000), Newman *et al.* (2006), Cole *et al.* (2008), Monz and D'Luhosch (2010), Monz and Twardock (2010)

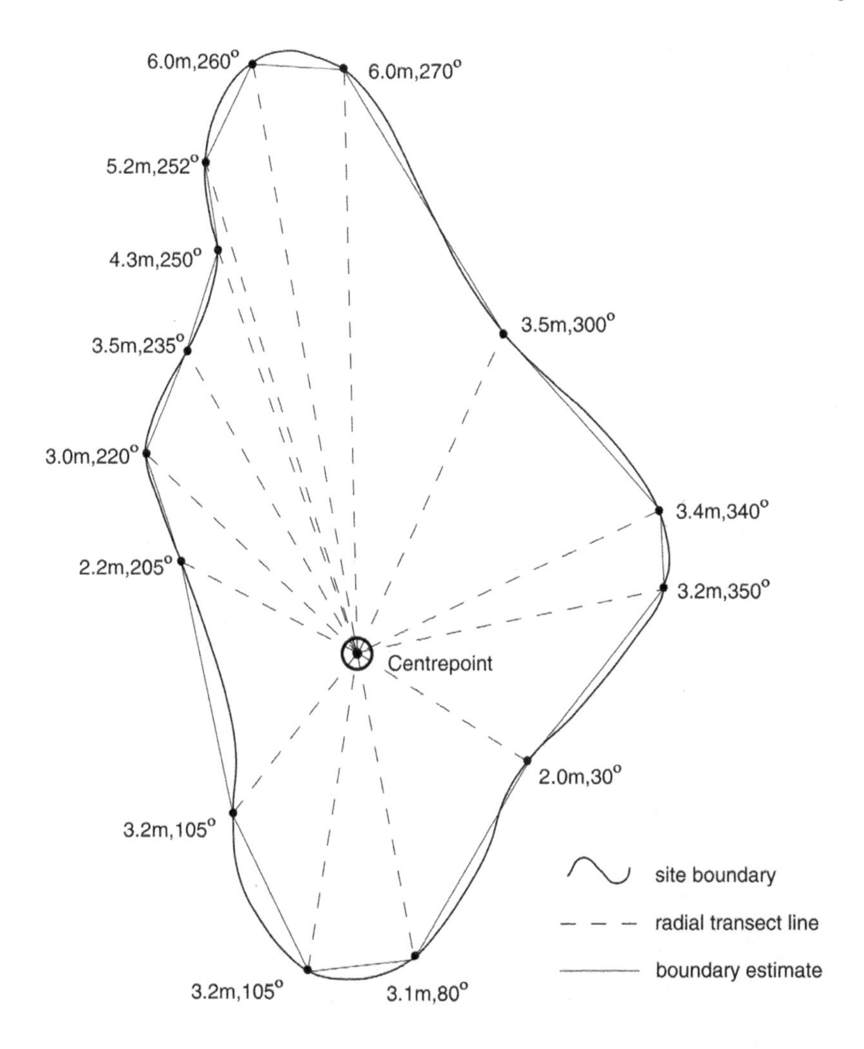

Figure 7.2 Variable radial transect method for measuring site area (Derived from Marion, 1991)

associated compass bearing, is then recorded. The result could be a list of as many as 15–20 lengths and bearings. Computerised arithmetic procedures can then be used to calculate the area (Marion, 1991). The geometric figure method is based on approximating the area of the campsite to common geometric figures and calculating its area accordingly.

Time limitations in the field and measurement error are frequent concerns expressed by managers regarding the radial transect and geometric figure methods (Monz & D'Luhosch, 2010). These concerns have led to recent investigations regarding the feasibility of digital photography, and subsequent software analysis

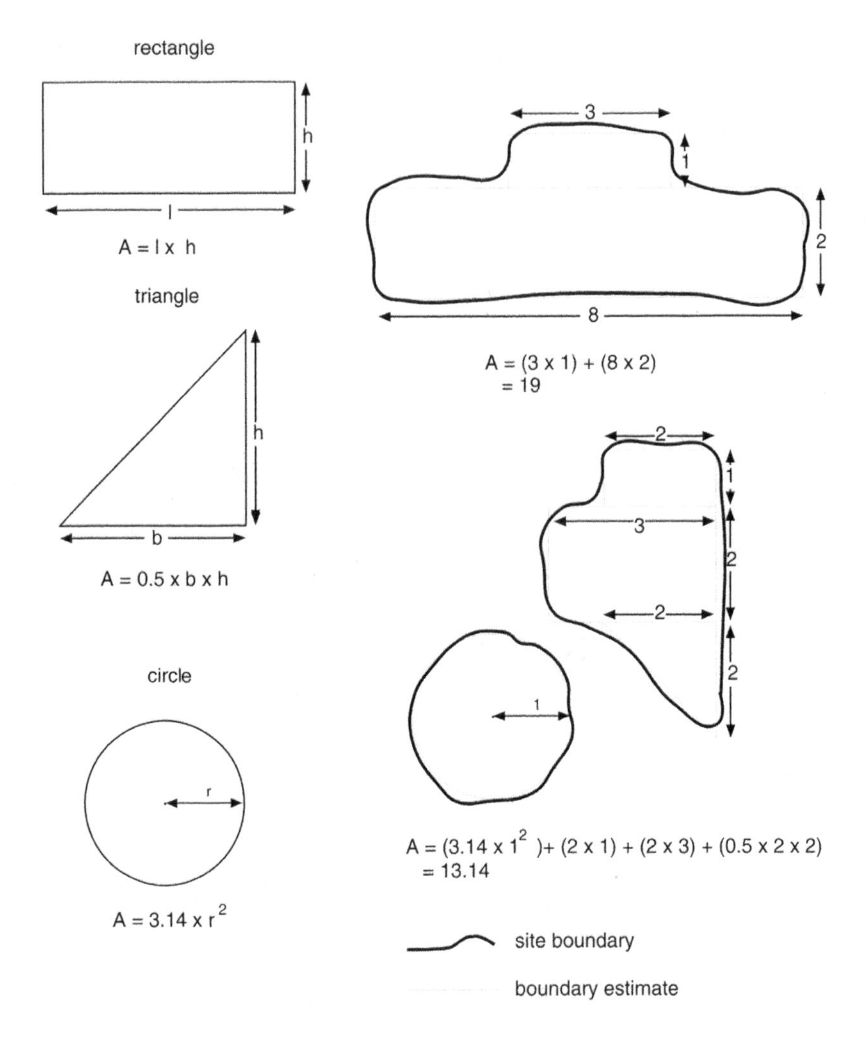

Figure 7.3 Geometric figure method for measuring site area (Derived from Marion, 1991)

of the captured images, as a viable alternative (Newman *et al.*, 2006a; Monz & D'Luhosch, 2010). Field experiments suggest a similar level of measurement error for this image-based technique and the traditional radial transect method (~6%) and for repeat measures using the latter method alone (4–8%) (Monz & D'Luhosch, 2010).

A digital camera with a wide-angle lens, held on a boom over the campsite, can be used to capture a single image of sites up to 25 m² and multiple images of larger sites. One image is preferable, as the multiple images from the larger sites require merging, which, although technically possible using programs such as Photoshop, can be time-consuming (Monz & D'Luhosch, 2010). Image analysis software (e.g. Able Image

Analyzer) is then used to determine campsite size (and vegetation cover; see section below on photographs). Before taking the photograph, the boundary of the campsite is delineated using flagging tape. Global positioning system (GPS) location data, to assist in spatial analyses and help relocate the campsite for future monitoring, can be collected separately or automatically through direct connection to the camera.

Irrespective of the method used, the centre-point of campsites must be marked in such a way that it can be accurately relocated in the future. Recording its co-ordinates using GPS technology is becoming widespread practice.

Photographs

The first approach, taking photos, provides a visual record of the site. Photos may be taken from a fixed point, and then subsequent photos used to determine changes over time. Another approach is to take a 360° panorama of the site by rotating a camera located at fixed point. Repeat photos may also be taken of vegetation quadrats (usually a square of 1 m × 1 m, laid over the top of the vegetation) to determine changes in vegetation cover. Digital image capture and analysis, with the camera mounted on a boom over the campsite, can provide information on vegetation cover (Newman *et al.*, 2006a; Monz & D'Luhosch, 2010). Photographs are recommended as a supplement to other forms of data collection rather than as a technique used on its own (Marion, 1991; Hammitt & Cole, 1998).

Condition class rating

The second approach, developed by Frissell (1978), is to determine the condition class that best describes the campsite being surveyed. The five classes range from class 1, with vegetation flattened but not permanently damaged, through to class 5, with obvious soil erosion (Table 7.3). This system is quick and easy to apply and provides a useful single value for each site (Marion, 1991). This value can be used to compare the levels of impacts for campsites across an area. Repeat monitoring can be used to indicate movements from one condition class to another. Because of the inaccuracy of this approach and information lost through aggregating impact

Table 7.3 Condition classes for monitoring campsites in natural areas

Class	*Description*
1	Ground vegetation flattened but not permanently damaged. Minimal physical change except for possibly a simple rock fireplace
2	Ground vegetation worn away around fireplace or centre of activity
3	Ground vegetation lost on most of the site, but humus and litter still present in all but a few areas
4	Bare mineral soil widespread. Tree roots, where present, exposed on the surface
5	Soil erosion obvious, as indicated by exposed tree roots and rocks (where present), and/or gullying

Derived from Frissell (1978) and Marion (1995)

information into a single rating, condition class ratings are generally only used to complement other forms of data collection.

Multiple indicator ratings

The third approach, variously referred to as a multiple-indicator approach (Leung & Marion, 1999c), multiparameter systems (Marion, 1991) and multiple parameter systems (Hammitt & Cole, 1998), involves collecting information on a number of indicators. A widely used indicator rating system is that developed by Parsons and MacLeod (1980) for Sequoia and Kings Canyon National Parks in the USA (Figure 7.1) and subsequently modified by Cole (1983a). It assigns ratings to a suite of indicators. Cole (1983a) used ratings of 1–3 for nine indicators – campsite area, barren core area, condition of ground cover vegetation, exposed bare mineral soil, damage to trees, exposed tree roots (a measure of erosion), extent of development, site cleanliness, and number of associated trails ('social' trails) (Table 7.4). A raw value and rating are recorded for each indicator. For example, for campsite area, the rating categories were (1) 0–50 m^2, (2) 51–100 m^2 and (3) >100 m^2; for a campsite area of 60 m^2, this area and a rating of '2' are recorded.

A summary impact score is then obtained by summing the raw or weighted ratings for the nine indicators. Parsons and MacLeod (1980) used raw individual ratings while Cole (1983a) weighted his ratings. Although it is simpler to have equal weightings, this implies that all types of change are of equal importance. For example, bare mineral soil and camp area might be much more important than cleanliness of the site. As such, they should be weighted more heavily. A weight for each indicator should be decided by managers, in consultation with stakeholders. When applied to the Bob Marshall Wilderness Complex (Figure 7.1), scores ranged from 20 to 60. This range was divided into four impact classes: light (ratings of 20–29), moderate (30–40), heavy (41–50) and severe (51–60) (Hammitt & Cole, 1998). This system has also been used to quantify the impacts of natural area tourism in Kibale National Park in Africa (Box 7.1, Figure 7.1).

When using this monitoring system it is essential to adapt it to the specific setting. Modification of the rating descriptions and the range of each summary impact score category may be necessary to ensure that approximately equal numbers of sites fall into each category. If these changes are not made then differentiation of sites for different levels of management attention will not be possible. For example, if all the sites fall into the severe impact category then it will be impossible to prioritise management actions. Also, for developed sites such as an 80-bay campground, the rating choices for campsite area will need to be very different from those for a back-country campsite.

Multiple indicator measurements

Taking individual measurements of a number of indicators is the best way to get accurate, replicable data for individual campsites. Table 7.5 lists and describes the most commonly used indicators. Recording visual indicators, plus measuring the campsite area, takes two workers 10–15 minutes per site (Marion, 1991). Although more time-consuming than the multiple indicator ratings system, it is still feasible

Table 7.4 Indicators commonly used in multiple indicator ratings systems

Indicator	*Description*	*Rating*
Campsite area	Estimation (by tape, pacing or eye) of the total area trampled	(1) 0–50 m^2 (2) 51–100 m^2 (3) >100 m^2
Barren core area	Estimation (by tape, pacing or eye) of the area denuded of vegetation	(1) 0–5 m^2 (2) 6–50 m^2 (3) >50 m^2
Condition of ground cover vegetation*	Relative measure (from walking around) of the extent of vegetation cover within the campsite compared with a similar, adjacent unimpacted area	(1) Site and control belong to the same coverage class** (2) Coverage on site is one class less than on control (3) Difference in coverage is two or more classes (e.g. 51–75% at control, 6–25% at site)
Exposed bare mineral soil*	Relative measure (from walking around) of the extent of bare soil within the campsite compared with a similar, adjacent unimpacted area	Rate as for condition of ground cover vegetation (above)
Damage to trees	Number of lower branches broken, boles hacked, carvings and nails	(1) No damage/a few broken lower branches (2) 1–7 mutilations (e.g. axe marks, carvings, cut stumps) (3) >7 mutilations
Exposed tree roots (a measure of erosion)	Number of exposed roots as a simple measure of soil erosion	(1) No trees with exposed roots (2) Three trees or less with exposed roots (3) Four or more trees with exposed roots
Extent of development	Number of human 'improvements' such as fire rings, seats, tent pads and windbreaks	(1) Nothing more than a scattered fire ring (2) Nothing more than 1 fire ring and rudimentary seats (3) More than 1 fire ring or seats, tables, windbreaks, levelled tent pads etc.
Site cleanliness	Amount of charcoal, blackened logs, litter, human waste and horse manure	(1) Nothing more than scattered charcoal from 1 fire site (2) Either scattered charcoal from >1 fire site or some litter (3) Horse manure, human waste, widespread litter, remnants of campfires
Number of associated trails ('social' trails)	Number of trails radiating from the campsite and their degree of development to provide a measure of impact on surrounding areas	(1) No trails discernible (2) One or two discernible trails, no more than one well developed trail (3) Either more than two discernible trails or more than one well developed trail

*Relies on assessment of an adjacent, similar, unimpacted site (i.e. a 'control' site); **Coverage classes are 0–5, 6–25, 26–50, 51–75, 76–100%
Derived from Parsons and MacLeod (1980), Cole (1983a)

Box 7.1 Applying a multiple indicator ratings system to monitoring campsites in Kibale National Park, Uganda

Kibale is one of 10 national parks in Uganda. It covers about 560 km^2 of high forest, grassland and swamps. The forest has a rich diversity of trees and animals, with 11 primate and 325 bird species. Ecotourism commenced in 1992 following the establishment of campsites and nature trails. Increasing visitor numbers led to concerns regarding campsite degradation (Obua & Harding, 1997).

A multiple indicator ratings system, as developed by Parsons and MacLeod (1980) and modified by Cole (1983a), was used to assess the extent of campsite degradation by visitors (Table 7.4). The nine indicators were vegetation loss, mineral soil increase, tree damage, root exposure, development, cleanliness, social trails, camp area and barren core area. All the ratings were based on visual estimates. Assessing vegetation loss and increase in mineral soil involved comparing the campsite (S_0) with nearby undisturbed forest (S_1). Five classes (0–5, 6–25, 26–50, 51–75 and 76–100%) were used to describe cover. Ratings were then allocated as follows: both the site (S_0) and control (S_1) belong to the same coverage class was rated 1; coverage on S_0 is one class lower than on S_1 was rated 2; the difference in coverage on S_0 and S_1 is two or more classes was rated 3.

A rating of 1–3 was given to each indicator according to the severity of the impact. Each rating was then multiplied by a weighting before being summed to give an overall impact score for each campsite. The ratings and weightings were those developed by Cole (1983a). For example, a campsite where the trees had no exposed roots was given a rating of 1, from a choice of 1–3. This rating was multiplied by 3, the weight assigned to exposed tree roots, before being summed with the other weighted ratings to give an overall impact score. The summary impact score for this campsite was 24, described by Obua and Harding (1997) as a low level of impact. This designation was based on the following categories: 0–29 low, 30–40 moderate, 41–50 high and 51–60 severe.

Obua and Harding (1997) concluded that more than three-quarters of the campsites had experienced some form of degradation. This finding was supported by work elsewhere showing that even low levels of use cause noticeable impacts (Hammitt & Cole, 1998).

to apply this approach to a reasonably large number of sites. Leung and Marion (1999c) surveyed 195 campsites in the Great Smoky Mountains National Park using this approach.

More comprehensive vegetation quadrats and soil measurements can also be added but such work is generally reserved for research or monitoring of a small number of sites due to the substantial field time required. This comprehensive assessment, plus collecting data on visual indicators, can take up to two hours per site (Hammitt & Cole, 1998). About 15–20 vegetation quadrats at each campsite and an adjacent control are sampled to determine groundcover loss, species composition and exposed soil. Soil parameters, such as organic matter and soil compaction, are similarly measured at the campsite and control, with multiple samples taken (Cole & Marion, 1988; Cole, 1989; Cole *et al.*, 2008). Skilled, experienced staff are needed.

Table 7.5 Indicators commonly used in multiple indicator measurement systems

Indicator (including measurement unit)	*Measured visually*	*Descriptions and comments*
Campsite size (m^2)	No	Measured using the variable transect or geometric figure methods (Figures 7.2, 7.3)
Groundcover loss (%)*	Yes/no	Determined using six cover classes (e.g. class 1, 0–5%) and derived by subtracting from the value for a nearby, similar, undisturbed control site to give percentage loss *or* 15–20 1 m × 1 m permanent quadrats on site and nearby control – infrequently used because of time involved
Exposed soil (%)*	Yes	Determined using six exposure classes (e.g. class 1, 0–5%) and derived by subtracting from the value for a nearby, similar, undisturbed control site *or* 15–20 1 m × 1 m permanent quadrats on site and nearby control – infrequently used because of time involved
Trees with exposed roots (number, %)	Yes	Number of trees with exposed roots or as a percentage of all trees on the site
Tree stumps (number, %)	Yes	Number of tree stumps or as a percentage of all trees on the site
Damaged trees (number, %)	Yes	Number of damaged trees or as a percentage of all trees on the site
Fire sites (pits or rings) (number)	Yes	Number of fire sites
Visitor-created social trails (number)	Yes	Number of visitor-created social trails
Cleanliness (rating, number)	Yes	Determined by categorising the amount of charcoal, blackened logs, litter, human waste and horse manure – not widely used
Campsite development (rating, number)	Yes	Determined by categorising the number of human 'improvements' such as fire rings, seats, tent pads and windbreaks – not widely used
Impacts to soil organic horizons (number)*	No	Measures include organic horizon cover, organic horizon depth and degree of disturbance of litter and duff from site and control; one sample from each of the vegetation quadrats established to measure groundcover loss – infrequently used because of time involved
Impacts to mineral soil (number)*	No	Measures include soil compaction (bulk density or resistance of the soil to penetration), water infiltration rates, moisture content, organic matter content and chemical composition; more than five samples from site and control needed – infrequently used because of the time involved in fieldwork and laboratory analyses

*Relies on measures of an nearby, similar, undisturbed site (i.e. a 'control' site)

Derived from Cole (1989), Marion (1991, 1995), Hammitt and Cole (1998), Leung and Marion (1999c), Smith and Newsome (2002)

Combined systems

A combined approach using ratings and multiple indicator measurements has been developed and applied by Leung and Marion (1999c) (Box 7.2). The visual indicators are quickly measured, with vegetation cover and exposed soil estimated for the site and a control rather than using quadrats. Soils are not sampled.

Box 7.2 A combined system for monitoring campsites in Great Smoky Mountains National Park, USA

Great Smoky Mountains National Park, with an area of 209,000 ha and located along the border of Tennessee and North Carolina (Figure 7.1), is one of the most visited national parks in the USA. About 470,000 overnight stays were reported in 1998, in 87 designated back-country campgrounds and 18 shelters. Each designated campground typically consists of two to four individual campsites. A study was conducted by Leung and Marion (1999c) to improve assessment procedures and understanding of back-country impacts.

A total of 377 back-country campsites were assessed, 308 legal and the remainder illegal. Site size, number of fire sites and damaged trees and stumps were assessed for all sites. A ratings system was used to give each site a condition class: class 1 sites were barely evident while class 5 sites had lost most vegetation and litter cover and were eroding.

To reduce field assessment time, only campsites rated class 3 or above were assessed further (195 sites). For these sites, eight indicators were selected and measured, based on their ecological and managerial significance, use in other studies and their level of measurement being suitable for multivariate analysis. They were grouped as *area disturbance indicators*, including campsite size (ft^2), fire sites (number of pits or rings) and visitor-created social trails (number); *soil and ground cover damage indicators*, including trees with exposed roots (%), absolute groundcover loss (%) and exposed soil (%); and *tree-related damage indicators*, including tree stumps (%) and damaged trees (%). The groundcover and exposed soil indicators required measurements at the site and at an environmentally similar but undisturbed nearby site.

Leung and Marion (1999c) concluded that two aspects of impact, spatial extent (areal measures) and intensity (percentage measures), could be used to differentiate four campsite types. The first type of campsite, low impact (38% of campsites), had minimal areas and levels of vegetation and soil disturbance. These sites were generally remote from trails and water sources such as streams. Moderately impacted sites (24%) had low to medium areas and levels of impact. Intensively impacted campsites (21%) had high levels of soil and ground cover damage but the area disturbed was small compared with extensively impacted campsites. The area of these sites appeared to be constrained by topography and dense vegetation. Extensively impacted campsites (8%), while only showing intermediate levels of soil and ground cover damage, had large areas of disturbance. For these sites there were no topographic or vegetative barriers restricting site expansion or proliferation of new campsites in adjacent areas.

In this study, the combination of a rapid but standardised set of field procedures with statistical analysis provided a useful assessment tool for a large number of campsites. For managers it seems more effective to design management strategies based on the campsite types differentiated in this study rather than individual impact indicators (Leung & Marion, 1999c). For example, strategies for intensively and extensively impacted sites are likely to be different. For the former, the focus will be controlling soil erosion and site restoration. For extensively impacted sites, the priorities are more likely to be controlling the expansion and proliferation of sites.

Choosing a monitoring strategy

Choice of a monitoring approach involves considering accuracy, precision, sensitivity and cost. Accuracy refers to how close a measurement is to its true value, while precision refers to how close repeated measures are to each other. Both are important and generally improved by using more time-consuming techniques. Sensitivity refers to how large a change must be before it can confidently be identified as a real change in conditions. A system using a small number of broad categories will have low sensitivity. For example, if a system for campsite size has three categories, with the third described as >100 m^2, a site of 150 m^2 could triple in size without the site moving to another impact category. The last consideration is cost, with staff time often being the limiting factor.

Deciding which approach to take also depends on the objectives of the monitoring programme, as well as the resources and time available. How many sites to survey, the sampling design, is an important consideration. Managers ideally would like an inventory and then repeat monitoring of all sites, to provide data on total number of sites, their distribution and levels of impact. Cole (1983a) recommended inventorying all sites and, if funding is insufficient to take measurements at all sites, using an estimate system at all sites rather than taking detailed measurements at only a few. Such an approach is a census rather than a sampling process, so concerns regarding selection of a representative sample of sites are largely irrelevant.

Campsite censuses have been undertaken in a number of iconic national parks. For example, a complete census of back-country campsites in Yosemite was undertaken in 1972, taking 28 people a full summer to complete (Boyers *et al.*, 2000, cited in Newman *et al.*, 2006a). Most back-country campsites in Grand Canyon National Park were surveyed over the period 1985–92 using a rolling approach, with ~8% of sites located and their condition assessed each year (Cole *et al.*, 2008). A census approach was similarly taken in Smoky Mountains National Park, by Leung and Marion (1999c), when they surveyed 377 back-country campsites.

Newman *et al.* (2006a) note that often a complete survey cannot be done because of the effort and resources required. They developed a geographic information system (GIS) for the Merced River corridor in Yosemite National Park, including details on distance from trail-heads, landscape slope, distance from water, distance visitors tend to travel off trail, and presence of designated no-camping zones, to help them predict where back-country campsites might be located. They then field surveyed a subset of the potential 'campable area' identified using this GIS and found that 8% of this area had evidence of a campsite. They concluded that their method had potential for providing a measure of the total area of campsites in the Merced corridor.

In their campsite monitoring work in Grand Canyon National Park, Cole *et al.* (2008) selected 24 campsites for detailed examination. Their rationale was as follows: the number of sites selected (i.e. sample size) was a balance between the number of sites that could be examined in one field season and the number required for reasonable statistical power. They chose four campsites in each of six combinations of use level (high and low) and vegetation type (pinyon–juniper, catclaw and desert scrub vegetation types), giving a total of 24 sites. Apart from examples such as this, there is no general guidance available about the 'best' number of sites to sample the campsites adequately in a particular park or park system. The resources available will always be a determining factor.

If resources are very limited, a rating system may be the best approach. More preferable is a combined ratings and multiple indicator measurement system, as employed by Marion (1995), Leung and Marion (1999c), Cole *et al.* (2008) and Monz and Twardock (2010). Little can be saved in time or money by collecting data on a small rather than large number of indicators, given that most of the time spent on monitoring in natural areas is devoted to travelling between sites. Most programmes seem to rely on eight or more indicators. From Leung and Marion's (1999c) work it appears that indicators of both the spatial extent (e.g. campsite area) and intensity of impacts (e.g. number of tree stumps) are needed.

Roads and trails

Much research effort has been directed to developing and applying monitoring techniques to trails in back-country and wilderness areas, similarly to campsite monitoring. Trails are taken here to include 'formal', managed (designated) trails and 'informal' trails created by visitors seeking access to points of interest or taking shortcuts. Trail monitoring techniques are relevant to other natural areas, although will need modification before they can be applied to sealed and gravel roads and tracks from four-wheel drive vehicles. Most of the following concerns the monitoring of trails because limited information is available on the monitoring of the roads and tracks used for natural area tourism, although a brief description of a monitoring system for gravel roads concludes this section.

A number of techniques are available, with the choice depending on the objectives guiding monitoring and the resources available. Three different approaches can be applied: trail attribute inventory, trail condition assessment and trail prescriptive management assessments (Marion & Leung, 2011; Marion *et al.*, 2011) (Table 7.6). All three approaches are increasingly relying on GPS devices to locate and map trails and their features. Also, using GPS allows electronic data entry, saving time and reducing data entry errors (Marion *et al.*, 2011). Such information can also be readily included in geographic information systems, where it may be combined with remotely sensed data, for example information on changing vegetation cover over time (Kim & Daigle, 2011).

All of the survey approaches include some measure of erosion, usually recorded as trail depth. The most accurate and time-consuming approach is measuring the cross-sectional area between the trail surface and a taut line stretched between two fixed points on each side of the trail (Leonard & Whitney, 1977, cited in Hammitt & Cole, 1998) (Figure 7.4). Olive and Marion (2009) modified this approach by adding sliding beads on the transect line to identify the locations for vertical measures, with the number varying depending on the complexity of the trail tread. Cross-sectional area is determined by measuring the distance between the two fixed points (L) and then taking at least 20 vertical measurements ($V_1 - V_{n+1}$). The formula following Figure 7.4, with these measurements included, gives the cross-sectional area.

Table 7.6 Summary of trail monitoring techniques (organised according to Marion *et al.*, 2011)

*Trail monitoring technique**	*Advantages*	*Disadvantages*	*Application*
(1) *Trail attribute inventories* – rapid inventory and mapping of trail characteristics such as location, types of use (e.g. mountain bikers, hikers) and maintenance features, such as culverts and water bars See Marion *et al.* (2011)	Provides rapid description of location and characteristics of trails	No assessment of conditions included	Provides general information for managers; inventory information on distribution and lineal extent
(2) *Trail condition assessments*			
(2a) Sampling-based rapid surveys – conditions recorded at systematicallylocated but non-permanent points, aiming for generalisation to the whole trail system See Cole (1983b)	Permits rapid assessment of general trail conditions	May not accurately reflect trail problems unless a large number of sample points are used	Common survey approach, specifically the approach of taking measurements at points located at a constant interval along a trail
(2b) Problem-based rapid surveys – evaluator searches for and documents the location and extent of trail problems in a trail system, usually based on searching the whole trail system (a census approach) See Leung and Marion (1999a)	Permits rapid assessment of trail problems; allows description of the condition of the whole trail system	Not able to provide information on average conditions; for example no statement can be made about the average level of muddiness for the system	Common survey approach, used to advise managers on extent and location of trail problems as well as the efficacy of current management measures
(2c) Condition class surveys (sectional evaluations) – impact ratings are given to sections of trails See Marion *et al.* (2011)	Highly efficient survey method that can be used to map and describe trail networks	No measures of changes in indicators obtained; rating classes are so broad that major changes may occur before allocation to another class occurs	Used to map and describe entire trail networks, especially informal trail networks
(3) *Trail prescriptive management assessments*			
(3a) Sustainability analyses – collect and analyse physical data on trails to inform management See Marion *et al.* (2011)	Efficient and cost-effective as remote sensing and GIS are increasingly used	Resolution of imaging may be insufficient for management purposes	Provides data for assessing trail features (e.g. grades, substrate) for planning and management purposes

* All techniques except 2c (Condition class surveys) use multiple indicators

Derived from Cole (1983b), Williams and Marion (1992), Cole *et al.* (1997), Hammitt and Cole (1998), Leung and Marion (1999b, 2000), Monz (2000), Marion and Leung (2001, 2011), Mende and Newsome (2006), Randall and Newsome (2008), Kincey and Challis (2010), Marion *et al.* (2011), Wimpey and Marion (2011)

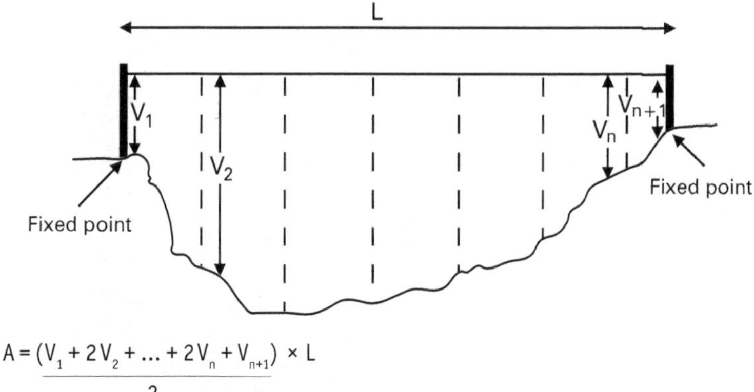

$$A = \frac{(V_1 + 2V_2 + \ldots + 2V_n + V_{n+1})}{2} \times L$$

where A = cross-sectional area

$V_1 - V_{n+1}$ = vertical distance, with V_1 measured at the first fixed point, through to V_{n+1} measured on the other side of the trail at the other fixed point

L = length of the horizontal line

Figure 7.4 Method for measuring trail cross-sectional area to determine trail erosion (Derived from Hammitt & Cole, 1998)

Phillips and Newsome (2002) and subsequently Randall and Newsome (2009) refined the measurement process by developing a pin frame 95 cm wide for determining cross-sectional area. The pin frame is an adjustable metal frame holding 20 pins, each 1.5 m long, which slide up and down within upper and lower cross-bars to provide 20 measurements for calculating cross-sectional area.

An even simpler and more rapid, but less accurate, way of measuring trail depth is to record the depth to which a trail has eroded below its construction level. Leung and Marion (1999a) used a range to describe the depth of each trail segment measured (31–60 cm, 61–90 cm and so on). In his rapid survey work, Cole (1983b) measured the maximum depth of the trail every 0.32 km.

Trail attribute inventories

Information collected as part of such inventories includes types of use, attraction features, cultural features, maintenance features (e.g. signs, culverts, water bars) and 'sustainability' attributes (e.g. trail grade, substrate) (Marion *et al.*, 2011). The location of these features can be readily recorded using GPS, enabling these data to be stored and manipulated spatially in geographic information systems, and then used for planning and management.

Trail condition assessments

Such assessments provide information on the type and severity of trail impacts. Some, but not all, of the following assessment approaches provide specific locations for trail impacts.

In *sampling-based rapid surveys*, data on trail condition are collected at points a predetermined, set distance apart, 50–500 m being most common, and following a randomly selected first point. A sampling interval of 100 m or less is usually recommended (Leung & Marion, 1999b). The indicators most commonly used are trail width, width of bare ground and maximum trail depth. Determination of the average width and depth of the trail system sampled and the portion with problems should be possible using these data. And, if the standard for the system is, for example, 'no more than 1% of all trails will be more than 30 cm deep', this can be easily monitored using this approach (Hammitt & Cole, 1998).

Such systematic sampling, with many points separated by a set distance, can be extremely time-consuming and may still miss features of interest. Pettebone *et al.* (2009) compared the usefulness of an alternative approach – spatially balanced probability-based sampling – with the more widely used and known systematic sampling, through application to monitoring the Glacier Gorge Trail in Rocky Mountain National Park, Colorado, USA (Figure 7.1). Sampling was stratified based on the level of visitor use, which was assumed to decline with decreasing distance from the trail-head. Fewer sample points using spatially balanced sampling provided results similar to those for the complete data-set from the systematic sampling. For example, trail width estimates from 25 sample points from the balanced sampling approach were similar to the trail width estimates from 83 sample points using systematic sampling (Pettebone *et al.*, 2009). Given that fieldwork is expensive, this alternative sampling strategy appears to have cost-saving benefits with no associated decline in the accuracy of the results obtained.

Problem-based rapid surveys let managers know where the problems are and can guide trail location, design and maintenance (Cole, 1983b). Such surveys require recording of the starting and end points of problems such as soil erosion/trail depth, excessive root exposure, excessive trail width, wet soil, running water on the trail and multiple trails (Cole, 1983b; Leung & Marion, 1999a). Features of the site such as slope, the type of use (usually pedestrians and less often horses), track surface and drainage measures are also noted. Generally this approach relies on censusing the whole trail system. Leung and Marion (1999b), however, sampled a representative portion and then extrapolated their findings to the whole trail system (Box 7.3).

Condition class surveys provide descriptions of trail systems according to condition classes. Individual sections are allocated to a class, ranging from class 1 where the vegetation is flattened through to class 5 with obvious soil erosion (Wimpey & Marion, 2011) (Table 7.7). The system is quick and easy to apply and provides a useful single value for each trail section. These values can then be collectively used to describe a trail system according to the length of trail in each of the condition classes. Such surveys have been most commonly applied to mapping and describing informal trail networks (Marion *et al.*, 2011).

Permanent point surveys are occasionally mentioned but such an approach is very time-consuming, because of the need to locate, permanently mark and re-locate each survey point. As such, they are rarely used. Where such permanent point surveys are employed, the points may be randomly distributed throughout the trail system or purposively located on portions of the trail where managers want to closely watch

Box 7.3 A rapid monitoring system for trails in Great Smoky Mountains National Park, USA

As described in Box 7.2, Great Smoky Mountains National Park, on the border of Tennessee and North Carolina (Figure 7.1), is one of the most visited national parks in the USA. It has 1496 km of official trails, with two-thirds of this system open for horse use. Permit data indicate more than 10,000 animals annually are part of groups on overnight stays. In 1985 the number of day hikers was estimated at 700,000 (Leung & Marion, 1999a).

Leung and Marion (1999b) used a problem-based rapid survey approach, the trail problem-assessment method (TPAM), to document the locations and extent of various trail problems, in particular excessively muddy, wide or eroded trails. They surveyed 35% of the Park's total trail length, arguing that this was a sufficiently large and representative sample to allow conclusions to be drawn of relevance to the whole trail system. A total of 23 indicators were assessed, grouped in three categories – inventory indicators (e.g. use type, such as horse riding and trail width), resource condition indicators (e.g. soil erosion, root exposure and wet soil) and design and maintenance indicators (e.g. grades >20% and the effectiveness of water bars).

The data from TPAM provided managers with locational information on all trail problems and maintenance features such as drainage dips and water bars. Trail maintenance crews could then be directed to sections with erosion and/or wet soil. The most common impact was soil erosion, with 4.6% of the trails surveyed with erosion exceeding 30 cm in depth. Running water, wet soil and rutted tracks were the main contributors to excessive trail width (affecting 0.7% of surveyed trails) and multiple trails (1.8% of trails).

The TPAM also allowed several other simple assessments. Lineal measurements of the extent of a problem could be used to determine for an indicator whether its standard had been exceeded. For example, if the standard for wet soil is 'no wet segment to exceed 30 m in length', then this monitoring system can give the number of segments where this value has been exceeded. This method also allows comparison between different trails and lengths of trail by converting the number of occurrences of a problem to m/km or a percentage.

By standardising the data, different trail lengths can be characterised and compared in terms of user type, trail width, erosion problems, and design and maintenance conditions. Such information is vital for planning and managing trails. From this characterisation, it was apparent that the eastern portion of the Great Smoky Mountains National Park had the poorest trails, with the most prevalent problems being wet soil and excessive width. Both problems were a result of poorly designed trails located close to streams on poorly drained soils. Poor trail conditions such as these lead to trail widening and multiple parallel tracks as hikers attempt to avoid wet areas. High horse use of the area had also exacerbated the problems (Leung & Marion, 1999b).

Table 7.7 Condition classes for monitoring informal walk trails in natural areas

Class	*Description*
1	Trail distinguishable, with slight loss of vegetation cover
2	Trail obvious, with vegetation cover lost
3	Vegetation cover and organic litter lost across most of the walk trail tread
4	Soil erosion in the tread beginning in places
5	Soil erosion common along the tread

Derived from Wimpey and Marion (2011)

active erosion or where some new trail design (e.g. a new trail surface or drainage structure) is being tried. The results from surveying the randomly distributed points can be generalised to the whole trail system if a sufficient number of points are sampled. For the purposive sampling this is not the case as only trail portions with features of interest have been surveyed.

Trail prescriptive management assessments

Marion *et al.* (2011) report that such assessments can record and evaluate maintenance requirements, features important for trail sustainability (e.g. substrate durability), capabilities with respect to different types of users (e.g. horses versus hikers) and relocation options. Sustainability analyses, as part of these management assessments, currently focus on the physical attributes of trails, such as trail grade and substrate (Marion *et al.*, 2011).

Using remote sensing and GIS

Remote sensing techniques are being increasingly investigated and applied to monitoring trails. Trails can be inventoried and mapped (i.e. a trail attribute inventory – see Table 7.6) using GPS devices; recreational-grade GPS is accurate to 3–8 m but survey-grade equipment to 1–3 m. The higher-accuracy units are most useful when mapping dense networks of informal trails (Marion & Leung, 2011). This information can then be added to GIS to create maps of trail networks and allow manipulation of these plus other spatial data for management purposes.

Leung (2008), Marion and Leung (2011) and Leung *et al.* (2011) report the use of GPS-based inventory and mapping of informal trails in eight major meadows of Yosemite Valley, Yosemite National Park (Figure 7.1) to develop spatial indices of trail impacts. Having input trail locations and meadow boundaries into GIS they calculated two fragmentation indices to give a measure of disturbance: weighted mean patch index (gives average size of undisturbed patches; gets smaller with increasing fragmentation/proliferation of trails) and largest five patches index (another measure of fragmentation; also gets smaller with increasing fragmentation) (Leung, 2008; Leung *et al.*, 2011). GIS-based inventory of trails also enables the aggregate lineal extent of trails to be calculated (Newsome & Davies, 2009; Marion & Leung, 2011).

The valuable contribution that GIS can make to the storage, manipulation and analysis of trail management data is increasingly being realised. Chiou *et al.* (2010) used GIS to collate and analyse trail information for managers. Working in Sitou Nature Education Area in central Taiwan (Figure 7.1) they obtained slope data from digital elevation models to derive the time cost of walking on each trail. GIS can also be used to locate and place sampling points. For example, Wimpey and Marion (2010) used a GIS for Arcadia National Park, Maine (Figure 7.1), to locate sampling points at 152.4 m intervals along a trail system as a basis for a sampling-based rapid survey. They then used a handheld GPS device to locate and sample these points in the field.

If data can be collected remotely, using airborne sensors for example, there is the potential for cost-savings and the ability to cover larger areas than is possible with field-based surveying. Lidar – airborne laser scanning – can be used to locate

Box 7.4 Lidar as a remote sensing technique for trail inventory and monitoring

Lidar (an acronym for light detection and ranging) uses an aircraft-mounted laser to produce digital surface models (which includes trees and buildings) and digital terrain models (bare-earth models). The high-resolution topographic data provided by lidar can be used to evaluate the sustainability of whole trail systems, by describing trail grade, slope ratio and substrate. The amounts and types of use a system is able to support can then be determined and, if necessary, relocation alternatives considered (Marion *et al.*, 2011).

Kincey and Challis (2010) used lidar to determine the extent and severity of soil erosion in an area of 3.8 km^2 in Brecon Beacons National Park, Carmarthenshire, Wales (Figure 7.1). About 15% of the park contains Sites of Special Scientific Interest, including the Waun Dhu raised bog, one of only 20 such raised bogs in Wales. Visitor-days have been estimated at 3.6 million. Sources of erosion include overgrazing by livestock, footpath erosion and illegal use of motorised vehicles.

These researchers produced both digital surface and digital terrain models, at a spatial resolution of 0.5 m, with a vertical accuracy of ~15 cm and an ability to differentiate changes in elevation to within 1–2 cm. Image interpretation, using ArcGIS 9.2 software, used several illumination angles to ensure that features aligned with only one illumination source were not overlooked. Analysis of the lidar data was complemented by examination of colour aerial photography and Ordnance Survey digital mapping.

Analysis showed over 46.8 km of erosion features associated with trails in the study area. The depth-of-erosion features ranged from minor features only 5–10 cm deep to incised gullies over 75 cm in depth. A field survey independently confirmed the lidar results, although the field survey missed a number of erosion areas that were identified using lidar. Kincey and Challis (2010) conclude that when lidar results are incorporated in a GIS they provide further analytical opportunities. For example, opportunities are created to calculate quantitative path measurements or to integrate lidar data with hydrological modelling to determine susceptibility to erosion.

trails in the landscape, map the nature and extent of soil erosion, and identify areas requiring urgent management attention (Kincey & Challis, 2010). It can also be used to develop terrain models, providing a basis for determining trail grades and slope ratios so the sustainability of trail segments and systems can be evaluated (i.e. sustainability analyses – see Table 7.6) (Marion *et al.*, 2011). Box 7.4 provides further details on the application of lidar to trail monitoring.

Choosing a monitoring strategy

The choice depends on the objectives of the survey and the resources available. Any combination of trail attribute inventory, condition assessment and prescriptive management survey can be undertaken (Marion *et al.*, 2011) (Table 7.6). In most cases GPS-based mapping of trail location will be essential. Since most of the survey expenses relate to getting staff into the field, adding a condition assessment component will add very little to the survey's cost. It is likely to become increasingly possible to obtain condition information using remote sensing platforms such as lidar. Entry of the data from trail surveys and monitoring in a GIS assists managers to evaluate the trail system and conditions, problem locations and/or sustainability (Marion & Leung, 2011; Marion *et al.*, 2011).

For condition assessment, the choice is between condition class surveys and sampling- and problem-based rapid surveys (Table 7.6). The first is usually undertaken only when time is very limited and general descriptions are required. The sampling-based approach requires a large number of sample points, which is potentially resource intensive; however, to be meaningful, the problem-based approach requires a search of the whole trail system, which also takes time. The latter approach seems more useful for managers, as it provides details on where problems begin and end and can also record information on the effectiveness of management actions, such as dips and banks for controlling water movement.

Marion and Leung (2001) noted that point sampling and problem assessment methods yield distinctly different types of quantitative information. Point sampling provides more accurate and precise measures of trail characteristics that are continuous and frequent (e.g. tread width or exposed soil). The problem assessment method is a useful approach for monitoring trail characteristics that can be easily predefined and are infrequent (e.g. excessive width or secondary treads), particularly where information on the location of specific trail impact problems is needed. The usefulness of these data may be offset by their reduced precision due to subjectivity in defining where impact problems begin and end on a trail (Leung & Marion, 2000).

Surveying roads and vehicle tracks

Gravel roads and other unsealed vehicle tracks can be monitored, similarly to walk trails, using rapid survey approaches. Walker (1991) developed a monitoring system for gravel roads in Wisconsin. First, he divided the roadway system into segments of one mile or more (i.e. at least 1.6 km) based on similar function, traffic volume and surface thickness. Each segment had its geometry (e.g. road width) and maintenance history recorded and a rating was generated based on the major factors likely to influence road performance – road crown, drainage and adequacy of the gravel layer. A road with a good crown falls 1–2 cm per 30 cm from its centre to its edges. Poor drainage is characterised by ponding, flooding, erosion and collapsed or silted culverts and bridges. Evidence of an inadequate gravel layer is rutting, corrugations (washboarding) and potholes. If the time and skills are available, the thickness of the gravel layer can be measured. A simple five-point rating scale is used to assign a value to each road segment (Table 7.8). The information and ratings can then be used to assign maintenance priorities to road segments. This system can easily be used to assess 30–65 km of gravel roads per day (Walker, 1991).

More recently, lidar has been used for assessment and planning of roads and trails by the US Forest Service and National Park Service units in Georgia and West Virginia (Marion *et al.*, 2011). Routes were collected using GPS and digitised from aerial photographs and then evaluated for grade and slope ratio using a lidar terrain model. The resultant information was used to assess grades for an off-road vehicle trail system and evaluate pre-existing roads for inclusion in this system (Marion *et al.*, 2011).

Table 7.8 Rating scale for monitoring gravel roads in natural areas

Class	*Description*
5 (Excellent)	Excellent surface condition and ride
4 (Good)	Slight corrugations
3 (Fair)	Good crown (7.5–15 cm). Ditches present on >50% of roadway. Moderate corrugations (2.5–7.5 cm) over 10–25% of the area. No or only slight rutting
2 (Poor)	Little or no crown (<7.5 cm). Adequate ditches on <50% of roadway. Culverts partly full of debris. Moderate to severe corrugations (>7.5 cm deep) over 25% of the area. Moderate rutting (2.5–7.5 cm) over 10–25% of the area. Moderate potholes (5–10 cm) over 10–25% of the area
1 (Failed)	No crown, or road is bowl-shaped, with extensive ponding. Little if any ditching. Filled or damaged culverts. Severe rutting (>7.5 cm) for over 25% of the area. Severe potholes (>10 cm) for over 25% of the area

Derived from Walker (1991)

Water bodies

Water, whether lakes, streams, estuaries or the ocean, attracts people, resulting in facilities such as resorts, campgrounds, roads and trails being located on the edges of water bodies and in some instances over the top of them (e.g. pontoons over coral reefs). These environments also provide habitat for a wealth of plants and animals. Water quality can be adversely affected by nutrients, pathogens and sediments moving into the water (see Chapter 3 for details on impacts). Monitoring is essential for human and ecosystem health.

Design of the sampling programme is an important consideration. From the data collected, managers need to be able to accurately measure, and differentiate from other causes, changes due to natural area tourism. For most natural area tourism, the impacts are likely to be localised. For a resort or floating pontoon over a coral reef, for example, sampling should be conducted in the immediate vicinity. An undisturbed but environmentally similar site should also be sampled, to provide a control. Warnken and Buckley (2000) make a strong case for resource monitoring at the impact and control site before a tourism facility is developed, during construction and then once it is operational. Teh and Cabanban (2007) similarly advocated an assessment of the biophysical conditions of the coral reefs at Pulau Banggi, Sabah (Figure 7.1), before tourism development commenced. Very often lack of money, forethought or time precludes pre-development monitoring.

The frequency of sampling and the number of samples taken can be decided only once some idea of data variability has been obtained (Hammitt & Cole, 1998). For example, bacterial contamination from human wastes varies greatly depending on the time of year and the flushing effects of precipitation. Sampling must be frequent enough to pick up these variations.

Monitoring techniques and standards for drinking water and for swimming are known and widely used. The usual measure is the number of coliform bacteria. These are found in human faeces and are regarded as a good indicator of bacterial contamination. Numerous physical and chemical water quality indicators can be measured, including phosphorus, nitrogen, pH and dissolved oxygen. Other aquatic elements that may be sampled include plankton, algae, macro-invertebrates and fish. Where road or trail construction and use may lead to increased sedimentation, monitoring of aquatic fauna to determine the impacts of suspended solids may be useful.

Reef Check is a marine monitoring programme of relevance to natural area tourism. It aims to document human impacts, including those originating from natural area tourism, on coral reefs worldwide (Hodgson, 2001; Reef Check, 2012a). There is some variability in how data are collected but central to efforts is data collection by recreational divers trained and led by marine scientists. Four spatial replicates along a 100 m transect line are given as a core method (Reef Check, 2012b). Information is collected on: the site itself; fish species, especially those targeted by spearfishers and aquarium collectors; invertebrates, particularly those taken for food or collected as curios; and reef substrate types (Habibi *et al.*, 2007; Shuman *et al.*, 2008). Reef Check California also documents urchin frequency and add extra fish survey transects to facilitate meaningful stock assessment (Shuman *et al.*, 2008). Reef Check Indonesia conducts reef surveys at two depths, with a shallow transect at 2–6 m and a deeper transect at 6–12 m, taking into account the tidal range (Habibi *et al.*, 2007).

Monitoring programmes have been designed specifically to determine the impacts of tourism on coral reefs. Tilot *et al.* (2008) used a combination of a video survey method and visual census to assess the impacts of tourism on the South Sinai coral reefs in the Egyptian Red Sea (Figure 7.1). Nine monitoring stations were selected, to represent contrasting topographies, different degrees of exposure to wave action and different levels of tourism impact. Each station had nine permanently marked transects 50 m in length, with three each at 3, 7 and 16 m depth. Transects were filmed with a digital video-camera in an underwater housing to determine the cover of hard and soft corals. Fish abundance at each station was estimated using an underwater visual census along four contiguous 50 m transects at depths of 3 and 10 m, within 5 m each side of the observer's path. Only commercially or ecologically important fish were counted (Tilot *et al.*, 2008).

Soundscapes

Sounds can have a profound effect on visitors' experiences, including the perceived quality of the landscape (Briggs *et al.*, 2012). Acoustic data can be used to set standards for acceptable noise levels in wilderness, where freedom from modern human impacts, including noise, is central to the landscape's character (USDA FS, 2009). Sound intensity and pitch can be measured and/or visitors asked to identify sounds that they hear and whether they find them pleasing or annoying (Newman *et al.*, 2006b).

The first approach has been applied in Yosemite National Park (Figure 7.1) to measure sound levels beyond ambient noise levels and the amount of time sounds are above speech interference thresholds (USDI NPS, 2010). Staff used digital audio-recorders to collect sound data. Recorders were located to sample the natural ambient baseline in developed, back-country and transition zones. At Lake Mead National Recreation Area Wilderness, Nevada, USA (Figure 7.1), Briggs *et al.* (2012) continuously recorded the intensity and pitch of the ambient sound for 30 days at three sites. On-site listening was also conducted to identify the sounds recorded.

The second approach – visitors identifying and assessing sounds – has been undertaken at Muir Woods National Monument, California, USA (Newman *et al.*, 2006b) (Figure 7.1). Visitors were asked to listen and then identify and rate sounds on a checklist.

Monitoring Visitors to Natural Areas

Monitoring of visitors to natural areas can provide information on visitor numbers, visitor and visit characteristics, and visitor outcomes (e.g. satisfaction) (Table 7.9). Such monitoring provides data for management, planning, resource allocation, public accountability, enhancing visitors' experiences, marketing and interpretation, and to meet legal requirements. Although visitor monitoring provides a wealth of information, managers regularly place the natural environment and its monitoring and management ahead of visitor management. The reasons are both philosophical and practical. Often, managers find it is easier to manage natural resources rather than visitors, especially as most of their training has centred on the natural environment. Also, managing visitors is difficult because of the lack of information on the relationships between visitors and the natural environment. Visitor monitoring may not be done because it generally has a lower priority than other activities, such as replacing old facilities or building new ones. The inadequate collection, storage and analysis of visitor data are regarded as a major impediment to managing visitor use in protected areas (Hadwen *et al.*, 2007; Griffin *et al.*, 2010).

Sampling design is an important consideration, as it is in monitoring resource impacts. The design chosen depends on the objectives of the monitoring programme. Generally, the population of concern will be all visitors to the natural area or system of natural areas. Visitor monitoring then relies on sampling part of this population and drawing conclusions about the whole population from this sample. The lack of scientific sampling strategies was identified as a greatest weakness in Australian visitor monitoring in protected areas (Pitts & Smith, 1993). This problem is unlikely to be restricted to Australia. Also, sampling must represent the full diversity of the user population, while acknowledging that selecting representative samples is notoriously difficult. This difficulty stems from users being widely dispersed, mobile and engaged in diverse activities.

If more general information from non-users as well as users of natural areas is required, then the population of interest will be much larger. Again, the same principle as outlined above applies: sampling must be designed to represent the diversity of the whole population. To assess future demand and the community's perceived benefits of natural areas, it is critical to sample non-users as well as users. The remainder of this section, however, focuses on visitors themselves.

Table 7.9 Summary of different areas of focus for visitor monitoring, monitoring techniques and main uses of the resultant data

Monitoring focus	*Data required*	*Monitoring technique*	*Main uses of data**
Visit/visitor numbers	Visit/visitor numbers for site, park or park system; mode of arrival; entry and exit points	Automated counters, telephone surveys, entrance records (e.g. ticket sales), manual counts, videoing, visitor books, tour records, aerial counts and photos, guesstimates	• Making resource allocation decisions • Public accountability
Visitor characteristics (visitor profiling)	Demographic and socio-economic attributes of visitors; reasons for visiting, attitudes, motivations, expectations and preferences	Questionnaires, telephone surveys, personal interviews, interactive techniques	• Marketing and interpretation • Planning – planning frameworks (e.g. LAC) and site design
Visit characteristics	Sites visited, seasonal use patterns, group size, length of stay, frequency of visits, activities undertaken	Questionnaires, telephone surveys, personal interviews, field observations, visitor (object) locators and trackers	• Planning – (as given above) • Making resource allocation decisions • Routine management, especially ameliorating impacts
Visitor outcomes	Satisfaction, experiences, disappointments, and future intentions and behaviours	Questionnaires, telephone surveys, personal interviews, interactive techniques	• Enhancing visitors' experiences • Planning (as above) • Routine management, especially ameliorating impacts

*Only the most important use for each monitoring focus is given, to make the information in this table as accessible and meaningful as possible, while recognising that visitor numbers through to visitor outcomes all provide data for planning and resource allocation

Derived from Pitts and Smith (1993), Cope *et al.* (2000), Griffin *et al.* (2010)

Visitor monitoring techniques range from counting visitors, through questionnaires, telephone surveys and personal interviews, to advisory committees and other interactive techniques, including the internet (Table 7.10).

Visitor counts

Counting visitors is the most widespread form of visitor monitoring in natural areas. Buckley *et al.* (2008), in his analysis of monitoring for the management of conservation and recreation in Australian protected areas, noted that most parks count visitors but few know what these visitors do. Techniques range from on-site automated counts and direct counts by observers (Cessford & Muhar, 2003) through to off-site estimates via telephone surveys (Griffin *et al.*, 2010). The total number of visits to a park or park system is of great interest to senior managers, as this information can help make a case to governments for continuing or increased funding. It may also be an annual corporate reporting requirement. The US Forest Service, in its National Visitor Use Monitoring Program, defines a recreation visit as 'one person entering and exiting a national forest for the purpose of recreation' (English *et al.*, 2001).

The numbers of visitors to parks, and to individual sites in a particular park, are often estimated using vehicle counters, located on roads and vehicle tracks and triggered by vehicle wheels passing over an air-filled tube (pneumatic tube counters) or an inducted loop (inducted loop axle counters). More recently, traffic classifiers have been used. These have double pneumatic tubes and record and classify vehicle types, speeds, direction of travel, time and date. The use of vehicle counters is widespread in the USA and Australia (Cessford & Muhar, 2003; Griffin *et al.*, 2010). The US Forest Service also employs hand tally recorders, using visual counts by staff at sites where it is not safe to have traffic counters (English *et al.*, 2001).

The ability of vehicle counters to provide accurate estimates of visitor numbers, however, continues to be questioned (Cessford & Muhar, 2003; English *et al.*, 2003; Griffin *et al.*, 2010). Deductions must be made for staff traffic and through traffic, otherwise visitor numbers can be overestimated. All counters require careful calibration and frequent checking. Regular observation-based surveys are required in tandem with automated vehicle counts to determine the numbers of people in the different types of vehicles. This information is then used to multiply the counts recorded for different vehicle types to obtain total visitor numbers.

Griffin *et al.* (2010), in their review of visitor data collection by protected area agencies in Australia, referred to problems with calibration, regular maintenance and the associated requirement for skilled staff, and the ability to differentiate visitors from other users who trigger the counters. An alternative approach to obtaining visitor numbers, based on telephone surveys, has been successfully implemented by several park management agencies in Australia (Griffin *et al.*, 2010). These researchers advocated using telephone surveys to obtain aggregate visitor numbers for park systems, with traffic counters used only for short periods at certain sites, to address specific questions regarding visitor use.

Table 7.10 Summary of visitor monitoring techniques

Visitor monitoring technique	*Advantages*	*Disadvantages*	*Application*
(1) Visitor counts – includes automated counters, telephone surveys, entrance records (e.g. ticket sales), manual counts, videoing, visitor books, tour records, aerial counts and photos, guesstimates	Provides simple measure of extent of use of a natural area; telephone surveys have the most potential to produce a reliable estimate	Most methods provide estimates only; automated counters are expensive to purchase and may be poorly calibrated, with significant margins of error	Telephone survey of broader community can provide reliable estimate of visit numbers for a park system; traffic counters can be used on most roads; aerial counts and photos useful for marine areas and difficult-to-access locations such as beach dunes
(2) Visitor patterns of use – includes log books, aerial surveys, trackers, questionnaires (see below)	Provides spatial and temporal information for planning and management	Can be difficult and expensive for large areas or where use is widely dispersed	Remote sensing and GPS technologies are contributing to rapid developments; useful for marine parks and other difficult-to-access areas
(3) Questionnaires and personal interviews – includes site-based, mail, internet or telephone data collection	Questionnaires provide comprehensive information on visitors, their activities, expectations, preferences and experiences; they are widely used, making results comparable with those obtained elsewhere	Can be expensive to design, administer and analyse	Best used where detailed information on visitors and their visit characteristics and outcomes are required for planning and impact management
(4) Interactive techniques –users are brought together or come together to provide data, often on more than one occasion	Efficient means of accessing a range of ideas at one time or seeking determination and agreement over time on indicators and standards (advisory committee or task force)	Time-consuming to organise and administer and data may be difficult to analyse if consensus is not reached; applies to both face-to-face and internet-based interactions	Using advisory committees only warranted for large, complex natural areas with multiple stakeholders; internet interactions require adequate, ongoing moderating and resourcing

Derived from Cole (1997), Moore *et al.* (2009), Griffin *et al.* (2010)

Parks Victoria, for example, conducts a biennial community telephone survey of Victorians (Victoria is one of the six states in Australia) and face-to-face interviews with interstate and international visitors to determine the number of visitors to parks and reserves in Victoria (Zanon, personal communication, 2012). Respondents are asked when they last visited a park and its name, the number of visits in the last four weeks, reasons for their visits and their demographic details. The agency complements this statewide approach with short, intensive periods of monitoring of vehicle numbers (using vehicle counters) at a small number of high-use sites. The site-based monitoring provides an estimate of yearly visits and the distribution of arrivals and departures. Parks Victoria relies predominantly on the community survey, rather than the vehicle count data, for decision making and reporting.

For walk trails, both automated and manual approaches are taken. Automated infra-red, photoelectric and seismic pad counters have been used on walk trails (Watson *et al.*, 1998; Cope *et al.*, 2000; Cessford & Muhar, 2003). Arnberger *et al.* (2005) counted trail users in the Danube Floodplains National Park, Austria (Figure 7.1) using two different methods: fixed-point time-lapse video-recording over a year; and counts by human observers over four days. At low use levels, the video-recording counts of single rapidly moving visitors (e.g. cyclists) were significantly lower than those obtained by the human observers, whereas at high use levels human observers counted fewer walkers and cyclists than the video interpreters. Video interpretation is suggested as the preferred approach, especially if the imaging can be modified to better capture fast-moving objects and workable digital image analysis software becomes available (Arnberger *et al.*, 2005). The need for manual analysis and privacy concerns are still issues with video and still-camera technologies (Warnken & Blumenstein, 2008)

Fairfax *et al.* (2012) used infra-red sensors connected to digital cameras in D'Aguilar National Park, south-east Queensland, Australia (Figure 7.1), to capture images of fast-moving objects on a multi-use trail. Four cameras provided over 7000 photographs over a period of 1000 days. Counts of cyclists, horses, walkers and motorised vehicles were obtained. About half of the photographs were false triggers, caused by environmental factors. Capture rates were highest for horses (100%) and lowest for cyclists (63%).

Less sophisticated methods of counting visitors are in widespread use. Numbers may be manually counted or guesstimated by staff or volunteers, or extrapolated from entry passes purchased or numbers self-registering at trail-heads. Data can also be collected from tour operators through log books and receipts. Information on numbers of visitors accessing certain sites or engaged in activities such as whale watching as part of an organised tour can then be collated. Similar data on sites used and numbers involved can be obtained from associations and clubs where permits are required for using natural areas. In Western Australia, for example, clubs using natural areas for orienteering (a cross-country navigational sport) require approval from land management agencies. This gives managers information on the areas accessed, plus they can estimate numbers knowing the levels of club membership.

Proxies can be used to assist in counting visitors (English *et al.*, 2003). Proxies are variables that can substitute for direct counts of visitors. Using proxies is a response to the realisation that comprehensive sampling of a park system to determine total visitor numbers can be prohibitively expensive. Proxies for visitation to protected areas include traffic counts, use of car parks, self-registration records, permits, and number of information brochures taken (Watson *et al.*, 2000; English *et al.*, 2003). Double sampling, where actual counts of visitors and measures of the proxies are made, can provide precise estimates of visitation (English *et al.*, 2003). It is important to establish the relationship between the visitation proxy variable and actual visitation, especially if the proxy measures are to be used on their own.

For marine areas and difficult-to-access places like sand dunes and beaches, aerial photography or visual observations from the land, sea or air allows visitor numbers to be counted. Numbers are often extrapolated from the number of vehicles or vessels recorded. In marine parks, aerial surveys can give rapid estimates of the number of boats and where boat use is concentrated. Aerial photography has routinely been used to monitor camping activity adjacent to the beaches of Fraser Island off the eastern coast of Australia (Hockings & Twyford, 1997) (Figure 7.1).

A combination of manual counts by observers and digital photography from a small plane (a four-seat fixed high-wing Cessna 172), combined with geospatial referencing using a GPS linked to a PalmPilot, were used to record visitor numbers and patterns of use in Ningaloo Marine Park and the associated shoreline (Smallwood *et al.*, 2011). Ningaloo is a large marine park extending for 300 km along the north-western coast of Australia (Figure 7.1), with extensive areas of dispersed use, making it expensive and challenging to monitor visitor numbers and patterns of use. Over the year-long survey period, 2906 aerial observations of boat activity were made and 15,373 people in recreational activities were observed on the shoreline, with 7696 observations of camps, boat trailers and vehicles, as well as boats that were not being used for recreation at the time of observation.

King and McGregor (2012) used periodic counts of visitors combined with a short survey to determine beach attendance numbers in southern California. The short survey was used to determine visitors' length of stay so that turn-over rates could be used to adjust the visual counts to obtain daily attendance estimates for these beaches. Official agency counts were up to five times higher and six times lower than the estimates obtained from this study (King & McGregor, 2012). These results highlight the value of carefully designed and executed survey and monitoring programmes.

Visitor patterns of use

Spatial information regarding visitor use is essential for both planning and management (Yuan & Fredman, 2008). For marine areas, aerial surveys, as applied by Smallwood *et al.* (2011) at Ningaloo Marine Park and described above, provide a cost-effective means of surveying large, difficult-to-access areas. Smallwood *et al.* (2011) recorded and geo-located shore- and boat-based activities, with sampling errors of 6.1 m for vessels and 4.3 m for groups observed on the shore. For observations near known landmarks that had been previously geo-referenced there was no sampling error. Aerial photographs, site visits, observation points and boat-based transect surveys are other means of locating and mapping the location of visitors. Thompson and Dalton (2010) used a laser locator with built-in compass and a handheld computer with built-in GPS and GIS software to survey visitors to the shoreline of Narragansett Bay, Rhode Island (Figure 7.1), allowing each observation to be immediately entered into a geo-database.

Mass-produced locational and communication devices – GPS and mobile phone technologies respectively – have opened up opportunities for automating the collection of movement data over time for visitors (Warnken & Blumenstein, 2008). Tracking may be location restricted or independent of location (Table 7.11). The position of mobile phones has been used to determine the location of visitors in Emajõe-Suursoo, Endla and Alam-Pedja Nature Protection Areas in Estonia (Roose, 2010) (Figure 7.1). Locational information was stored, as depersonalised log files, by the mobile phone operators. Residents and local employees were filtered out of the data-set. Large discrepancies existed between direct counts of visitors and mobile phone counts, suggesting more research is needed on filters and the behaviour of phone users (Roose, 2010).

Table 7.11 Visitor (object) counting and tracking technologies

Detection mode	*Requirements*	*Focus of detection*	*Examples*
Location restricted			
Passive sensing of 'signals' from visitors ('objects')	Natural signal emitted by an object	Visitors on trails, using interpretive facilities	Track counters, wireless sensor networks with motion detection, video or camera surveillance (motion-sensor activated), infra-red image analyses
Detecting a reflected signal	Detection of unique signal	Vehicles, boats	Radar, laser and sonar
Detection of specific signal (i.e. tagging)	Unique ID tag fixed to object with one or more local receivers to detect and position ID tag	Visitors, vehicles or boats fitted with tags	Radio frequency identification (RFID) tags, swipe cards, toll bridge systems, tracking of mobile phones
Location independent			
GPS-based	Access to GPS signal (satellites) and mechanism to store or transmit data to central unit	Visitors and vehicles fitted with GPS receivers and data-logging or data-transmitting devices	Fleet management systems, tracking of fishing vessels

Derived from Warnken and Blumenstein (2008)

Questionnaires and personal interviews

Questionnaires are widely used to collect information on visitors, their activities, expectations, preferences and experiences in relation to natural areas. Personal interviews are less common, primarily because they are more time-consuming. Attention to sampling design is important in both approaches because conclusions are usually being drawn for the whole population of users, with temporal comparisons being made. To make these conclusions as robust as possible, a large sample size is needed, as well as proper representation of all groups within the population. Babbie (1992) suggested if we want to be 95% confident that the results are within 5% of their values for the population, then a sample size of at least 400 should be the aim. If a response rate of only 50–60% is likely, as is often the case with questionnaire-based surveys (Neuman, 1994), then at least 670–800 questionnaires should be distributed.

A variety of means of distributing questionnaires exists. Cole *et al.* (1997) surveyed visitors on-site, asking two members from each group leaving wilderness areas in the Cascade Mountains of Washington and Oregon to fill out a short (10-minute) questionnaire. Roggenbuck *et al.* (1993) interviewed all parties entering and leaving four wilderness areas (Cohutta, Georgia; Caney Creek, Arkansas; Upland Island, Texas; Rattlesnake, Montana) (Figure 7.1) and, having obtained contact names and addresses, sent party members a mail-back questionnaire. Chin *et al.* (2000), in their survey of visitors to Bako National Park in Borneo, handed out questionnaires at the park office, accommodation areas and canteen, areas most frequented by visitors, and then collected them once completed (Box 7.5). Morin *et al.* (1997) used trail-head registration details from Nuyts Wilderness in Western Australia to mail out questionnaires. Cavana and Moore (1988) provided boxes at entrances and exits to Fitzgerald River National Park (Figure 7.1) at which questionnaires could be collected and returned.

For questionnaires and interviews, the more personalised the level of contact, the higher is the response rate. Personal interviews have the highest response rate, followed by questionnaires handed out and back on-site, while response rates to mail-out/mail-back questionnaires are much lower. Telephone surveys have response rates between those for interviews and questionnaires. Surveys should be designed and conducted to get the highest possible response rate, given the resources available. A higher response rate gives a larger sample, meaning that the survey findings can be more confidently generalised to the whole visitor population. It also means that one or more groups are unlikely to have been omitted (omissions potentially bias the results). Response rates of 50–60% are regarded as acceptable, although a higher rate is preferable (Neuman, 1994).

To ensure that, as far as possible, samples are independent, only one member per group is asked to complete a questionnaire. To minimise bias, the individual who has most recently had a birthday can be selected (Horneman *et al.*, 2002). With groups, however, multiple members can be asked to complete questionnaires. For this reason, it is useful (and in some cases essential) to include a question about the type of travel group so potential differences in responses can be explained.

Box 7.5 Using questionnaires to monitor visitors to Bako National Park, Borneo

Bako National Park, 37 km east of the capital city of Kuching (Figure 7.1), was the first national park gazetted in Sarawak. Attractions include its diverse coastal and forest ecosystems and abundant wildlife. Three species of monkey are found in the Park – the proboscis monkey (*Nasalis larvatus*) inhabiting mangrove swamps and found only in Borneo, and the silver leaf (*Presbytis cristata*) and common long-tailed macaques (*Macaca fascicularis*). Facilities include the park office, information centre, canteen, accommodation and camping ground, and the jetty where visitors arrive by longboat. Visitor activities include enjoying nature, hiking, sightseeing, observing wildlife, relaxing and photography (Chin *et al.*, 2000).

Visitors were surveyed using a questionnaire handed out and returned in the Park. It sought information on visitor and visit characteristics, activities undertaken, and visitor perceptions of impacts and management strategies. Visitors were young – most were aged between 16 and 40 – and their activities centred on enjoying nature. The environmental conditions of greatest concern to visitors were litter, damage to natural vegetation and erosion along walk trails. Very few visitors were concerned about visitor numbers or the size of groups encountered. These results suggest that suitable indicators, all of them environmental rather than social, could be litter, damage to natural vegetation and erosion along walk trails. In terms of management strategies to address these problems, those surveyed favoured most of the strategies suggested, from education to limiting the overall number of visitors (Chin *et al.*, 2000). Least favoured was providing more visitor facilities, probably because visitors to natural areas prefer little to no development (Buckley & Pannell, 1990).

How questionnaires and interviews are conducted, presented and worded all affect visitor responses. Many useful social research texts exist, such as Frankfort-Nachmias and Nachmias (1992) and Neuman (1994), which should be consulted before commencing this type of work. A number of design choices need to be made, an important one being whether questions are open- or closed-ended. Open-ended questions often provide more information, but they can be more difficult and time-consuming to analyse than closed-ended ones. Closed-ended questions are easier to analyse but may bias the results by limiting the choice of response visitors have. Also, if questions are not carefully worded, they can lead or even threaten respondents.

Questionnaires and personal interviews rely on visitors' perceptions as a data source. Using perceptions must be accompanied by several understandings. First, they are a valid data source given that a central part of natural area management is addressing visitors' perceptions of the experiences they have and are seeking. Second, differences in perceptions may exist between managers and visitors, for example regarding the acceptability of impacts. Martin and McCool (1989), working in the USA, found that managers were more sensitive than visitors to bare ground, while visitors found tree damage and fire rings more objectionable than managers. These findings suggest that the perceptions of both visitors and managers are essential inputs to natural area management.

Selecting indicators and standards

Pivotal to the successful management of natural area tourism is understanding the experiences visitors are seeking and that they obtain, an understanding obtained through research and monitoring. This section on indicators and standards (and the following one, on visitor satisfaction) is underpinned by an interest in understanding visitor experiences and managing to enhance them. Measuring experiences has remained an elusive task, although it is assisted by the basic tenet of outdoor recreation management that the quality of visitor experience depends on the conditions of the setting (Clark & Stankey, 1979; Cole & Hall, 2009). Although managers can control the setting (i.e. the environmental, social and managerial conditions at a site) and hence the nature and quality of visitor experiences, there are also elements beyond the control of managers that can also powerfully affect visitors' experiences. Included are weather, within-group relationships, and interactions with other groups and their members. However, by managing settings, managers are increasing the probability that positive experiences are realised by visitors (Cole & Hall, 2009).

As such, the following information on questionnaire content as a basis for monitoring (and subsequent actions by managers) focuses on the indicators and standards for the desired conditions in natural areas, and satisfaction with attributes of these conditions (e.g. helpful staff, clean facilities). Through monitoring, management actions can be triggered by standards being exceeded or visitors not being satisfied by the performance of particular attributes. This section on questionnaires concludes with details on monitoring visitors' future behavioural intentions. Such intentions provide a guide to managers regarding whether visitors have had 'satisfactory' experiences and hence are likely to return or recommend to others the natural area they have visited.

Questionnaires can be used to identify what resource and social conditions influenced the quality of visitors' experiences (Roggenbuck *et al.*, 1993). The answers to this question give possible indicators, such as the amount of litter, vegetation loss and the number of people in other groups. Visitors can then be asked to suggest maximum acceptable levels, before their experience would be changed, for a list of indicators. Box 7.6 describes this approach as applied by Morin *et al.* (1997) in Nuyts Wilderness Area in south-western Australia (Figure 7.1). These authors used a single questionnaire; other researchers, such as Manning *et al.* (1996a), have used two, the first to collect information on conditions and possible indicators, and the second to determine standards for the indicators identified in the first phase.

Having standards is a crucial aspect of managing natural areas for tourism. Without standards it is impossible to tell if impacts have reached unacceptable levels, jeopardising the values that attracted visitors in the first place. The planning frameworks detailed in Chapter 4 rely on standards for the indicators identified. It is only when these standards have been exceeded and the impacts become unacceptable that the management strategies detailed in Chapter 5 become necessary.

Managers are often hesitant to set standards, generally because data on impacts and their causes are lacking and they are concerned about the political consequences of poorly informed management decisions. As with many aspects of environmental

Box 7.6 Using a questionnaire to identify indicators and standards for Nuyts Wilderness Area, Western Australia

Nuyts Wilderness, with an area of only 4,500 ha, lies on the south coast of Western Australia (Figure 7.1). It is used for day trips and overnight stays, with activities including appreciating nature, walking for exercise, solitude, viewing wildlife and photography. A questionnaire was mailed to visitors who had filled in the log book at the trail-head over a six-month period. Included were questions on visit and visitor characteristics, preferences regarding current conditions and management strategies, and maximum levels of acceptable impact for a number of environmental and social indicators (Morin *et al.*, 1997).

Most visitors were aged 26–60 and their activities centred on enjoying nature. The environmental conditions of greatest concern were the amount of litter, inadequate disposal of human waste, lack of wildlife, erosion of trails leading to the beaches, vegetation loss at the beaches and tree damage at the campsite. Fewer visitors were concerned about visitor numbers or the size of groups encountered than were concerned about environmental conditions such as vegetation loss and tree damage. These results suggest a number of potential indicators. Most management strategies were supported, with education and rehabilitation of degraded areas most favoured.

Standards were requested for a list of impacts, including trees damaged by humans, vegetation loss/bare ground, number and size of groups encountered, litter, human-made structures (e.g. signs) and fire rings. Standards acceptable to 50% of visitors were calculated from the results (Roggenbuck *et al.*, 1993). The standard for trees damaged, acceptable to 50% of visitors, was two trees per site. Therefore, if damage to more than two trees occurred this would be an unacceptable impact and warrant management intervention. The standard for group size was six people or fewer per group. For several indicators, the standard suggested by respondents depended on location. For the area of bare ground, up to 11 m^2 was acceptable at a campsite but only up to 3 m^2 was acceptable adjacent to the beaches (Morin *et al.*, 1997). These data provided managers with indicators and standards for an ongoing monitoring programme.

management, it reasonable to make a 'best guess', selecting a standard based on the best available information and then modifying it as new information becomes available. The political dimension can be managed by ensuring standards are determined with visitor input. It is also important to consult visitors because, as noted above, they may have different standards to managers (Martin & McCool, 1989). To help deal with uncertainty, the Tourism Optimisation Management Model relies on an acceptable range for indicators, rather than a single standard. For example, for the indicator 'proportion of visitors who were very satisfied with their overall visit' the acceptable range is 95–100% of respondents (Manidis Roberts Consultants, 1997).

Visual images are increasingly being used to explore visitors' perceptions and evaluations of a range of park conditions. Such images include sketches and drawings, photographs, computer-edited photographs (image capture technology: ICT), computer-generated photographs and videos. The quality of ICT rendering now makes these images almost indistinguishable from original photographs (Manning & Freimund, 2004). Visual approaches, in comparison with numerical or narrative approaches, have the benefit of enabling respondents to clearly visualise the conditions.

Manning *et al.* (1998) used computer-edited photographs in their questionnaires to help people visualise different standards. This research in Acadia National Park in Maine (Figure 7.1) focused on the use of carriage roads built in the early 1990s and today enjoyed by thousands of hikers and bikers. Their research had two phases – the first to identify indicators and the second, where photographs were used, to determine standards. The two indicators identified in the first stage were numbers of visitors on the carriage roads and problem behaviours such as bikes passing without warning. A series of photographs showing different numbers of visitors, from 0 to 30, along a 100 m section of the carriage way was used to determine the standard for visitor numbers. Visitors were shown the photographs in random order and asked to rate their acceptability on a scale from –4 (very unacceptable) to +4 (very acceptable). The results indicated that visitors found it acceptable to see up to 14 people on the 100 m section.

Computer-generated photographs can be used in 'stated choice' surveys to present a range of resource, social and managerial conditions. Stated choice analysis asks respondents to choose between two alternative sets of indicators and standards. The choices made by respondents between a number of alternative sets of conditions mimics the trade-offs made in real life. This method is regarded as providing a realistic reflection of human behaviour (van Riper *et al.*, 2011). Application of stated choice analysis to the Northern Forest Region of the USA involved six variables (resource conditions on-trail, resource conditions off-trail, number of people on-trail, number of people off-trail, on-trail management to mitigate ecological impacts, management to keep visitors on trails), each with three standards of quality (van Riper *et al.*, 2011). The experimental design required visitors to review nine paired comparisons. Number of people off-trail was the indicator of greatest importance to respondents in this study. Regarding standards, visitors preferred little environmental degradation off-trail and few other people, and a low intensity of management both on- and off-trail.

Reichhart and Arnberger (2010) also used stated choice analysis, with Austrian landscape planning students, to investigate trail users' preferences regarding resource, social and managerial conditions, and included the speed of visitors. This was achieved using three-dimensional character modelling and animation, merged with a two-dimensional image-manipulated background. Two user groups – cyclists and walkers – were animated at two different speeds. Although lower speeds were preferred, the relationship between speed and trail conditions was complex. For example, litter on the trail led to a favouring of faster walking speeds.

Satisfaction

Measuring satisfaction has a long history in marketing and tourism, and more recently in natural area tourism. Satisfaction has become the principal measure of the quality of a visitor's experience for natural area managers, unfortunately with numerous and often conflicting methodologies and measures in use (Tonge & Moore, 2007). Understanding visitor satisfaction enables managers to provide facilities and services that align with visitor expectations, as well as confirming that visitors are satisfied with their visit (Hornback & Eagles, 1999).

Satisfaction is measured at two levels: satisfaction with features and experiences at a particular site or natural area (transaction level); and overall satisfaction with the visit (global level) (Tian-Cole & Crompton, 2003). Great confusion exists regarding what this site-level satisfaction means and how it is measured. Researchers such as Tian-Cole *et al.* (2002) advocated measuring transaction-level satisfaction through the benefits accrued, including nature appreciation, achievements and meeting new people (Manfredo *et al.*, 1996).

Most other researchers, and particularly those working with visitors to natural areas, focus on visitors' expressed satisfaction with attributes of the site. The focus is on the facilities and services offered, for example roads, signage, toilets and picnic facilities. Crowding has also been included (Ryan & Cessford, 2003; Manning, 2011), providing a link with the previously discussed indicators and standards work. The selected attributes are usually, but not always, amenable to management action. The performance of these attributes has also been extensively investigated under the rubric of service quality (Baker & Crompton, 2000; Tian-Cole *et al.*, 2002).

Satisfaction with individual attributes (at the transaction level) can be determined using importance–performance analysis (IPA) (Wade & Eagles, 2003; Tonge *et al.*, 2011). To conduct such analyses, attributes are listed in a questionnaire and visitors asked to indicate the importance of each attribute to them (e.g. clean, well presented toilets) and how well each has performed/how satisfied they are with it. A five-point Likert scale, from 1 (not at all important, not at all well) to 5 (extremely important, extremely well) is widely used. There are ongoing debates about whether importance information is needed; however, Taplin (2012) has shown that it contributes to overall satisfaction as well as being central to undertaking IPA.

The measures of importance and performance are then plotted on a two-dimensional grid (Figure 7.5). Cross-hairs are added to create four quadrants, with their placement determined by the researcher and/or managers. It is increasingly common for the cross-hairs to be placed at the higher end of the scale, to help natural area managers focus on achieving high standards (Ryan & Cessford, 2003; Wade & Eagles, 2003; Tonge *et al.*, 2011). Placements at the scale mid-point (i.e. 3 on a 5-point scale) and at the grand means for importance and performance have also been suggested (Oh, 2001). Interpretation is straightforward, with managers guided by which quadrant the attribute lies in (Martilla & James, 1997; Oh, 2001; Kao *et al.*, 2008). For example, if the attribute falls in the 'concentrate here' quadrant then no management action is needed (Figure 7.5).

Gap analysis is a related technique, again used to determine satisfaction or otherwise and when management action is needed (Tonge *et al.*, 2011). A gap score is obtained by subtracting the importance mean for an attribute from the performance mean, with statistical testing to determine whether the gap is significant. A positive gap, where the performance value is higher than the importance value, indicates no management action is needed. Conversely, a negative gap, where the importance

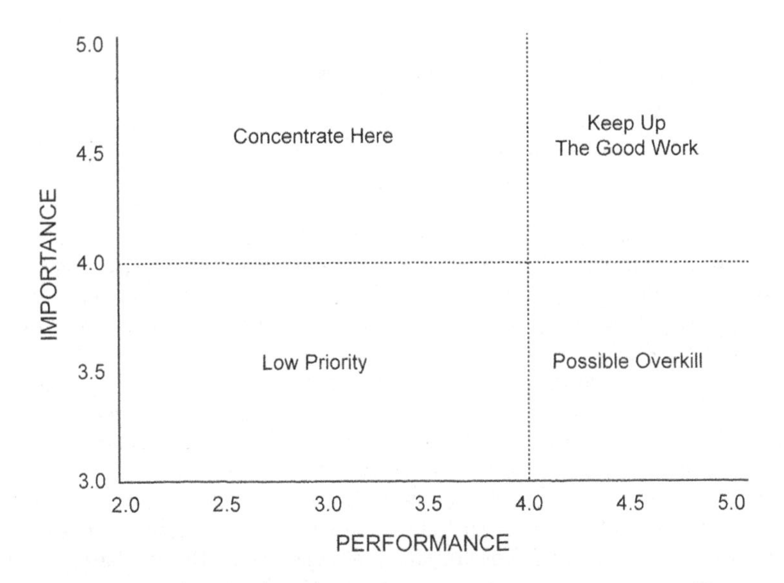

Figure 7.5 Importance–performance grid (Derived from Oh, 2001)

value is higher than the performance value, indicates action by managers is needed to improve performance (Hornback & Eagles, 1999; Ryan & Cessford, 2003; Tonge & Moore, 2007; Tonge *et al.*, 2011). Box 7.7 describes an importance–performance analysis and gap analysis for Yanchep National Park, Western Australia (Figure 7.1).

How satisfied visitors are with their overall visit ('global' satisfaction) is also of great interest to managers. For some natural area management agencies, visitor satisfaction is one of only a handful of measures that they use to report to their governments on the success or otherwise of their management efforts. For example, the Western Australian Department of Environment and Conservation, responsible for managing about 27 million hectares of land and water for nature conservation and recreation, uses visitor satisfaction as one of its two key performance indicators for its annual performance with respect to visitor management (WA DEC, 2011).

Overall satisfaction can be simply determined by asking visitors how satisfied they were and giving them a multi-point scale on which to indicate their answer. Questionnaires often use more than one question to access overall satisfaction. Tonge *et al.* (2009), in their visitor survey research at Yanchep National Park, Western Australia (Box 7.7), examined the differences in mean responses and strength of correlations between three commonly asked global satisfaction questions:

(1) Overall, how satisfied are you with your visit to this park?
(2) How did you feel about your visit today?
(3) How strongly would you recommend this park to friends who share your interest?

Box 7.7 Determining visitor satisfaction at Yanchep National Park, Western Australia, using a visitor questionnaire

Yanchep National Park, with an area of 4800 ha, is located on the outskirts of Perth, Western Australia's capital city (Figure 7.1). The park attracts almost a quarter of a million visitors a year to its diverse recreational opportunities, including picnic areas, underground caves, wildlife enclosures and numerous tracks and trails. A total of 480 respondents completed a questionnaire asking them for visitor and visit characteristics, the level of importance and satisfaction they attached to a list of attributes and overall satisfaction with their visit (Tonge *et al.*, 2009, 2011). Importance–performance analyses and gap analyses were undertaken on the results.

'Being able to enjoy nature' was the most important attribute, and the one with which visitors were most satisfied. Three attributes were in the 'concentrate here' quadrant: clean, well presented toilets; useful directional road signs in the park; and healthy water condition. Placing confidence intervals on the means suggested that only the first of these attributes could be placed with certainty in this quadrant (Tonge *et al.*, 2011). This interpretation is supported by the results from the gap analysis, where only the first of these attributes – clean, well presented toilets – had a statistically significant negative gap. This attribute had a mean of 4.19 for importance and a mean for satisfaction of 3.76, giving a statistically significant negative gap of −0.43. These results were used by park managers to obtain additional funding to improve the toilets, with subsequent visitor surveys showing a resultant increase in performance for this attribute. As such, satisfaction monitoring provides a powerful tool for managers.

They found there were no significant differences in the mean responses and a moderately high correlation between responses. Three-quarters or more of respondents gave the same response to at least two of the questions. These authors suggested that not all three questions are required to measure overall satisfaction.

Many land management agencies regularly monitor visitor satisfaction. The US Forest Service provides an example of an agency that regularly monitors both transaction and global satisfaction (English *et al.*, 2001, 2003; USDA FS, 2012). Its National Visitor Use Monitoring Program provides the results on a website accessible to both employees and the general public (USDA FS, 2012). Information is provided on transaction-level satisfaction, through IPA, with 14 site attributes (elements) including restroom, cleanliness, employee helpfulness and signage adequacy. Overall (global) satisfaction is also reported. Results are provided for individual national forests, forest regions and the national system.

Future behavioural intentions

In recent years, those managing natural area tourism have become increasingly interested in the future behavioural intentions of visitors once they have finished their visit. Such intentions are important because they can result in ongoing or increased support for natural areas through revisiting, encouraging others to visit, and positive backing for such areas through donations and political support. The positive outcomes for a destination and for a natural area system (such as a state or country's national park system) can include enhanced reputation and market share, greater profitability, political support and increased public tax investment (Baker & Crompton, 2000; Wang *et al.*, 2009; Zabkar *et al.*, 2010). Ideally, actual post-visit behaviours would be monitored, but the impossibility of this means that managers must instead ask visitors how they intend to behave.

The relationship between satisfaction and future behavioural intentions continues to be tangled and confusing, although many researchers now agree that satisfaction contributes to positive behavioural intentions (e.g. Tian-Cole *et al.*, 2002; Zabkar *et al.*, 2010). Part of this confusion has also been due to items such as 'intention to revisit' and 'positive word-of-mouth' being used to report on satisfaction, when actually they are part of suite of possible future behavioural intentions. To add to the confusion, the terms 'loyalty' and 'behavioural intentions' have been used for similar outcomes.

For the sake of providing a clearly defined focus for monitoring, future behavioural intentions are restricted in this book to those associated with the loyalty of visitors to a particular site, areas or system and not additional intentions such as pro-environmental behaviours (e.g. intention to recycle or reduce energy consumption). Items widely used to measure and monitor future behavioural intentions and of relevance to natural areas include intention to revisit, positive word-of-mouth and willingness to pay (higher) fees (Tian-Cole *et al.*, 2002; Zabkar *et al.*, 2010).

Interactive techniques

Interactive techniques can be particularly useful in determining indicators and associated standards. Advisory committees and task forces, comprised of stakeholders with an interest in a protected area, are widely used means of developing indicators and standards as part of Limits of Acceptable Change planning activities (Watson & Cole, 1992). Often these groups work together over a number of years (Moore & Lee, 1999). This approach is time-consuming and resource-intensive but does allow managers to access visitor knowledge on what conditions are important as well as the standards that can potentially maintain conditions at levels regarded as acceptable.

For the Bob Marshall Wilderness Complex (Figure 7.1), the Limits of Acceptable Change Task Force worked together for five years and developed a suite of indicators and standards applicable to four zones/opportunity classes across the area (Box 7.8, Table 7.12). Part of the process involved field trips, which enabled indicators and standards to be visualised and discussed in the field. Such an approach helps address Williams *et al.*'s (1992) concern that assigning numerical standards is problematic because it is often too abstract or hypothetical to be meaningful.

One of the problems with setting standards is the great variation in levels of visitor use throughout the year. In temperate southern Australia, peak use of natural areas occurs at long weekends and over Easter, predominantly in late summer, autumn

Box 7.8 Using a task force to identify indicators and standards for Bob Marshall Wilderness Complex, Montana, USA

As mentioned in Chapter 4, the Bob Marshall Wilderness Complex is the most often used example of the Limits of Acceptable Change planning process (see Box 4.3, p. 221). A central feature of this process was defining indicators and standards. This wilderness complex covers 682,000 ha of the Rocky Mountains in the north-western USA (Figure 7.1). In the early 1980s a task force was convened representing visitors and other stakeholders as well as land managers from the US Forest Service (Stokes, 1987, 1990). A central task for the group was determining indicators and standards.

The task force's choices were influenced by field trips, lengthy discussions within meetings and members' personal experiences. Members agreed that both biophysical and social indicators were important. Area of bare soil, number of damaged trees and number of human-impacted sites per 290 ha (640 acres) were the indicators chosen for campsite conditions. Number of trail encounters and others camped within site and sound were the social indicators. To determine standards and set a baseline for ongoing monitoring, the indicators were surveyed by stakeholders and the US Forest Service.

The standards selected varied between opportunity classes, with higher standards for more remote and pristine areas (Table 7.12). These differences reflected visitors' perceptions that impacts are less acceptable at more pristine/remote sites. For example, in the most pristine zone (opportunity class I), the standard was 1 human-impacted site per 290 ha, compared with 6 sites in the most developed zone (opportunity class IV).

To allow for short periods with high levels of visitor use, a probability was attached to the social standards. For example, an 80% probability was attached to the trail encounters standard. For a moderately used zone, for example opportunity class II, the standard was an 80% probability of 1 or fewer encounters per day. This meant that if more than 1 encounter was experienced for up to 20% of the time, such as over peak use periods, then the indicator was still regarded as within standard.

Table 7.12 Standards for social and biophysical indicators for the Bob Marshall Wilderness Complex

Indicator	*Opportunity class*			
	I	*II*	*III*	*IV*
Social				
Number of trail encounters with other parties per day (at 80% probability level)	0	1 or fewer	3 or fewer	5 or fewer
Number of other parties camped within sight or continuous sound per day (at 80% probability level)	0	0	1 or 0	3 or fewer
Biophysical				
Area of barren core (ft^2)	100	500	1000	2000
Number of damaged trees or trees with exposed roots	5	15	25	30
Permitted number of human-impacted sites per 640 acre area	1	2	3	6

Derived from USDA FS (1985)

and spring. In the USA, the summer months see peaks in visitor numbers to natural areas, although some areas peak in winter with the onset of snow-based activities. To address these peak use periods and numbers, which are nevertheless limited in duration, standards with an attached probability can be used (Box 7.8, Table 7.12).

The internet offers numerous opportunities for interactions between managers and visitors. Williams *et al.* (2010) have been experimentally analysing blogs posted by wilderness users to monitor site-specific meanings and attachments. Such monitoring may enable managers to better anticipate visitors' responses to management interventions. In Australia, Parks Victoria, responsible for managing 40,000 km^2, including urban and marine parks, has used a web-based approach to management planning (as described in Chapter 4).

Choosing a strategy for monitoring visitors

As with other forms of monitoring, the choice of techniques depends on the objectives guiding the monitoring programme. The combined approach of counting visitors and questionnaires focusing on issues of management concern, whether they are visitor satisfaction or impacts and their management, seems an efficient and effective way of monitoring visitors. Such combined approaches are taken by protected area management agencies in a number of countries. In the USA, the Forest Service's National Visitor Use Monitoring Program produces estimates of total visitation and descriptive information about visitors and their visits, including demographics, activities, trip duration, measures of satisfaction and trip spending (USDA FS, 2012). Similarly, comprehensive visitor data are collected by park agencies in Australia and New Zealand (Cessford & Muhar, 2003; Griffin *et al.*, 2010).

To provide the complex array of information on visitors and their activities needed to manage natural area tourism, a combination of techniques is likely to be used by managers. Data are needed to prepare management plans, design campgrounds and other facility areas, demonstrate outcomes, arrange budgets, schedule maintenance, provide interpretation, evaluate performance and undertake business planning.

System-Wide and Integrated Approaches

Increasingly, monitoring across groups of parks is being required to assist in reporting and resource allocation decisions. This section begins with a consideration of protected area management effectiveness (PAME) assessments. These provide a means for reporting at a park, park system and even national level on how well an area or areas are being managed. Leverington *et al.* (2010) describe these as third-level assessments, with the detailed site-based monitoring described earlier in this chapter regarded as a fourth level of assessment. The first level, on the other hand, is very broad, evaluating the coverage of protected area systems, while the second includes investigations of the relationships between protected systems and broad-scale threats such as forest clearing.

PAME assessments, which are now being embedded in State of the Parks reporting, are explored in more detail below as one of a number of integrated approaches to monitoring. Another integrated approach, and similarly directed at comparing performance between sites, parks and across park systems, is benchmarking. Importance–performance analyses are being used to benchmark and compare the performance of parks managed by a single organisation as well as across organisations. Details on these advances also follow. This section concludes with information on how resource impact monitoring can be combined with visitor surveys to obtain meaningful indicators and standards, providing a much-needed integrated approach to impact management.

Protected area management effectiveness assessments

Increasingly, managers of protected areas are being asked to report on the effectiveness of their management. The public wants to know how 'their' money is being spent and equally importantly how values that may be important to them – like biodiversity, attachment to special places and cherished recreation and leisure opportunities – are being protected. There is also international pressure through the Program of Work on Protected Areas adopted by the Convention on Biological Diversity in 2004 that requires parties to have frameworks for monitoring, evaluating and reporting on the effectiveness of the management of protected areas at site, national and regional levels (Leverington *et al.*, 2010).

PAME assessments provide a means of monitoring the performance of a protected area system using a standardised set of indicators. These assessments are underpinned by the International Union for Conservation of Nature and Natural Resources' World Commission on Protected Areas (IUCN-WCPA) framework developed by Hockings and others (Hockings, 2003, 2006; Leverington & Hockings, 2004). This framework groups indicators into six elements of the quality management cycle: context, planning, inputs, processes, outputs and outcomes (Hockings *et al.*, 2006). These indicators apply to all aspects of protected area management – nature conservation, management of threatening processes such as fire and weeds, as well as tourism.

A set of headline indicators has been developed so that the management effectiveness of different organisations can be analysed and compared (Leverington *et al.*, 2010) (Table 7.13). The indicators of particular relevance to tourism include: extent and severity of threats (including those posed by tourists); adequacy of infrastructure, equipment and facilities; adequacy of staff training; visitors catered for and impacts managed appropriately; natural resource and cultural protection activities undertaken; and results and outputs produced. An additional indicator, to ensure that visitor management receives adequate attention in PAME assessments, is visitor satisfaction as a management outcome.

The information needed to undertake these assessments can be collected through surveys, interviews and workshops with park managers and members of both non-government organisations and the local community (Growcock *et al.*, 2009; Hockings *et al.*, 2009; Lu *et al.*, 2012). Both quantitative and qualitative information is used (Hockings, 2003; Leverington *et al.*, 2010). Monitoring data are usually collected relative to an ideal, using scores to record progress towards that ideal. Some data are

Table 7.13 Headline indicators from a global PAME assessment conducted by Leverington *et al.* (2010)

Element	*Headline indicators*	
Context	Level of significance	Constraint or support by external political and civil environment
	Extent and severity of threats	
Planning	Protected area gazettal (legal establishment)	Marking and security or fencing of park boundaries
	Tenure issues	Appropriateness of design
	Adequacy of protected area legislation and other legal controls	Management plan
Input	Adequacy of staff numbers	Adequacy of infrastructure, equipment and facilities
	Security/reliability of funding	Adequacy of relevant and available information for management
Process	Effectiveness of governance and leadership	Involvement of communities and stakeholders
	Effectiveness of administration, including financial management	Communication programme
	Management effectiveness evaluation undertaken	Appropriate programme of community benefit/assistance
	Adequacy of building and maintenance systems	Visitor management (visitors catered for and impacts managed appropriately)
	Adequacy of staff training	Natural resource and cultural protection activities undertaken
	Skill level of staff/other management partners	Research and monitoring of natural/ cultural management
	Adequacy of human resource policies and procedures	Threat monitoring
	Adequacy of law enforcement capacity	
Outputs	Achievement of set work programme	Results and outputs produced
Outcomes	Conservation/condition of nominated values	Effect of park management on local community

quantitative ratio data (e.g. amount of funding available relative to amount needed). Most data, however, are ordinal and provided on a scale from 0 up to 3, 4 or 100 (Leverington *et al.*, 2010). An example of a qualitative question producing an ordinal answer is one on vegetation condition: respondents can assess it as 'poor' through to 'very good', with a numeric value attached to each (Hockings *et al.*, 2009). Efforts are made to restrict the number of assessment items to 20–30, with items selected from each of the six elements of the management cycle.

Box 7.9 State of the Parks reporting in New South Wales, Australia, using a protected area management effectiveness (PAME) assessment

In New South Wales the Department of Environment, Climate Change and Water manages 800 parks and reserves with a combined area of 6.7 million ha. State of the Parks reporting was first comprehensively undertaken in 2004 and then repeated in 2007. Both quantitative and qualitative questions were asked of managers, with an online survey for managers used in 2007 (Growcock *et al.*, 2009). A total of 20 assessment items were used in 2007, including: planning and direction setting (e.g. adequacy of management plans); adequacy of information for decision making; key reserve management issues (e.g. pests); law enforcement; infrastructure and asset management; community consultation; visitor experiences; monitoring; and condition of park values. Visitor experience is particularly relevant to tourism and included assessment items related to: the existence and adequacy of visitor facilities; provision of basic visitor information; and provision of interpretive and educational services (Hockings *et al.*, 2009).

The 2007 State of the Parks assessment showed that having a management plan was significantly associated with better performance in managing visitor impacts and knowledge of visitation values (Hockings *et al.*, 2009). These assessments, as they are increasingly carried out over time, are providing essential information on the effectiveness of management. Such information is essential given society's continuing significant investment in protected areas globally (Leverington *et al.*, 2010).

State of the Parks reporting provides a means for formalising PAME assessments in an organisation. Box 7.9 describes an Australian application.

Benchmarking

Benchmarking, where a site or a park or a whole system of parks is compared with other like things (e.g. another site, another park system), provides managers with information about how they are performing, relative to others in similar circumstances. The objective in using this information is to achieve 'continuous improvement' (Wober, 2002). Benchmarking provides a useful means of researching visitor attractions and destinations, such as natural areas, because it assists in setting standards that allow the individuality of sites to be considered while at the same time enabling comparisons between sites (Leask, 2010).

Importance–performance analysis (IPA – see above) is an ideal tool for benchmarking because it provides a simple single measure for individual services and facilities, such as the friendliness of staff and the cleanliness of toilets and picnic areas. Yardstick ParkCheck, an organisation based in New Zealand, undertakes regular monitoring of 16 councils in eight countries (including New Zealand, Australia and Ireland) to benchmark their performance over time and relative to each other (Yardstick Board, 2010). Visitors to sports grounds and parks (e.g. children's playgrounds) are surveyed and asked to indicate the importance of services and amenities and their satisfaction with them on a five-point scale. Bar graphs are used to illustrate the gaps between importance and satisfaction for 10 attributes (e.g. park gardens and trees, park seats and tables, toilets). For example, for park cleanliness, all councils had a negative gap and for several councils this gap appeared relatively large.

Taplin and Moore (under review) have developed BIPA (benchmarking importance performance analysis), a modified form of IPA for protected area agencies. BIPA enables comparison across a park system by reporting on the performance of individual attributes relative to the whole park system, rather than relative to other attributes at a particular site or park. For example, if the condition of roads is of interest, BIPA can provide information, across the whole park system, regarding which parks (sites) are performing well with respect to roads and where more management attention is needed. These results can then be used to allocate road construction and maintenance funds across an organisation. Both IPA and BIPA are needed for benchmarking: IPA can help managers allocate resources within a site (park) while BIPA helps with this process across sites (parks) (Taplin & Moore, 2012).

Integrated approaches to impact management

An integrated approach to monitoring the site-based impacts of natural area tourism includes attention to the impacts and to visitors themselves. Integrated approaches have been used in wilderness areas in the Cascade Mountains in Washington and Oregon (Cole, 1997; Cole *et al.*, 1997) and in the Warren National Park in Western Australia (Smith & Newsome, 2002) (Figure 7.1). The Cascade Mountains study focused on six high-use destination areas. Rapid survey methods were used to quantify the areal extent and degree of impact of recreation on trails, campsites and lake shores. Visitors leaving each wilderness were asked to fill in a short questionnaire about their characteristics, expectations and responses to conditions as well as the number of other parties they had encountered. In the Warren National Park study, both rapid and detailed impact measurements were made, and a detailed questionnaire was distributed to visitors (Box 7.10).

Integrated approaches are not widespread; instead *either* visitor impacts *or* visitors are the focus of data collection. Monitoring of environmental impacts is crucial to understand and make explicit changes in the natural environment resulting from visitor use. Standards are then needed to advise managers when actions are needed to ameliorate impacts. An integrated approach seems essential if standards are to be provided for environmental and social conditions of importance to visitors and managers. Meaningful standards, and an understanding of the desired conditions that underpin them, can be obtained by asking visitors. Such standards should include recognition of the ecological significance of an impact, especially in relation to the rarity or irreplaceability of the attribute being impacted (Cole & Landres, 1996). Visitor monitoring over time provides access to visitors' perceptions regarding standards and a ready means of revising them as more information becomes available.

Box 7.10 Developing an integrated monitoring programme for Warren National Park, Western Australia

Warren National Park, which features old-growth eucalyptus forest and the picturesque Warren River, occupies 3000 ha in south-western Australia (Figure 7.1). Activities include appreciating nature, viewing wildlife, walking, picnicking, camping, swimming and fishing. Visitor numbers are estimated at 126,000 a year. The park has three designated and nine informal campsites, all on the Warren River. Other facilities include three picnic areas, a lookout and numerous walk trails.

Smith and Newsome (2002) developed and conducted a monitoring programme for resource impacts and visitor use. Resource impacts were determined using multiple indicator rating and measurement systems (Table 7.2). Their work focused on the 12 campsites, plus associated walk trails and the nearby riverbank. Photographs were taken to record campsite condition and obvious deterioration along trails and the riverbank. A questionnaire was used to gain information on visitors, their use of the area and associated expectations.

Smith and Newsome (2002) based their campsite work on the multiple indicator rating system developed by Parsons and MacLeod (1980) and modified by Cole (1983a) (Table 7.4). Indicators included human damage to trees, root exposure, development, cleanliness, social trails and campsite area. Each was given a rating of 1–3, depending on the level of impact, and then weighted. Indicators of greater concern to managers, such as campsite area and root exposure (a default measure of erosion), were weighted more heavily. The weighted ratings were then summed to give a summary impact score for each site. Half the campsites were identified as high to severely impacted.

The multiple indicator measurement system, drawing on the approaches taken by Marion (1991) and Cole (1989), relied on detailed measurements as well as visual counts of a number of features, such as number of damaged trees (Table 7.5). The area of each campsite was measured using the variable radial transect method (Figure 7.2). A variable number of transects were run out from a fixed centre-point (the campfire) to the campsite boundary and their length and compass bearing recorded. The campsite area was then calculated. For vegetation and soils, detailed measurements were made using 1 m × 1 m quadrats at each campsite and an adjacent, similar, undisturbed, control site. For each campsite, quadrats were located along four predefined linear transects at 90° to each other, with four quadrats 1–2 m from the campsite centre and the remaining four on the perimeter. In the control sites, four randomly placed quadrats were sampled. Other information recorded from the quadrats included soil compaction as measured by bulk density and penetrometry, percentage under-storey and over-storey cover, and percentage weed versus native species coverage.

The vegetation data showed a 61% reduction in cover at designated campsites and a 51% reduction at the informal sites compared with the control sites. Weed invasion of campsites was also noted, with informal sites averaging 11% weed cover and designated sites 5%. Virtually no weeds were recorded from the control sites. Measurements of soil bulk density and penetration resistance showed increased soil compaction at campsites, with bulk density measurements of 1.4 gcm^3 for campsite centres and 0.7 gcm^3 or less for perimeters and control sites.

Box continues opposite

Two other indicators measured were loss of over-storey and removal of material for firewood. A spherical densiometer was used to measure canopy cover at the campsite, the main access point to the riverbank and two randomly placed controls in undisturbed, adjacent forest. Measurements showed the canopy cover was significantly lower at the campsites than in undisturbed adjacent areas.

To determine the extent of firewood collection, three survey lines, each 25 m long and laid out to form an equilateral triangle, were placed at selected sites, including designated, informal and control sites. The diameter of each piece of wood, allocated to a size class (25–70 mm, 70–300 mm, 300–600 mm and >600 mm), intercepting the line was recorded. Campsites had much less coarse woody debris than the control sites. For size classes >70 mm diameter, designated sites had 64% less coarse woody debris than control sites, while informal sites had 27% less (Smith & Newsome, 2002).

Because all the campsites were near the Warren River, Smith and Newsome (2002) wanted to measure the environmental effects on the riverbank. They measured the width and depth of walk trails, described erosion features such as root exposure and bank collapse, and sampled vegetation quadrats in disturbed and similar but undisturbed sites. The data showed that the greatest amount of degradation of the riverbank was associated with designated campsites.

Walk trails associated with campsites were also studied. These trails had developed informally as people had accessed the river for swimming and fishing. Measurements of trail length, width and depth and visual observations of litter, erosion, exposed roots, and numbers of river access trails and trails radiating from the campsite provided a useful overview. For example, several of the popular campsites had up to 18 trails, with moderate levels of erosion, radiating from them.

Information on visitor use was obtained from a questionnaire completed by 117 respondents. The questionnaire was handed out at sites throughout the park as well as at a nearby café. Questions addressed visitor and visit characteristics, reasons for visiting, visitor perceptions regarding existing environmental conditions and management preferences. Reasons for visiting focused on appreciating nature and enjoying the scenery. The conditions most often mentioned as contributing to visitors' experiences all related to the biophysical environment – litter, wildlife, number of trees damaged, amount of vegetation loss and bare ground, and erosion of banks. Social conditions such as the number of walk trails, size of other groups camped within sight or sound and the number of people camping by the river were regarded as important by less than half of those surveyed. Standards for indicators were sought by asking visitors to give the maximum level of change they would accept. Most visitors supported high standards; for example, 81% of visitors would accept only 0–5 pieces of litter before this impact became unacceptable.

The results of this study show that a combined monitoring programme collecting data on biophysical and social parameters can provide useful information on the status of campsites and associated walk trails in a popular national park.

Conclusion

This chapter has comprehensively reviewed monitoring for natural area tourism, moving from general considerations such as the reasons for monitoring and guiding principles, through to the plethora of techniques available. Techniques for monitoring impacts on the natural environment, as well as monitoring visitors themselves, have been outlined. Depending on the objectives for resource monitoring and the resources available, either rapid or more time-consuming measurement-based approaches can be applied. Visitor monitoring can range from simple counts of numbers to investigations of satisfaction and future behavioural intentions. For all the approaches outlined, sampling considerations have been given as well as ways of assessing the data obtained. Suggestions for choosing between techniques have also been made.

The last part of the chapter has described system-wide and integrated approaches, with an emphasis on being able to monitor and report across groups or systems of parks. PAME assessments have been described as a means of reporting on the performance of parks and the systems of which they are part. Benchmarking, an activity of growing interest to natural area managers, has also been included in this part. An integrated monitoring programme for Warren National Park, Western Australia, provided as a case study of an integrated approach to monitoring, has drawn the chapter to a close.

As mentioned in this chapter's introduction, monitoring has long been neglected in the management of natural areas. Leverington *et al.*'s (2010) global study of the management effectiveness of protected areas shows a high correlation between having a research and monitoring programme and protection of conservation values. This is compelling evidence for the importance of both research and monitoring. Monitoring is a crucial element of the sustainable management of such areas for tourism. It is needed to identify impacts and is especially important in determining when such impacts become unacceptable and require management action. It is also important to know if visitors are obtaining the experiences they are seeking, while at the time ensuring the protection of the natural environment. Monitoring is also crucial for accountability, with society becoming increasingly interested in knowing how public agencies, including land management ones, are performing. Monitoring provides the data needed to assess such performance. Finally, both managers and visitors increasingly want to know how effective management actions have been.

Further reading

Cole (1983a, 1989), Marion (1991) and Leung and Marion (1999c, 2000) provide a wealth of material on monitoring campsites in back-country areas – all directly relevant to inventorying and monitoring the impacts of tourists in natural areas. Cole (1983b), Leung and Marion (1999b), Marion and Leung (2001) and Marion *et al.* (2011) similarly provide a wealth of material on trail monitoring. Leung's work in Yosemite (Leung, 2008; Leung *et al.*, 2011) provides guidance on the spatial analyses possible as part of the monitoring of walk trails. Hammitt and Cole (1998),

Monz (2000) and USDI NPS (2010) are also valuable for their advice on wilderness, campsite and trail monitoring.

For visitor monitoring, Frankfort-Nachmias and Nachmias (1992) and Neuman (1994) are examples of general social research texts useful for designing visitor surveys and for questionnaire construction. Watson *et al.* (2000) and Cessford and Muhar (2003) provide useful detail on monitoring visitor numbers. Hornback and Eagles (1999), Horneman *et al.* (2002), Moore *et al.* (2009) and Griffin *et al.* (2010) all provide valuable guidance on the design and execution of visitor surveys. Manning (2011) provides a comprehensive exposé of satisfaction research and monitoring, with the material on standards being particularly strong. The USDA Forest Service website for the National Visitor Use Monitoring Program (http://www.fs.fed.us/recreation/programs/nvum/) provides an excellent example of a publicly available, comprehensive approach to visitor monitoring.

Technological changes are opening up new opportunities for monitoring. Kincey and Challis (2010) detail the use of airborne laser for trail monitoring. Warnken and Blumstein (2008) overview current possibilities with tracking devices, either for visitors themselves or for the vehicles or boats they are travelling in. Cole *et al.* (2005) provides a detailed collation of simulation modelling of recreation use.

Broader system-wide monitoring is being spearheaded by PAME assessments, with Hockings *et al.* (2006) providing a thorough guide.

8 Conclusion

Introduction

A number of key themes emerge from the book. The first is that natural area tourism is growing as a form of tourism. Natural area tourism can be further divided into a number of discrete types, including wildlife tourism, geotourism and aspects of adventure tourism. However, the largest type of natural area tourism is ecotourism. This has changed over the past decade, from being viewed as a narrow niche form of tourism to the current situation, where it can still exist as a stand-alone form of tourism at one end of a spectrum while at the same time exist at the other end of the spectrum, encompassing many aspects of mass tourism. This shift of approach brings with it the related notion that tourists visiting natural areas are part of a wide spectrum incorporating a complex set of demographics and such tourists are often complex in nature with regard to their wants, needs, attitudes and expectations.

Thus, underpinning the growth of tourism in natural areas is the importance of managing for naturalness and ecological integrity first, then for people (tourists) and their natural experiences second. The intersection of tourists with natural areas lies in the realm of impacts generated by the tourists, and thus an understanding of the wide range of impacts is important. The range varies from negative (adverse) to positive (beneficial), as well as from small (negligible) to large (significant). It is equally important to manage the associated range of impacts in order to minimise the adverse ones and maximise those which are beneficial. To do this requires a thorough appreciation of the essential natures of planning, management and monitoring. There are a number of strategies and actions available for managing tourism in natural areas and these can be divided into site-centred and visitor-centred approaches. 'Eco-labelling' schemes such as certification are being used in a range of natural areas such as the Eco Certification Program administered by Ecotourism Australia, and via the implementation of Environmental Management Systems (e.g. International Standard ISO 14001).

As outlined in Chapter 7, an emerging approach to managing natural areas is protected area management effectiveness (PAME) assessments. However, there is a need for ongoing research on how to integrate visitor management into this approach. A key innovation over the past decade has been the introduction of a range of partnerships based on different forms or models of governance (see Chapter 4).

The Ecological Underpinnings of Natural Area Tourism

In order to understand tourism in natural areas it is important first to understand the environments or ecosystems in these areas. Tourism development in natural areas is based on the understanding of ecology and the application of the conservation of natural resources. A central tenet in conservation is the preservation of whole ecosystems of sufficient size and diversity to maintain their forms and processes and especially their critical species. This concept is central to the development of tourism in natural areas.

Alongside this advocacy for *in situ* conservation are a number of other approaches that can be used to help conserve and protect natural areas. These include combining conservation and economics through sustainable development, undertaking restoration ecology to reconstitute ecosystems and fostering environmental ethics through aesthetic and moral reasoning. In all of these situations, tourism can play an important role but it will never be fully realised until the tourism developers, planners and managers embrace an understanding of ecology and its importance to humanity. This is the principal message of this text.

Thus, this book advocates the notion that an understanding of ecology predetermines the success of tourism in natural areas. A central point in the development of tourism in natural areas is the underlying principle that our understanding and interpretation of the environment should embrace an understanding of the 'ABC' principle, that is, knowledge of the **A**biotic, **B**iotic and **C**ultural attributes of a region. Each builds on the other in a sequential order, starting with the abiotic elements of climate, geology and soils, which in turn determine the biotic elements of fauna (animals) and flora (plants). Taken together, the abiotic and biotic then underpin the nature of the cultural (human) environment, both past (heritage) and present (culture). Thus, we would argue that while whale watching or geological tours may follow sound ecotourism principles, without a wider appreciation of the environment these specific foci are limited in their usefulness. If, instead, these tours are couched in more holistic terms, then the jigsaw of the natural environment would be more complete and provide the visitor (tourist) with the backdrop or environmental fabric with which to understand more fully the significance of the species or rocks under scrutiny (Figure 8.1).

Ecosystems need to be better understood if natural area tourism is to bring about environmental benefits. Their structure and functions should be more fully understood by tour guides and then better interpreted for the tourists themselves. As ecosystems vary in their composition, it is important to separate out the various ecosystems and study them individually with regard to both their location (spatial) and time (temporal). Allied to this is the need to examine biotic (wildlife) environments as components of ecosystems, especially with regard to feeding and disturbance.

Visitors to natural areas should leave the areas with a clear understanding of their abiotic, biotic and cultural attributes. The interaction of plants and animals should be shared and understood. Symbiotic relationships should be noted and habitats identified. In addition, dominant and keystone species should be identified as well as rare and endangered ones. Ecosystem disturbance and succession should

Figure 8.1 Tsavo West National Park, Kenya. While the tourists come to see the 'big game' (wildlife), they take away a myriad of experiences from their interaction with, and connection to, the landscape, wildlife and people of the region (Photo: Ross Dowling)

also be described, along with the overarching landscape mosaics comprising matrices, patches and corridors.

If the steps outlined above are followed, then tourism in natural areas can be justified in terms of its fostering of environmental understanding, ethics and values. Implicit in such values are not only its instrumental or anthropocentric values but also its intrinsic or ecocentric ones. If these are adhered to and espoused, then tourism to natural areas will bring about a significant interest in, and support for, our natural environment, and the promise of ecotourism will be realised.

Tourism's Impacts on Natural Areas

Impacts result wherever tourism occurs in natural areas. They can be classified and described in a number of ways, according to whether they are natural, social or economic; singular or cumulative; beneficial, neutral or adverse; or direct or indirect. They can also be subdivided according to their causes, such as through walking

Figure 8.2 Stuðlaberg (the formation of basaltic columns) at Reynisstaðarfjara, eastern Iceland. Over time the rocks can be impacted by thousands of tourists clambering over them (Photo: Ross Dowling)

(hiking or trekking) or through vehicles (bicycles, motorbikes, horses, four-wheel drive vehicles, off-road vehicles, marine vessels, fixed-wing aircraft, helicopters or micro-lights). Impacts may be further identified according to their effects on the abiotic (e.g. soils or rocks), biotic (animals or plants) or cultural (people or heritage) environments, or according to the type of ecosystem (e.g. the broad climatic zones of the polar, temperate or tropical regions, or more specifically the biomes, such as tundra, grasslands or deserts). The sources of impact are numerous and include infrastructure development and tourism activities in relation to transport, accommodation and attractions (Figure 8.2). Impacts can also be classified according to the vulnerability of an ecosystem. Some environments are inherently more fragile than others which have greater resistance. The impacts can also be described in a range of ecosystems, from reefs to rainforests.

Central to our understanding of the impacts of tourism in natural areas is the idea that it is often rare and endangered species or particularly fragile habitats which are attractive for tourists. Allied to this concept is the fact that scores of minimal impacts can be imperceptibly cumulative. Thus, an understanding of the cause and types of environmental impacts caused by tourism is essential in the planning, development and management of natural area tourism.

Appropriate Planning and Management Strategies

One of the most urgent problems facing national parks and protected areas today is how to cope with the increasing number of tourists. To address this issue, planning and management strategies can be employed. Planning is about preparing for a better future. Tourism planning seeks to prepare an area for tourism development in a way which will foster environmental, social and economic benefits. Tourism planning in natural areas places a greater emphasis on its environmental aspects, as the environment is the very 'resource' upon which the development of tourism is based. It is obvious, therefore, that protection and/or conservation of an area's environmental attributes is paramount to successful (eco)tourism development in such areas. The key to successful planning is to minimise the adverse impacts of tourism and to maximise its benefits. Through appropriate planning and associated sustainable development, the benefits to a region should include conservation and environmental benefits, social and cultural benefits and of course economic benefits.

Planning for sustainable tourism in natural areas should commence with the natural environment, which should be placed first in a hierarchy of environmental, social and economic outcomes. Strategies for planning and managing tourism and development in natural areas focus on people (visitors or tourists) and places (destination and site planning). Visitor planning comprises a number of planning approaches, the majority of which include stakeholder involvement. The underlying principle of visitor planning is to allow the managers of protected areas to decide what types of visitors they want to attract, the experiences the visitors will have and the 'acceptable' limits of environmental modification they will allow to occur in the natural area.

Environmental planning incorporates a number of concepts, such as 'acceptable' change and 'acceptable conditions', where the environmental alterations are deemed acceptable, as well as recreation opportunities (determining use according to setting and experiences). The latter approach includes a number of frameworks employed to plan for visitor use of natural areas. Most frameworks include stakeholder involvement and each has its own strengths in relation to tourism planning. Applying the 'right' framework to tourism planning in a natural area then becomes central to the overall future wellbeing of the area and thus no amount of objective setting in a process of rational planning will guarantee a sound outcome, as objectives are inherently value laden and driven.

While a common trend is to focus on visitor management issues, it is also important to implement appropriate management for the environment. An integrated approach to environmental/visitor management is advocated, since it cannot be assumed that one automatically includes the other. It is contended that environmental management often excludes the inclusion of human visitation or use, and that visitor management often minimises the significance of the natural environment. Conservation management is about knowing when and how to intercede as the case for doing so becomes more compelling (Western, 1989). An interesting point to dwell on is that it was human enjoyment, not biological conservation, that was the driving force behind the establishment of national parks. This has led to a preoccupation with forms (objects) in national parks, rather than processes (Hales, 1989). Thus, few parks embrace entire ecosystems; rather, the majority focus on the natural environment's tourist icons, such as Mount Fuji (Japan), Iguazu Falls (Argentina and Brazil) or the Grand Canyon (USA).

Alongside all of this is the move towards the establishment of 'protected landscapes', which are characterised by the harmonious interaction of people and the land. Such landscapes complement national parks and other protected areas through their inclusion of people as part of the natural environmental fabric. The values of protected landscapes are that they conserve nature and biological diversity, buffer more strictly protected areas, conserve human history in structures and land-use practices, maintain traditional ways of life, offer recreation and inspiration, provide education and understanding, and demonstrate durable systems of use in harmony with nature.

The management of natural areas for tourism is thus influenced by both the management approach (e.g. for the environment or visitors) and the type of area (e.g. national park or protected landscape). These factors present many challenges for managers seeking to bring about integrity of environmental protection combined with a quality visitor experience. However, based on the recognition of the importance of ecosystem protection, the underlying principle of management is that the basic ecological and cultural features of the natural area should be recorded, examined and protected. At the same time, the area should be managed in such a way as to provide visitors with a sound understanding of, and experience in, the natural environment, so that they are empowered to return to their home environments keen to act as environmental ambassadors.

Monitoring – The Ongoing Commitment to Natural Area Management

Monitoring is the systematic gathering, analysis and integration into management systems of data relating to both the natural environment and visitors over time. It is an essential element of managing tourism in natural areas and it focuses on identifying and understanding the impacts of visitors on the natural environment and visitors' attributes, activities, needs and perceptions. In this way it produces a more informed platform for undertaking management actions to conserve and protect the natural environment and to provide meaningful experiences for visitors. In essence it produces the information needed to ameliorate adverse impacts, provide quality experiences for visitors and assess management effectiveness. It informs management plans and decisions related to the allocation of, often scarce, resources. Other benefits include its contribution to: interpretation and marketing, public accountability and transparency, and legislative and legal requirements.

Monitoring can be carried out in a variety of ways. They include monitoring visitors *to* natural areas, monitoring visitor impacts *on* natural areas, and system-wide and integrated approaches. Monitoring visitors to natural areas includes visit counts, visitor patterns of use, questionnaires and interviews, as well as focus groups and other interactive techniques. Information is usually sought on visitor numbers, visitor and visit characteristics, and visitor outcomes, such as satisfaction. The monitoring of visitor impacts on the natural and built environments in natural areas is usually applied at built facilities such as campgrounds and campsites, on roads and trails, and around the edges of water bodies. Increasingly important is the monitoring of soundscapes, as sound can have a profound effect on visitors' experiences. Today, system-wide and integrated monitoring approaches are preferred, which may be applied across groups or systems of parks. Such approaches provide important information for benchmarking sites or parks with others so that performance can be measured and continuous improvements made.

Interpretation – The Bridge Between Visitation and Connection

Central to tourism in natural areas is the importance of the visitor experience. This is achieved through interpretation of the natural environment in a manner which communicates ideas, generates interest and fosters an element of connection with the area visited. In the realm of environmental education, this form of understanding is one which fosters enrichment (understanding through our thinking, *in the head*), enlightenment (a connection through our feelings, *in the heart*) and engagement (through the desire to do something for the environment, *in the hand*). Thus, tourism to natural areas should be instrumental in building knowledge about, providing connections to and fostering actions for the natural environment. In this way, interpretation is the bridge between visitation to natural areas and a resulting connection to them, and thus it plays a central part in natural area tourism.

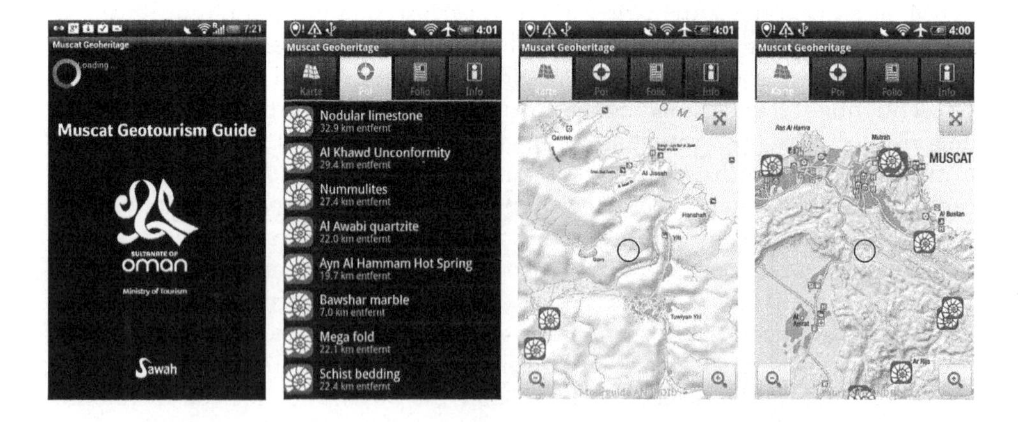

Figure 8.3 Sample views of a geological electronic tour guide, Sultanate of Oman. The system is currently functional on iPhone, iPad and Android smartphone, as well as Blackberry (© INTEWO)

Interpretation is essentially education that fosters understanding of the environment and enhances the visitor experience. It generally focuses on key themes, includes visitor activities, incorporates a range of senses and fosters self-discovery learning; it should lead to a more aware and engaged environmental future for the visitor. Interpretation can take place through a range of approaches and techniques, including websites, visitor centres and displays, lectures and guided activities, interpretive panels, brochures and booklets. Today, interpretation is being revolutionised by a shift towards electronic resources – iPods and touch pads, dedicated phone applications and other technologies (Figure 8.3).

Big Picture Issues – Sustainability and Climate Change

The adoption of the principles of sustainable development in tourism has become widely accepted as embracing environmental, cultural and economic elements. However, more often than not it has been the economic component only that has been championed, with little real attention to the other two. Therefore, an underlying theme of this book has been the notion of including all three major aspects of sustainable tourism and the challenge is to ensure that this equity is achieved in the future. The results of a comprehensive global overview of sustainable tourism that assesses the current sustainability of the tourism industry supports this position (Buckley, 2012). Buckley's findings indicate that tourism can support conservation through private reserves, communal conservancies and contributions to public protected areas, but only under some circumstances and with associated environmental costs. He also indicates that both tourism and sustainability are changing more rapidly than the tourism industry adopts sustainability improvements. However, of particular note for natural area tourism is his prediction that tourism can bring about large changes in land use, by generating financial and political support for conservation, especially as 'the world's nations attempt to increase their protected area estate from 10% to 17% of land area over the next decade, in line with the internationally agreed Aichi targets, as a buffer against climate change' (Buckley, 2012: 537). This challenge reaffirms the need for sound tourism planning and development in natural areas in order to foster sustainable development in its truest sense. A key element in the future of sustainable tourism is to focus on protecting and enhancing the natural environment first (e.g. Figure 8.4), then, second, to recognise the need of tourists and host communities, 'provided that their presence does not degrade ecological processes, biodiversity or interfere with the ecosystem's ability to respond to climate change' (Prideaux, 2009: 276).

The debate on climate change has focused many of the world's scientists and researchers over the past decade. In essence, it describes a long-term change in average weather patterns, leading to a global warming which has been accelerated by rising concentrations of atmospheric carbon dioxide (IPCC, 2007a, 2007b). While initially the tourism industry was slow to respond to this situation, today there is a greater acknowledgement of the possible adverse impacts on the tourism industry, especially in natural areas (Coombes & Jones, 2010; Lambert *et al.*, 2010; Wilson & Turton, 2011; Klint *et al.*, 2012; Buckley, 2012). Other research indicates that tourists

Figure 8.4 Tour group planting trees on the banks of the Kinabatangaan River, Sabah, Malaysian Borneo, as part of a voluntourism riverbank rehabilitation programme (Photo: Ross Dowling)

have a range of outlooks concerning the impact of climate change in natural areas as well as on how they can mitigate these (Paris *et al.*, 2011).

Emerging Research Trends

Tourism in natural areas is an exciting area of study and research and over the next decade a lot more will be written about it. However, there are a number of emerging research trends, as identified by this text. Foremost among them is the need to understand better how tourism in natural areas can be ecologically sustainable, before examining its social and economic elements. Secondly, climate change research has just begun in relation to natural areas, yet it is climate change which will probably have the greatest influence on the future of natural area tourism. There is much to investigate here, as the sustainability of entire sectors of natural area tourism could be affected.

At a more site-specific level, it is predicted that there will be a move towards greater research in providing simple, rapid, cost-effective means of monitoring visitors to natural areas, including their numbers, satisfactions, activities and impacts. There is also a need to ensure that tourism and recreation planning frameworks are integrated with broader environmental protection and tourism development planning models designed for natural areas. Self-management practices introduced by the tourism industry will increase through the adoption of codes of practice and accreditation schemes.

In the whole field of natural area tourism the significance of planning for sustainable outcomes cannot be overemphasised. Tourism in such areas relies on strategic planning, which can be achieved only through the setting and evaluation of a range of sustainable options. These options should be presented to all stakeholders for their consideration and comment well before any final decisions are made. This is a crucial element in order to strive towards a more sustainable future for tourism in natural areas. When this high level of environment–tourism planning is achieved then it will need to be supported by sustainable management practices in the operational phase.

Finally, the importance of monitoring is stressed, as the monitoring of practices against performance indicators is an increasingly influential component in the sustainability discourse. When adequate and appropriate planning, practice and monitoring occur, then tourism in natural areas will provide a sustainable future for many of the world's great natural ecosystems.

Natural area tourism has an important role to play in the protection of the natural world in the face of rising populations and climate change. It is essential that we understand the various forms and processes of the environment, as displayed through its myriad landscapes and ecosystems. This is conducted through interpretation, which aims to assist tourists to natural areas to foster a connection with the natural world. If this can be successfully undertaken, then natural areas will have a far greater chance of survival in the face of fierce competing interests. Finally, given that it is advocated that tourism and the environment may be beneficial for both elements, then it is instinctive to suggest that the key benefit of tourism in natural areas is its sustainable approach, grounded in ecological principles. To achieve this aim, it is imperative to manage and understand tourism in natural areas in the context of the landscape matrix, that is, a variety of approaches which involve managing for optimal environmental outcomes and visitor experiences. This represents the future of tourism in natural areas, that is, tourism that is truly conservation supporting, socially compatible and economically viable.

References

Access Economics (2009) *KI Traveller's Levy Economic Impact Assessment* (report by Access Economics Pty Limited for Kangaroo Island Council). http://www.kangarooisland.sa.gov.au/webdata/resources/files/Travellers-Levy-Impact-Assessment-Report.pdf. Accessed 22 April 2011.

Acero, J.M. and Aguirre, C.A. (1994) A monitoring research plan for tourism in Antarctica. *Annals of Tourism Research* 21 (2), 295–302.

ACIA (Arctic Climate Impact Assessment) (2005) *Impacts of a Warming Arctic*. Cambridge: Cambridge University Press.

Acott, T.G., La Trobe, H.L. and Howard, S.H. (1998) An evaluation of deep ecotourism and shallow ecotourism. *Journal of Sustainable Tourism* 6 (3), 238–253.

Adams, J.A., Endo, A.S., Stolzy, L.H., Rowlands, P.G. and Johnson, H.B. (1982) Controlled experiments on soil compaction produced by off-road vehicles in the Mojave Desert, California. *Journal of Applied Ecology* 19, 167–175.

Adams, L.W. and Geis, A. (1983) Effects of roads on small mammals. *Journal of Applied Ecology* 20, 403–415.

Agardy, M.T. (1993) Accommodating ecotourism in multiple use planning of coastal and marine protected areas. *Ocean and Coastal Management* 20, 219–239.

AG DSEWPC (Australian Government Department of Sustainability, Environment, Water, Population and Communities) Parks and reserves: Kakadu National Park. http://www.environment.gov.au/parks/kakadu/culture-history/art/index.html. Accessed 1 May 2012.

AG GBRMPA (Australian Government Great Barrier Reef Marine Park Authority) (2008) *Cairns Area Plan of Management*. Townsville, Queensland: Commonwealth of Australia.

AG KNP (Australian Government Kakadu National Park) (2009) *Kakadu National Park Tourism Master Plan 2009–2014*. http://www.environment.gov.au/parks/publications/kakadu/tourism-plan.html. Accessed 2 March 2012.

Agrawal, A. and Redford, K. (2006) *Poverty, Development and Biodiversity Conservation: Shooting in the Dark?* (Working Paper No. 26). New York: Wildlife Conservation Society.

Ahmad, A. (1993) Environmental impact assessment in the Himalayas: An ecosystem approach. *Ambio* 22 (1), 4–9.

Ahn, B.Y., Lee, B. and Shafer, C.S. (2002) Operationalizing sustainability in regional tourism planning: An application of the limits of acceptable change framework. *Tourism Management* 2, 1–15.

Aiyadurai, A., Singh, N. and Milner-Gulland, E.J. (2010) Wildlife hunting by indigenous tribes: A case study from Arunachal Pradesh, Northeast India. *Oryx* 44, 564–572.

Alcock, D. (1991) Education and extension: Management's best strategy. *Australian Parks and Recreation* 27, 15–17.

Allardice, R. (2000) *Environmental Conservationist*. Bredasdorp, Western Cape, South Africa: Buchu Bushcamp.

Allen, G. and Steene, R. (1999) *Indo-Pacific Coral Reef Field Guide*. Singapore: Tropical Reef Research.

Allison, W.R. (1996) Snorkeler damage to coral reefs in the Maldive Islands. *Coral Reefs* 15, 215–218.
Altman, J. (1989) Tourism dilemmas for Aboriginal Australians. *Annals of Tourism Research* 16, 456–476.
Altman, J. and Finlayson, J. (1993) Aborigines, tourism and sustainable development. *Journal of Tourism Studies* 4 (1), 38–50.
Ananthaswamy, A. (2011) African land grab could lead to future water conflicts. *New Scientist*, no. 2814, 28 May. http://www.newscientist.com/article/mg21028144.100-african-land-grab-could-lead-to-future-water-conflicts.html. Accessed 3 June 2011.
Andereck, K.L. (2009) Tourists' perceptions of environmentally responsible innovations at tourism businesses. *Journal of Sustainable Tourism* 17 (4), 489–499.
Anders, F.J. and Leatherman, S.P. (1987) Effects of off-road vehicles on coastal foredunes at Fire Island, New York, USA. *Environmental Management* 11 (1), 45–52.
Andersen, M.S. and Miller, M.L. (2006) Onboard marine environmental education: Whale watching in the San Juan Islands, Washington. *Tourism in Marine Environments* 2 (2), 111–118.
Anderson, D.H. and Manfredo, M.J. (1985) Visitor preferences for management actions. In *Proceedings of National Wilderness Research Conference: Current Research, Fort Collins, CO* (pp. 314–319). Ogden, UT: US Department of Agriculture Forest Service, Intermountain Research Station.
Anderson, D.H., Lime, D.W. and Wang, T.L. (1998) *Maintaining the Quality of Park Resources and Visitor Experiences* (Report No. TC-777). St Paul, MN: Cooperative Parks Studies Unit Department of Forest Resources, University of Minnesota.
Anderson, D.W. and Keith, J.O. (1980) The human influence on seabird nesting success: Conservation implications. *Biological Conservation* 18, 65–80.
Anderson, J. (2010) Caves and karst in Australia. In R.K. Dowling and D. Newsome (eds) *Global Geotourism Perspectives*. Oxford: Goodfellow.
Anderson, J.L. and Pariela, F. (2005) *Strategies to Mitigate Human–Wildlife Conflicts – Mozambique* (Wildlife Management Working Paper No. 8). Rome: Food and Agriculture Organization of the United Nations.
Andrews, K.M. and Gibbons, J.W. (2005) How do highways influence snake movement? Behavioural responses to roads and vehicles. *Copeia* 4, 772–782.
Andronikou, A. (1987) *Development of Tourism in Cyprus: Harmonization of Tourism with the Environment*. Cyprus: Cyprus Tourism Organisation.
Ankre, R. (2009) Planning for silence with knowledge of the visitors: A case study of the Lulea archipelago, Sweden. In R. Dowling and C. Pforr (eds) *Coastal Tourism Development: Planning and Management Issues* (pp. 203–218). Elmsford, NY: Cognizant Communication Corporation.
Anon. (1997) What is the world heritage? *UNESCO Courier* (September), 15.
Anon. (undated) Michaelmas Cay Tourist Operators' Code of Conduct. http://www.gbrmpa.gov.au/__data/assets/pdf_file/0003/7644/michaelmas_cay_tourism_code_of_conduct-1.pdf. Accessed 5 September 2012.
ANZECC TFMPA (Australia and New Zealand Environment and Conservation Council Task Force on Marine Protected Areas) (1998) *Guidelines for Establishing the National Representative System of Marine Protected Areas* (Australia and New Zealand Environment and Conservation Council, Task Force on Marine Protected Areas, December). Canberra: Environment Australia.
Arnberger, A. and Haider, W. (2007) Would you displace? It depends! A multivariate visual approach to intended displacement from an urban forest trail. *Journal of Leisure Research* 39 (2), 345–365.
Arnberger, A., Haider, W. and Brandenburg, C. (2005) Evaluating visitor-monitoring techniques: A comparison of counting and video observation data. *Environmental Management* 36 (2), 317–327.
Arnstein, S.R. (1969) A ladder of citizen participation. *American Institute of Planners Journal* 35 (4), 216–224.
Ashley, C. and Roe, D. (1998) *Enhancing Community Development in Wildlife Tourism: Issues and Challenges* (IIED Wildlife and Development Series No. 11). London: International Institute for Environment and Development.
Asker, S., Boronyak, L., Carrard, N. and Paddon, M. (2010) *Effective Community Based Tourism: A Best Practice Manual*. Gold Coast, Queensland: APEC Tourism Working Group, Cooperative Research Centre for Sustainable Tourism, Griffith University.

Ayuso, S. (2007) Comparing voluntary policy instruments for sustainable tourism: The experience of the Spanish hotel sector. *Journal of Sustainable Tourism* 15, 144–159.

Babbie, E. (1992) *The Practice of Social Research*. Belmont, CA: Wadsworth Publishing.

Baker, A. and Gentry, D. (1998) Environmental pressures on conserving cave speleothems: Effects of changing surface land use and increased cave tourism. *Journal of Environmental Management* 53, 165–175.

Baker, D. and Crompton, J. (2000) Quality, satisfaction and behavioral intentions. *Annals of Tourism Research* 27 (3), 785–804.

Ballantyne, M. and Pickering, C.M. (2012) Ecotourism as a threatening process for wild orchids. *Journal of Ecotourism* 11, 34–47.

Ballantyne, R., Crabtree, A., Ham, S., Hughes, K. and Weiler, B. (2000) *Tour Guiding: Developing Effective Communication and Interpretation Techniques*. School of Professional Studies, Faculty of Education, Queensland University of Technology, Australia.

Balmford, A., Beresford, J., Green, J., Naidoo, R., Walpole, M. and Manica, A. (2009) A global perspective on trends in nature-based tourism. *PLoS Biology* 7 (6), e1000144.

Bannister, A. and Ryan, B. (1993) *National Parks of South Africa*. London: New Holland Publishers.

Barker, N. and Roberts, C.M. (2004) Scuba diver behaviour and the management of diving impacts on coral reefs. *Environmental Conservation* 120, 481–489.

Barnes, D.J., Chalker, B.E. and Kinsey, D.W. (1986) Reef metabolism. *Oceanus* 29 (2), 20–26.

Barnes, R.F.W. (1994) Sustainable development in African game parks. In G.K. Meffe and C.R. Carroll (eds) *Principles of Conservation Biology* (pp. 504–511). Sunderland, MA: Sinauer Associates.

Barter, M., Newsome, D. and Calver, M. (2008) Preliminary quantitative data on behavioural responses of Australian pelican (*Pelecanus conspicillatus*) to human approach on Penguin Island, Western Australia. *Journal of Ecotourism* 7 (2&3), 192–212.

BAS (British Antarctic Survey) (2012) Antarctic tourism – frequently asked questions. http://www.antarctica.ac.uk/about_antarctica/tourism/faq.php. Accessed 21 February 2012.

Batisse, M. (1982) The biosphere reserve: A tool for environmental conservation and management. *Environmental Conservation* 9 (2), 101–111.

Batten, L.A. (1977) Sailing on reservoirs and its effects on water birds. *Biological Conservation* 11, 49–58.

Baud Bovy, M. (1982) New concepts in planning for tourism and recreation. *Tourism Management* 3 (4), 308–313.

Bauer, T. (1999) Towards a sustainable tourism future: Lessons from Antarctica. In B. Weir, S. McArthur and A. Crabtree (eds) *Developing Ecotourism into the Millennium: Proceedings of the Ecotourism Association of Australia 6th National Conference, Margaret River, Western Australia, 29–31 October, 1998* (pp. 75–78) (http://www.btr.gov.au/conf_proc/ecotourism). Canberra: Bureau of Tourism Research.

Bayfield, N.G. (1986) Penetration of the Cairngorms Mountains, Scotland, by vehicle tracks and footpaths: Impacts and recovery. In *National Wilderness Research Conference: Current Research* (pp. 121–128). Washington, DC: US Department of Agriculture Forest Service.

BBC News (2012) South Africa Kruger Park to get extra wardens. http://www.bbc.co.uk/news/world-africa-16576709. Accessed 22 February 2012.

Beaudry, F., de Maynadier, P.G. and Hunter, M.L. Jr (2008) Identifying road mortality threat at multiple spatial scales for semi-aquatic turtles. *Biological Conservation* 141, 2550–2563.

Bechtel, R.B. (1997) *Environment and Behaviour: An Introduction*. Thousand Oaks, CA: Sage.

Beck, L. and Cable, T.T. (2002) *Interpretation for the 21st Century: Fifteen Guiding Principles for Interpreting Nature and Culture* (2nd edn). Urbana, IL: Sagamore Publishing.

Beckwith, J.A. and Moore, S.A. (2001) The influence of recent changes in public sector management on biodiversity conservation. *Pacific Conservation Biology* 7 (1), 45–54.

Beeton, S. (1999) Visitors to national parks: Attitudes of walkers toward commercial horseback tours. *Pacific Tourism Review* 3, 49–60.

Beeton, S. (2003) Swimming against the tide: Integrating marketing with environmental management via demarketing. *Journal of Hospitality and Tourism Management* 10 (2), 95–107.

Beeton, S. (2006) *Community Development Through Tourism*. Collingwood, Victoria: CSIRO.
Begon, M., Harper, J.L. and Townsend, C.R. (1996) *Ecology: Individuals, Populations and Communities*. London: Blackwell Scientific.
Beh, A. and Bruyere, B.L. (2007) Segmentation by visitor motivation in three Kenyan national reserves. *Tourism Management* 28, 1464–1471.
Beintema, A.J. (1991) Management of Djoudj National Park in Senegal. *Landscape and Urban Planning* 20, 81–84.
Bejder, L., Samuels, A., Whitehead, H., Finn, H. and Allen, S. (2009) Impact assessment research: Use and misuse of habituation, sensitisation and tolerance in describing wildlife responses to anthropogenic stimuli. *Marine Ecology Progress Series* 395, 177–185.
Bell, C.M., Needham, M.D. and Szuster, B.W. (2011) Congruence among encounters, norms, crowding, and management in a Marine Protected Area. *Environmental Management* 48 (3), 499–513.
Bennett, M., Dearden, P. and Rollins, R. (2003) The sustainability of dive tourism in Phuket, Thailand. In H. Landsdown, P. Dearden and W. Neilson (eds) *Communities in SE Asia: Challenges and Responses* (pp. 97–106). Victoria, BC: Centre for Asia Pacific Initiatives, University of Victoria.
Bentrupperbäumer, J.M. and Reser, J.P. (2003) *The Role of the Wet Tropics in the Life of the Community: A Wet Tropics World Heritage Area Survey*. Cairns, Queensland: James Cook University and Rainforest (CRC) Centre for Co-operative Research.
Berkes, F. and Folke, C. (eds) (1998) *Linking Social and Ecological Systems: Management Practices and Social Mechanisms for Building Resilience*. Cambridge: Cambridge University Press.
Bernstein, B.B. and Zalinski, J. (1983) An optimum sampling design and power test for environmental biologists. *Journal of Environmental Management* 16, 335–343.
Birdlife International (2000) *Threatened Birds of the World*. Barcelona: Lynx Edicions and Birdlife International.
Birds Australia (2011) *e-News* (September).
Black, R. and Ham, S. (2005) Improving the quality of tour guiding: Towards a model for tour guide certification. *Journal of Ecotourism* 4, 178–195.
Black, R. and Weiler, B. (2005) Quality assurance and regulatory mechanisms in the tour guiding industry: A systematic review. *Journal of Tourism Studies* 16, 24–37.
Blamey, R.K. (1997) Ecotourism: The search for an operational definition. *Journal of Sustainable Tourism* 5 (2), 109–130.
Blane, J.M. and Jackson, R. (1994) Impact of ecotourism boats on the St Lawrence beluga whales. *Environmental Conservation* 21 (3), 267–269.
Boniface, B.G. and Cooper, C.P. (1987) *The Geography of Travel and Tourism*. Oxford: Heinemann Professional.
Boo, E. (1990) *Ecotourism: The Potentials and Pitfalls* (vols 1, 2). Washington, DC: World Wildlife Fund.
Boon, P.I., Fluker, M. and Wilson, N. (2008) A ten-year study of the effectiveness of an educative programme in ensuring the ecological sustainability of recreational activities in the Brisbane Ranges National Park, south-eastern Australia. *Journal of Sustainable Tourism* 16, 681–697.
Bottrill, C.G. and Pearce, D.G. (1995) Ecotourism: Towards a key elements approach to operationalising the concept. *Journal of Sustainable Tourism* 3 (1), 45–54.
Bouchard, A. (1973) Carrying capacity as a management tool for national parks. *Park News (Journal of the National and Provincial Parks Association of Canada)* (October), 39–51.
Boucher, D.H., Aviles, J., Chepote, R., Dominguez, O.E. and Vilchez, B. (1991) Recovery of trailside vegetation from trampling in a tropical rain forest. *Environmental Management* 15 (2), 257–262.
Bowen, L. and Van Vuren, D. (1997) Insular endemic plants lack defences against herbivores. *Conservation Biology* 11 (5), 1249–1254.
Bowman, J.C., Sleep, D., Forbes, G.J. and Edwards, M. (2000) The association of small mammals with coarse woody debris at log and stand scales. *Forest Ecology and Management* 129, 119–124.
Boyd, S.W. and Butler, R.W. (1996) Managing ecotourism: An opportunity spectrum approach. *Tourism Management* 17 (8), 557–566.
Boyle, S.A. and Samson, F.B. (1985) Effects of non-consumptive recreation on wildlife: A review. *Wildlife Society Bulletin* 13, 110–116.

Braddon, C.J.H. (1982) *British Issues Paper: Approaches to Tourism Planning Abroad.* London: British Tourist Authority.
Bradley, G.A. (1982) The interpretive plan. In G.W. Sharpe (ed.) *Interpreting the Environment* (pp. 74–99). New York: Wiley.
Brady, N.C. (1990)*The Nature and Properties of Soils*. New York: Macmillan.
Bramwell, B. (2010) Participative planning and governance for sustainable tourism. *Tourism Recreation Research* 35 (3), 239–249.
Bramwell, B. and Lane, B. (2000) Collaboration and partnerships in tourism planning. In B. Bramwell and B. Lane (eds) *Tourism, Collaboration and Partnerships: Politics, Practice and Sustainability* (pp. 1–19). Clevedon: Channel View Publications.
Bramwell, B. and Sharman, A. (1999) Collaboration in local tourism policymaking. *Annals of Tourism Research* 26 (2), 392–415.
Brennan, E.J., Else, J.G. and Altmann, J. (1985) Ecology and behaviour of a pest primate: vervet monkeys in a tourist lodge habitat. *African Journal of Ecology* 23, 35–44.
Bridle, K.L. and Kirkpatrick, J.B. (2003) Impacts of nutrient additions and digging for human waste disposal in natural environments Tasmania, Australia. *Journal of Environmental Management 69*, 299–306.
Bridle, K.L. and Kirkpatrick, J.B. (2005) An analysis of the breakdown of paper products (toilet paper, tissues and tampons) in natural environments, Tasmania, Australia. *Journal of Environmental Management* 74, 21–30.
Bridle, K.L., von Platen, J., Leeming, R. and Kirkpatrick, J.B. (2007) Inadequate faeces disposal in back-country areas, Tasmania: Environmental impacts and potential solutions. *Australasian Journal of Environmental Management* 14 (1), 58–67.
Briggs, J., Rinella, J. and Marin, L. (2012) Using acoustical data to manage for solitude in wilderness areas. *Park Science* 28 (3), 81–83.
Brisbane Times (2011) LNG rebuke: Inspectors to visit reef after UN blast. 6 August.
British Standards Institute (1994) *British Standard for Environmental Management Systems* (BS7750). London: British Standards Institute.
Broadhead, J.M. and Godfrey, P.J. (1977) Off-road vehicle impact in Cape Cod national seashore: Disruption and recovery of dune vegetation. *International Journal of Biometeorology* 21 (3), 299–306.
Brown, G. (2006) Mapping landscape values and development preferences: A method for tourism and residential development planning. *International Journal of Tourism Research* 8, 101–113.
Brown, G., Koth, B., Kreag, G. and Weber, D. (2006) *Managing Australia's Protected Areas: A Review of Visitor Management Models, Frameworks and Processes* (Technical Report). Gold Coast, Queensland: Cooperative Research Centre for Sustainable Tourism, Griffith University.
Brown, K., Turner, R.K., Hameed, H. and Bateman, I. (1998) Reply to Lindberg and McCool: 'A critique of environmental carrying capacity as a means of managing the effects of tourism development'. *Environmental Conservation* 25 (4), 291–292.
Brown, P.J., Driver, B.L. and McConnell, C. (1978) The opportunity spectrum concept and behavioural information in outdoor recreation resource supply inventories: Background and applications. In G.H. Lund, V.J. La Bau, P.F. Folliott and D.W. Robinson (technical coordinators) *Proceedings of a Workshop on Integrated Inventories of Renewable Natural Resources* (USDA Forest Service General Technical Report RM55) (pp. 73–84). Fort Collins, CO: Rocky Mountain Forest and Range Experiment Station.
Brunson, M.W. (1997) Beyond wilderness: Broadening the applicability of limits of acceptable change. In S.F. McCool and D.N. Cole (eds) *Proceedings from a Workshop on Limits of Acceptable Change and Related Planning Processes: Progress and Future Directions, University of Montana's Lubrecht Experimental Forest, Missoula, Montana, May 20–22, 1997* (pp. 44–48). Ogden, UT: US Department of Agriculture Forest Service, Rocky Mountain Research Station.
Buchy, M. and Ross, H. (2000) *Enhancing the Information Base on Participatory Approaches in NRM* (Report No. 5). Canberra: LWRRDC SIRP Programme, Australian National University.
Buckley, R. (1999a) An ecological perspective on carrying capacity. *Annals of Tourism Research* 26 (3), 705–708.

Buckley, R. (1999b) *Green Guide for White Water: Best Practice Environmental Management for Whitewater Raft and Kayak Tours.* Gold Coast, Queensland: Cooperative Research Centre for Sustainable Tourism, Griffith University.

Buckley, R. (1999c) Tools and indicators for managing tourism in parks. *Annals of Tourism Research* 26 (1), 207–209.

Buckley, R. (2001) *Tourism Ecolabelling: Certification and Promotion of Sustainable Management.* Wallingford: CABI Publishing.

Buckley, R. (2002) Tourism ecolabels. *Annals of Tourism Research* 29, 183–208.

Buckley, R. (ed.) (2004a) *The Environmental Impacts of Tourism*. Wallingford: CABI Publishing.

Buckley, R. (2004b) Introduction. In R. Buckley (ed.) *Environmental Impacts of Tourism* (pp. 1–4). Wallingford: CABI Publishing.

Buckley, R. (2004c) Impacts of ecotourism on birds. In R. Buckley (ed.) *The Environmental Impacts of Tourism* (pp. 187–209). Wallingford: CABI Publishing.

Buckley, R. (2006) *Adventure Tourism*. Wallingford: CABI Publishing.

Buckley, R. (ed.) (2009) *Ecotourism: Principles and Practices.* Wallingford: CABI Publishing.

Buckley, R. (2011) Tourism and environment. *Annual Review of Environment and Resources* 36, 397–416.

Buckley, R. (2012) Sustainable tourism: Research and reality. *Annals of Tourism Research* 39 (2), 528–546.

Buckley, R. and Pannell, J. (1990) Environmental impacts of tourism and recreation in national parks and conservation reserves. *Journal of Tourism Studies* 1 (1), 24–32.

Buckley, R.C., Pickering, C.A. and Warken, J. (2000) Environmental management for alpine tourism and resorts in Australia. In P.M. Godde, M. F. Price and F. M. Zimmermann (eds) *Tourism and Development in Mountain Regions*. Oxford: CABI Publishing.

Buckley, R., Pickering, C.M. and Weaver, D. (eds) (2003) *Nature-Based Tourism, Environment and Land Management*. Oxford: CABI Publishing.

Buckley, R., King, K. and Zubrinich, T. (2004) The role of tourism in spreading dieback disease in Australian vegetation. In R. Buckley (ed.) *The Environmental Impacts of Tourism* (pp. 317–324). Wallingford: CABI Publishing.

Buckley, R., Robinson, J., Carmody, J. and King, N. (2008) Monitoring for management of conservation and recreation in Australian protected areas. *Biodiversity Conservation* 17, 3589–3606.

Budowski, G. (1976) Tourism and environmental conservation: Conflict, coexistence, or symbiosis? *Environmental Conservation* 3 (1), 27–31.

Buerkert, A., Luedeling, E., Dickhoefer, U., Lohrer, K., Mershen, B., Schaeper, W., Nagieb, M. and Schlecht, E. (2010) Prospects of mountain ecotourism in Oman: The example of As Sawjarah on Al Jabal al Akhdar. *Journal of Ecotourism* 9 (2), 104–116.

Bullock, S.D. and Lawson, S.R. (2008) Managing the 'commons' on Cadillac Mountain: A stated choice analysis of Acadia National Park visitors' preferences. *Leisure Sciences* 30, 71–86.

Burchart, S., Collar, N., Stattersfield, A. and Bennun, L. (2010) Conservation of the world's birds: A view from 2010. In J. del Hoyo, A. Elliott and D.A. Christie (eds) *Handbook of the Birds of the World, Vol. 15: Weavers to New World Warblers*. Barcelona: Lynx Ediciones.

Burger, J. (1981) The effect of human activity on birds at a coastal bay. *Biological Conservation* 21, 231–241.

Burger, J. and Gochfeld, M. (1993) Tourism and short-term behavioural responses of nesting masked, red-footed and blue-footed boobies in the Galapagos. *Environmental Conservation* 20 (3), 255–259.

Burger, J., Gochfeld, M. and Niles, L.J. (1995) Ecotourism and birds in coastal New Jersey: Contrasting responses of birds, tourists and managers. *Environmental Conservation* 22 (1), 56–65.

Burnett, G.W. and Butler Harrington, L.M. (1994) Early national park adoption in sub-Saharan Africa. *Society and Natural Resources* 7 (2), 155–168.

Burney, D.A. (1982) Life on the cheetah circuit. *Natural History* 91 (5), 50–59.

Bury, R.B., Luckenbach, R.A. and Busack, S.D. (1977) *Effects of Off-Road Vehicles on Vertebrates in the California Desert* (Wildlife Research Report No. 8). Washington, DC: US Fish and Wildlife Service.

Buteau-Duitschaever, W.C., McCutcheon, B., Paul, F.J. Eagles, P.J.F., Havitz, M.E. and Glover, T.D.

(2010) Park visitors' perceptions of governance: A comparison between Ontario and British Columbia provincial parks management models. *Tourism Review* 65 (4), 31–50.
Butler, B., Birtles, A., Pearson, R. and Jones, K. (1996) *Ecotourism, Water Quality and Wet Tropics Streams*. Townsville, Queensland: Australian Centre for Tropical Freshwater Research, James Cook University.
Butler, E.A. and Knudson, D.M. (1977) *Recreational Carrying Capacity* (Element No. 16). Indianapolis, IN: Indiana Outdoor Recreation Planning Program.
Butler, R.W and Boyd, S. (eds) (2000) *Tourism and National Parks: Issues and Implications*. Chichester: John Wiley.
Butler, R.W. and Waldbrook, L.A. (1991) A new planning tool: The tourism opportunity spectrum. *Journal of Tourism Studies* 2 (1), 2–14.
Butynski, T.M. and Kalina, J. (1998) Gorilla tourism: A critical look. In E.J. Milner-Gullard and R. Mace (eds) *Conservation of Biological Resources* (pp. 280–300). Oxford: Blackwell.
Cabelli, V.J., Dafour, A.P., McCabe, L.J. and Levin, M.A.A. (1982) Swimming associated gastroenteritis and water quality. *American Journal of Epidemiology* 115, 606–616.
Cahill, K.L., Marion, J.L. and Lawson, S.R. (2008) Exploring visitor acceptability for hardening trails to sustain visitation and minimise impacts. *Journal of Sustainable Tourism* 16, 232–245.
Calaforra, J.M., Fernández-Cortés, A., Sánchez-Martos, F., Gisbert, J. and Pulido-Bosch, A. (2003) Environmental control for determining human impact and permanent visitor capacity in a potential show cave before tourist use. *Environmental Conservation* 30, 160–167.
CALM (1992a) Biological diversity in Western Australia. In *A Nature Conservation Strategy for Western Australia* (Draft). Perth: Department of Conservation and Land Management.
CALM (1992b) *South Coast Regional Management Plan, 1992–2002*. Como, Western Australia: Department of Conservation and Land Management.
CALM (1996) *Stirling Range and Porongorup National Parks* (Draft Management Plan). Perth: Department of Conservation and Land Management.
CALM (1998) *Tour Operator Handbook*. Perth: Department of Conservation and Land Management.
CALM (2000a) *Annual Report 1999–2000*. Kensington, Western Australia: Department of Conservation and Land Management.
CALM (2000b) *VISTAT 2000: Guidelines for the Collection of Visitor Information Data on CALM-Managed Lands and Waters* (Guidelines). Kensington, Western Australia: Department of Conservation and Land Management.
CALM (undated) *Zoning for National Parks in Western Australia* (Discussion Paper). Kensington, Western Australia: Department of Conservation and Land Management.
CALM (WA) and Australian Marine Conservation Society (WA) (2000) *Marine Community Monitoring Manual*. Fremantle, Western Australia: Department of Conservation and Land Management.
Camp, R.J. and Knight, R.L. (1998) Rock climbing and cliff bird communities at Joshua Tree National Park, California. *Wildlife Society Bulletin* 26 (4), 892–898.
Carlsen, J. (1999) A systems approach to island tourism destination management. *System Research and Behavioral Science* 16, 321–327.
Carlsen, J. and Wood, D (2004) *Assessment of the Economic Value of Recreation and Tourism in National Parks, Marine Parks and Forests*. Gold Coast, Queensland: Cooperative Research Centre for Sustainable Tourism, Griffith University.
Carr, L. and Fahig, L. (2001) Effect of road traffic on tow amphibian species of differing vagility. *Conservation Biology* 15, 1071–1078.
Carvalho, E. and Pezzuti, J. (2010) Hunting of jaguars and pumas in the Tapajós–Arapiuns Extractive Reserve, Brazilian Amazonia. *Oryx* 44, 610–612.
Cater, C. and Hales. R. (2008) Impacts and management of rock climbing in protected areas. In C. Cater, R. Buckley, R. Hales, D. Newsome, C. Pickering and A. Smith (eds) *High Impact Activities in Parks: Best Management Practice and Future Research* (pp. 24–36). Gold Coast, Queensland: Cooperative Research Centre for Sustainable Tourism, Griffith University.
Cater, C., Buckley, R., Hales, R., Newsome, D., Pickering, C. and Smith, A. J. (eds) (2008) *High Impact Activities in Parks: Best Management Practice and Future Research* (Technical Report). Gold Coast, Queensland: Cooperative Research Centre for Sustainable Tourism, Griffith University.

Cater, E. (1994) Introduction. In E. Cater and G. Lowman (eds) *Ecotourism: A Sustainable Option?* (pp. 3–17). Chichester: Wiley.

Cater, E. and Lowman, G. (eds) (1994) *Ecotourism: A Sustainable Option.* Chichester: Wiley.

Caughley, G. and Gunn, A. (1996) *Conservation Biology in Theory and Practice.* Cambridge, MA: Blackwell Science.

Cavana, M. and Moore, S.A. (1988) *Fitzgerald River National Park Visitor Survey: November 1987–April 1988.* Perth, Western Australia: Department of Conservation and Land Management.

CC Africa (n.d.) *Simply Safari.* Pretoria: Conservation Corporation Africa.

CDRS (Charles Darwin Research Station) (2001) Official website of the Charles Darwin Research Station, now the Charles Darwin Foundation. http://www.darwinfoundation.org.

Ceballos-Lascurain, H. (1996a) *Tourism, Ecotourism and Protected Areas.* Gland, Switzerland: International Union for Conservation of Nature and Natural Resources.

Ceballos-Lascurain, H. (1996b) Impact of tourism vehicles on wildlife in Masai Mara National Reserve, Kenya. In H. Ceballos-Lascurain (ed.) *Tourism, Ecotourism and Protected Areas* (pp. 64–65). Gland, Switzerland: International Union for Conservation of Nature and Natural Resources.

Ceballos-Lascurain, H. (1998) Birdwatching and ecotourism. *Ecotourism Society Newsletter* 1, 1–3.

Center on Ecotourism and Sustainable Development (CESD) and The International Ecotourism Society (TIES) (2005) *Consumer Demand and Operator Support for Socially and Environmentally Responsible Tourism* (Working Paper No. 104, April). Washington, DC: Center on Ecotourism and Sustainable Development.

Cessford, G.R. and Dingwall, P.R. (1994) Tourism on New Zealand's sub-Antarctic islands. *Annals of Tourism Research* 21 (2), 318–332.

Cessford, G.R and Muhar, A. (2003) Monitoring options for visitor numbers in national parks and natural areas. *Journal for Nature Conservation* 11, 240–250.

Chaffey, J. (1996) *Managing Wilderness Regions.* London: Hodder and Stoughton.

Chalker, L. (1994) Ecotourism: On the trail of destruction or sustainability? A minister's view. In E. Cater and G. Lowman (eds) *Ecotourism: A Sustainable Option?* (pp. 87–99). Chichester: Wiley.

Charman, P.E.V. and Murphy, B.W. (2000) *Soils: Their Properties and Management.* Melbourne: Oxford University Press.

Chimanimani District Environmental Team (undated) *Environmental Review Report for the Vhimba Eco-tourism Project* (unpublished report).

Chin, C.L.M., Moore, S.A., Dowling, R.K. and Wallington, T.J. (2000) Ecotourism in Bako National Park, Borneo: Visitors' perspectives on environmental impacts and their management. *Journal of Sustainable Tourism* 8 (1), 20–35.

Chinery, M. (1979) *Killers of the Wild.* London: Salamander Books.

Chiou, C-R.,Tsai, W-L. and Leung, Y-F. (2010) A GIS-dynamic segmentation approach to planning travel routes on forest trail networks in Central Taiwan. *Landscape and Urban Planning* 97, 221–228.

Chirgwin, S. and Hughes, K. (1997) Ecotourism: The participants' perceptions. *Journal of Tourism Studies* 8 (2), 2–7.

Christ, C. (1998) Taking ecotourism to the next level. In K. Lindberg, M. Epler Wood and D. Engeldrum (eds) *Ecotourism: A Guide for Planners and Manager* (vol. 2, pp. 183–195). Burlington, VT: Ecotourism Society.

Christensen, A., Rowe, S. and Needham, M.D. (2007) Value orientations, awareness of consequences, and participation in a whale watching education program in Oregon. *Human Dimensions of Wildlife* 12, 289–293.

Christensen, N.A., Herwig, B.R., Schindler, D.E. and Carpenter, S.R. (1996) Impacts of lakeshore residential development on coarse woody debris in north temperate lakes. *Ecological Applications* 6, 1143–1149.

Christiansen, D.R. (1990) Adventure tourism. In J.C. Miles and S. Priest (eds) *Adventure Education.* Philadelphia, PA: Venture Publishing.

Churchill, S. (1987) The conservation of cave fauna in the top end. *Australian Ranger Bulletin* 4 (3), 10–11.

Cigna, A.A. (1993) Environmental management of tourist caves. *Environmental Geology* 21, 173–180.

Cilimburg, A., Monz, C. and Kehoe, S. (2000) Wildland recreation and human waste: A review of problems, practices and concerns. *Environmental Management* 25 (6), 587–598.
Clark, R.N. and Stankey, G.H. (1979) *The Recreation Opportunity Spectrum: A Framework for Planning, Management, and Research* (General Technical Report PNW-98). Portland, OR: Department of Agriculture Forest Service, Pacific Northwest Forest and Range Experiment Station.
Clark, R.N., Hendee, J.C. and Campbell, F. (1971) Values, behaviour, and conflict in modern camping culture. *Journal of Leisure Research* 3 (3), 143–159.
Clark, R.W., Brown, W.S., Stechert, R. and Zamudio, K.R. (2010) Roads, interrupted dispersal and genetic diversity in timber rattlesnakes *Conservation Biology* 24, 1059–1069.
Clarke, J.E. (2000) Protected area management planning. *Oryx* 34 (2), 85–87.
Cochrane, J. (1996) The sustainability of ecotourism in Indonesia: Fact or fiction? In M.J.G. Parnwell and R.L. Bryant (eds) *Environmental Change in South-East Asia: People, Politics and Sustainable Development*. London: Routledge.
Cochrane, J. (2006) Indonesian national parks: Understanding leisure users. *Annals of Tourism Research* 33 (4), 979–997.
Cohen, E. (1972) Toward a sociology of international tourism. *Social Research* 39, 164–182.
Cole, D.N. (1981a) Managing ecological impacts at wilderness campsites: An evaluation of techniques. *Journal of Forestry* 79, 86–89.
Cole, D.N. (1981b) Vegetational changes associated with recreational use and fire suppression in the Eagle Gap Wilderness, Oregon: Some management implications. *Biological Conservation* 20, 247–270.
Cole, D.N. (1982) Controlling the spread of campsites at popular wilderness destinations. *Journal of Soil and Water Conservation* 37, 291–295.
Cole, D.N. (1983a) *Assessing and Monitoring Backcountry Trail Conditions* (Research Paper INT-303). Ogden, UT: US Department of Agriculture Forest Service, Intermountain Forest and Range Experiment Station.
Cole, D.N. (1983b) *Monitoring the Condition of Wilderness Campsites* (Research Paper INT-302). Ogden, UT: US Department of Agriculture Forest Service, Intermountain Forest and Range Experiment Station.
Cole, D.N. (1989) *Wilderness Campsite Monitoring Methods: A Sourcebook* (General Technical Report INT-259). Ogden, UT: US Department of Agriculture Forest Service, Intermountain Research Station.
Cole, D.N. (1990a) Trampling disturbance and recovery of cryptogamic crusts in Grand Canyon National Park. *Great Basin Naturalist* 50 (4), 321–325.
Cole, D.N. (1990b) Ecological impacts of wilderness recreation and their management. In J.C. Hendee, G.H. Stankey and R.C. Lucas (eds) *Wilderness Management* (pp. 425–466). Golden, CO: North American Press.
Cole, D.N. (1992) Modelling wilderness campsites: Factors that influence amount of impact. *Environmental Management* 16 (2), 255–264.
Cole, D.N. (1993) Wilderness recreation management: We need more than bandages and toothpaste. *Journal of Forestry* 91 (2), 22–24.
Cole, D.N. (1994) *The Wilderness Threats Matrix: A Framework for Assessing Impacts* (Research Paper INT-475). Ogden, UT: US Department of Agriculture Forest Service, Intermountain Research Station.
Cole, D.N. (1995) Wilderness management principles: Science, logical thinking or personal opinion? *Trends* 32 (1), 6–9.
Cole, D.N. (1996) *Wilderness Recreation Use Trends, 1965 Through 1994* (Research Paper INT-RP-488). Ogden, UT: US Department of Agriculture Forest Service Intermountain Research Station.
Cole, D.N. (1997) Visitors, conditions, and management options for high-use destination areas in wilderness. Making protection work. *Proceedings of the Ninth Conference on Research and Resource Management in Parks and on Public Lands, Albuquerque, New Mexico, March 17–21, 1997* (pp. 29–35). Albuquerque, NM: George Wright Society Biennial Conference.
Cole, D.N. (2000) Paradox of the primeval: Ecological restoration in wilderness. *Ecological Restoration* 18 (2), 77–86.

Cole, D.N. (2004) Impacts of hiking and camping on soils and vegetation: A review. In R. Buckley (ed.) *Environmental Impacts of Ecotourism* (pp. 41–60). Oxford: CABI Publishing.

Cole, D.N. (compiler) (2005) *Computer Simulation Modeling of Recreation Use: Current Status, Case Studies, and Future Directions* (General Technical Report RMRS-GTR-143). Fort Collins, CO: US Department of Agriculture Forest Service, Rocky Mountain Research Station.

Cole, D.N. and Hall, T.E. (2009) Perceived effects of setting attributes on visitor experiences in wilderness: Variation with situational context and visitor characteristics. *Environmental Management* 44, 24–36.

Cole, D.N. and Landres, P.B. (1995) Indirect effects of recreation on wildlife. In R.L. Knight and K.J. Gutzwiller (eds) *Wildlife and Recreationists: Coexistence Through Management and Research*. Washington, DC: Island Press.

Cole, D.N. and Landres, B. (1996) Threats to wilderness ecosystems: Impacts and research needs. *Ecological Applications* 6 (1), 168–184.

Cole, D.N. and Marion, J.L. (1986) Wilderness campsite impacts: Changes over time. *Proceedings of the National Wilderness Research Conference: Current Research, July 23–26, 1985, Fort Collins, CO* (pp. 144–151). Ogden, UT: US Department of Agriculture Forest Service, Intermountain Research Station.

Cole, D.N. and Marion, J.L. (1988) Recreational impacts in some riparian forests of the eastern United States. *Environmental Management* 12 (1), 99–107.

Cole, D.N. and McCool, S.F. (1998) The limits of acceptable change process: Modifications and clarifications. In S.F. McCool and D.N. Cole (eds) *Proceedings of a Workshop on Limits of Acceptable Change and Related Planning Processes: Progress and Future Directions, University of Montana's Lubrecht Experimental Forest, Missoula, Montana, May 20–22, 1997* (General Technical Report INT-GTR-371) (pp. 61–68). Ogden, UT: US Department of Agriculture Forest Service, Rocky Mountain Research Station.

Cole, D.N. and Monz, C. (2004) Spatial patterns of recreation impact on experimental campsites. *Journal of Environmental Management* 70, 73–84.

Cole, D.N. and Stankey, G.H. (1998) Historical development of limits of acceptable change: Conceptual clarifications and possible extensions. In S.F. McCool and D.N. Cole (eds) *Proceedings of a Workshop on Limits of Acceptable Change and Related Planning Processes: Progress and Future Directions, University of Montana's Lubrecht Experimental Forest, Missoula, Montana, May 20–22, 1997* (General Technical Report INT-GTR-371) (pp. 5–9). Ogden, UT: US Department of Agriculture Forest Service, Rocky Mountain Research Station.

Cole, D.N., Petersen, M.E. and Lucas, R.C. (1987) *Managing Wilderness Use: Common Problems and Potential Solutions* (General Technical Report INT230). Ogden, UT: US Department of Agriculture Forest Service, Intermountain Research Station.

Cole, D.N., Watson, A.E. and Roggenbuck, J.W. (1995) *Trends in Wilderness Visitors and Visits: Boundary Waters Canoe Area, Shining Rock, and Desolation Wildernesses* (Research Raper INT-RP-483). Ogden, UT: US Department of Agriculture Forest Service, Intermountain Research Station.

Cole, D.N., Watson, A.E., Hall, T.E. and Spildie, D.R. (1997) *High-Use Destinations in Wilderness: Social and Biophysical Impacts, Visitor Responses, and Management Options* (Research Paper INT-RP-496). Ogden, UT: US Department of Agriculture Forest Service, Intermountain Research Station.

Cole, D.N., Manning, R. and Lime, D. (2005) Addressing visitor capacity of parks and rivers. *Parks and Recreation* 40 (3), 8–12.

Cole, D.N., Foti, P. and Brown, M. (2008) Twenty years of change in campsites in the backcountry of Grand Canyon National Park. *Environmental Management* 41, 959–970.

Colinvaux, P.A. (1993) *Ecology*. New York: Wiley.

Collins, M. (1990) *The Last Rain Forests*. London: Mitchell Beazley.

Colmar Brunton (2010) *TOMM Kangaroo Island Committee. Visitor Exit Survey 2009/2010*. http://rdahc.com.au/sites/default/files/4161643_TOMM_Visitors%20Exit%20Survey_Final%20Report_27-09-11.pdf. Accessed 22 April 2011.

Commonwealth of Australia (1991) *Kakadu National Park Plan of Management* (Plan of Management). Canberra: Australian National Parks and Wildlife Service.

Connell, S.D. and Gillanders, B.M. (2007) *Marine Ecology*. Melbourne: Oxford University Press.
Considine, M. (1988) The corporate management framework as administrative science: A critique. *Australian Journal of Public Administration* 47 (1), 4–18.
Coombes, E. and Jones, P. (2010) Assessing the impact of climate change on visitor behaviour and habitat use at the coast: A UK case study. *Global Environmental Change* 20, 303–313.
Cope, A., Doxford, D. and Probert, C. (2000) Monitoring visitors to UK countryside resources: The approaches of land and recreation resource management organisations to visitor monitoring. *Land Use Policy* 17, 59–66.
Cordell, H.K., James, G.A. and Tyre, G.L. (1974) Grass establishment on developed recreation sites. *Journal of Soil and Water Conservation* 29, 268–271.
Corlett, R. (2009) *The Ecology of Tropical East Asia*. Oxford: Oxford University Press.
Corrick, M.G. and Fuhrer, B.R. (1996) Introduction. In M.G. Corrick and B.R. Fuhrer (eds) *Wildflowers of Southern Western Australia* (pp. 9–15). Melbourne: Monash University.
Crabtree, A. (2000) Interpretation: Ecotourism's fundamental but much neglected tool. International *Ecotourism Society Newsletter* (third quarter), 1–3.
Crabtree, B. and Bayfield, N. (1998) Developing sustainability indicators for mountain ecosystems: A study of the Cairngorms, Scotland. *Journal of Environmental Management* 52, 1–14.
Craven, S.A. (1996) Carbon dioxide variations in Cango Cave, South Africa. *Cave and Karst Science* 23 (3), 89–92.
Craven, S.A. (1999) Speleothem deterioration at Cango Cave, South Africa. *Cave and Karst Science* 26 (1), 29–34.
Crosbie, K. and Splettstoesser, J. (2009) Antarctic tourism introduction. In P.T. Maher, E. Stewart and M. Lück (eds) *Polar Tourism: Human, Environmental and Governance Dimensions* (pp. 105–120). Elmsford, NY: Cognizant Communication Corporation.
Cubit, S. (1990) Horse riding in national parks: Some critical issues. *Australian Parks and Recreation* 26 (1), 24–28.
Curry, S. (1985) Tourism circuit planning: Carrying capacity at Ngorongoro Crater, Tanzania. In *The Impact of Tourism and Recreation on the Environment: A Miscellany of Readings* (selected papers from a Seminar at the University of Bradford, 27 June–1 July 1983, Occasional Paper No. 8). Bradford: University of Bradford.
Curtin, S. (2010) Managing the wildlife tourism experience: The importance of tour leaders. *International Journal of Tourism Research* 12, 219–236.
Curtis, J., Ham, S. and Weiler, B. (2010) Identifying beliefs underlying visitor behaviour: A comparative elicitation study based on the theory of planned behavior. *Annals of Leisure Research* 13 (4), 564–589.
Cushman, J.H. and Meentemeyer, R.K. (2008) Multi-scale patterns of human activity and the incidence of an exotic forest pathogen. *Journal of Ecology* 96, 766–776.
Cushman, J.H., Cooper, M., Meentemeyer, R.K. and Benson, S. (2007) Human activity and the spread of *Phytophthora ramorum*. In *Proceedings of the Sudden Oak Death Third Science Symposium. March 5–9, Santa Rosa, California* (General Technical Report PSW-GTR-214) (pp. 179–180). Ogden, UT: US Department of Agriculture Forest Service, Pacific Southwest Research Station.
Daby, D. (2003) Effects of seagrass removal for tourism purposes in a Mauritian bay. *Environmental Pollution* 125, 313–324.
Daily, G.C., Ehrlich, P.R. and Haddad, N.M. (1993) Double keystone bird in a keystone species complex. *Proceedings of the National Academy of Science USA* 90, 592–594.
Dale, D. and Weaver, T. (1974) Trampling effects on vegetation of the trail corridors of North Rocky Mountain Forest. *Journal of Applied Ecology* 11, 767–772.
Daniel, R., Taylor, J. and Guest, D. (2006) Distribution and occurrence of *Phytophthora cinnamomi* at Middle Head and North Head, Sydney Harbour. *Australasian Plant Pathology* 35, 569–571.
Daniels, M. and Marion, J. (2006) Visitor evaluations of management actions at a highly impacted Appalachian Trail camping area. *Environmental Management* 38, 1006–1019.
Davidson, J.M., Wickland, A.C., Patterson, H.A., Falk, K.R. and Rizzo, D.M. (2005) Transmission of *Phytophthora ramorum* in mixed evergreen forest in California. *Phytopathology* 95, 587–596.

Davies, C. and Newsome, D. (2009) *Assessing the Impact of Mountain Bike Activity in John Forrest National Park, Western Australia* (Technical Report). Gold Coast, Queensland: Cooperative Research Centre for Sustainable Tourism, Griffith University.

Davis, D., Tisdell, C. and Hardy, M. (2000) *The Role of Economics in Managing Wildlife Tourism* (Wildlife Tourism Research Report, Status Assessment of Wildlife Tourism in Australia Series). Gold Coast, Queensland: Cooperative Research Centre for Sustainable Tourism, Griffith University.

Davis, G., Wanna, J., Warhurst, J. and Weller, P. (1993) *Public Policy in Australia*. St Leonards, New South Wales: Allen and Unwin.

Davis, P.B. (1999) Beyond guidelines: A model for Antarctic tourism. *Annals of Tourism Research* 26 (3), 516–533.

Deacon, A. (1992) People pressure threatens rivers. *Custos* 21 (4), 18.

Dearden, P., Bennett, M., and Rollins, R. (2006) Implications for coral reef conservation of diver specialization. *Environmental Conservation* 33 (4), 353–363.

Dearden, P., Bennett, M. and Rollins, R. (2007) Perceptions of diving impacts and implications for reef conservation. *Coastal Management* 35 (2), 305–317.

DEAT (1996) *White Paper: The Development and Promotion of Tourism in South Africa*. Pretoria: Department of Environmental Affairs and Tourism (DEAT).

DEAT (1998) *Tourism in Gear: Tourism Development Strategy 1998–2000*. Pretoria: Department of Environmental Affairs and Tourism (DEAT).

del Hoyo, J., Elliott, A. and Christie, D.A. (eds) (2010) *Handbook of the Birds of the World, Vol. 15: Weavers to New World Warblers*. Barcelona: Lynx Edicions.

De Lacy, T. (1994) The Uluru/Kakadu model–Anangu Tjukurrpa: 50,000 years of Aboriginal law and land management changing the concept of national parks in Australia. *Society and Natural Resources* 7 (5), 479–498.

De Lacy, T. and Whitmore, M. (2006) Tourism and recreation. In M. Lockwood, G. Worboys and A. Kothari (eds) *Managing Protected Areas: A Global Guide* (pp. 497–527). London: Earthscan.

Dell, B., Hopkins, A.J.M. and Lamont, B.B. (1986) *Resilience in Mediterranean-Type Ecosystems*. Boston, MA: Kluwer Academic.

Dellue, B. and Dellue, G. (1984) Lascaux II: A faithful copy. *Antiquity* 58, 194–196.

De Luca, T.H., Patterson, W.A. (IV), Freimund, W.A. and Cole, D.N. (1998) Influence of llamas, horses, and hikers on soil erosion from established recreation trails in Western Montana. *Environmental Management* 22 (2), 255–262.

Deng, J., Qiang, S., Walker, G.J. and Zhang, Y. (2003) Assessment on and perception of visitors' environmental impacts of nature tourism: A case study of Zhangjiajie National Forest Park, China. *Journal of Sustainable Tourism* 11 (6), 529–548.

Department of Tourism (1995) *Best Practice Tourism: A Guide to Energy and Waste Minimisation*. Canberra: Commonwealth of Australia.

Dickinson, G. and Murphy, K. (1998) *Ecosystems*. London: Routledge.

Dieback Working Group (DWG) (2009) *Managing* Phytophthora *Dieback in Bushland*. Perth: Dieback Working Group.

Diedrich, A., Huguet, P.B. and Subirana, J.T. (2011) Methodology for applying the limits of acceptable change process to the management of recreational boating in the Balearic Islands, Spain (western Mediterranean). *Ocean and Coastal Management* 54, 341–351.

Ding, P. and Pigram, J.J. (1995) Environmental audits: An emerging concept in sustainable tourism development. *Journal of Tourism Studies* 6 (2), 2–10.

Ding, P. and Pigram, J.J. (1996) An approach to monitoring and evaluating the environmental performance of Australian beach resorts. *Australian Geographer* 27 (1), 77–86.

Dingwall, P.R. (1998) Implementing an environmental management regime in Antarctica. In A.E. Watson, G.H. Aplet and J.C. Hendee (eds) *Personal, Societal and Ecological Values of Wilderness: Proceedings of Sixth World Wilderness Congress on Research, Management and Allocation, 1997 October, Bangalore, India* (RMRSP4) (pp. 1–5). Ogden, UT: Department of Agriculture Forest Service, Rocky Mountain Research Station.

Dixit, S.K. and Narula, V.K. (2010) Ecotourism in Madhav National Park: Visitors' perspectives on environmental impacts. *South Asian Journal of Tourism and Heritage* 3, 109–115.

Dixon, G. and Hawes, M. (2011) *Walking Track Management Strategy for Tasmania's National Parks and Reserves 2011–2020*. Hobart, Tasmania: Parks and Wildlife Service, Department of Primary Industries, Parks, Water and Environment.

Dixon, I., Douglas, M., Burrows, D., Dowe, J., Grice, T., Ludwig, J., Mitchell, S., Setterfield, S. and Werren, G. (2004) Riparian Health Workshop: The Development of a Tropical Rapid Appraisal of Riparian Condition (TRARC). Summary of a Workshop Held at James Cook University Townsville, 29 October 2003 (Internal Report for the Tropical Savannas CRC, Charles Darwin University, April).

Dolnicar, S. (2010) Identifying tourists with smaller environmental footprints. *Journal of Sustainable Tourism* 18 (6), 717–734.

Dovers, S. (2003) Processes and institutions for resource and environmental management: Why and how to analyse? In S. Dovers and S. Wild River (eds) *Managing Australia's Environment* (pp. 3–12). Sydney: Federation Press.

Dowling, R.K. (1996) The implementation of ecotourism in Australia. In P. Dhamabutra, V. Stithyudhakarn and S. Hirunburana (eds) *Proceedings of the 2nd International Conference on the Implementation of Ecotourism* (pp. 29–50). Bangkok: Institute of Ecotourism, Srinakharinwirot University.

Dowling, R.K. (2001) Environmental tourism. In N. Douglas, N. Douglas and R. Derrett (eds) *Special Interest Tourism: Contexts and Cases* (pp. 283–306). Brisbane: Wiley.

Dowling, R.K. (2011) Geotourism's global growth. *Geoheritage* 3, 1–13.

Dowling, R.K. (2013) The history of ecotourism. In R. Ballantyne and J. Packer (eds) *The International Handbook on Ecotourism*. Cheltenham: Edward Elgar (in press).

Dowling, R.K. and Newsome, D. (eds) (2006) *Geotourism*. Oxford: Elsevier/Heinemann.

Dowling, R.K. and Newsome, D. (eds) (2010) *Global Geotourism Perspectives*. Oxford: Goodfellow Publishers.

Dowling, R.K. and Weiler, B. (1997) Ecotourism in South East Asia: A report of the 2nd International Ecotourism Conference 'The Implementation of Ecotourism: Planning and Managing for Sustainablity'. *Tourism Management* 18 (1), 51–53.

Driscoll, D., Milkovits, G. and Freudenberger, D. (2000) *Impact and Use of Firewood in Australia*. Canberra: CSIRO Sustainable Ecosystems.

Driver, B.L. (1975) Quantification of outdoor recreationists' preferences. In B. Van der Smissen and J. Myers (eds) *Research: Camping and Environmental Education* (pp. 165–187). Pennsylvania, PA: Pennsylvania State University, University Park.

Driver, B.L. (1996) Benefits-driven management of natural areas. *Natural Areas Journal* 16, 94–99.

Driver, B.L. (2008) *Managing to Optimise the Beneficial Outcomes of Recreation*. State College, PA: Venture Publishing.

Driver, B.L. and Brown, P.J. (1978) The opportunity spectrum concept and behavioral information in outdoor recreation resource supply inventories: A rationale. In G.H. Lund, V.J. La Bau, P.F. Folliott and D.W. Robinson (eds) *Integrated Inventories of Renewable Natural Resources* (General Technical Report RM-55) (pp. 24–31). Fort Collins, CO: US Department of Agriculture Forest Service, Rocky Mountain Forest and Range Experiment Station.

Driver, B.L. and Tocher, S.R. (1970) Toward a behavioural interpretation of recreational engagements, with implications for planning. In B.L. Driver (ed.) *Elements of Outdoor Recreation Planning* (pp. 9–13). Ann Arbor, MI: University of Michigan Press.

Driver, B.L., Brown, P. and Peterson, G. (ed.) (1991) *Benefits of Leisure*. State College, PA: Venture Publishing.

Drumm, A. (1998) New approaches to community-based ecotourism management. In K. Lindberg, M. Epler Wood and D. Engeldrum (eds) *Ecotourism: A Guide for Planners and Managers* (vol. 2, pp. 197–213). Burlington, VT: Ecotourism Society.

DSEWPC (Department of Sustainability, Environment, Water, Population and Communities: Australian Government) (2011) CAPAD (Collaborative Australian Protected Area Database) 2008. http://www.environment.gov.au/parks/nrs/science/capad/2008/index.html. Accessed 18 November 2011.

DSEWPC (Department of Sustainability, Environment, Water, Population and Communities)

(2012) Indigenous protected areas. http://www.environment.gov.au/indigenous/ipa/index.html. Accessed 19 January 2012.

Duffus, D.A. and Dearden, P. (1990) Non-consumptive wildlife-oriented recreation: A conceptual framework. *Biological Conservation* 53, 213–231.

Dunham, K.M., Ghiurghi, A., Cumbi, R. and Urbano, F. (2010) Human–wildlife conflict in Mozambique: A national emphasis on wildlife attacks on humans. *Oryx* 44, 185–193.

du Toit, J.T., Roger, K.H. and Biggs, H.C. (2003) *The Kruger Experience: Ecology and Management of Savanna Heterogeneity* (pp. 332–348). Washington, DC: Island Press.

Dye, T. (1992) *Understanding Public Policy* (7th edn). Englewood Cliffs, NJ: Prentice Hall.

Ecotourism Australia (2012) Ecotourism Australia. http://www.ecotourism.org.au/index.asp. Accessed 21 February 2012.

EAA (2000) *EcoGuide Program Guidebook*. Brisbane: Ecotourism Association of Australia.

EAA and ATOA (1996) *National Ecotourism Accreditation Program*. Brisbane: Ecotourism Association of Australia and Australian Tourism Operators Network.

Eagles, P.F.J. (1984) *The Planning and Management of Environmentally Sensitive Areas*. London: Longman.

Eagles, P.F.J. (1992) The motivations of Canadian ecotourists. *Journal of Travel Research* 31 (2), 3–7.

Eagles, P.F.J. (1995) Understanding the market for sustainable tourism. In S.F. McCool and A.E. Watson (eds) *Linking Tourism, the Environment, and Sustainability – Topical Volume of Compiled Papers from a Special Session of the Annual Meeting of the National Recreation and Park Association, 1994 October 12–14* (General Technical Report INT-GTR-323) (pp. 25–33). Ogden, UT: US Department of Agriculture Forest Service, Rocky Mountain Research Station.

Eagles, P.F.J. (2008) Governance models for parks, recreation, and tourism. In K.S. Hanna, D.A. Clark and D.S. Slocombe (eds) *Transforming Parks and Protected Areas: Policy and Governance in a Changing World* (pp. 39–61). New York: Routledge.

Eagles, P.F.J. (2009) Governance of recreation and tourism partnerships in parks and protected areas. *Journal of Sustainable Tourism* 17 (2), 231–248.

Eagles, P.F.J. and McCool, S.F. (2002) *Tourism in National Parks and Protected Areas: Planning and Management*. Wallingford: CABI Publishing.

Eagles, P.F.J., McCool, S.F. and Haynes, C.D. (2002) *Sustainable Tourism in Protected Areas: Guidelines for Planning and Management*. Gland, Switzerland: International Union for Conservation of Nature and Natural Resources.

Ecotourism Society (1998) *Ecotourism Statistical Fact Sheet*. North Bennington, VT: Ecotourism Society.

Edington, J.M. and Edington, M.A. (1977) *Ecology and Environmental Planning*. London: Chapman and Hall.

Edington, J.M. and Edington, M.A. (1986) *Ecology, Recreation and Tourism*. Cambridge: Cambridge University Press.

Eldridge, D.J. and Rosentreter, R. (1999) Morphological groups: A framework for monitoring microphytic crusts in arid landscapes. *Journal of Arid Environments* 41, 11–25.

English, D.B.K., Kocis, S.M., Zarnoch, S.J. and Arnold, J.R. (2001) *Forest Service National Visitor Use Monitoring Process: Research Method Documentation*. http://www.fs.fed.us/recreation/programs/nvum/. Accessed 6 April 2012.

English, D.B.K., Kocis, S.M., Arnold, J.R., Zarnoch, S.J. and Warren, L. (2003) The effectiveness of visitation proxy variables in improving recreation use estimates for the USDA Forest Service. *Journal for Nature Conservation* 11, 332–338.

Environment Canada and Park Service (1991) *Selected Readings on the Visitor Activity Management Process*. Ottawa: Environment Canada.

Enzenbacher, D.J. (1992) Antarctic tourism and environmental concerns. *Marine Pollution Bulletin* 25, 258–265.

Enzenbacher, D.J. (1993) Tourists in Antarctica: Numbers and trends. *Tourism Management* (April), 142–146.

Enzenbacher, D.J. (1995) The regulation of Antarctic tourism. In C.M. Hall and M.E. Johnston (eds) *Polar Tourism: Tourism in the Arctic and Antarctic Regions* (pp. 179–215). Chichester: Wiley.

EPA (1995) *Environmental Management Systems: Best Practice in Environmental Management in Mining*. Perth, Western Australia: Environmental Protection Agency.

Etherington, J.R. (1975) *Environment and Plant Ecology*. New York: Wiley.

European Environment Agency (2004) *Impacts of Europe's Changing Climate: An Indicator Based Assessment*. Copenhagen: EEA.

Fa, J.E. (1988) Supplemental food as an extranormal stimulus in Barbary Macaques (*Macaca sylvanus*) at Gibraltar – Its impact on activity budgets. In J.E. Fa and C.H. Southwick (eds) *Ecology and Behaviour of Food Enhanced Primate Groups* (pp. 53–78). New York: Alan R. Liss.

Fairfax, R.J., Dowling, R. and Neldner, V.J. (2012) The use of infra-red sensors and digital cameras for documenting visitor use patterns: A case study from D'Aguilar National Park, south-east Queensland, Australia. *Current Issues in Tourism* (in press).

Farrell, T. and Marion, J. (2001) Identifying and assessing ecotourism visitor impacts at selected protected areas in Costa Rica and Belize. *Environmental Conservation* 28, 215–225.

Farris, M.A. (1998) The effects of rock climbing on the vegetation of three Minnesota cliff systems. *Canadian Journal of Botany* 76, 1981–1990.

Fauna and Flora International Update (2011) Antigua's wildlife thrives thanks to an innovative rat eradication campaign. *Fauna and Flora International Update* (May), no. 18.

Fennell, D. (1999) *Ecotourism: An Introduction*. London: Routledge.

Fennell, D. (2001) A content analysis of ecotourism definitions. *Current Issues in Tourism* 4 (5), 403–421.

Fennell, D. (2008) *Ecotourism* (3rd edn). London: Routledge.

Fennell, D. and Eagles, P.F.J. (1990) Ecotourism in Costa Rica: A conceptual framework. *Journal of Park and Recreation Administration* 8 (1), 23–34.

Fennell, D. and Weaver, D. (2005) The ecotourism concept and tourism–conservation symbiosis. *Journal of Sustainable Tourism* 13 (4), 373–390.

Floyd, M.F., Jang, H. and Noe, F.P. (1997) The relationship between environmental concern and acceptability of environmental impacts among visitors to two U.S. national park settings. *Journal of Environmental Management* 51, 391–412.

Font, X. and Harris. C. (2004) Rethinking standards from green to sustainable. *Annals of Tourism Research* 31, 986–1007.

Forbes, B.C. (1992) Tundra disturbance studies 1: Long term effects of vehicles on species richness and biomass. *Environmental Conservation* 19 (1), 48–58.

Forman, R.T.T. (1995) *Land Mosaics: The Ecology of Landscapes and Regions*. Cambridge: Cambridge University Press.

Fowler, G. (1999) Behavioural and hormonal responses of Magellanic penguins (*Spheniscus megellanicus*) to tourism and nest site visitation. *Biological Conservation* 90, 143–149.

France, L. (ed.) (1998) *Earthscan Reader in Sustainable Tourism*. London: Earthscan.

Frankfort-Nachmias, C. and Nachmias, D. (1992) *Research Methods in the Social Sciences*. New York: St Martin's Press.

Freedman, B., Zelazny, V., Beaudette, D., Fleming, T., Fleming, S., Forbes, G., Gerrow, J. S., Johnson, G. and Woodley, S. (1996) Biodiversity implications of changes in the quantity of dead organic matter in managed forests. *Environmental Review* 4, 238–265.

Freitag-Ronaldson, S. and Foxcroft, L.C. (2003) Anthropogenic influences at the ecosystem level. In J.T. du Toit, K.H. Roger and H.C. Biggs (eds) *The Kruger Experience: Ecology and Management of Savanna Heterogeneity* (pp. 332–348). Washington, DC: Island Press.

Friedlander, A.M., Aeby, G., Brown, E., Clark, A., Coles, S. and Dollar, S. (2005) The state of coral reef ecosystems of the main Hawaiian Islands. In J. Waddell (ed.) *The State of Coral Reef Ecosystems in the United States and Pacific Freely Associated States* (pp. 222–269). Silver Spring, MD: NOAA/NCCOS Center for Coastal Monitoring and Assessment.

Friis, C. and Reenberg, A. (2010) *Land Grab in Africa: Emerging Land System Drivers in a Teleconnected World*. Copenhagen: GLP International Project Office, University of Copenhagen.

Frissell, S.S. (1978) Judging recreation impacts on wilderness campsites. *Journal of Forestry* 76, 481–483.

Frohoff, T.G. (2000) *Behavioral Indicators of Stress in Odontocetes During Interactions With Humans: A Preliminary Review and Discussion* (SC/52WW2). Cambridge: International Whaling Commission Scientific Committee.

Furbank, D. (2010) *Briefing Note on Beach Nesting Shorebirds*. Adelaide: Yorke Peninsula Natural Resource Management Group, Department of Environment and Natural Resources, Government of South Australia.
Furley, P.A. and Newey, W.W. (1983) *The Geography of the Biosphere*. London: Butterworth.
Gabrielsen, G.W. and Smith, E.N. (1995) Physiological responses of wildlife to disturbance. In R.L. Knight and K J. Gutzwiller (eds) *Wildlife and Recreationists: Coexistence Through Management and Research*. Washington, DC: Island Press.
Gajraj, A.M. (1988) A regional approach to environmentally sound tourism development. *Tourism Recreation Research* 13 (2), 5–9.
Gales, N. (2000) Presentation on wildlife tourism at Marine Tourism Conference, Notre Dame University. Perth, Western Australia: Forum Advocating Cultural and Ecotourism Inc.
Galicia, E. and Baldassarre, G.A. (1997) Effects of motorised tour boats on the behaviour of non-breeding American flamingos in Yucatan, Mexico. *Conservation Biology* 11 (5), 1159–1165.
Gardner, T. (1994) Visitor monitoring programs in urban parks and natural areas – Taking a strategic approach. *Australian Parks and Recreation* (spring), 27–31.
Garland, G.G. (1987) Rates of soil loss from mountain footpaths: An experimental study in the Drakensberg mountains, South Africa. *Applied Geography* 1, 121–131.
Garland, G.G. (1988) Experimental footpath soil losses and path erosion risk assessment in the Natal Drakensberg. Unpublished PhD thesis, University of Natal, Durban, South Africa.
Garland, G.G. (1990) Technique for assessing erosion risk from mountain footpaths. *Environmental Management* 14 (6), 793–798.
Garland, G.G., Hudson, C. and Blackshaw, J. (1985) An approach to the study of path erosion in the Natal Drakensberg, a mountain wilderness area. *Environmental Conservation* 12 (4), 337–342.
Garrison R.W. (1997) Sustainable nature tourism: California's regional approach. In *World Ecotour '97 Abstracts Volume* (pp. 180–182). Rio de Janiero: BIOSFERA.
Garrod, B. and Gössling, S. (2008). Introduction. In B. Garrod and S. Gössling (eds) *New Frontiers in Marine Tourism: Diving Experiences, Sustainability, Management* (pp. 3–28). Amsterdam: Elsevier.
Garven, I. (undated) The Australian Alps National Parks Co-operative Management. In *Proceedings of the Recreation Management Workshop and Third Working Meeting of Officers from Management Agencies* (Report Series 9). Canberra: Australian National Parks and Wildlife Service.
Getz, D. (1987) Tourism planning and research: Traditions, models and futures. Paper presented to the Australian Travel Research Workshop, Bunbury, Western Australia, 3–6 November.
Gibeau, M.L. and Heuer, K. (1996) Effects of transportation corridors on large carnivores in the Bow River Valley, Alberta. In G.L. Evink, P. Garrett, D. Zeigler and J. Berry (eds) *Trends in Addressing Transportation Related Wildlife Mortality*. Tallahassee, FL: State of Florida Department of Transportation.
Giese, M. (2000) Polar wandering. *Wingspan* 10 (1), 8–13.
Gillen, K. and Watson, J. (1993) Controlling *Phytophthora cinnamomi* in the mountains of south-western Australia. *Australian Ranger* 27, 18–20.
Gillieson, D. (1996) *Caves: Processes, Development, Management*. Oxford: Blackwell.
Gilpin, A. (1990) *An Australian Dictionary of Environment and Planning*. Melbourne: Oxford University Press.
Giuliano, W. (1994) The impact of hiking and rock climbing in mountain areas. *Environmental Conservation* 21 (3), 278–279.
Goeft, U. and Alder, J. (2000) Mountain bike rider preferences and perceptions in the south-west of Western Australia. *CALM Science* 3 (2), 261–275.
Goeldner, C. and Ritchie, J. (2006) *Tourism: Principles, Practices, Philosophies* (10th edn). Hoboken, NJ: John Wiley.
Gomez-Limon, F.J. and de Lucio, J. (1995) Recreational activities and loss of diversity in grasslands in Alta Manzanares Natural Park, Spain. *Biological Conservation* 74, 99–105.
Goodwin, H., Kent, I., Parker, K. and Walpole, M. (1998) *Tourism, Conservation and Sustainable Development: Case Studies from Asia and Africa* (Wildlife and Development Series No. 10). London: International Institute for Environment and Development.

Goodwin, S.E. and Shriver, W.G. (2011) Traffic noise and occupancy patterns of forest birds. *Conservation Biology* 25 (2), 406–411.
Goosem, M. (2000) Effects of tropical rainforest roads on small mammals: Edge effects in community composition. *Wildlife Research* 27, 151–163.
Gordon, D.M. (1987) *Disturbance to Mangroves in Tropical-arid Western Australia: Hypersalinity and Restricted Tidal Exchange as Factors Leading to Mortality* (Technical Series 12). Perth, Western Australia: Environmental Protection Authority.
Graefe, A.R., Kuss, F.R. and Vaske, J.J. (1990) *Visitor Impact Management: The Planning Framework, Vol. 2*. Washington, DC: National Parks and Conservation Association.
Graham, A.W. and Hopkins, M.S. (1993) Ecological prerequisites for managing change in tropical rainforest ecosystems. In N. McIntyre (ed.) *Track to the Future: Managing Change in Parks and Recreation*. Melbourne: Royal Institute of Parks and Recreation.
Grant, C., Davidson, T., Funston, P. and Pienaar, D. (2002) Challenges faced in the conservation of rare antelope: A case study on the northern basalt plains of the Kruger National Park. *Koedoe* 45, 45–62.
Grant, T.J. and Doherty, P.F. (2009) Potential mortality effects of off-highway vehicles on the flat-tailed horned lizard (*Phrynosoma mcalli*): A manipulative experiment. *Environmental Management* 43, 508–513.
GBRMPA (Great Barrier Reef Marine Park Authority) (1997) *Guidelines for Managing Visitation to Seabird Breeding Islands*. Townsville: GBRMPA.
GBRMPA (2009) *Great Barrier Reef Tourism Climate Change Action Strategy 2009–2012*. Townsville: GBRMPA.
Green, J.B. and Paine, J. (1997) State of the world's protected areas at the end of the twentieth century. IUCN World Commission on Protected Areas Symposium on 'Protected Areas in the 21st Century: From Islands to Networks', 24–29th November, 1997, Albany, Australia.
Green, R.J. (1999) Negative effects of wildlife viewing on wildlife in natural areas. In K. Higginbottom and M. Hardy (eds) *Wildlife Tourism – Discussion Document* (unpublished report). Gold Coast, Queensland: Cooperative Research Centre for Sustainable Tourism, Griffith University.
Green, R.J. and Higginbottom, K. (2000) The effects of non-consumptive wildlife tourism on free-ranging wildlife: A review. *Pacific Conservation Biology* 6, 183–197.
Green, R.J. and Higginbottom, K. (2001) *The Negative Effects of Wildlife Tourism on Wildlife* (Wildlife Tourism Research Report Series No. 5, Status Assessment of Wildlife Tourism in Australia Series). Gold Coast, Queensland: Cooperative Research Centre for Sustainable Tourism, Griffith University.
Green Globe (2012) Green Globe Certification Standard. http://greenglobe.com/register/green-globe-certification-standard/. Accessed 29 February 2012.
Greeves, G.W. and Leys, J.F. (2000) Soil erodibility. In P.E.V. Charman and B.W. Murphy (eds) *Soils: Their Properties and Management* (pp. 205–218). Melbourne: Oxford University Press.
Greig, A. (2011) Interpretation Project Officer, Department of Environment and Conservation. Forum Advocating Cultural and Eco-Tourism. *FACET News* 16, 2–5.
Greiner, R., Stoeckl, N. and Schweigert, R. (2004) Estimating community benefits from tourism: The case of Carpentaria Shire. Paper presented to the 48th Annual Conference of the Australian Agricultural and Resource Economics Society, Melbourne, 11–13 February.
Greller, A.M., Goldstein, M. and Marcus, L. (1974) Snowmobile impact on three alpine tundra plant communities. *Environmental Conservation* 1 (2), 101–110.
Griffin, T., Darcy, S., Moore, S. and Crilley, G. (2008) Visitor data needs of protected area agencies. In *APAC08 Proceedings (Australian Protected Areas Congress 2008): Protected Areas in a Century of Change, 24–28 November 2008, Twin Waters, Queensland*. Brisbane: Queensland Government Environmental Protection Agency.
Griffin, T., Moore, S., Crilley, G., Darcy, S. and Schweinsberg, S. (2010) *Protected Area Management: Collection and Use of Visitor Data. Vol. 1: Summary and Recommendations.* Gold Coast, Queensland: Cooperative Research Centre for Sustainable Tourism, Griffith University.
Griffiths, M. and van Schaik, C.P. (1993) The impact of human traffic on the abundance and activity periods of Sumatran rain forest wildlife. *Conservation Biology* 7 (3), 623–626.

Groom, J.D., McKinney, L.B., Ball, L.C. and Winchell, C.S. (2007) Quantifying off-highway vehicle impacts on density and survival of a threatened dune-endemic plant. *Biological Conservation* 135, 119–134.

Grove, S., Meggs, J. and Goodwin, A. (2002) *A Review of Biodiversity Conservation Issues Relating to Coarse Woody Debris Management in the Wet Eucalypt Production Forests of Tasmania*. Hobart: Forestry Tasmania.

Growcock, A.J. (2006) Impacts of camping and trampling on Australian alpine and subalpine vegetation and soils. PhD thesis, Griffith University.

Growcock, A.J. and Pickering, C.M. (2011) Impacts of small group short term experimental camping on alpine and subalpine vegetation in the Australian Alps. *Journal of Ecotourism* 10, 86–100.

Growcock, A.J., Sutherland, E. and Stathis, P. (2009) Challenges and experiences in implementing a management effectiveness evaluation program in a protected area system. *Australasian Journal of Environmental Management* 16, 218–226.

Gunn, C.A. (1979) *Tourism Planning*. New York: Crane-Russak.

Gunn, C.A. (1987) Environmental designs and land use. In J.R.B. Ritchie and C.R. Goeldner (eds) *Travel, Tourism and Hospitality Research: A Handbook for Managers and Researchers* (pp. 229–247). New York: Wiley.

Gunn, C.A. (1988a) *Tourism Planning* (2nd edn). New York: Taylor and Francis.

Gunn, C.A. (1988b) *Vacationscape: Designing Tourist Regions* (2nd edn). New York: Van Nostrand Reinhold.

Gunn, C.A. (1994) *Tourism Planning: Basics, Concepts, Cases* (3rd edn). Washington, DC: Taylor and Francis.

Gunn, C.A. and Var, T. (2002) *Tourism Planning: Basic, Concepts, Cases* (4th edn). New York: Routledge.

Gunther, K.A., Beil, M.J. and Robison, H.L. (1998) Factors influencing the frequency of road killed wildlife in Yellowstone National Park. In G.L. Evink, P. Garrett, D. Zeigler and J. Berry (eds) *Proceedings of the International Conference on Wildlife Ecology and Transportation*. Tallahasse, FL: State of Florida Department of Transportation.

Haaland, H. and Aas, O. (2010) Eco-tourism certification – does it make a difference? A comparison of systems from Australia, Costa Rica and Sweden. *Scandinavian Journal of Hospitality and Tourism* 10, 375–385.

Haase, D., Lamers, M. and Amelung, B. (2009) Heading into uncharted territory? Exploring the institutional robustness of self-regulation in the Antarctic tourism sector. *Journal of Sustainable Tourism* 17, 411–430.

Habibi, A., Setiasih, N. and Sartin, J. (eds) (2007) *A Decade of Reef Check Monitoring: Indonesian Coral Reefs, Conditions and Trends*. Indonesian Reef Check Network, http://www.jkri.or.id. Accessed 11 September 2012.

Hadwen, W.L., Arthington, A.H., Boon, P.I., Lepesteur, M. and McComb, A.J. (2005) *Rivers, Streams, Lakes and Estuaries: Hot Spots for Cool Recreation and Tourism in Australia*. Gold Coast, Queensland: Cooperative Research Centre for Sustainable Tourism, Griffith University.

Hadwen, W.L., Hill, W. and Pickering, C.M. (2007) Icons under threat: Why monitoring visitors and their ecological impacts on protected areas matters. *Ecological Management and Restoration* 8 (3), 177–181.

Hadwen, W.L., Arthington, A.H. and Boon, P.I. (2008) *Detecting Visitor Impacts in and Around Aquatic Ecosystems Within Protected Areas*. Gold Coast, Queensland: Cooperative Research Centre for Sustainable Tourism, Griffith University.

Hagan, G.M. (1973) Carbonate banks and hypersaline basin development, Shark Bay, Western Australia. PhD thesis, Department of Geology, University of Western Australia, Perth.

Hales, D. (1989) Changing concepts of national parks. In D. Western and M.C. Pearl (eds) *Conservation for the Twenty-First Century* (pp. 139–144). New York: Oxford University Press.

Hall, C.M. (1991) *Introduction to Tourism in Australia: Impacts, Planning and Development*. Melbourne: Longman Cheshire.

Hall, C.M. (1998) *Introduction to Tourism: Development, Dimensions and Issues* (3rd edn). South Melbourne, Australia: Addison Wesley Longman.

Hall, C.M. (2000a) *Tourism Planning: Policies, Processes and Relationships*. Harlow: Prentice Hall, Pearson Education.
Hall, C.M. (2000b) Policy. In J. Jafari (ed.) *Encyclopedia of Tourism* (pp. 445–448). London: Routledge.
Hall, C.M. and Härkönen, T. (eds) (2006) *Lake Tourism: An Integrated Approach to Lacustrine Tourism Systems*. Clevedon: Channel View.
Hall, C.M. and Jenkins, J. (1995) *Tourism and Public Policy*. London: Routledge.
Hall, C.M. and Johnston, M.E. (1995) *Polar Tourism: Tourism in the Arctic and Antarctic Regions*. Chichester: Wiley.
Hall, C.M. and Lew, A. (2009) *Understanding and Managing Tourism Impacts: An Integrated Approach*. London: Routledge.
Hall, C.M. and McArthur, S. (1998) *Integrated Heritage Management: Principles and Practice*. London: The Stationery Office.
Hall, C.M. and Wouters, M. (1995) Issues in Antarctic tourism. In C.M. Hall and M.E. Johnston (eds) *Polar Tourism: Tourism in the Arctic and Antarctic Regions* (pp. 147–166). Chichester: Wiley.
Hall, C.M., Jenkins, J. and Kearsley, G. (eds) (1997) *Tourism Planning and Policy in Australia and New Zealand: Cases, Issues and Practice*. Sydney: Irwin Publishers.
Hall, C.N. and Kuss, F.R. (1989) Vegetation alteration along trails in Shenandoah National Park, Virginia. *Biological Conservation* 48, 211–227.
Hall, D. and Kinnaird, V. (1994) Ecotourism in Eastern Europe. In E. Cater and G. Lowman (eds) *Ecotourism: A Sustainable Option?* (pp. 111–137). Chichester: Wiley.
Hall, M. and Wouters, M. (1994) Managing nature tourism in the sub-Antarctic. *Annals of Tourism Research* 21 (2), 355–374.
Ham, S.H. (1992) *Environmental Interpretation: A Practical Guide for People with Big Ideas and Small Budgets*. Golden, CO: Fulcrum/North American Press.
Ham, S.H. and Krumpe, E. (1996) Identifying audiences and messages for non-formal environmental education – A theoretical framework for interpreters. *Journal of Interpretation Research* 1 (1), 11–23.
Ham, S.H. and Weiler, B. (2002) Interpretation as a centrepiece in sustainable wildlife tourism. In R. Harris, T. Griffin and P. Williams (eds) *Sustainable Tourism: A Global Perspective* (pp. 35–44). London: Butterworth-Heinemann.
Ham, S.H. and Weiler, B. (2005) *Interpretation Evaluation Toolkit: Methods and Tools for Assessing the Effectiveness of Face-to-Face Interpretive Programs*. Gold Coast, Queensland: Sustainable Tourism Cooperative Research Centre.
Ham, S.H. and Weiler, B. (2006) *Development of a Research-Based Tool for Evaluating Interpretation*. Gold Coast, Queensland: Sustainable Tourism Cooperative Research Centre.
Ham, S.H., Weiler, B., Hughes, M., Brown, T.J., Curtis, J. and Poll, M. (2008) *Asking Visitors to Help: Research to Guide Strategic Communication for Protected Area Management*. Gold Coast, Queensland: Sustainable Tourism Cooperative Research Centre.
Ham, S.H., Brown, T., Curtis, J., Weiler, B., Hughes, M. and Poll, M. (2009) *Promoting Persuasion in Protected Areas*. Gold Coast, Queensland: Sustainable Tourism Cooperative Research Centre.
Hamberg, L., Malmivaara-Lämsä, M., Lehvävirta, S., O'Hara, R.B. and Kotze, D.J. (2010) Quantifying the effects of trampling and habitat edges on forest understory vegetation – A field experiment. *Journal of Environmental Management* 91, 1811–1820.
Hamilton-Smith, E. (1987) Karst kreatures: The fauna of Australian Karst. *Australian Ranger Bulletin* 4 (3), 9–10.
Hammitt, W.E. and Cole, D.N. (1998) *Wildland Recreation: Ecology and Management*. New York: Wiley.
Hannak, J.S., Kompatscher, S., Stachowitsch, M. and Herler, J. (2011) Snorkelling and trampling in shallow-water fringing reefs: Risk assessment and proposed management strategy. *Journal of Environmental Management* 92, 2723–2733.
Hardiman, N. and Burgin, S. (2010) Recreational impacts on the fauna of Australian coastal marine ecosystems. *Journal of Environmental Management* 91, 2096–2108.
Hardman, J. (1996) Ecotourism in Khao Sok National Park. In P. Dhamabutra, V. Stithyudhakarn and S. Hirunburana (eds) *Proceedings of the 2nd International Conference on the Implementation of Ecotourism* (pp. 216–224). Bangkok: Institute of Ecotourism, Srinakharinwirot University.

Harmon, M.E., Franklin, J.F., Swanson, F.J., Sollins, P., Gregory, S.V., Lattin, J.D., Anderson, N.H., Cline, S.P., Aumen, N.G., Sedell, J.R., Lienkaemper, G.W., Cromack, K. and Cummins, K.W. (1986) Ecology of coarse woody debris in temperate ecosystems. *Advances in Ecological Research* 15, 133–302.

Harrington, R., Owen-Smith, N., Vilgoen, P.C., Biggs, H.C., Mason, D.R. and Funston, P. (1999) Establishing the causes of Roan Antelope decline in the Kruger National Park, South Africa. *Biological Conservation* 90, 69–78.

Harriott, V. J. (2002) *Marine Tourism Impacts and Their Management on the Great Barrier Reef* (CRC Reef Research Centre, Technical Report No. 46). Townsville, Queensland: CRC Reef Research Centre.

Harris, J. (1993) Horse riding impacts in Victoria's Alpine National Park. *Australian Ranger* 27 (spring), 14–17.

Harrison, D. (1997) Ecotourism in the South Pacific: The case of Fiji. In *World Ecotour '97 Abstracts Volume* (p. 75). Rio de Janiero: BIOSFERA.

Harrison, L.C. and Husbands, W. (eds) (1996) *Practicing Responsible Tourism: International Case Studies in Tourism Planning, Policy, and Development*. New York: Wiley.

Harvey, R.G. (1997) *Polycyclic Aromatic Hydrocarbons*. New York: Wiley.

Hawes, M. (1996) A walking track management strategy for the Tasmanian Wilderness World Heritage Area. *Australian Parks and Recreation* (winter), 18–26.

Hawkins, J.P. and Roberts, C.M. (1993) Effects of recreational scuba diving on coral reefs: Trampling on reef flat communities. *Journal of Applied Ecology* 30, 25–30.

Haynes, C. (2010) Realities, simulacra and the appropriation of Aboriginality in Kakadu's tourism. In I. Keen (ed.) *Indigenous Participation in Australian Economies: Historical and Anthropological Perspectives* (ch. 10). Canberra: Australian National University E Press.

Hecnar, S.J. and M'Closkey, R.T. (1998) Effects of disturbance on five-lined skink, *Eumeces fasciatus*, abundance and distribution. *Biological Conservation* 85, 213–222.

Hendee, J.C. and Schoenfeld, C. (1990) Wildlife in wilderness. In J.C. Hendee, G.H. Stankey and R.C. Lucas (eds) *Wilderness Management* (2nd edn, revised) (pp. 215–239). Golden, CO: North American Press.

Hendee, J.C., Stankey, G.H. and Lucas, R.C. (1990a) *Wilderness Management* (2nd edn, revised). Golden, CO: North American Press.

Hendee, J.C., Stankey, G.H. and Lucas, R.C. (1990b) Managing for appropriate wilderness conditions: The carrying capacity issue. In J.C. Hendee, G.H. Stankey and R.C. Lucas (eds) *Wilderness Management* (2nd edn, revised) (pp. 215–239). Golden, CO: North American Press.

Henderson, J.C. (2000) The survival of a forest fragment: Bukit Timah Nature Reserve, Singapore. In X. Font and J. Tribe (eds) *Forest Tourism and Recreation: Case Studies in Environmental Management* (pp. 23–39). Oxford: CABI Publishing.

Higginbottom, K. (2004) *Wildlife Tourism: Impacts, Management and Planning*. Altona, Victoria: Common Ground.

Higginbottom, K. and Buckley, R. (1999) Viewing of free-ranging land dwelling wildlife. In K. Higginbottom and M. Hardy (eds) *Wildlife Tourism – Discussion Document* (unpublished report). Gold Coast, Queensland: Cooperative Research Centre for Sustainable Tourism, Griffith University.

Higham, J.E.S. (1998) Tourists and albatrosses: The dynamics of tourism at the Northern Royal Albatross Colony, Taiaroa Head, New Zealand. *Tourism Management* 19 (6), 521–531.

Higham, J.E.S. and Lück, M. (eds) (2008) *Marine Wildlife and Tourism Management: Insights From the Natural and Social Sciences*. Wallingford: CABI International.

Hill, R. and Pickering, C.M. (2009) Differences in resistance of three subtropical vegetation types to experimental trampling. *Journal of Environmental Management* 90, 1305–1312.

Hill, W. and Pickering, C.M. (2006) Vegetation associated with different walking track types in the Kosciuszko alpine area, Australia. *Journal of Environmental Management* 78, 24–34.

Hobbs, R.J., Cole, D.N., Yung, L., Zavaleta, E.S., Aplet, G.H., Chapin, F.S. III, Landres, P.B., Parsons, D.J., Stephenson, N.L., White, P.S., Graber, D.M., Higgs, E.S., Millar, C.I., Randall, J.M., Tonnessen, K.A. and Woodley, S. (2009) Guiding concepts for parks and wilderness stewardship in an era of global environmental change. *Frontiers in Ecology and Environment* 8 (9), 483–490.

Hockin, D., Ounsted, M., Gorman, M., Keller, V. and Barker, M.A. (1992) Examination of the effects of disturbance on birds with reference to its importance in ecological assessments. *Journal of Environmental Management* 36, 253–286.
Hockings, M. (1994) A survey of the tour operators role in marine park interpretation. *Journal of Tourism Studies* 5 (1), 16–28.
Hockings, M. (2003) Systems for assessing the effectiveness of management in protected areas. *BioScience* 53 (9), 823–832.
Hockings, M. and Twyford, K. (1997) Assessment and management of beach camping impacts within Fraser Island World Heritage Area, south-east Queensland. *Australian Journal of Environmental Management* 4, 26–39.
Hockings, M., Stolton, S., Dudley, N., Leverington, F. and Courrau, J. (2006) *Evaluating Effectiveness: A Framework for Assessing the Management of Protected Areas.* Gland, Switzerland: International Union for Conservation of Nature and Natural Resources.
Hockings, M., Cook, C., Carter, R. and James, R. (2009) Accountability, reporting, or management improvement? Development of a state of the parks assessment system in New South Wales, Australia. *Environmental Management* 43, 1013–1025.
Hodgson, G. (1999) A global assessment of human effects on coral reefs. *Marine Pollution Bulletin* 38 (5), 345–355.
Hodgson, G. (2001) Reef Check: The first step in community-based management. *Bulletin of Marine Science* 69 (2), 861–868.
Hof, M. and Lime, D.W. (1998) Visitor experience and resource protection framework in the national park system: Rationale, current status, and future direction. In S.F. McCool and D.N. Cole (eds) *Proceedings – Limits of Acceptable Change and Related Planning Processes: Progress and Future Directions* (General Technical Report INT-GTR-371) (pp. 61–68). Ogden, UT: US Department of Agriculture Forest Service, Rocky Mountain Research Station.
Holden, A. (2000) *Environment and Tourism* (1st edn). London: Routledge.
Holden, A. (2008) *Environment and Tourism* (2nd edn). London: Routledge.
Holden, A. (2012) *Environment and Tourism* (3rd edn). London: Routledge.
Holling, C.S. (1978) *Adaptive Environmental Assessment and Management*. Chichester: Wiley.
Hollis, T. and Bedding, J. (1994) Can we stop the wetlands from drying up? *New Scientist* 2 July, 30–35.
Holloway, J.C. (2008) *The Business of Tourism* (8th edn). Harlow: Prentice Hall.
Holmes, D.O. and Dobson, H.E. (1976) *Ecological Carrying Capacity Research: Yosemite National Park. (1). The Effect of Human Trampling and Urine on Subalpine Vegetation – A Survey of Past and Present Backcountry Use and the Ecological Carrying Capacity of Wilderness*. Springfield, VA: Department of Commerce, National Technical Information Service.
Honey, M. (1999) *Ecotourism and Sustainable Development: Who Owns Paradise?* Washington, DC: Island Press.
Honey, M. (2009) Community conservation and early ecotourism. *Environment Magazine* January/February, 46–56.
Hornback, K. and Eagles, P. (1999) *Guidelines for Public Use Measurement and Reporting at Parks and Protected Areas*. Gland, Switzerland: International Union for Conservation of Nature and Natural Resources.
Horneman, L., Beeton, R. and Hockings, M. (2002) *Monitoring Visitors to Natural Areas: A Manual With Standard Methodological Guidelines*. Gatton, Queensland: University of Queensland.
Horton, L.R. (2009) Buying up nature: Economic and social impacts of Costa Rica's ecotourism boom. *Latin American Perspectives* no. 166, 36 (3), 93–107.
Howard, J.L. (1999) How do scuba diving operators in Vanuatu attempt to minimise their impact on the environment? *Pacific Tourism Review* 3, 61–69.
Hughes, M. (2011) The value of protected areas: Benefits from human use. *Forum Advocating Cultural and Ecotourism: News* 16, 16.
Hughes, M. and Morrison-Saunders, A. (2005) Influence of on-site interpretation intensity on visitors to natural areas. *Journal of Ecotourism* 4 (3), 161–177.
Hughes, M., Zulfa, M. and Carlsen, J. (2008) *A Review of Recreation in Public Drinking Water Catchment*

Areas in the Southwest Region of Western Australia. Bentley, Western Australia: Curtin Sustainable Tourism Centre, Curtin University.

Hunter, C. and Green, H. (1995) *Tourism and the Environment: A Sustainable Relationship?* (Issues in Tourism Series.) London: Routledge.

Hurley, K. (1990) *Visitor Impact Management (VIM)* (Discussion paper prepared for the Victorian Department of Conservation and Natural Resources). Washington, DC: Department of Conservation and Natural Resources.

Huston, M.A. (1994) *Biological Diversity: The Coexistence of Species in Changing Landscapes*. Cambridge: Cambridge University Press.

Hutchings, P., Kingsford, M. and Hoegh-Gulberg, O. (2008) *The Great Barrier Reef: Biology, Environment and Management*. Collingwood, Victoria: CSIRO Publishing.

Huxtable, D. (1987) *The Environmental Impact of Firewood Collection for Campfires and Appropriate Management Strategies*. Salisbury, South Australia: South Australian College of Advanced Education.

Huyser, O., Ryan, P.G. and Cooper, J. (2000) Changes in population size, habitat use and breeding biology of lesser sheathbills (*Chionis minor*) at Marion Island: Impacts of cats, mice and climate change? *Biological Conservation* 92, 299–310.

Hvenegaard, G.T. (1994) Ecotourism: A status report and conceptual framework. *Journal of Tourism Studies* 5 (2), 24–35.

Hylgaard, T. and Liddle, M.J. (1981) The effect of human trampling on a sand dune ecosystem dominated by *Empetrum nigrum*. *Journal of Applied Ecology* 18, 559–569.

IAATO (International Association of Antarctic Tour Operators) (1992) *Guidelines of Conduct for Antarctic Tour Operators*. Providence, RI: International Association of Antarctic Tour Operators.

IAATO (2007) *IAATO Overview of Antarctic Tourism – 2006–2007 Antarctic Season* (Information Paper 121. Antarctic Treaty Consultative Meeting). Providence, RI: International Association of Antarctic Tour Operators.

IAATO (2012) Guidelines and resources. http://iaato.org/guidelines-and-resources. Accessed 21 February 2012.

IAPP (International Association for Public Participation) (2004) IAP2 Public participation spectrum. http://www.iap2.org.au/resources/spectrum. Accessed 5 April 2011.

IFAW (International Fund for Animal Welfare) (1996) *Report on the Workshop on the Scientific Aspects of Managing Whale Watching, Montecastello di Vibio, Italy, 1995*. Crowborough: IFAW.

Indonesia Attractions (2002) http://www.asia-planetnet/Indonesia/Komodo.htm. Accessed 4 February 2012.

Inman, C., Ranjeva, J.P., Gustavo, S., Mesa, N. and Prado, A. (2002) *Destination: Central America. A Conceptual Framework for Regional Tourism Development* (Working Paper CEN607). New York: Latin American Center for Competitiveness and Sustainable Development (CLACDS).

Inskeep, E. (1987) Environmental planning for tourism. *Annals of Tourism Research* 14 (1), 118–135.

Inskeep, E. (1988) Tourism planning: An emerging specialisation. *Journal of the American Planning Association* 54 (3), 360–372.

Inskeep, E. (1990) Sustainable tourism development in South Asia – Some case studies. Paper presented to the Globe 90 Conference on Global Opportunities for Business and the Environment – Tourism Stream, held in Vancouver, Canada, 19–23 March.

Inskeep, E. (1991) *Tourism Planning: An Integrated and Sustainable Development Approach*. New York: Van Nostrand Reinhold.

International Labour Organisation (2010) *Reducing Poverty Through Tourism*. (Sectoral Activities Programme. Working Paper 266. Prepared by Dain Bolwell and Wolfgang Weinz.) Geneva: ILO.

IPCC (Intergovernmental Panel on Climate Change) (2001) *Climate Change 2001: The Scientific Basis*. (Contribution of Working Group I to the Third Assessment Report of the Intergovernmental Panel on Climate Change, ed. T.J. Houghton *et al.*) Cambridge: Cambridge University Press.

IPCC (2007a) *Climate Change 2007: Synthesis Report*. Cambridge: Cambridge University Press.

IPCC (2007b) *Climate Change 2007: The Physical Science Basis – Summary for Policy Makers*. Cambridge: Cambridge University Press.

Itami, R., Raulings, R., MacLaren, G., Hirst, K., Gimblett, R., Zanon, D. and Chladek, P. (2003) RBSim2: Simulating complex interactions between human movement and the outdoor recreation environment. *Journal for Nature Conservation* 11, 278–286.

IUCN (International Union for Conservation of Nature and Natural Resources) (1980) *World Conservation Strategy: Living Resource Conservation for Sustainable Development.* Gland, Switzerland: International Union for the Conservation of Nature and Natural Resources, United Nations Environment Programme and the World Wildlife Fund.

IUCN (2012) *IUCN Protected Areas Categories System.* Gland, Switzerland: International Union for Conservation of Nature and Natural Resources.

Iveson, J.B. and Hart, R.P. (1983) Salmonella on Rottnest Island: Implications for public health and wildlife management. *Journal of the Royal Society of Western Australia* 66, 15–23.

Jackson, I. (1986) Carrying capacity for tourism in small tropical Caribbean islands. *UNEP's Industry and Environment Newsletter* 9 (1), 7–10.

Jacobson, S.K. and Lopez, A.F. (1994) Biological impacts of ecotourism and nesting turtles in Tortuguero National Park, Costa Rica. *Wildlife Society Bulletin* 22 (3), 414–419.

Jafari, J. (ed.) (2000) *Encyclopedia of Tourism.* London: Routledge.

Jafari, J. (2001) The scientification of tourism. In V.L. Smith and M. Brent (eds) *Hosts and Guests Revisited: Tourism Issues of the 21st Century* (pp. 28–41). New York: Cognizant Communication.

Jarvinen, O. and Vaisanen, R.A. (1977) Long term changes of North European land bird fauna. *Oikos* 29, 225–228.

Jenkins, C.N. and Joppa, L. (2009) Expansion of the global terrestrial protected area system. *Biological Conservation* 142, 2166–2174.

Jim, C.Y. (1989) Visitor management in recreation areas. *Environmental Conservation* 16 (1), 19–32 and 40.

Johannes, R.E. (1977) Coral reefs. In J. Clark (ed.) *Coastal Ecosystem Management* (pp. 593–594). New York: Wiley.

Johns, B.G. (1996) Responses of chimpanzees to habituation and tourism in the Kibale Forest, Uganda. *Biological Conservation* 78, 257–262.

Johnson, D. (2006) Providing ecotourism excursions for cruise passengers. *Journal of Sustainable Tourism* 14 (1), 43–54.

Johnstone, I.M., Coffey, B. and Howard-Williams, C. (1985) The role of recreational boat traffic in interlake dispersal of macrophytes: A New Zealand case study. *Journal of Environmental Management* 20, 263–279.

Jones, D. and Bond, A. (2010) Road barrier effect on small birds removed by vegetated overpass in south east Queensland, Australia. *Ecological Management and Restoration* 11, 65–67.

Jones, S. (2005) Community-based ecotourism: The significance of social capital. *Annals of Tourism Research* 32, 303–324.

Jordon, C.F. (1995) *Conservation: Replacing Quantity with Quality as a Goal for Global Management.* New York: Wiley.

Kakadu Board of Management and Parks Australia (1998) *Kakadu National Park Plan of Management.* Jabiru, NT: Commonwealth of Australia.

Kang, M. and Gretzel, U. (2012) Effects of podcast tours on tourist experiences in a national park. *Tourism Management* 33, 440–455.

Kao, M.C., Patterson, I., Scott, N. and Li, C.K. (2008) Motivations and satisfactions of Taiwanese tourists who visit Australia. *Journal of Travel and Tourism Marketing* 24 (1), 17–33.

Kellert, S.R. (1993) The biological basis for human values of nature. In S.R. Kellert and E.O. Wilson (eds) *The Biophilia Hypothesis* (pp. 42–69). Washington, DC: Island Press.

Kelly, P.E. and Larson, D.W. (1997) Effects of rock climbing on populations of presettlement eastern white cedar (*Thuja occidentalis*) on cliffs of the Niagara escarpment, Canada. *Conservation Biology* 11 (5), 1125–1132.

Kenteris, M., Gavalas, D. and Economou, D. (2011) Electronic mobile guides: A survey. *Personal and Ubiquitous Computing* 15 (1), 97–111.

Kevan, P.G., Forbes, B.C., Kevan, S.M. and Behan-Pelletier, V. (1995) Vehicle tracks on high Arctic tundra: Their effects on the soil, vegetation and soil arthropods. *Journal of Applied Ecology* 32, 655–667.

Khan, M. (1997) Tourism development and dependency theory: Mass tourism vs ecotourism. *Annals of Tourism Research* 24 (4), 988–991.

Kiernan, K. (1987) Soils and cave management. *Australian Ranger Bulletin* 4 (3), 67.

Kiernan, T. (2011) Rediscover parks, lecture, 14 September, Yanchep National Park, FACET, Perth Western Australia.

Kim, M-K. and Daigle, J.J. (2011) Detecting vegetation cover change on the summit of Cadillac Mountain using multi-temporal remote sensing datasets: 1979, 2001, and 2007. *Environmental Monitoring and Assessment* 180, 63–75.

Kincey, M. and Challis, K. (2010) Monitoring fragile upland landscapes: The application of airborne lidar. *Journal for Nature Conservation* 18, 126–134.

King, D.A. and Stewart, W.P. (1996) Ecotourism and commodification: Protecting people and places. *Biodiversity and Conservation* 5, 293–305.

King, P. and McGregor, A. (2012) Who's counting: An analysis of beach attendance estimates and methodologies in southern California. *Ocean and Coastal Management* 58, 17–25.

Kingsford, R.T. (2000) Ecological impacts of dams, water diversions and river management on floodplain wetlands in Australia. *Austral Ecology* 25, 109–127.

Kinlaw, A. (1999) A review of burrowing by semi-fossorial vertebrates in arid environments. *Journal of Arid Environments* 41, 127–145.

Kinnaird, M.F. and O'Brien, T.G. (1996) Ecotourism in the Tangkoko Duasudara Nature Reserve: Opening Pandora's box? *Oryx* 30 (1), 65–73.

Kirkby, M.J. (1980) The problem. In M.J. Kirkby and R.C.P. Morgan (eds) *Soil Erosion* (pp. 1–16). Chichester: Wiley.

Klein, M.L., Humphrey, S.R. and Percival, H.F. (1995) Effects of ecotourism on distribution of waterbirds in a wildlife refuge. *Conservation Biology* 9 (6), 1454–1465.

Klint, L.M., Jiang, M., Law, A., De Lacy, T. and Filep, S. (2012) Dive tourism in Luganville, Vanuatu: Shocks, stressors and vulnerability to climate change. *Tourism in Marine Environments* 8 (1&2), 91–109.

Knight, R.L. and Gutzwiller, K.J. (1995) *Wildlife and Recreationists: Coexistence Through Management and Research*. Washington, DC: Island Press.

Knox, B., Ladiges, P. and Evans, B. (1994) Living in communities. In B. Knox, P. Ladiges and B. Evans (eds) *Biology* (pp. 954–977). Sydney: McGraw-Hill.

KNPBM (Kakadu National Park Board of Management) (2007) *Kakadu National Park Management Plan 2007–2014*. Darwin, Northern Territory: Director of National Parks, Australian Government.

Kociolek, A., Clevenger, A., St Clair, C. and Prappe, D. (2010) Effects of road networks on bird populations. *Conservation Biology* 25, 241–249.

Koens, J.F., Dieperink, C. and Miranda, M. (2009) Ecotourism as a development strategy: Experiences from Costa Rica. *Environmental Development Sustainability* 11, 1225–1237.

Konu, H. and Kajala, L. (2012) *Segmenting Protected Area Visitors Based on Their Motivations* (Nature Protection Publications of Metsahallitus, Series A 194). Savonlinna, Finland: Metsahallitus.

Kozlowski, J.M. (1984) Threshold analysis to the definition of environmental capacity in Poland's Tatry National Park. In J. McNeely and J. Miller (eds) *National Parks, Conservation and Development, Proceedings of the World Congress on National Parks, Bali, Indonesia, 11–22 October 1982* (pp. 450–462). Washington, DC: Smithsonian Institution Press.

Kozlowski, J.M. (1986) *Threshold Approach in Urban, Regional and Environmental Planning*. St Lucia, Queensland: University of Queensland Press.

Kozlowki, J.M. (1990) Sustainable development in professional planning: A potential contribution of the EIA and UET concepts. *Landscape and Urban Planning* 19 (4), 307–332.

Kozlowski, J.M., Rosier, J. and Hill, G. (1988) Ultimate environmental threshold (UET) method in a marine environment (Great Barrier Reef Marine Park in Australia). *Landscape and Urban Planning* 15, 327–336.

Krebs, C. (2008) *The Ecological World View*. Collingwood, Victoria: CSIRO Publishing.

Krippendorf, J. (1987) *The Holiday Makers: Understanding the Impact of Leisure and Travel*. Oxford: Heinemann.

Kruger, F.J., van Wilgen, B.W., Weaver, A.v.B. and Greyling, T. (1997) Sustainable development and the environment: Lessons from the St Lucia environmental impact assessment. *South African Journal of Science* 93, 23–33.

Krumpe, E. and McCool, S.F. (1997) Role of public involvement in the limits of acceptable change wilderness planning system. In S.F. McCool and D.N. Cole (eds) *Proceedings from a Workshop on Limits of Acceptable Change and Related Planning Processes: Progress and Future Directions, University of Montana's Lubrecht Experimental Forest, Missoula, Montana, May 20–22, 1997* (General Technical Report INT-GTR-371) (pp. 16–20). Ogden, UT: University of Montana's Lubrecht Experimental Forest, Rocky Mountain Research Station.

Kuhre, W.L. (1995) *ISO 14001 Certification*. Upper Saddle River, NJ: Prentice Hall.

Kuitunen, M., Rossi, E. and Stenroos, A. (1998) Do highways influence density of land birds? *Environmental Management* 22 (2), 297–302.

Kuss, F.R. and Morgan, J.M. (1984) Using the USLE to estimate the physical carrying capacity of natural areas for outdoor recreation planning. *Journal of Soil and Water Conservation* 39, 383–387.

Kuss, F.R., Graefe, A.R. and Vaske, J.J. (1990) *Visitor Impact Management: A Review of Research, Vol. 1*. Washington, DC: National Parks and Conservation Association.

Kutiel, P., Eden, E and Zheveley, Y. (2000) Effect of experimental trampling and off-road motorcycle traffic on soil and vegetation of stabilized coastal dunes, Israel. *Environmental Conservation* 27 (1), 14–23.

Lambert, E., Hunter, C., Pierce, G.J. and MacLeod, C.D. (2010) Sustainable whale-watching tourism and climate change: Towards a framework of resilience. *Journal of Sustainable Tourism* 18, 409–427.

Landres, P., Cole, D. and Watson, A. (1994) A monitoring strategy for the national wilderness preservation system. In J.C. Hendee and V.G. Martin (eds) *International Wilderness Allocation, Management, and Research* (pp. 19–27). Fort Collins, CO: International Wilderness Leadership (WILD) Foundation.

Lawrence, T., Wickins, D. and Phillips, N. (1997) Managing legitimacy in ecotourism. *Tourism Management* 18 (5), 307–316.

Lawson, S.R. (2006) Computer simulation as a tool for planning and management of visitor use in protected natural areas. *Journal of Sustainable Tourism* 14 (6), 600–617.

Lawson, S.R., Hallo, J.C. and Manning, R.E. (2008) Measuring, monitoring and managing visitor use in parks and protected areas using computer-based simulation modeling. In R. Gimblett and H. Skov-Petersen (eds) *Monitoring, Simulation, and Management of Visitor Landscapes* (pp. 175–188). Tucson, AZ: University of Arizona Press.

Leask, A. (2010) Progress in visitor attraction research: Towards more effective management. *Tourism Management* 31, 155–166.

Lee, W.H. and Moscardo, G. (2005) Understanding the impact of ecotourism resort experiences on tourists' environmental attitudes and behavioural intentions. *Journal of Sustainable Tourism* 13 (6), 546–565.

Leiper, N. (1981) Towards a cohesive, curriculum in tourism: The case for a distinct discipline. *Annals of Tourism Research* 8 (1), 69–84.

Leiper, N. (1990) Tourist attraction systems. *Annals of Tourism Research* 17 (3), 367–384.

Leiper, N. (2004) *Tourism Management* (3rd edn). Sydney: Pearson Education Australia.

Lemelin, R.H. (2006) The gawk, the glance, and the gaze: Ocular consumption and polar bear tourism in Churchill, Manitoba Canada. *Current Issues in Tourism* 9 (6), 516–534.

Lemelin, R.H., Dawson, J., Stewart, E.J., Maher, P. and Liick, M. (2010) Last chance tourism: Boom, doom, and gloom. *Current Issues in Tourism* 13 (5), 477–493.

Lemos, M.C. and Agrawal, A. (2006) Environmental governance. *Annual Review of Environment and Resources* 31, 297–325.

Lenanton, R.C.J. (1977) Fishes from the hypersaline waters of the stromatolite zone of Shark Bay, Western Australia. *Copeia* 2, 1–33.

Leonard, M. and Holmes, D. (1987) Recreation management and multi-resource planning for the Mt Cole Forest, Victoria. In *Forest Management in Australia: Proceedings of the 1987 Conference of the Institute of Foresters of Australia, September 28–October 2, 1987, Perth, Western Australia* (pp. 399–415). Perth: Institute of Foresters of Australia.

Leujak, W. and Ormond, R. (2007) Visitor perceptions and the shifting social carrying capacity of south Sinai's coral reefs. *Environmental Management* 39, 472–489.
Leujak, W. and Ormond, R. (2008) Quantifying acceptable levels of visitor use on Red Sea reef flats. *Aquatic Conservation* 18, 930–944.
Leung, Y-F. (2008) *Visitor Experience and Resource Protection Data Analysis Protocol: Social Trails 2007 Field Season.* Yosemite National Park, CA: National Park Service US Department of the Interior.
Leung, Y-F. and Marion, J.L. (1995) *A Survey of Campsite Conditions in Eleven Wilderness Areas of the Jefferson National Forest* (USDI National Biological Service Report). Blacksburg, VA: Virginia Tech Cooperative Park Studies Unit, Virginia Tech University.
Leung, Y-F. and Marion, J.L. (1996) Trail degradation as influenced by environmental factors: A state of the knowledge review. *Journal of Soil and Water Conservation* 51, 130–136.
Leung, Y-F. and Marion, J.L. (1999a) Spatial strategies for managing visitor impacts in national parks. *Journal of Park and Recreation Administration* 17 (4), 2–38.
Leung, Y-F. and Marion, J.L. (1999b) Assessing trail conditions in protected areas: Application of a problem assessment method in Great Smoky Mountains National Park, USA. *Environmental Conservation* 26 (4), 270–279.
Leung, Y-F. and Marion, J.L. (1999c) Characterizing backcountry camping impacts in Great Smoky-Mountains National Park, USA. *Journal of Environmental Management* 57, 193–203.
Leung, Y-F. and Marion, J. (2000) Wilderness campsite conditions under an unregulated camping policy: An eastern example. In D.N. Cole *et al.* (eds) *Proceedings: Wilderness Science in a Time of Change; Vol. 5: Wilderness Ecosystems, Threats, and Management, May 23–27, 1999, Missoula, MT (Proceedings RMRS-P-15-Vol-5)* (pp. 148–152). Ogden, UT: USDA Forest Service, Rocky Mountain Research Station.
Leung, Y-F. and Monz, C. (2006) Visitor impact monitoring: Old issues, new challenges. *George Wright Forum* 23 (2), 7–10.
Leung, Y-F., Marion, J.L. and Farrell, T.A. (2008) Recreation ecology in sustainable tourism and ecotourism: A strengthening role. In S.F. McCool and R.N. Moisey *Tourism, Recreation and Sustainability: Linking Culture and the Environment*. Oxford: CABI Publishing.
Leung, Y-F., Newburger, T., Jones, M., Kuhn, B. and Woiderski, B. (2011) Developing a monitoring protocol for visitor-created informal trails in Yosemite National Park, USA. *Environmental Management* 47 (1), 93–106.
Leverington, F. and Hockings, M. (2004) Evaluating the effectiveness of protected area management: The challenge of change. In C.V. Barber, K.R. Miller and M. Boness (eds) *Securing Protected Areas in the Face of Global Change: Issues and Strategies* (pp. 169–214). Gland, Switzerland: International Union for Conservation of Nature and Natural Resources.
Leverington, F., Costa, K.L., Pavese, H. and Hockings, M. (2010) A global analysis of protected area management effectiveness. *Environmental Management* 46 (5), 685–698.
Liddle, M.J. (1975) A theoretical relationship between the primary production of vegetation and its ability to tolerate trampling. *Biological Conservation* 8, 251–255.
Liddle, M.J. (1997) *Recreation Ecology: The Ecological Impact of Outdoor Recreation and Ecotourism*. London: Chapman and Hall.
Liddle, M.J. and Kay, A.M. (1987) Resistance, survival and recovery of trampled corals on the Great Barrier Reef. *Biological Conservation* 42, 1–18.
Liddle, M.J. and Scorgie, H.R.A. (1980) The effects of recreation on freshwater plants and animals: A review. *Biological Conservation* 17, 183–206.
Lime, D. and Stankey, G.H. (1972) Carrying capacity: Maintaining outdoor recreation quality. *Proceedings of the Northeastern Forest Experiment Station Recreation Symposium* (pp. 174–83). Upper Darby, PA: Northeastern Forest Experiment Station.
Lindberg, K. (1998) Economic aspects of tourism. In K. Lindberg, M. Epler Wood and D. Engeldrum (eds) *Ecotourism: A Guide for Planners and Managers, Vol. 2* (pp. 87–117). Vermont: The Ecotourism Society.
Lindberg, K. (2001) Economic impacts. In D.B. Weaver (ed.) *Encyclopedia of Ecotourism* (pp. 363–377). Wallingford: CABI.

Lindberg, K. and Hawkins, D. (eds) (1993) *Ecotourism: A Guide for Planners and Managers, Vol. 1.* Burlington, VT: Ecotourism Society.

Lindberg, K. and McCool, S.F. (1998) A critique of environmental carrying capacity as a means of managing the effects of tourism development. *Environmental Conservation* 25 (4), 291–292.

Lindberg, K., McCool, S. and Stankey, G. (1997) Rethinking carrying capacity. *Annals of Tourism Research* 24, 461–465.

Lindberg, K., Epler Wood, M. and Engeldrum, D. (eds) (1998a) *Ecotourism: A Guide for Planners and Managers, Vol. 2.* Burlington, VT: Ecotourism Society.

Lindberg, K., Furze, B., Staff, M. and Black, R. (1998b) *Ecotourism in the Asia Pacific Region: Issues and Outlook.* Washington, DC: FAO Forestry Policy and Planning Division, US Department of Agriculture, Forest Service, and the Ecotourism Society.

Lindenmayer, D.B., Claridge, A.W., Gilmore, A.M., Michael, D. and Lindenmayer, B.D. (2002) The ecological roles of logs in Australian forests and the potential impacts of harvesting intensification on log-using biota. *Pacific Conservation Biology* 8, 121–140.

Lindsay, J.J. (1986) Carrying capacity for tourism development in national parks of the United States. *UNEP's Industry and Environment Newsletter* 9 (1), 17–20.

Lipscombe, N.R. (1987) *Park Management Planning: A Guide to the Writing of Management Plans.* Wagga Wagga, New South Wales: Riverina-Murray Institute of Higher Education.

Lipscombe, N.R. (1993) Recreation planning: Where have all the frameworks gone? In *Proceedings of the Royal Australian Institute of Parks and Recreation, Cairns, Queensland, September 1993.* Cairns: Centre for Leisure Research, Griffith University.

Liu, H., Feng, C., Luo, Y., Wang, Z. and Gu, H. (2010) Potential challenges of climate change to orchid conservation in a wild orchid hotspot in Southwestern China. *Botanical Review* 76, 174–192.

Liu, J.-Y. (1995) Studies on the trail erosion and its monitoring of Tataka area, Yushan National Park, Central Taiwan. *Quarterly Journal of Experimental Forestry, National Tiawan University* 9 (3), 1–19.

LNT (Leave No Trace) (2012) Leave No Trace. http://www.lnt.org/programs/index.php. Accessed 2 February 2012.

Lockwood, M. (2010) Good governance for terrestrial protected areas: A framework, principles and performance outcomes. *Journal of Environmental Management* 91, 754–766.

Lockwood, M., Worboys, G. and Kothari, A. (2006) *Managing Protected Areas: A Global Guide.* London: Earthscan.

Louv, R. (2005) *Last Child in the Woods.* Chapel Hill, NC: Algonquin.

Lu, D.J., Kao, C.W. and Chao C.L. (2012) Evaluating the management effectiveness of five protected areas in Taiwan using WWF's RAPPAM. *Environmental Management* 50 (2), 272–82.

Lucas, R.C. (1990a) The wilderness experience and managing the factors that influence it. In J.C. Hendee, G.H. Stankey and R.C. Lucas (eds) *Wilderness Management* (pp. 469–499). Golden, CO: North America Press.

Lucas, R.C. (1990b) Wilderness use and users: Trends and projections and wilderness recreation management: An overview. In J.C. Hendee, G.H. Stankey and R.C. Lucas (eds) *Wilderness Management* (pp. 355–398). Golden, CO: North American Press.

Luckenbach, R.A. and Bury, R.B. (1983) Effects of off-road vehicles on the biota of the Algodunes, Imperial County, California. *Journal of Applied Ecology* 20, 265–286.

Lucrezi, S. and Schlacher, T.A. (2010) Impacts of off-road vehicles on burrow architecture of ghost crabs (genus *Ocypode*) on sandy beaches. *Environmental Management* 45, 1352–1362.

Lück, M., Maher, P.T. and Stewart, E. (2010) Setting the scene: polar cruise tourism in the 21st century. In M. Lück, P.T. Maher and E. Stewart (eds) *Cruise Tourism in the Polar Regions: Promoting Environmental and Social Sustainability?* (pp. 1–10). London: Earthscan.

Lull, H.J. (1959) *Soil Compaction on Forest and Rangelands* (USDA Forest Service Miscellaneous Publication 768). Washington, DC: US Department of Agriculture, Forest Service.

Ma, X-L., Ryan, C. and Bao, J-G. (2009) Chinese national parks: Differences, resource use and tourism product portfolios. *Tourism Management* 30, 21–30.

Mackay, R. (1995) Visitor impact management: Determining a social and environmental carrying capacity for Jenolan Caves. In H. Richins, J. Richardson and A. Crabtree (eds) *Ecotourism and Nature-Based Tourism: Taking the Next Steps: Proceedings of the Ecotourism Association of Australia,*

National Conference, 18–23 November 1995, Alice Springs, Northern Territory (pp. 223–228). Newcastle, New South Wales: Department of Leisure and Tourism Studies, University of Newcastle.

MacLeod, C.D. (2009) Global climate change, range changes and potential implications for the conservation of marine cetaceans: A review and synthesis. *Endangered Species Research* 7, 125–136.

MacNally, R., Parkinson, A., Horrocks, G., Conole, L. and Tzaros, C. (2001) Relationships between terrestrial vertebrate diversity, abundance and availability of coarse woody debris on south-eastern Australian floodplains. *Biological Conservation* 99, 191–205.

Madej, M.A., Weaver, W.E. and Hagans, D.K. (1994) Analysis of bank erosion on the Merced River, Yosemite Valley, Yosemite National Park, California, USA. *Environmental Management* 18 (2), 234–250.

Mader, S.S. (1991) *Inquiry into Life*. Dubuque, IA: C. Brown Publishers.

Madin, E. and Fenton, D.M. (2004) Environmental interpretation in the Great Barrier Reef Marine Park: An assessment of programme effectiveness. *Journal of Sustainable Tourism* 12 (2), 121–137.

Magnus, Z., Kriwoken, L.K., Mooney, N.J. and Jones, M.E. (2004) *Reducing the Incidence of Wildlife Roadkill: Improving the Visitor Experience in Tasmania* (Technical Report). Gold Coast, Queensland: Cooperative Research Centre for Sustainable Tourism, Griffith University.

Maher, P.T., Stewart, E. and Lück, M. (eds) (2009) *Polar Tourism: Human, Environmental and Governance Dimensions*. Elmsford, NY: Cognizant Communication.

Malanson, G.P. (1993) *Riparian Landscapes*. Cambridge: Cambridge University Press.

Malisz, B. (1963) Threshold theory. *Biuletyn IUA*, 16/7 (Warsaw).

Malmivaara-Lämsä, M., Hamberg, L., Löfström, I., Vanha-Majamaa, I. and Niemelä, J. (2008) Trampling tolerance of understorey vegetation in different hemiboreal urban forest site types in Finland. *Urban Ecosystems* 11 (1), 1–16.

Mamit, J.M. (2011) The synergy between community based tourism and ecotourism in Malaysia. Paper presented at the Global Eco Asia-Pacific Conference, Seize the Potential, 7–11 November, Sydney, Australia.

Mancini, M. (2000) *Conducting Tours* (3rd edn). Independence, KY: Delmar Cengage Learning.

Mancini, M. (2008) *Selling Destinations: Geography for the Travel Professional* (5th edn). Independence, KY: Delmar Cengage Learning.

Manidis Roberts Consultants (1995) *Determining an Environmental and Social Carrying Capacity for Jenolan Caves Reserve: Applying a Visitor Impact Management System*. Sydney: Manidis Roberts Consultants.

Manidis Roberts Consultants (1997) *Developing a Tourism Optimisation Management Model (TOMM)* (Final Report). Sydney: Manidis Roberts Consultants.

Mann, J. and Barnett, H. (1999) Lethal tiger shark (*Gaeocerdo cuvier*) attack on a bottlenose dolphin (*Tursiops sp*) calf: Defense and reactions by the mother. *Marine Mammal Science* 15 (2), 568–575.

Manning, R.E. (1979) Strategies for managing recreational use of national parks. *Parks* 4 (1), 13–15.

Manning, R.E. (2004) Use rationing and allocation. In R. Buckley (ed.) *Environmental Impacts of Tourism* (pp. 273–286). Wallingford: CABI Publishing.

Manning, R.E. (2009) *Parks and People: Managing Outdoor Recreation at Acadia National Park*. Burlington, VT: University of Vermont Press.

Manning, R.E. (2011) *Studies in Outdoor Recreation: Search and Research for Satisfaction*. Corvallis, OR: Oregon State University Press.

Manning, R.E. and Freimund, W.A. (2004) Use of visual research methods to measure standards of quality for parks and outdoor recreation. *Journal of Leisure Research* 36 (4), 557–579.

Manning, R.E., McCool, S.F. and Graefe, A.R. (1995) Trends in carrying capacity. In J.L. Thompson, D.W. Lime, B. Gartner and W.M. Sames (compilers) *Proceedings of the Fourth International Outdoor Recreation and Tourism Trends Symposium and the 1995 National Recreation Resource Planning Conference, St Paul, Minnesota, May 14–17, 1995* (pp. 334–341). St Paul, MN: University of Minnesota College of Natural Resources and Minnesota Extension Service.

Manning, R.E., Ballinger, N.L., Marion, J. and Roggenbuck, J. (1996a) Recreation management in natural areas: Problems and practices, status and trends. *Natural Areas Journal* 16 (2), 142–146.

Manning, R.E., Lime, D.W. and Hof, M. (1996b) Social carrying capacity of natural areas: Theory and application in the U.S. national parks. *Natural Areas Journal* 16 (2), 118–127.

Manning, R.E., Jacobi, C., Valliere, W. and Wang, B. (1998) Standards of quality in parks and recreation. *Parks and Recreation* (July), 88–94.

Marchant, S. and Higgins, P.J. (1990) *Handbook of Australian, New Zealand and Antarctic Birds, Vol. 1*. Melbourne: Oxford University Press.

Marechal, L., Semple, S., Majolo, B., Qarro, M., Heistermann, M. and MacLarnon, A. (2011) Impacts of tourism on anxiety and physiological stress levels in wild male Barbary macaques. *Biological Conservation* 144, 2188–2193.

Margules, C.R. and Pressey, R.L. (2000) Systematic conservation planning. *Nature* 405, 243–253.

Marion, J.L. (1991) *Developing a Natural Resource Inventory and Monitoring Program for Visitor Impacts on Recreation Sites: A Procedural Manual* (Natural Resources Report, NPS/NRVT/NRR-91/06). Denver, CO: US Department of the Interior National Park Service.

Marion, J.L. (1995) Capabilities and management utility of recreation impact monitoring programs. *Environmental Management* 18 (5), 763–771.

Marion, J.L. and Leung, Y-F. (2001) Trail resource impacts and an examination of alternative assessment techniques. *Journal of Park and Recreation Administration* 19 (3) (Special issue on trails and greenways).

Marion, J.L. and Leung, Y-F. (2004a) Environmentally sustainable trail management. In R. Buckley (ed.) *Environmental Impacts of Tourism* (pp. 229–243). Wallingford: CABI Publishing.

Marion, J.L. and Leung, Y-F. (2004b) Managing impacts of camping. In R. Buckley (ed.) *Environmental Impacts of Tourism* (pp. 245–258). Wallingford: CABI Publishing.

Marion, J.L. and Leung Y-F. (2011) Indicators and protocols for monitoring impacts of formal and informal trails protected areas. *Journal of Tourism and Leisure Studies* 17 (2), 215–236.

Marion, J.L. and Reid, S.E. (2007) Minimising visitor impacts to protected areas: The efficacy of low impact education programmes. *Journal of Sustainable Tourism* 15, 5–27.

Marion, J.L., Roggenbuck, J.W. and Manning, R.E. (1993) *Problems and Practices in Backcountry Recreation Management: A Survey of National Park Service Managers* (Natural Resources Report NPS/NRVT/NRR-93/12). Denver, CO: US Department of the Interior National Park Service.

Marion, J.L., Wimpey, J.F. and Park, L.O. (2011) The science of trail surveys: Recreation ecology provides new tools for managing wilderness trails. *Park Science* 28 (3), 60–65.

Marr Consulting Services (2008) Green your business: Toolkit for tourism operators. Tourism Industry Association of Canada, Canadian Tourism Commission and Parks Canada. http://www.marrcc.com/toolkit.html. Accessed 1 March 2012.

Marron, C-H. (1999) The impact of ecotourism. *Flora and Fauna News* 11, 14.

Martilla, J.A. and James, J.C. (1977) Importance–performance analysis. *Journal of Marketing* 41 (1), 77–79.

Martin, B.S. and Uysal, M. (1990) An examination of the relationship between carrying capacity and the tourism lifecycle: Management and policy implications. *Journal of Environmental Management* 31 (4), 327–333.

Martin, S.R. and McCool, S.F. (1989) Wilderness campsite impacts: Do managers and visitors see them the same? *Environmental Management* 13 (5), 623–629.

Mason, P.A. and Legg, S.J. (1999) Antarctic tourism: Activities, impacts, management issues, and a proposed research agenda. *Pacific Tourism Review* 3, 71–84.

Mason, S.A. and Moore, S.A. (1998) Using the Sorensen network to assess the potential effects of tourism on two Australian marine environments. *Journal of Sustainable Tourism* 6 (2), 143–154.

Mastran, T.A., Dietrich, A.M., Gallagher, D.L. and Grizzard, T.J. (1994) Distribution of polyaromatic hydrocarbons in the water column and sediments of a drinking water reservoir with respect to boating activity. *Water Research* 28, 2353–2366.

Mata, C., Hervás, I., Herranz, J., Suárez, F. and Malo, J.E. (2008) Are motorway wildlife passages worth building? Vertebrate use of road-crossing structures on a Spanish motorway. *Journal of Environmental Management* 88, 407–415.

Mathieson, A. and Wall, G. (1982) *Tourism: Economic, Physical and Social Impacts*. London: Longman.

Mau, R. (2008) Managing for conservation and recreation: The Ningaloo whale shark experience. *Journal of Ecotourism* 7 (2&3), 213–225.

Mazerolle, M.J. (2004) Amphibian road mortality in response to nightly variations in traffic intensity. *Herpetologica* 60, 45–53.

Mbaiwa, J.E., Bernard, F.E. and Orford, C.E. (2002) Limits of acceptable change for tourism in the Okavango Delta. Paper presented at the Environmental Monitoring of Tropical and Sub-Tropical Wetlands International Conference, Maun Lodge, Maun, Botswana, 4–7 December.

McArthur, S. (1994a) Acknowledging a symbiotic relationship: Better heritage management via better visitor management. *Australian Parks and Recreation* (spring), 12–17.

McArthur, S. (1994b) Guided nature-based tourism: Separating 'fact from fiction'. *Australian Parks and Recreation* 30 (4), 31–38.

McArthur, S. (1996) Beyond the limits of acceptable change – Developing a model to monitor and manage tourism in remote areas. In *Towards a More Sustainable Tourism Down Under 2 Conference, Centre for Tourism, University of Otago, Dunedin, New Zealand* (pp. 223–229). Otago, Dunedin, New Zealand: Centre for Tourism.

McArthur, S. (1998a) Embracing the future of ecotourism, sustainable tourism and the EAA in the new millennium. In *Proceedings of Sixth Annual Conference of the Ecotourism Association of Australia, 29 October–1 November, 1998, Margaret River, Western Australia* (pp. 1–14). Brisbane: Ecotourism Association of Australia.

McArthur, S. (1998b) Introducing the undercapitalized world of interpretation. In K. Lindberg, M. Epler Wood and D. Engeldrum (eds) *Ecotourism: A Guide for Planners and Managers, Vol. 2*. Burlington, VT: Ecotourism Society.

McArthur, S. (2000a) Visitor management in Action: An analysis of the development and implementation of visitor management models at Jenolan Caves and Kangaroo Island. PhD thesis, University of Canberra.

McArthur, S. (2000b) Beyond carrying capacity: Introducing a model to monitor and manage visitor activity in forests. In X. Font and J. Tribe (eds) *Forest Tourism and Recreation: Case Studies in Environmental Management* (pp. 259–78). Wallingford: CABI Publishing.

McArthur, S. and Hall, C.M. (1996) Interpretation: Principles and practice. In C.M. Hall and S. McArthur (eds) *Heritage Management in Australia and New Zealand* (pp. 88–106). Melbourne: Oxford University Press.

McCaw, L. and Gillen, K. (1993) Fire. In C. Thomson, G. Hall and G. Friend (eds) *Mountains of Mystery: A Natural History of the Stirling Range* (pp. 143–148). Perth, Western Australia: Department of Conservation and Land Management.

McCawley, R. and Teaff, J.D. (1995) Characteristics and environmental attitudes of coral reef divers in the Florida Keys. In S.F. McCool and A.E. Watson (eds) *Linking Tourism, the Environment, and Sustainability: Topical Volume of Compiled Papers from a Special Session of the Annual Meeting of the National Recreation and Park Association, 1994*. Washington, DC: US Department of Agriculture Forest Service.

McClaran, M.P. and Cole, D.N. (1993) *Packstock in Wilderness: Use, Impacts, Monitoring, and Management* (General Technical Report INT-301). Ogden, UT: United States Department of Agriculture Forest Service, Intermountain Research Station.

McCool, S.F. (1986) Putting wilderness research and technology to work in the Bob Marshall Wilderness Complex. *Proceedings from the National Wilderness Research Conference: Current Research, July 23–26, 1985, Fort Collins, Colorado, July 23–26, 1985* (General Technical Report INT-212). Fort Collins, CO: US Department of Agriculture Forest Service, Intermountain Research Station.

McCool, S.F. and Cole, D.N. (1998) Experiencing limits of acceptable change: Some thoughts after a decade of implementation. In S.F. McCool and D.N. Cole (eds) *Proceedings from a Workshop on Limits of Acceptable Change and Related Planning Processes: Progress and Future Directions, University of Montana's Lubrecht Experimental Forest, Missoula, Montana, May 20–22, 1997* (General Technical Report INTGTR-371) (pp. 72–78). Ogden, UT: US Department of Agriculture Forest Service, Rocky Mountain Research Station.

McCool, S.F. and Patterson, M.E. (2000) Trends in recreation, tourism and protected area planning. In W.C. Gartner and D.W. Lime (eds) *Trends in Outdoor Recreation, Leisure and Tourism* (pp. 111–120). Wallingford: CABI International.

McCool, S.F., Clark, R.N. and Stankey, G.H. (2007) *An Assessment of Frameworks Useful for Public Land Recreation Planning* (General Technical Report, PNW-GTR-705). Portland, OR: US Department of Agriculture Forest Service, Pacific Northwest Research Station.

McKay, H. (2006) *Applying the Limits of Acceptable Change Process to Visitor Impact Management in New Zealand's Natural Areas* (Internal Report). Christchurch, NZ: Lincoln University. http://www.tourismresearch.govt.nz/Documents/Scholarships/HeatherMcKayLimitsofAcceptableChange.pdf. Accessed 8 August 2011.

McKercher, B. (1996) Differences between tourism and recreation in parks. *Annals of Tourism Research* 23 (3), 563–575.

McLain, R.J. and Lee, R.G. (1996) Adaptive management: Promises and pitfalls. *Environmental Management* 20 (4), 437–448.

McLaren, D. (1998) *Rethinking Tourism and Ecotravel: The Paving of Paradise and What You Can Do To Stop It*. Sterling, VA: Kumarian Press.

McNamara, K.E. and Prideaux, B. (2010) Reading, learning and enacting: Interpretation at visitor sites in the Wet Tropics rainforest, Australia. *Environmental Education Research* 16, 2.

McNeely, J.A. (1990) The future of national parks. *Environment* 32 (1), 1620 and 3641.

McNeely, J.A. and Thorsell, J.W. (1989) Jungles, mountains and islands. How tourism can help conserve the natural heritage. *World Leisure and Recreation* 31 (4), 29–29.

McNeely, J.A., Harrison, J. and Dingwall, P. (1994) Introduction: Protected areas in the modern world. In J.A. McNeely, J. Harrison and P. Dingwall (eds) *Protecting Nature: Regional Reviews of Protected Areas* (pp. 1–28). Gland, Switzerland: International Union for Conservation of Nature and Natural Resources.

Medio, D., Ormond, R.F.G. and Pearson, M. (1997) Effect of briefings on rates of damage to corals by SCUBA divers. *Biological Conservation* 79, 91–95.

Mende, P. and Newsome, D. (2006) The assessment, monitoring and management of hiking trails: A case study from the Stirling Range National Park, Western Australia. *Conservation Science Western Australia* 5 (3), 27–37.

Meredith, C. (1997) *Best Practice in Performance Reporting in Natural Resource Management: ANZECC Working Group on National Parks and Protected Area Management Benchmarking and Best Practice Program* (Report). Port Melbourne, Victoria: Biosis Research.

Merigliano, L., Cole, D.N. and Parsons, D.J. (1997) Applications of LAC-type processes and concepts to nonrecreation management issues in protected areas. In S.F. McCool and D.N. Cole (eds) *Proceedings from a Workshop on Limits of Acceptable Change and Related Planning Processes: Progress and Future Directions, University of Montana's Lubrecht Experimental Forest, Missoula, Montana, May 20–22, 1997* (General Technical Report INT-GTR-371) (pp. 37–43). Ogden, UT: US Department of Agriculture Forest Service, Rocky Mountain Research Station.

Mignucci-Giannoni, A.A., Monyoya-Ospina, R.A., Jimenez-Marrero, N.M., Rodriguez-Lopez, MA., Williams, E.H. and Bonde, R.K. (2000) Manatee mortality in Puerto Rico. *Environmental Management* 25 (2), 189–198.

Milazzo, M., Badalamenti, F., Vega Fernandez, T. and Chemello, R. (2005) Effects of fish feeding by snorkelers on the density and size distribution of fishes in a Mediterranean marine protected area. *Marine Biology* 146, 1213–1223.

Milazzo, M., Anastasi, I. and Willis, T.J. (2006) Recreational fish feeding affects coastal fish behaviour and increases frequency of predation on damselfish *Chromis chromis* nests. *Marine Ecology Progress Series* 310, 165–172.

Mill, R.C. and Morrison, A.M. (1985) *The Tourism System: An Introductory Text*. Upper Saddle River, NJ: Prentice-Hall.

Miller, G.T. (1998a) Science, matter, energy and ecology: Connections in nature. In G.T. Miller (ed.) *Sustaining the Earth* (pp. 27–56 and 62–66). Belmont, CA: Wadsworth Publishing.

Miller, G.T. (1998b) *Sustaining the Earth*. Belmont, CA: Wadsworth Publishing.

Miller, G.T. and Spoolman, S. (2008) *Living in the Environment: Principles, Connections, and Solutions* (16th edn). Belmont, CA: Wadsworth Publishing.

Mintel (2008) *Wildlife Tourism International*. London: Mintel International Group Ltd.

Mlinaric, I.B. (1985) Tourism and the environment: A case for Mediterranean cooperation. *International Journal of Environmental Studies* 25 (4), 239–245.

Mohammadi, M., Khalifah, Z. and Hosseini, H. (2010) Local people perceptions toward social, economic and environmental impacts of tourism in Kermanshan (Iran). *Asian Social Science* 6 (11), 220–225.

Moncrieff, D. (1997) *A Tourism Optimisation Management Model for Dryandra Woodland* (unpublished report). Perth, Western Australia: Department of Conservation and Land Management.

Moncrieff, D. and Lent, L. (1996) Tune into the future of interpretation. *Ranger* 34, 10–11.

Monz, C.A. (1998) Monitoring recreation resource impacts in two coastal areas of western North America: An initial assessment. In A.E. Watson, G.H. Aplet and J.C. Hendee (eds) *Personal, Societal, and Ecological Values of Wilderness: Proceedings of the Sixth World Wilderness Congress on Research, Management and Allocation, October 1997, Bangalore, India* (pp. 117–122). Ogden, UT: US Department of Agriculture Forest Service, Rocky Mountain Research Station.

Monz, C.A. (2000) Recreation resource assessment and monitoring techniques for mountain regions. In P.M. Godde, M.F. Price and F.M. Zimmermann (eds) *Tourism and Development in Mountain Regions* (pp. 47–68). Oxford: CABI Publishing.

Monz, C.A. and D'Luhosch, P. (2010) Monitoring campsite conditions with digital image analysis. *International Journal of Wilderness* 16 (1), 26–31.

Monz, C.A. and Twardock, P. (2010) A classification of backcountry campsites in Prince William Sound, Alaska, USA. *Journal of Environmental Management* 91, 1566–1572.

Monz, C.A., Cole, D.N., Leung, Y-F. and Marion, J.L. (2010a) Sustaining visitor use in protected areas: Future opportunities in recreation ecology research based on the USA experience. *Environmental Management* 45, 551–562.

Monz, C., Marion, J., Goonan, K., Manning, K., Wimpey, J. and Carr, C. (2010b) Assessment and monitoring of recreation impacts and resource conditions on mountain summits: Examples from the Northern Forest, USA. *Mountain Research and Development* 30 (4), 332–343.

Moore, S.A. and Lee, R.G. (1999) Understanding dispute resolution processes for American and Australian public wildlands: Towards a conceptual framework for managers. *Environmental Management* 23 (4), 453–465.

Moore, S.A. and Polley, A. (2007) Defining indicators and standards for tourism impacts in protected areas: Cape Range National Park, Australia. *Environmental Management* 39, 291–300.

Moore, S.A. and Walker, M. (2008) Progressing the evaluation of management effectiveness for protected areas: Two Australian case studies. *Journal of Environmental Policy and Planning* 10, 405–421.

Moore, S.A., Smith, A.J. and Newsome, D. (2003) Environmental performance reporting for natural area tourism: Contributions by visitor impact management frameworks and their indicators. *Journal of Sustainable Tourism* 11 (4), 348–375.

Moore, S.A., Crilley, G., Darcy, S., Griffin, T., Taplin, R., Tonge, J., Wegner, A. and Smith, A. (2009) *Designing and Testing a Park-Based Visitor Survey*. Gold Coast, Queensland: Cooperative Research Centre for Sustainable Tourism, Griffith University.

Morgan, J.M. and Kuss, F.R. (1986) Soil loss as a measure of carrying capacity in recreation environments. *Environmental Management* 10 (2), 263–270.

Morin, S.L., Moore, S.A. and Schmidt, W. (1997) Defining indicators and standards for recreation impacts in Nuyts Wilderness, Walpole-Nornalup National Park, Western Australia. *CALM Science* 2 (3), 247–266.

Morrison, A., Hsieh, S. and Wang, C.Y. (1992) Certification in the travel and tourism industry: The North American experience. *Journal of Tourism Studies* 3 (2), 32–40.

Morrison, R.J. and Munroe, A.J. (1999) Waste management in the small island developing states of the South Pacific: An overview. *Journal of Environmental Management* 6 (4), 232–246.

Moscardo, G. (2000) Interpretation. In J. Jafari (ed.) *Encyclopaedia of Tourism* (pp. 327–328). London: Routledge.

Moscardo, G., Ballantyne, R. and Hughes, K. (2007) *Designing Interpretive Signs: Principles in Practice*. Golden, CO: Fulcrum.

Mosisch, T.D. and Arthington, A.H. (1998) The impacts of power boating and water skiing on lakes and reservoirs. *Lakes and Reservoirs: Research and Management* 3, 1–17.

Mosisch, T.D. and Arthington, A.H. (2004) The impacts of recreational power boating on freshwater ecosystems. In R. Buckley (ed.) *Environmental Impacts of Ecotourism* (pp. 125–154). New York: CABI Publishing.

Moss, D. and McPhee, D.P. (2006) The impacts of recreational four-wheel driving on the abundance of ghost crab (*Ocypode cordimanus*) on subtropical beaches in SE Queensland. *Coastal Management* 34, 133–140.

Mowforth, M. and Munt, I. (1998) *Tourism and Sustainability: New Tourism in the Third World*. London: Routledge.

Muir, F. and Chester, G. (1993) Case study: Managing tourism of a seabird nesting island. *Tourism Management* April, 99–105.

Munro, J.K., Morrison-Saunders, A. and Hughes, M. (2008) Environmental interpretation evaluation in natural areas. *Journal of Ecotourism* 7, 1–14.

Murphy, K.J., Willby, N.J. and Eaton, J.W. (1995) Ecological impacts and management of boat traffic on navigable inland waterways. In D.M. Harper and A.J.D. Ferguson (eds) *The Ecological Basis for River Management* (pp. 427–442). Chichester: Wiley.

Murphy, P.E. (1985) *Tourism: A Community Approach*. New York: Methuen.

Myers, N., Mittermeier, R.A., Mittermeier, C.G., da Fonseca, G.A.B. and Kent, J. (2000) Biodiversity hotspots for conservation priorities. *Nature* 403, 853–858.

Naidoo, R. and Adamowickz, W. (2005) Biodiversity and nature based tourism at forest reserves in Uganda. *Environment and Development Economics* 10, 159–78.

Naiman, R.J. (1988) Animal influences on ecosystem dynamics. *Bioscience* 38, 7502.

Naturetrek (2012) http://www.naturetrek.co.uk/wildlife_holidays_in_madagascar. Accessed 6 February 2012.

Nayar, A. (2012) African land grabs hinder sustainable development. *Nature* doi:10.1038/nature.2012.9955.

NBTAC (1997) *Nature Based Tourism Strategy for Western Australia*. Perth, WA: Nature Based Tourism Advisory Committee of the Western Australian Tourism Commission.

NEAP (2000) *National Ecotourism Accreditation Program* (2nd edn). Brisbane: Nature and Ecotourism Accreditation Program.

Neff, J.M. (1979) *Polycyclic Aromatic Hydrocarbons in the Aquatic Environment, Fates and Biological Effects*. London: Applied Science Publishers.

Nelson, B. (1980) *Seabirds: Their Biology and Ecology*. London: Hamlyn.

Netherwood, A. (1996) Environmental management systems. In R. Welford (ed.) *Corporate Environmental Management: Systems and Strategies* (pp. 35–58). London: Earthscan.

Neuman, W.L. (1994) *Social Research Methods: Qualitative and Quantitative Approaches*. Boston, MA: Allyn and Bacon.

Neves, F.M. and Bemuenuti, C.E. (2006) The ghost crab *Ocypode quadrata* as a potential indicator of anthropogenic impact along the Rio Grande do Sul coast, Brazil. *Biological Conservation* 133, 431–435.

Newman, P., Monz, C., Leung, Y-F. and Theobald, D.M. (2006a) Monitoring campsite proliferation and conditions: Recent methodological considerations. *George Wright Forum* 23, 28–35.

Newman, P., Manning, R.E., Pilcher, E., Trevino, K. and Savidge, M. (2006b) Understanding and managing soundscapes in national parks: Part 1 – indicators of quality. In D. Siegrist, C. Clivaz, M. Hunziker and S. Iten (eds) *Exploring the Nature of Management. Proceedings of the Third International Conference on Monitoring and Management of Visitor Flows in Recreational and Protected Areas* (pp. 198–200). Rapperswil, Switzerland: University of Applied Sciences.

Newsome, D. (2000) The role of an accidentally introduced fungus in degrading the health of the Stirling Range National Park ecosystem in South Western Australia: Status and prognosis. In *International Congress on Ecosystem Health, Sacramento, California, USA, August 1999* (Special volume: Managing for healthy ecosystems).

Newsome, D. (2003) The role of an accidentally introduced fungus in degrading the health of the Stirling Range National Park ecosystem in South Western Australia: Status and prognosis. In

D.J. Rapport, W.L. Lasley, D.E. Rolston, N.O. Nielsen, C.O. Qualset and A.B. Damania (eds) *Managing for Healthy Ecosystems* (pp. 375–388). Boca Raton, FL: Lewis Publishers.

Newsome, D. (2010) The problem of mountain biking as leisure and sporting activity in protected areas. In Y-C. Hsu (ed.) *Proceedings of the Conference on Visions and Strategies for World's National Parks and Issues Confronting the Management of World's Parks*. Hualien, Taiwan: National Dong Hwa University.

Newsome, D. and Davies, C. (2009) A case study in estimating the area of informal trail development and associated impacts caused by mountain bike activity in John Forrest National Park, Western Australia. *Journal of Ecotourism* 8 (3), 237–253.

Newsome, D. and Dowling, R. (eds) (2010) *Geotourism: The Tourism of Geology and Landscape*. Oxford: Goodfellow Publishers.

Newsome, D. and Lacroix, C. (2011) Changing recreational emphasis and the loss of 'natural experiences' in protected areas: An issue that deserves consideration, dialogue and investigation. *Journal of Tourism and Leisure Studies* 17 (2), 315–333.

Newsome, D. and Rodger, K. (2008a) Impacts of tourism on pinnipeds and implications for tourism management. In J. Higham and M. Lück (eds) *Marine Wildlife and Tourism Management* (pp. 182–205). Oxford: CABI Publishing.

Newsome, D. and Rodger, K. (2008b) To feed or not to feed: A contentious issues in wildlife tourism. In D. Lunney, A. Munn and W. Meikle (eds) *Too Close For Comfort: Contentious Issues in Human–Wildlife Encounters* (pp. 255–270). Sydney: Royal Zoological Society of New South Wales.

Newsome, D. and Rodger, K. (2012a) Wildlife tourism. In A. Holden and D. Fennell (eds) *A Handbook of Tourism and the Environment* (pp. 55–70). London: Routledge.

Newsome, D. and Rodger, K. (2012b) Vanishing fauna of tourism interest. In R.H. Lemelinn, J. Dawson and E.J. Stewart (eds) *Last Chance Tourism: Adapting Tourism Opportunities in a Changing World* (pp. 55–70). London: Routledge.

Newsome, D., Milewski, A., Phillips, N. and Annear, R. (2002) Effects of horse riding on ecosystems in Australian national parks: Implications for management. *Journal of Ecotourism* 1, 52–74.

Newsome, D., Cole, D, and Marion, J. (2004) Environmental impacts associated with recreational horse riding. In R. Buckley (ed.) *The Environmental Impacts of Tourism* (pp. 61–82). Wallingford: CABI Publishing.

Newsome, D., Dowling, R. and Moore, S. (2005) *Wildlife Tourism*. Clevedon: Channel View Publications.

Newsome, D., Smith, A. and Moore, S. (2008) Horse riding in protected areas: A critical review and implications for research and management. *Current Issues in Tourism* 11 (2), 144–166.

Newsome, D., Lacroix, C. and Pickering, C. (2011) Adventure racing events in Australia: Context, assessment and implications for protected area management. *Australian Geographer* 42 (4), 403–418.

Newsome, D., Dowling, R. and Leung, Y-F. (2012) The nature and management of geotourism: A case study of two established iconic geotourism destinations. *Tourism Management Perspectives* 2–3, 19–27.

Newton, I. (1995) The contribution of some recent research on birds to ecological understanding. *Journal of Animal Ecology* 64, 675–696.

Niesward, G.H. and Pizor, P.J. (1977) *Current Planning Capacity: A Practical Carrying-Capacity Approach to Land-Use Planning* (Bulletin No. 413). New Brunswick, NJ: Co-operative Extension Service, Rutgers University.

Nilsen, P. and Tayler, G. (1998) A comparative analysis of protected area planning and management frameworks. In S.F. McCool and D.N. Cole (eds) *Proceedings of a Workshop on Limits of Acceptable Change and Related Planning Processes: Progress and Future Directions, University of Montana's Lubrecht Experimental Forest, Missoula, Montana, May 20–22, 1997* (pp. 49–57). Ogden, UT: US Department of Agriculture Forest Service, Rocky Mountain Research Station.

Nolan, H.J. (1980) Tourist attractions and recreation resources providing for natural and human resources. In D.E. Hawkins, E.L. Shafer and J.M. Rovelstad (eds) *Tourism Planning and Development Issues* (pp. 27–282). Washington, DC: George Washington University.

Norman, J. (2011) Kimberley and the dinosaur. *Habitat* (May), 25.
North Carolina State University and Department of Parks, Recreation and Tourism Management (1994) *Conflicts on Multiple-Use Trails: Synthesis of the Literature and State of the Practice.* Washington, DC: Federal Highway Administration Intermodal Division. Sourced via the New Zealand mountain bike website, http://www.mountainbike.co.nz.
NPS (National Parks Service, US Department of the Interior) (2012) Campground regulations. http://www.nps.gov/yose/planyourvisit/campregs.htm. Accessed 1 February 2012.
Nsele Reserve (2011) nsele-reserve.co.za/habituation. Accessed 6 February 2012.
Nuampukdee, R. (2002) Impacts of forest hiking activity on vegetation and some physical properties of soil in Khao Yai National Park. Unpublished Masters thesis, Kasetsart University, Bangkok, Thailand.
Nyaupane, G.P. (2007) Ecotourism versus nature-based tourism: Do tourists really know the difference? *Anatolia* 18 (1), 161–165.
Obua, J. and Harding, D.M. (1997) Environmental impact of ecotourism in Kibale National Park, Uganda. *Journal of Sustainable Tourism* 5 (3), 213–223.
Odum, E.P. (1975) *Ecology*. London: Holt Rinehart and Winston.
OEH NSW (Office of Environment and Heritage NSW) (2011) *NSW Alpine Resorts Environment Report 2010–11*. Sydney: Office of Environment and Heritage.
Office of National Tourism (1998) *Tourism in Antarctica. Tourism Facts.* Canberra: Department of Industry, Science and Tourism, Commonwealth of Australia.
Oh, H. (2001) Revisiting importance-performance analysis. *Tourism Management* 22 (6), 617–627.
Olindo, P. (1991) The old man of nature tourism: Kenya. In P. Olindo and T. Whelan (eds) *Nature Tourism* (pp. 23–38). Washington, DC: Island Press.
Olive, N.D. and Marion, J.L. (2009) The influence of use-related, environmental, and managerial factors on soil loss from recreational trails. *Journal of Environmental Management* 90, 1483–1493.
Olliff, T., Legg, K. and Keading, B. (1999) *Effects of Winter Recreation on Wildlife of the Greater Yellowstone Area: A Literature Review and Assessment.* Wyoming: Greater Yellowstone Coordinating Committee.
Oma, V.P.M., Clayton, D.M., Broun, J.B. and Keating, C.D.M. (1992) *Coastal Rehabilitation Manual.* South Perth, Western Australia: Department of Agriculture.
Ong, T.F. and Musa, G. (2011) An examination of recreational divers' underwater behaviour by attitude–behaviour theories. *Current Issues in Tourism* 14 (8), 779–795.
Ongerth, J.E., Hunter, G.D. and DeWalle, F.B. (1995) Watershed use and *Giardia* cyst presence. *Water Research* 29 (5), 1295–1299.
Onyeanusi, A.E. (1986) Measurements of impact of tourist off-road driving on grasslands in Masai Mara National Reserve, Kenya: A simulation approach. *Environmental Conservation* 13 (4), 325–329.
Orams, M.B. (1995a) Towards a more desirable form of ecotourism. *Tourism Management* 16 (1), 3–8.
Orams, M.B. (1995b) Using interpretation to manage nature based tourism. *Journal of Sustainable Tourism* 4 (2), 81–93.
Orams, M.B. (1996) A conceptual model of tourist–wildlife interaction: The case for education as a management strategy. *Australian Geographer* 27 (1), 39–51.
Orams, M.B. (1999) *Marine Tourism: Development, Impacts and Management*. London: Routledge.
O'Reilly, A.M. (1986) Tourism carrying capacity concept and issues. *Tourism Management* 7 (4), 254–258.
O'Riordan, T., Shadrake, A. and Wood, C. (1989) Interpretation, participation and national park planning. In D. Uzzell (ed.) *Heritage Interpretation – The Natural and Built Environment* (pp. 179–189). London: Belhaven.
PADI (2011) *PADI Statistics 2011*. http://www.padi.com/scuba/about-padi/PADI-statistics/default.aspx. Accessed 20 October 2012.
Page, S. and Dowling, R.K. (2002) *Ecotourism*. London: Pearson Education.
Palmer, L. (2007) Interpreting 'nature': The politics of engaging with Kakadu as an Aboriginal place. *Cultural Geographies* 14, 255–273.

Paris, C.M., Hopkins, S. and Westbrook, T. (2011) Tourists' perceptions of climate change in Cairns, Australia. *e-Review of Tourism Research* 9 (6), 279–291.

Parsons, D.J. (1986) Campsite impact data as a basis for determining wilderness use capacities. In *National Wilderness Research Conference: Current Research* (General Technical Report INT-212) (pp. 449–455). Washington, DC: US Department of Agriculture, Forest Service.

Parsons, D.J. and MacLeod, S.A. (1980) Measuring impacts of wilderness use. *Parks* 5 (3), 8–12.

Pastorelli, J. (2002) *Enriching the Experience: An Interpretive Approach to Tour Guiding*. Melbourne: Hospitality Press.

PATA (1991) *PATA Code for Environmentally Responsible Tourism: An Environmental Ethic for the Travel and Tourism Industry*. San Francisco, CA: Pacific Asia Tourism Association.

PDNPA (Peak District National Parks Authority) (2012) Facts and figures about the Peak District National Park. http://www.peakdistrict.gov.uk/news/mediacentrefacts. Accessed 24 January 2012.

Pearce, D.G. (1989) *Tourist Development* (2nd edn). Harlow: Longman Scientific and Technical.

Pearce, D.G. and Kirk, R.M. (1986) Carrying capacities for coastal tourism. *UNEP's Industry and Environment Newsletter* 9 (1), 3–7.

Pearce-Higgins, J.W. and Yalden, D.W. (1997) The effect of resurfacing the Pennine Way on recreational use of blanket bog in the Peak District National Park, England. *Biological Conservation* 82, 337–343.

Peeters, P. (2012) A clear path towards sustainable mass tourism? Rejoinder to the paper 'Organic, incremental and induced paths to sustainable mass tourism convergence' by David B. Weaver. *Tourism Management* 33 (5), 1038–1041.

Peterson, G.L. (1974) Evaluating the quality of the wilderness environment: Congruence between perceptions and aspirations. *Environment and Behaviour* 6, 169–193.

Pettebone, D., Newman, P. and Theobald, D. (2009) A comparison of sampling designs for monitoring recreational trail impacts in Rocky Mountain National Park. *Environmental Management* 43, 523–532.

Petts, G.E. (1984) *Impounded Rivers: Perspectives for Ecological Management*. Chichester: Wiley.

Phillips, N. (2000) A field experiment to quantify the environmental impacts of horse riding in D'Entrecasteaux National Park, Western Australia. Unpublished honours thesis, Department of Environmental Science, Murdoch University, Western Australia.

Phillips, N. and Newsome, D. (2002) Understanding the impacts of recreation in Australian protected areas: Quantifying damage caused by horse riding in D'Entrecasteaux National Park, Western Australia. *Pacific Conservation Biology* 7, 256–273.

Pickering, C.M. (2010) Ten factors that affect the severity of environmental impacts of visitors in protected areas. *Ambio* 39, 70–77.

Pickering, C.M. and Ballantyne, M. (2012) Orchids: An example of charismatic megaflora tourism. In A. Holden and D. Fennell (eds) *Handbook of Tourism and the Environment*. London: Routledge.

Pickering, C.M. and Barros, A. (2012) Mountain environments and tourism. In A. Holden and D. Fennell (eds) *Handbook of Tourism and the Environment* (pp. 183–191). London: Routledge.

Pickering, C.M. and Growcock, A. (2009) Impacts of experimental trampling on tall alpine herbfields and subalpine grasslands in the Australian Alps. *Journal of Environmental Management* 91 (2), 532–540.

Pickering, C.M. and Hill, W. (2007) Impacts of recreation and tourism on plant biodiversity and vegetation in protected areas in Australia. *Journal of Environmental Management* 85, 791–800.

Pickering, C.M. and Mount, A. (2010) Do tourists disperse weed seed? A global review of unintentional human-mediated terrestrial seed dispersal on clothing, vehicles and horses. *Journal of Sustainable Tourism* 18, 239–256.

Pickering, C.M., Johnston, S., Green, K. and Enders, G. (2003) Impacts of nature tourism on the Mount Kosciusko alpine area, Australia. In R. Buckley, C.M. Pickering and D. Weaver (eds) *Nature-Based Tourism, Environment and Land Management* (pp. 123–135). New York: CABI Publishing.

Pickering, C., Castley, G., Hill, W. and Newsome, D. (2010a) Environmental, safety and management issues of unauthorised trail technical features for mountain bicycling. *Landscape and Urban Planning* 97, 58–67.

Pickering, C.M., Hill, W., Newsome, D. and Leung, Y-F. (2010b) Comparing hiking, mountain biking and horse riding impacts on vegetation and soils in Australia and the United States of America. *Journal of Environmental Management* 91, 551–562.

Pickering, C.M., Rossi, S. and Barros, A. (2011) Assessing the impacts of mountain biking and hiking on subalpine grassland in Australia using an experimental protocol. *Journal of Environmental Management* 92 (12), 3049–3057.

Pitts, D. and Smith, J. (1993) A visitor monitoring strategy for Kakadu National Park. In *Track to the Future: Managing Change in Parks and Recreation: National Conference of the Royal Australian Institute of Parks and Recreation, September 1993, Cairns* (p. 12). Cairns: RAIPR, Dickson.

Plathong, S., Inglis, G.J. and Huber, M.E. (2000) Effects of self-guided snorkeling trails in a tropical marine park. *Conservation Biology* 14, 1831–1830.

Plimmer, N. (1995) Antarctic tourism: Issues and outlook. *Australian Journal of Hospitality Management* 2 (1), 31–38.

Plog, S.C. (1973) Why destination areas rise and fall in popularity. *Cornell Hotel and Restaurant Administration Quarterly* November, 13–16.

Poland, R.H.C., Hall, G.B. and Smith, M. (1996) Turtles and tourists: a hands-on experience of conservation for sixth formers from King's College, Taunton, on the lonian island of Zakynthos. *Journal of Biological Education* 30 (2), 120–128.

Pollock, A. (2007) *The Climate Change Challenge: Implications for the Tourism Industry*. Toronto: Icarus.

Pomerantz, G.A., Decker, D.J., Goff, G.R. and Purdy, K.G. (1988) Assessing impact of recreation on wildlife: A classification scheme. *Wildlife Society Bulletin* 16, 58–62.

Powell, R. and Ham, S. (2008) Can ecotourism interpretation really lead to pro-conservation knowledge, attitudes and behaviour? Evidence from the Galapagos Islands. *Journal of Sustainable Tourism* 16, 467–489.

Press, A.J. and Hill, M.A. (1994) Kakadu National Park: An Australian experience in comanagement. In D. Western, R.M. Wright and S.C. Strum (eds) *Natural Connections: Perspectives in Community-Based Conservation* (pp. 135–157). Washington, DC: Island Press.

Press, T. and Lawrence, D. (1995) Kakadu National Park: Reconciling competing interests. In T. Press, D. Lea, A. Webb and A. Graham (eds) *Kakadu: Natural and Cultural Heritage and Management* (pp. 1–14). Darwin: Australian Nature Conservation Agency and North Australia Research Unit, ANU.

Press, T., Lea, D., Webb, A. and Graham, A. (eds) (1995) *Kakadu: Natural and Cultural Heritage and Management*. Darwin: Australian Nature Conservation Agency and North Australia Research Unit, ANU.

Preston, F., Whitehead, I. and Byrne, N. (1986) *Impacts of Camping Use on Snow Plains in the Baw Baw National Park*. National Parks Service, Conservation Forests and Lands.

Prideaux, B. (2009) *Resort Destinations: Evolution, Management and Development*. Oxford: Butterworth-Heinemann.

Prideaux, B. and Cooper, M. (2009) *River Tourism*. Wallingford: CABI International.

Primack, R.B. (1998) *Essentials of Conservation Biology*. Sunderland, MA: Sinauer Associates.

Primack, R.B. and Corlett, R. (2006) *Tropical Rain Forests: An Ecological and Biogeographical Comparison*. Oxford: Blackwell.

Priskin, J. (2004) Four-wheel drive vehicle impacts in the central coastal region of Western Australia. In R. Buckley (ed.) *Environmental Impacts of Ecotourism* (pp. 339–348). New York: CABI Publishing.

Prosser, G. (1986) The limits of acceptable change: An introduction to a framework for natural area planning. *Australian Parks and Recreation* 22 (2), 5–10.

Pusey, B. and Arthington, A. (2003) Importance of the riparian zone to the conservation and management of freshwater fish: A review. *Marine and Freshwater Research* 54, 1–16.

PV (Parks Victoria) (2012) wePlan Alpine. http://www.weplan.parks.vic.gov.au/alpine. Accessed 10 April 2012.

Pyle, R.M. (2003) Nature matrix: Reconnecting people and nature. *Oryx* 37, 206–214.

QG (Queensland Government Department of Environment and Resource Management) (2012) About Michaelmas and Upolu. http://www.derm.qld.gov.au/parks/michaelmas-upolu-cays/about.html. Accessed 1 February 2012.

Quartermain, R. and Telford, R. (2011) Private sector accommodation in national parks and protected areas. Paper presented at the Global Eco Asia-Pacific Conference, 'Seize the Potential', 7–11 November, Sydney, Australia.

Queensland Government (1997) *Queensland Ecotourism Plan.* Brisbane: Queensland Department of Tourism, Small Business and Industry.

Ramsar (2012) About the Ramsar Convention. http://www.ramsar.org/cda/en/ramsar-about-about-ramsar/main/ramsar/1-36%5E7687_4000_0__. Accessed 1 March 2012.

Randall, C. and Rollins, R.B. (2009) Visitor perceptions of the role of tour guides in natural areas. *Journal of Sustainable Tourism* 17 (3), 357–374.

Randall, M. and Newsome, D. (2008) Assessment, evaluation and a comparison of planned and unplanned walk trails in coastal south-western Australia. *Conservation Science Western Australia* 7 (1), 19–34.

Randall, M. and Newsome, D. (2009) Changes in the soil micro-topography of two coastal hiking trails in south-western Australia. *Conservation Science Western Australia* 7 (2), 279–299.

Randall, C. and Rollins, R. (2009) Visitor perceptions of the role of tour guides in natural areas. *Journal of Sustainable Tourism* 17, 357–374.

Rapport, D., Costanza, R. and McMichael, A.J. (1998) Assessing ecosystem health. *Tree* 13, 397–402.

Reader's Digest (1997) *Illustrated Guide to the Game Parks and Nature Reserves of Southern Africa.* Cape Town: Reader's Digest Association.

Reef Check (2012a) Reef Check. http://www.reefcheck.org. Accessed 25 March 2012.

Reef Check (2012b) Reef Check long term monitoring. http://www.reefcheck.org/conservation/long_term_monitoring.php. Accessed 3 April 2012.

Regel, J. and Putz, K. (1997) Effect of human disturbance on body temperature and energy expenditure in penguins. *Polar Biology* 18, 246–253.

Reichhart, T. and Arnberger, A. (2010) Exploring the influence of speed, social, managerial and physical factors on shared trail preferences using a 3D computer animated choice experiment. *Landscape and Urban Planning* 96, 1–11.

Reid, S.E. and Marion, J. (2004) Effectiveness of a confinement strategy for reducing campsite impacts in Shenandoah National Park. *Environmental Conservation* 31 (4), 274–282.

Reijnen, R., Foppen, R., ter Braak, C. and Thissen, J. (1995) The effects of car traffic on breeding bird populations in woodland. III. Reduction of density in relation to the proximity to main roads. *Journal of Applied Ecology* 32, 187–202.

Reynolds, P.C. and Braithwaite, D. (2001) Towards a conceptual framework for wildlife tourism. *Tourism Management* 22, 31–42.

Rice, N. (1996) A precautionary approach to whale watching is needed. *African Wildlife* 50 (6), 22.

Richter, L.K. (1989) *The Politics of Tourism in Asia.* Honolulu, HI: University of Hawaii Press.

Rickard, W.E. and Brown, J. (1974) Effects of vehicles on Arctic tundra. *Environmental Conservation* 1 (1), 55–62.

Ripple, W.J. and Beschta, R.L. (2007) Restoring Yellowstone's aspen with wolves. *Biological Conservation* 138, 514–519.

Ritter, D. (1997) Limits of acceptable change planning in the Selway-Bitterroot Wilderness: 1985 to 1997. In *Proceedings – Limits of Acceptable Change and Related Planning Processes: Progress and Future Directions [from a Workshop]; May 20–22, 1997, Missoula, Montana* (General Technical Report INT-GTR-371) (pp. 25–28). Missoula, MT: US Department of Agriculture Forest Service, Rocky Mountain Research Station.

Robbins, M. and Boesch, C. (2011) *Among African Apes*. Berkeley, CA: University of California Press.

Rodger, K., Moore, S.A. and Newsome, D. (2007) Wildlife tourism in Australia: Characteristics, the place of science and sustainable futures. *Journal of Sustainable Tourism* 15 (2), 160–179.

Rodger, K., Smith, A., Newsome, D. and Moore, S.A. (2011) Developing and testing a rapid assessment framework to guide the sustainability of the marine wildlife tourism industry. *Journal of Ecotourism* 10, 149–164.

Rodriguez, A. (1999) Kapawi: A model of sustainable development in Ecuadorean Amazonia. *Cultural Survival Quarterly* (summer), 43–44.

Roe, D., Leader-Williams, N. and Dalal-Clayton, B. (1997) *Take Only Photographs, Leave Only Footprints: The Environmental Impacts of Wildlife Tourism* (Wildlife and Development Series No. 10). London: International Institute for Environment and Development.

Rogers, J.A. and Smith, H.T. (1995) Set-back distances to protect nesting bird colonies from human disturbance in Florida. *Conservation Biology* 9 (1), 89–99.

Rogers, J.A. and Smith, H.T. (1997) Buffer zone distances to protect foraging and loafing waterbirds from human disturbance in Florida. *Wildlife Society Bulletin* 25 (1), 139–145.

Roggenbuck, J.W. (1992) Use of persuasion to reduce resource impacts and visitor conflicts. In M.J. Manfredo (ed.) *Influencing Human Behaviour: Theory and Applications in Recreation, Tourism, and Natural Resources Management* (pp. 149–208). Champaign, IL: Sagamore Publishing.

Roggenbuck, J.W. and Lucas, R.C. (1987) Wilderness use and user characteristics: A state-of-knowledge review. In *Proceedings of National Wilderness Research Conference: Issues, State-of-Knowledge, Future Directions* (General Technical Report INT-220) (pp. 204–245). Ogden, UT: US Department of Agriculture, Forest Service, Intermountain Research Station.

Roggenbuck, J.W., Williams, D.R. and Watson, A.E. (1993) Defining acceptable conditions in wilderness. *Environmental Management* 17 (2), 187–197.

Roman, G.S.J., Dearden, P. and Rollins, R. (2007) Application of zoning and 'limits of acceptable change' to manage snorkelling tourism. *Environmental Management* 39, 819–830.

Romeril, M. (1985) Tourism and the environment: Towards a symbiotic relationship (introductory paper). *International Journal of Environmental Studies* 25 (4), 215–218.

Romeril, M. (1989a) Tourism and the environment – accord or discord? *Tourism Management* 10 (3), 204–208.

Romeril, M. (1989b) Tourism – the environmental dimension. In C.P. Cooper (ed.) *Progress in Tourism, Recreation and Hospitality Management* (pp. 103–113). London: Belhaven Press.

Roose, A. (2010) Designing visitor monitoring system in Estonian nature reserves combining passive mobile positioning with other counting methods. In M. Goossen, B. Elands and R. van Marwijk (eds) *Recreation, Tourism and Nature in a Changing World. Proceedings of the Fifth International Conference on Monitoring and Management of Visitor Flows in Recreational and Protected Areas, May 30–June 3, 2010, Wageningen, The Netherlands* (pp. 132–133). Wageningen: Wageningen University.

Rosen, P.C. and Lowe, C.H. (1994) Highway mortality of snakes in the Sonoran Desert of southern Arizona. *Biological Conservation* 68, 143–148.

Rosier, J., Hill, G. and Kozlowski, J.M. (1986) Environmental limitations: A framework for development on Heron Island, Great Barrier Reef. *Journal of Environmental Management* 23, 59–73.

Ross, H., Grant, C., Robinson, C.J., Izurieta, A., Smyth, D. and Rist, P. (2009) Co-management and indigenous protected areas in Australia: Achievements and ways forward. *Australasian Journal of Environmental Management* 16, 242–252.

Rowe, S.J. (1994) Ecocentrism: The chord that harmonizes humans and earth. *The Trumpeter* 11 (2), 106–107.

Ruschmann, D.vd.M. (1992) Ecological tourism in Brazil. *Tourism Management* 13 (1), 125–128.

Ryan, C. (1998) Kakadu National Park (Australia). In M. Shackley (ed.) *Visitor Management: Case Studies from World Heritage Sites* (pp. 121–138). Oxford: Butterworth-Heinemann.

Ryan, C. and Cessford, G. (2003) Developing a visitor satisfaction monitoring methodology: Quality gaps, crowding and some results. *Current Issues in Tourism* 6 (6), 457–507.

SAG (South Australian Government) (2006) *Recovery Plan for Twelve Threatened Orchids in the Lofty Block Region of South Australia*. Adelaide. South Australian Government.

Salm, R.V. (1986) Coral reefs and carrying capacity: The Indian Ocean experience. *Industry and Environment* 1 (9), 11–14.

Sangjun, N., Tanakanjana, N., Pattanavobool, A. and Bhumpakphan, N. (2006) Impacts of recreation activities on sanbar deer behavior and habitat utilisation in Khao Yai National Park. *Thai Journal of Forestry* 25, 30–43.

Santos-Delgado, R. (2011) Ecotourism, housing and community: Impacts of participation in *Gawad Kalinga* villages, Cam Sur, Philippines. Paper presented at the Global Eco Asia-Pacific Conference, 'Seize the Potential', 7–11 November, Sydney, Australia.

Saunders, D. (2010) Bird conservation, tourism and the value of monitoring. In J. Kirkwood and

J. O'Connor (compilers) *The State of Australia's Birds 2010: Islands and Birds. Wingspan (Supplement)* 20, 12–13.
Saunders, R. and Hough, D. (1997) Changing role of parks and park management. *Ranger* 41–42, 4–8.
Saundry, P. (2009) *IUCN Protected Area Management Categories.* Encyclopedia of Earth. http://www.eoearth.org. Accessed 20 April 2012.
Sautter, E.T. and Leisen, B. (1999) Managing stakeholders: A tourism planning model. *Annals of Tourism Research* 26 (2), 312–328.
Schellhorn, M. (2010) Development for whom? Social justice and the business of ecotourism. *Journal of Sustainable Tourism* 18 (1), 115–135.
Scheyvens, R. (1999) Ecotourism and the empowerment of local communities. *Tourism Management* 20, 245–249.
Scheyvens, R. (2002) *Tourism for Development: Empowering Communities*. Harlow: Prentice-Hall, Pearson Education.
Schlacher, T.A. and Thompson, L. (2008) Physical impacts caused by off-road vehicles to sandy beaches: Spatial quantification of car tracks on an Australian barrier island. *Journal of Coastal Research* 24, 234–242.
Schlacher, T.A., Richardson, D. and McLean, I. (2008) Impacts of off-road vehicles (ORVs) on macrobenthic assemblages on sandy beaches. *Environmental Management* 41, 878–892.
Schneider, D.M., Godschalk, D.R. and Axler, N. (1978) *The Carrying Capacity Concept as a Planning Tool* (Report No. 338). Chicago, IL: American Planning Association.
Schoegel, C. (2007) Sustainable tourism. *Journal of Sustainable Forestry* 25 (3&4), 247–264.
Schonhardt, S. (2011) Indonesia: Is Komodo Island a real life Jurassic Park? http://www.globalpost.com/dispatch/Indonesia. Accessed 4 February 2012.
Scott, R.L. (1998) Wilderness management and restoration in high use areas of Olympic National Park, Washington, USA. In A.E. Watson, G.H. Aplet and J.C. Hendee (eds) *Personal, Societal and Ecological Values of Wilderness: Sixth World Wilderness Congress Proceedings on Research, Management, and Allocation, 1997 October* (Proceedings RMRS-P-4) (pp. 144–147). Ogden, UT: US Department of Agriculture Forest Service, Rocky Mountain Research Station.
Seaton, P.T., Hu, H., Perner, H. and Pritchard, H. (2010) Ex-situ conservation of orchids in a warming world. *Botanical Review* 7, 193–203.
Secrett, C. (undated) *Rainforest, Protecting the Planet's Richest Resource*. London: Russell Press/Friends of the Earth.
Sekercioglu, C. (2002) Impacts of bird watching on human and avian communities. *Environmental Conservation* 29, 282–289.
Shackley, M. (1996) *Wildlife Tourism*. London: International Thomson Business Press.
Shackley, M. (1998a) 'Stingray City' – Managing the impact of underwater tourism in the Cayman Islands. *Journal of Sustainable Tourism* 6 (4), 328–338.
Shackley, M. (1998b) *Visitor Management – Case Studies from World Heritage Sites*. Oxford: Butterworth-Heinemann.
Shafer, C.S. and Inglis, G.J. (2000) Influence of social, biophysical, and managerial conditions on tourism experiences within the Great Barrier Reef World Heritage Area. *Environmental Management* 26 (1), 73–87.
Shafer, E.L., Jr, Hamilton, J.F. and Schmidt, E. (1969) Natural landscape preferences: A predictive model. *Journal of Leisure Research* 1, 1–9.
Sharpe, G.W. (1982) *Interpreting the Environment*. New York: Wiley.
Sharpley, R. (2009) *Tourism Development and the Environment: Beyond Sustainability*. London: Earthscan.
Shearer, B.L. (1994) The major plant pathogens occurring in the native ecosystems of south-western Australia. *Journal of the Royal Society of Western Australia* 77, 113–122.
Shearer, B.L., Crane, C. and Cochrane, A. (2004) Quantification of the susceptibility of the native flora of the southwest botanical province, Western Australia, to *Phytophthora cinnamomi*. *Australian Journal of Botany* 52, 435–443.
Shelby, B. and Heberlein, T.A. (1984) A conceptual framework for carrying capacity determination. *Leisure Sciences* 6 (4), 433–451.

Sherwood, B., Cutler, D. and Burton, J. (2002) *Wildlife and Roads: The Ecological Impact*. London: Imperial College Press.

Shew, R.L., Saunders, P.R. and Ford, J.D. (1986) Wilderness managers' perceptions of recreational horse use in the northwestern United States. In *Proceedings – National Wilderness Research Conference: Current Research, July 23–26, 1985, Fort Collins, Colorado* (pp. 320–332). Fort Collins, CO: Intermountain Research Station, Ogden, Utah.

Shindler, B. and Shelby, B. (1993) Regulating wilderness use: An investigation of user group support. *Journal of Forestry* 91 (2), 41–44.

Short, A.D. and Woodroffe, C.D. (2009) *The Coast of Australia*. Melbourne: Cambridge University Press.

Shultis, J.D. and Way, P.A. (2006) Changing conceptions of protected areas and conservation: Linking conservation, ecological integrity and tourism management *Journal of Sustainable Tourism* 14 (3), 223–237.

Shuman, C.S., Dawson, C., Wisniewski, C., Golden, W., Knight, C., Mihaly, J. and Hodgson, G. (2008) *Reef Check California 2006–2007: Citizen Monitoring to Improve Marine Conservation*. Pacific Palisades, CA: Reef Check Foundation.

Sigal, L.L. and Nash, T.H. (1983) Lichen communities on conifers in southern California mountains: An ecological survey relative to oxidant air pollution. *Ecology* 64, 1343–1354.

Sindiga, I. (1999) Alternative tourism and sustainable development in Kenya. *Journal of Sustainable Tourism* 7 (2), 108–127.

Singh, S., Timothy, D. and Dowling, R.K. (eds) (2003) *Tourism in Destination Communities*. Wallingford: CABI.

Singh, T.V. (1992) *Tourism Environment: Nature, Culture, Economy*. New Delhi: Inter India Publications.

Singh, T.V. and Kaur, J. (1986) The paradox of mountain tourism: Case references from the Himalaya. *UNEP's Industry and Environment Newsletter* 9 (1), 21–26.

Smallwood, C.B., Beckley, L.E., Moore, S.A. and Kobryn, H.T. (2011) Assessing patterns of recreational use in large marine parks: A case study from Ningaloo Marine Park, Australia. *Ocean and Coastal Management* 54, 330–340.

Smallwood, C.B., Beckley, L.E.B. and Moore, S.A. (2012) An analysis of visitor movement patterns using travel networks in a large marine park, north-western Australia. *Tourism Management* 33, 517–528.

Smit, I.P.J., Grant, C.C. and Devereux, B.J. (2007) Do artificial waterholes influence the way herbivores use the landscape? Herbivore distribution patterns around rivers and artificial surface water sources in a large African savanna park. *Biological Conservation* 136, 85–99.

Smith, A.J. (1998) Environmental impacts of recreation and tourism in Warren National Park, Western Australia and appropriate management planning. Unpublished honours thesis, Department of Environmental Science, Murdoch University, Western Australia.

Smith, A.J. (2003) Campsite impact monitoring in the temperate eucalypt forests of Western Australia: An integrated approach. Unpublished doctoral thesis, Murdoch University, Western Australia.

Smith, A.J. and Newsome, D. (2002) An integrated approach to assessing, managing and monitoring campsite impacts in Warren National Park, Western Australia. *Journal of Sustainable Tourism* 10 (4), 343–359.

Smith, A.J. and Newsome, D. (2005) *Research into the Factors Leading to and the Management of Impact Creep*. Gold Coast, Queensland: Cooperative Research Centre for Sustainable Tourism, Griffith University.

Smith, A.J. and Newsome, D. (2006) *An Investigation into the Concept of and Factors Leading to Impact Creep and Its Management*. Gold Coast, Queensland: Cooperative Research Centre for Sustainable Tourism, Griffith University.

Smith, A.J., Newsome, D. and Enright, N. (2012) Does provision of firewood reduce woody debris loss around campsites in south-west Australian forests? *Australasian Journal of Environmental Management* 19 (2), 108–121.

Smith, D.W., Petersen, R.O. and Houston, D.B. (2003) Yellowstone after wolves. *Bioscience* 53, 330–340.

Smith, R.H. and Neal, J.E. (1993) *Wood Residues in Regenerated Karri Stands* (CALM Internal Report). Manjimup, Western Australia: Science and Information Division, Department of Conservation and Land Management.

Smith, V.L. and Eadington, W.R. (eds) (1992) *Tourism Alternatives: Potentials and Problems in the Development of Tourism*. Pennsylvania, PA: International Academy for the Study of Tourism, University of Pennsylvania Press.

Smith, V.L. and Moore, S. (1990) Identifying park users and their expectations: A fundamental component in management plans. *Australian Parks and Recreation* 26 (1), 34–41.

Snowcroft, P.G. and Griffin, J.G. (1983) Feral herbivores suppress Mamane and other browse species on Mauna Kea, Hawaii. *Journal of Range Management* 36, 638–645.

SNV (Stichting Nederlandse Vrijwilligers; Foundation of Netherlands Volunteers) (2009) *The Market for Responsible Tourism Products in Latin America and Nepal*. Amsterdam: Netherlands Development Organisation.

Solbrig, O., Medina, E. and Silva. J. (1996) Determinants of tropical savannas. In O. Solbrig, E. Medina and J. Silva (eds) Biodiversity and Savanna Ecosystem Processes: A Global Perspective (pp. 31–41). Berlin: Springer.

Spellerberg, I. (2002) *Ecological Effects of Roads*. Enfield, NH: Science Publishers.

Splettstoesser, J. (1999) Antarctica tourism: Successful management of a vulnerable environment. In T.V. Singh and S. Singh (eds) *Tourism Development in Critical Environments* (pp. 137–148). New York: Cognizant Communication Corporation.

Stabler, M.J. (ed.) (1997) *Tourism and Sustainability – Principles to Practice*. Oxford: CAB International.

Standards Australia (1997) *Integrating Quality and Environmental Management Systems ISO 9001 and ISO 14001*. Homebush, New South Wales: Standards Australia.

Standards Australia and Standards New Zealand (1996) *Environmental Management Systems: General Guidelines on Principles, Systems and Supporting Techniques*. Homebush, New South Wales: Standards Australia and Standards New Zealand.

Stankey, G.H. (1978) Wilderness carrying capacity. In J.C. Hendee, G.H. Stankey and R.C. Lucas (eds) *Wilderness Management* (Miscellaneous Publication No. 1365) (pp. 168–188). Washington, DC: US Department of Agriculture, Forest Service.

Stankey, G.H. (1980) *A Comparison of Carrying Capacity Perceptions Among Visitors to Two Wildernesses* (Research Paper INT-242). Ogden, UT: US Department of Agriculture Forest Service, Intermountain Forest and Range Experiment Station.

Stankey, G.H. (1988) Issues and approaches in the management of recreational use in natural areas. *Recreation Australia* 8 (3), 1–6.

Stankey, G.H. (1989) Conservation, recreation and tourism. Paper prepared for the Institute of Australian Geographers' 23rd Conference, University of Adelaide, 13–16 February.

Stankey, G.H. (1997) Institutional barriers and opportunities in application of the limits of acceptable change. In S.F. McCool and D.N. Cole (eds) *Proceedings of a Workshop on Limits of Acceptable Change and Related Planning Processes: Progress and Future Directions* (pp. 10–15). Ogden, UT: University of Montana's Lubrecht Experimental Forest, Rocky Mountain Research Station.

Stankey, G.H. (2003) Adaptive management at the regional scale: Breakthrough innovation or mission impossible? A report on an American experience. In B.P. Wilson and A. Curtis (eds) *Agriculture for the Australian Environment* (pp. 159–177). Albury, New South Wales: Johnstone Centre, Charles Sturt University.

Stankey, G.H. and Brown, P.J. (1981) A technique for recreation planning and management in tomorrow's forests. In *Proceedings of XVII IUFRO World Congress, Japan* (pp. 63–73).

Stankey, G.H. and Lime, D.W. (1973) *Recreational Carrying Capacity: An Annotated Bibliography* (General Technical Report INT-3). Ogden, UT: US Department of Agriculture, Forest Service, Intermountain Forest and Range Experiment Station.

Stankey, G.H. and Manning, R.E. (1986) Carrying capacity of recreational settings. In *A Literature Review* (INT 4901 Publication No. 166) (pp. 47–57). Washington, DC: President's Commission on Americans Outdoors.

Stankey, G.H., McCool, S.F. and Stokes, G.L. (1984) Limits of acceptable change: A new framework for managing the Bob Marshall Wilderness complex. *Western Wildlands* 10 (3), 33–37.

Stankey, G.H., Cole, D.N., Lucas, R.C., Petersen, M.E. and Frissell, S. (1985) *The Limits of Acceptable Change (LAC) System for Wilderness Planning* (General Technical Report INT-176). Ogden, UT: US Department of Agriculture Forest Service, Intermountain Forest and Range Experiment Station.

Stankey, G.H., McCool, S.F. and Stokes, G.L. (1990) Managing for appropriate wilderness conditions: The carrying capacity issue. In J.C. Hendee, G.H. Stankey and R.C. Lucas (eds) *Wilderness Management* (pp. 215–239). Golden, CO: North American Press.

Starkey, R. (1996) The standardization of environmental management systems. In R. Welford (ed.) *Corporate Environmental Management: Systems and Strategies* (pp. 59–91). London: Earthscan.

State Ministry for Environment Republic of Indonesia and United Nations Development Programme (1997) *Agenda 21: Indonesia*. Jakarta: State Ministry for Environment, Republic of Indonesia, and United Nations Development Programme.

Stattersfield, A. Crosby, M., Long, A. and Wege, D. (1998) *Endemic Bird Areas of the World: Priorities for Biodiversity Conservation*. Washington, DC: Smithsonian Institute Press.

Steiner, A.J. and Leatherman, S.P. (1981) Recreational impacts on the distribution of ghost crabs (*Ocypode quadrata*). *Biological Conservation* 29, 111–122.

Stephenson, P.J. (1993) The impacts of tourism on nature reserves in Madagascar: Perinet, a case study. *Environmental Conservation* 20 (3), 262–265.

Steven, R., Pickering, C. and Castley, G. (2011) A review of the impacts of nature based recreation on birds. *Journal of Environmental Management* 92, 2287–2294.

Stewart, R.R., Ball, I.R. and Possingham, H.P. (2007) The effect of incremental reserve design and changing reservation goals on the long-term efficiency of reserve systems. *Conservation Biology* 21 (2), 346–354.

Stewart, W.P. (1989) Fixed itinerary systems in backcountry management. *Journal of Environmental Management* 29, 163–171.

Stewart, W.P. and Sekartjakrarini, S. (1994) Disentangling ecotourism. *Annals of Tourism Research* 21 (4), 840–841.

Stohlgren, T.J. and Parsons, D.J. (1992) Evaluating wilderness recreational opportunities: Application of an impact matrix. *Environmental Management* 16 (3), 397–403.

Stokes, G.L. (1987) Involving the public in wilderness management decision making: The Bob Marshall Wilderness Complex – a case study. In *Economic and Social Development: A Role for Forests and Forestry Professionals, Proceedings of the 1987 Society of American Foresters National Convention Minneapolis, Minnesota October 18–21* (pp. 157–161). Bethesda, MA: Society of American Foresters.

Stokes, G.L. (1990) The evolution of wilderness management: The Bob Marshall Wilderness Complex. *Journal of Forestry* 88 (10), 15–21.

Stonehouse, B. (1994) Ecotourism in Antarctica. In E. Cater and G. Lowman (eds) *Ecotourism: A Sustainable Option?* (pp. 195–212). Chichester: Wiley.

Stonehouse, B. and Crosbie, K. (1995) Tourism impacts and management in the Antarctic peninsula area. In C.M. Hall and M.E. Johnston (eds) *Polar Tourism: Tourism in the Arctic and Antarctic Regions* (pp. 217–233). Chichester: Wiley.

Stonehouse, B. and Snyder, J.M. (2010) *Polar Tourism: An Environmental Perspective*. Bristol: Channel View Publications.

Storrie, A. and Morrison, S. (1998) *The Marine Life of Ningaloo Marine Park and Coral Bay*. Perth: Department of Conservation and Land Management, Western Australia.

Strickland-Munro, J., Moore, S. and Freitag-Ronaldson, S. (2010) The impacts of tourism on two communities adjacent to the Kruger National Park, South Africa. *Development Southern Africa* 27, 663–678.

Stronza, A. and Gordillo, J. (2008) Community views of ecotourism. *Annals of Tourism Research* 35 (2), 448–468.

Sun, D. and Liddle, M.J. (1993a) A survey of trampling effects on vegetation and soil in eight tropical and subtropical sites. *Environmental Management* 17 (4), 497–510.

Sun, D. and Liddle, M.J. (1993b) Plant morphological characteristics and resistance to simulated trampling. *Environmental Management* 17 (4), 511–521.

Sun, D. and Walsh, D. (1998) Review of studies on environmental impacts of recreation and tourism in Australia. *Journal of Environmental Management* 53, 323–338.

Sundberg, J. (2004) Identities in the making: Conservation, gender and race in the Maya Biosphere Reserve, Guatemala. *Gender, Place and Culture* 11 (1), 43–66.

Swearingen, T.C. and Johnson, D.R. (1995) Visitor's responses to uniformed park employees. *Journal of Park and Recreation Administration* 13, 73–85.

Swinnerton, G.S. (1995) Conservation through partnership: Landscape management within national parks in England and Wales. *Journal of Park and Recreation Administration* 13 (4), 47–60.

Szuster, B.W., Needham, M.D. and McClure, B.P. (2011) Scuba diving perceptions and evaluations of crowding underwater. *Tourism in Marine Environments* 7 (3&4), 153–165.

Tao, C-H., Eagles, P. and Smith, S. (2004) Profiling Taiwanese tourists using a self-definition approach. *Journal of Sustainable Tourism* 12 (2), 149–168.

Tanakanjana, N. (2008) Recreation opportunity classification and challenges in maintaining recreation diversity in Thailand's national parks. In S. Weber and D. Harmon (eds) *Rethinking Protected Areas in a Changing World: Proceedings of the 2007 GWS Biennial Conference on Parks, Protected Areas, and Cultural Sites* (pp. 151–156). Hancock, Michigan: George Wright Society.

Taplin, R.H. (2012) The value of self-stated attribute importance to overall satisfaction. *Tourism Management* 33 (2), 295–304.

Taplin, R. and Moore, S.A. (2012) Benchmarking for visitor management in parks. In P. Fredman, M. Stenseke, H. Liljendahl, A. Mossing and D. Laven (eds) *Outdoor Recreation in Change – Current Knowledge and Future Challenges. Proceedings of the 6th International Conference on Monitoring and Management of Visitors in Recreational and Protected Areas. Stockholm, Sweden, August 21–24, 2012* (pp. 234–235). Östersund, Sweden: Mid Sweden University.

TAT (1995) *Policies and Guidelines: Development of Ecotourism (1995–1996) of the Tourism Authority of Thailand.* Bangkok: Tourism Authority of Thailand.

TAT (1999) Development of Ecotourism Areas project: Khao Sok National Park, Surat Thani Province. In C. Kandel and M. Marcolina (compilers) *Environment, Culture and Heritage: Best Practice Papers for 1999* (pp. 39–44). Bangkok: Pacific Asia Tourism Association Office of Environment and Culture.

Taylor, R. and De La Harpe, R. (1995) *Great Game Parks of Africa: St Lucia Wetland Park.* Cape Town: Struik Publishers.

Teh, L. and Cabanban, A.S. (2007) Planning for sustainable tourism in southern Pulau Banggi: An assessment of biophysical conditions and their implications for future tourism development. *Journal of Environmental Management* 85, 999–1008.

Temple, K.L., Camper, A.K. and Lucas, R.C. (1982) Potential health hazard from human wastes in wilderness. *Journal of Soil and Water Conservation* (November–December), 357–359.

Thomas, G. and Morgans, D. (2011) Tourism in protected areas: Sustainable nature-based tourism in Queensland's national parks. Paper presented at the Global Eco Asia-Pacific Conference, 'Seize the Potential', 7–11 November, Sydney, Australia.

Thompson, L.M.C. and Schlacher, T.A. (2008) Physical damage to coastal dunes and ecological impacts caused by vehicle tracks associated with beach camping on sandy shores: A case study from Fraser Island, Australia. *Journal of Coastal Conservation* 12, 67–82.

Thompson, M.J. and Henderson, R.E. (1998) Elk habituation as a credibility challenge for wildlife professionals. *Wildlife Society Bulletin* 26 (3), 477–483.

Thompson, R. and Dalton, T. (2010) Measuring public access to the shoreline: The boat-based offset survey method. *Coastal Management* 38 (4), 378–398.

Thrash, I. (1998) Impact of water provision on herbaceous vegetation in Kruger National Park, South Africa. *Journal of Arid Environments* 38, 437–450.

Tian-Cole, S. and Crompton, J. (2003) A conceptualization of the relationships between service quality and visitor satisfaction, and their links to destination selection. *Leisure Studies* 22 (1), 65–80.

Tian-Cole, S., Crompton, J.L. and Wilson, V.L. (2002) An empirical investigation of the relationships between service quality, satisfaction and behavioural intentions among visitors to a wildlife refuge. *Journal of Leisure Research* 34 (1), 1–24.

TIES (The International Ecotourism Society) (2005) *Ecotourism Facts and Statistics*. Washington, DC: The International Ecotourism Society.

Tilden, F. (1957) *Interpreting Our Heritage*. Chapel Hill, NC: University of North Carolina Press.

Tilden, F. (1977) *Interpreting Our Heritage* (3rd edn). Chapel Hill, NC: University of North Carolina Press.

Tilden, F. and Craig, R.B. (2008) *Interpreting Our Heritage* (4th edn). Chapel Hill, NC: University of North Carolina Press.

Tilot, V., Leujak, W., Ormond, R.F.G., Ashworth, J.A. and Mabrouk, A. (2008) Monitoring of South Sinai coral reefs: Influence of natural and anthropogenic factors. *Aquatic Conservation: Marine and Freshwater Ecosystems* 18, 1109–1126.

Timothy, D.J. (1999) Participatory planning: A view of tourism in Indonesia. *Annals of Tourism Research* 26 (2), 371–391.

TL (Tread Lightly) (2012) Welcome to Tread Lightly! http://www.treadlightly.org/page.php/home/Home.html. Accessed 3 February 2012.

Todd, S.E. and Williams, P.W. (1996) From white to green: A proposed environmental management system framework for ski areas. *Journal of Sustainable Tourism* 4 (3), 147–173.

Tonge, J. and Moore, S.A. (2007) Importance–satisfaction analysis for marine park hinterlands: A Western Australian case study. *Tourism Management* 28, 768–776.

Tonge, J., Wegner, A., Moore, S.A., Taplin, R. and Smith A.J. (2009) *Designing and Testing a Park-based Visitor Survey for Protected Areas in Western Australia*. Report prepared for the Western Australia Department of Environment and Conservation, Murdoch University.

Tonge, J., Moore, S.A. and Taplin, R. (2011) Visitor satisfaction analysis as a tool for park managers: A review and case study. *Annals of Leisure Research* 14 (4), 289–303.

Tourism Coordinates (1995) *Pilbara/Gascoyne Islands Ecotourism Management Strategy*. Pilbara: Pilbara Development Commission.

Tourism WA (2011) *Naturebank: Exceptional Ecotourism Development Opportunities in Western Australia*. Perth: Tourism Western Australia.

Townsend, C. (2003). Marine ecotourism through education: A case study of divers in the British Virgin Islands. In B. Garrod and J. C. Wilson (eds) *Marine Ecotourism: Issues and Experiences* (pp. 138–154). Clevedon: Channel View Publications.

Tratalos, J.A. and Austin, T.J. (2001) Impacts of recreational SCUBA diving on coral communities of the Caribbean island of Grand Cayman. *Biological Conservation* 102, 67–75.

Trauer, B. (1998) Green tourism in the hotel and resort sector: International and New Zealand perspectives. *Australian Parks and Leisure* (December), 5–9.

Tremblay, P. (2007) Economic contribution of Kakadu National Park to tourism in the Northern Territory. Gold Coast, Queensland: Cooperative Research Centre for Sustainable Tourism, Griffith University.

Trent, D.B. (1991) Case studies of two ecotourism destinations in Brazil. In J.A. Kusler (ed.) *Ecotourism and Resource Conservation, Vol. 1*. Miami Beach, FL: Ecotourism and Resource Conservation.

TripAdvisor (2010) TripAdvisor unveils 2011 travel trends forecast, 9 November. http://ir.tripadvisor.com/releasedetail.cfm?ReleaseID=631769. Accessed 22 September 2012.

Tritter, J.Q. and McCallum, A. (2006) The snakes and ladders of user involvement: Moving beyond Arnstein. *Health Policy* 76 (2), 156–168.

Trombulak, S. and Frissell, C. (2000) Review of ecological effects of roads on terrestrial and aquatic communities. *Conservation Biology* 14 (1), 18–30.

Tubb, K.N. (2003) An evaluation of the effectiveness of interpretation within Dartmoor National Park in reaching the goals of sustainable tourism development. *Journal of Sustainable Tourism* 11, 476–498.

Tuite, C.H., Hanson, P.R. and Owen, M. (1984) Some ecological factors affecting winter wildfowl distribution on inland waters in England and Wales, and the influence of water-based recreation. *Journal of Applied Ecology* 21, 41–62.

Turton, S.M. (2005) Managing environmental impacts of recreation and tourism in rainforests at the Wet Tropics of Queensland World Heritage Area. *Geographical Research* 43, 140–151.

Tyler, D. and Dangerfield, J.M. (1999) Ecosystem tourism: A resource based philosophy for ecotourism. *Journal of Sustainable Tourism* 7 (2), 146–158.
UNDP/WTO (1986) *Bhutan Tourism Development Master Plan*. Madrid: United Nations Development Programme and the World Tourism Organisation.
UNEP (United Nations Environment Program) (1986) *Carrying Capacity for Tourism Activities* (Special Issue). *UNEP Industry and Environment Newsletter* 9 (1), 1–2.
UNEP (2007) *Tourism in the Polar Regions: The Sustainability Challenge*. Paris: United Nations Environment Program.
UNEP (2010) *Are You a Green Leader? Business and Biodiversity: Making the Case for a Lasting Solution*. Paris: United Nations Environment Program.
UNEP (2011) Towards a green economy: Pathways to sustainable development and poverty eradication. In L. Pratt (coordinating author) *Part Two: Investing in Energy and Resource Efficiency – Tourism*. (pp. 413–452). New York: United Nations Environment Programme.
UNEP (2012) Aichi biodiversity targets. http://www.cbd.int/sp/targets. Accessed 7 May 2012.
UNESCO (United Nations Educational, Scientific and Cultural Organization) (2000) The World Heritage List. http://whc.unesco.org/en/list/. Accessed 1 March 2012.
UNESCO (2009) *The World's Heritage: A Complete Guide to the Most Extraordinary Places*. New York: United Nations Educational, Scientific and Cultural Organization and HarperCollins.
UNESCO (2011) *Global Geoparks Network*. Paris: Division of Ecological and Earth Sciences, United Nations Educational, Scientific and Cultural Organization.
UNESCO (2012a) World Heritage list. URL: http://whc.unesco.org/en/list. Accessed 2012.
UNESCO (2012b) Ecological sciences for sustainable development. http://www.unesco.org/new/en/natural-sciences/environment/ecological-sciences/biosphere-reserves/. Accessed 1 March 2012.
United Nations (1977) *Threshold Analysis Handbook* (Document No. ST/ESA/64). New York: Department of Economic and Social Affairs, United Nations.
University of Queensland (2011) Marxan. http://www.uq.edu.au/marxan/index.html?page=80354. Accessed 18 November 2011.
UNWTO (United Nations World Tourism Organization) (2005) Declaration of tourism and the millennium development goals. *Sustainable Development of Tourism e-bulletin*, 10. http://www.world-tourism.org/sustainable/ebulletin/dec2005eng.htm. Accessed 18 November 2011.
UNWTO (2010) *Tourism 2020 Vision*. Madrid: United Nations World Tourism Organization.
UNWTO (2011) *UNWTO Tourism Highlights: 2011 Edition*. Madrid: United Nations World Tourism Organization.
UNWTO (2012a) *World Tourism Barometer, Volume 10, November 2012*. Madrid: United Nations World Tourism Organization.
UNWTO (2012b) Global code of ethics for tourism. http://www.unwto.org/ethics/index.php. Accessed 17 February 2012.
UNWTO and UNEP (United Nations Environmental Program) (2008) *Climate Change and Tourism, Responding to Global Challenges*. Madrid: United Nations World Tourism Organization.
USDA FS (United States Department of Agriculture Forest Service) (1985) *Bob Marshall Great Bear Scapegoat Wildernesses Action Plan for Managing Recreation (The Limits of Acceptable Change)*. Washington, DC: USDA FS (Flathead National Forest).
USDA FS (2009) *Technical Guide for Monitoring Selected Conditions Related to Wilderness Character* (General Technical Report WO-80). Washington, DC: USDA FS.
USDA FS (2012) Recreation, heritage and wilderness programs. National Visitor Use Monitoring Program. http://www.fs.fed.us/recreation/programs/nvum/. Accessed 9 April 2012.
USDI NPS (United States Department of Interior National Park Service) (2010) *Field Monitoring Guide. Visitor Use and Impacts Monitoring Program. Yosemite National Park*. Washington, DC: US Department of Interior National Park Service.
Van Der Walt, W. (1998) Ecotourism development in South Africa. Paper presented at the Ecotourism Association of Australia Annual Conference on Ecotourism Development in the Millennium, Margaret River, Western Australia, 28–31 October.
Van der Zande, A.N., Berkhuizen, J.C., van Latesteijn, H.C., ter Keurs, W.J. and Poppelaars, A.J.

(1984) Impact of outdoor recreation on the density of a number of breeding bird species in woods adjacent to urban residential areas. *Biological Conservation* 30, 1–39.

Van Riet, W.F. and Cooks, J. (1990a) Planning and design of Berg-en-Dal, a new camp in Kruger National Park. *Environmental Management* 14 (3), 359–365.

Van Riet, W.F. and Cooks, J. (1990b) An ecological planning model. *Environmental Management* 14 (3), 339–348.

Van Riper, C.J., Manning, R.E., Monz, C.A. and Goonan, K.A. (2011) Tradeoffs among resource, social, and managerial conditions on mountain summits of the northern forest. *Leisure Sciences* 33, 228–249.

Van Vuren, D. and Coblentz, B.E. (1987) Some ecological effects of feral sheep on Santa Cruz Island, California, USA. *Biological Conservation* 41, 252–268.

Van Wagner, C.E. (1968) The line intersect method in forest fuel sampling. *Forest Science* 14, 20–26.

van Wagtendonk, J.W. (1986) The determination of carrying capacities for the Yosemite Wilderness. In *Proceedings – National Wilderness Research Conference: Current Research* (General Technical Report INT-212) (pp. 456–461). Washington, DC: US Department of Agriculture Forest Service.

Vaske, J.J., Decker, D.J. and Manfredo, M.J. (1995) Human dimensions of wildlife management: An integrated framework for coexistence. In R.L. Knight and L.J. Gutzwiller (eds) *Wildlife and Recreationists: Coexistence Through Management and Research* (pp. 33–49). Washington, DC: Island.

Veal, A.J. (2006) *Research Methods for Leisure and Tourism: A Practical Guide*. New York: Prentice Hall.

Veron, J.E.N. (1986) Distribution of reef building corals. *Oceanus* 29 (2), 27–32.

Vistad, O.I. (2003) Experience and management of recreational impact on the ground – A study among visitors and managers. *Journal for Nature Conservation* 11, 363–369.

Vivianco, L.A. (2001) Spectacular Quetzals, ecotourism, and environmental futures in Monte Verde, Costa Rica. *Ethnology* 40 (2), 79–92.

Vogler, F. and Reisch, C. (2011) Genetic variation on the rocks – The impact of climbing on the population ecology of a typical cliff plant. *Journal of Applied Ecology* 48, 899–905.

Vogt, C.A. and Williams, D.R. (1999) Support for wilderness recreation fees: The influence of fee purpose and day versus overnight use. *Journal of Park and Recreation Administration* 17 (3), 85–99.

von Platen, J. (2003). Human waste disposal in remote natural terrestrial areas: Is there a public health risk? Unpublished honours thesis, University of Tasmania.

von Ruschkowki, E. and Mayer, M. (2011) From conflict to partnership? Interactions between protected areas, local communities and operators of tourism enterprises in two German national park regions. *Journal of Tourism and Leisure Studies* 17 (2), 147–181.

Waayers, D., Newsome, D. and Lee, D. (2006) Observations of non-compliance behaviour by tourists to a voluntary code of conduct: A pilot study of turtle tourism in the Exmouth region, Western Australia. *Journal of Ecotourism* 5 (3), 211–222.

WA DEC (Western Australian Department of Environment and Conservation) (2011) *2010–2011 Annual Report*. Kensington, Western Australian: Department of Environment and Conservation, http://www.dec.wa.gov.au/content/view/6708/2422/. Accessed 4 May 2012.

Wade, D.J. and Eagles, P.F.J. (2003) The use of importance–performance analysis and market segmentation for tourism management in parks and protected areas: An application to Tanzania's national parks. *Journal of Sustainable Tourism* 2 (3), 196–212.

Wagar, J.A. (1964) *The Carrying Capacity for Wildlands for Recreation* (Forest Science Monograph 7). Washington, DC: Society of American Foresters.

Walker, D.M. (1991) Evaluation and rating of gravel roads. *Transportation Research Record* 2, 120–125.

Walker, R.H., Emslie, R.H., Owen-Smith, R.N. and Scholes, R.J. (1987) To cull or not to cull: Lessons from a southern African drought. *Journal of Applied Ecology* 24, 381–401.

Wallace, C.C., Babcock, R.C., Harrison, P.L., Oliver, J.K. and Willis, B.L. (1986) Sex on the reef: Mass spawning of corals. *Oceanus* 29 (2), 38–42.

Wallace, G.N. (1993) Wildlands and ecotourism in Latin America. *Journal of Forestry* 91 (2), 37–40.

Walpole, M. (2001) Feeding dragons in Komodo National Park: A tourism tool with conservation implications. *Animal Conservation* 4, 67–73.

Waltert, B., Wiemken, V., Rusterholz, H.-P., Boller, T. and Baur, B. (2002) Disturbance of forest by

trampling: Effects on mycorrhizal roots of seedlings and mature trees of *Fagus sylvatica*. *Plant and Soil* 243, 143–154.
Wang, X., Zhang, J., Gu, C. and Zhen, F. (2009) Examining antecedents and consequences of tourist satisfaction: A structural modelling approach. *Tsinghua Science and Technology* 14 (3), 397–406.
Wardell, M. and Moore, S.A. (2004) *Collection, Storage and Application of Visitor Use Data in Protected Areas: Guiding Principles and Case Studies*. Gold Coast, Queensland: Cooperative Research Centre for Sustainable Tourism, Griffith University.
Wardle, D.A. (1992) A comparative assessment of factors which influence microbial biomass carbon and nitrogen levels in soil. *Biological Reviews* 67, 321–358.
Warnken, J. and Blumenstein, M. (2008) *Monitoring Visitor Use in Australian Terrestrial and Marine Protected Areas: Practical Applications of Methodologies*. Gold Coast, Queensland: Cooperative Research Centre for Sustainable Tourism, Griffith University.
Warnken, J. and Buckley, R. (2000) Monitoring diffuse impacts: Australian tourism developments. *Environmental Management* 25 (4), 453–461.
Warnken, J. and Byrnes, T. (2004) Impacts of tour boats in marine environments. In R. Buckley (ed.) *Environmental Impacts of Ecotourism* (pp. 99–124). Wallingford: CABI Publishing.
Washburne, R.F. and Cole, D.N. (1983) *Problems and Practices in Wilderness Management: A Survey of Managers* (Research Paper, INT-304). Washington, DC: US Department of Agriculture, Forest Service.
Watson, A.E. and Cole, D. (1992) LAC indicators: An evaluation of progress and list of proposed indicators. In L. Merigliano (ed.) *Ideas for Limits of Acceptable Change Process* (pp. 65–84). Washington, DC: US Department of Agriculture, Forest Service.
Watson, A.E. and Niccolucci, M.J. (1995) Conflicting goals of wilderness management: Natural conditions vs. natural experiences. In *Proceedings of the Second Symposium on Social Aspects and Recreation Research, February 23–25, 1994, San Diego, CA* (pp. 11–15). San Diego, CA: US Department of Agriculture Forest Service, Pacific Southwest Research Station.
Watson, A.E. and Roggenbuck, J.W. (1997) Selecting human experience indicators for wilderness: Different approaches provide different results. In D.L. Kulhavey and M.H. Legg (eds) *Wilderness and Natural Areas in Eastern North America: Research, Management and Planning* (pp. 264–269). Nacogdoches, TX: Center for Applied Studies, Arthur Temple College of Forestry, Stephen F. Austin State University.
Watson, A.E. and Williams, D.R. (1995) Priorities for human experience research in wilderness. *Trends/Wilderness Research* 32 (1), 14–18.
Watson, A.E., Niccolucci, M.J. and Williams, D.R. (1993) *Hikers and Recreational Stock Users: Predicting and Managing Recreation Conflicts in Three Wildernesses* (General Technical Report INT-468). Ogden, UT: US Department of Agriculture Forest Service, Intermountain Research Station.
Watson, A.E., Niccolucci, M.J. and Williams, D.R. (1994) The nature of conflict between hikers and recreational stock users in the John Muir wilderness. *Journal of Leisure Research* 26 (4), 372–385.
Watson, A.E., Cronn, R. and Christensen, N.A. (1998) *Monitoring Inter-group Encounters in Wilderness* (Research Paper RMRS-RP-14). Fort Collins, CO: US Department of Agriculture Forest Service, Rocky Mountain Research Station.
Watson, A.E., Cole, D.N., Turner, D.L. and Reynolds, P.S. (2000) *Wilderness Recreation Use Estimation: A Handbook of Methods and Systems* (General Technical Report RMRS-GTR-56). Ogden, UT: US Department of Agriculture Forest Service Rocky Mountains Research Station.
Watson, J. (1997) Regional planning and protected areas in south Western Australia. *Parks* 7 (1), 2–8.
WCED (World Commission on Environment and Development) (1987) *Our Common Future* (Report of the Brundtland Commission). Oxford: Oxford University Press.
WDPA (World Database on Protected Areas) (2011a) Statistics (2010 BIP Indicator: Coverage of Protected Areas). http://www.wdpa.org/Statistics.aspx. Accessed 18 November 2011.
WDPA (World Database on Protected Areas) (2011b) WDPA Marine. http://www.wdpa-marine.org/#/country/AU. Accessed 18 November 2011.
WDPA (2012) World Data Base on Protected Areas. http://www.wdpa.org. Accessed 20 April 2012.
Wearing, S.L. and Neil, J. (1999) *Ecotourism: Impacts, Potentials and Possibilities*. Oxford: Butterworth.

Wearing S.L. and Neil, J. (2009) *Ecotourism: Impacts, Potential and Possibilities* (2nd edn). Oxford: Butterworth-Heinemann.

Wearing, S.L., Archer, D. and Beeton, S. (2007) *The Sustainable Marketing of Tourism in Protected Areas*. Brisbane: STCRC.

Weaver, D.B. (1991) Alternative to mass tourism in Dominica. *Annals of Tourism Research* 18 (3), 414–432.

Weaver, D.B. (1998) *Ecotourism in the Less Developed World.* Oxford: CAB International.

Weaver, D.B. (2000) Tourism and national parks in ecologically vulnerable areas. In R.W. Butler and S. Boyd (eds) *Tourism and National Parks: Issues and Implications* (pp. 107–134). Chichester: Wiley.

Weaver, D.B. (2005) Comprehensive and minimalist dimensions of ecotourism. *Annals of Tourism Research* 32 (2), 439–455.

Weaver, D.B. (2008) *Ecotourism* (2nd edn). Milton, Queensland: Wiley Australia.

Weaver, D.B. (2011) Organic, incremental and induced paths to sustainable mass tourism convergence. *Tourism Management* 32 (2), 439–455.

Weaver, D.B. and Lawton, L. (2010) *Tourism Management* (4th edn). Milton, Queensland: Wiley Australia.

Weaver, T. and Dale, D. (1978) Trampling effects of hikers, motorcycles and horses in meadows and forests. *Journal of Applied Ecology* 15, 451–457.

Weaver, D. and Oppermann, M. (2000) *Tourism Management*. Brisbane: John Wiley.

Webb, R.H. (1982) Off-road motorcycle effects on a desert soil. *Environmental Conservation* 9 (3), 197–208.

Webb, R.H. and Wiltshire, H.G. (1983) *Environmental Impacts of Off-Road Vehicles: Impacts and Management in Arid Regions*. New York: Springer-Verlag.

Webb, R.H., Ragland, H.C., Godwin, W.H. and Jenkins, D. (1978) Environmental effects of soil property changes with off-road vehicle use. *Environmental Management* 2 (3), 219–233.

WEF (World Economic Forum) (2009) *The Travel and Tourism Competitiveness Report 2009: Managing in a Time of Turbulence*. Geneva: World Economic Forum.

Weiler, B. and Davis, D. (1993) An exploratory investigation of the nature-based tour leader. *Tourism Recreation Research* 18 (1), 55–60.

Weiler, B. and Ham. S. (2001) Tour guides and interpretation. In D. Weaver (ed.) *Encyclopedia of Ecotourism* (pp. 549–564). Wallingford: CABI International.

Weiler, B. and Ham. S. (2010) Development of a research instrument for evaluating the visitor outcomes of face-to-face interpretation. *Visitor Studies* 13, 187–205.

Weiler, B. and Smith, L. (2009) Does more interpretation lead to greater outcomes? An assessment of the impacts of multiple layers of interpretation in a zoo context. *Journal of Sustainable Tourism* 17, 91–105.

Weir, J., Dunn, W., Bell, A. and Chatfield, B. (1996) *An Investigation into the Impact of 'Dolphin Swim Ecotours' in Southern Port Phillip Bay*. Hampton, Victoria: Dolphin Research Project Inc.

Wellings, P. (1995) Management considerations. In T. Press, D. Lea, A. Webb and A. Graham (eds) *Kakadu: Natural and Cultural Heritage and Management* (pp. 238–270). Darwin: Australian Nature Conservation Agency and North Australia Research Unit, ANU.

Welsh, H.H. Jr and Ollivier, L.M. (1998) Stream amphibians as indicators of ecosystem stress: A case study from California's redwoods. *Ecological Applications* 8 (4), 1118–1132.

Western, D. (1989) Why manage nature? In D. Western and M.C. Pearl (eds) *Conservation for the Twenty-First Century* (pp. 133–137). New York: Oxford University Press.

Whelan, T. (ed.) (1991) *Nature Tourism: Managing for the Environment.* Washington, DC: Island Press.

Whinam, J. and Chilcott, N. (1999) Impacts of trampling on alpine environments in central Tasmania. *Journal of Environmental Management* 57, 205–220.

White, D., Kendall, K.C. and Picton, H.D. (1999) Potential energetic effects of mountain climbers on foraging grizzly bears. *Wildlife Society Bulletin* 27 (1), 146–151.

Whitford, W.G., Rapport, D.J. and deSoyza, A.G. (1999) Using resistance and resilience measurements for 'fitness' tests in ecosystem health. *Journal of Environmental Management* 57, 21–29.

Whittaker, D. and Knight, R.L. (1998) Understanding wildlife responses to humans. *Wildlife Society Bulletin* 26 (2), 312–317.

Whyte, I.J., van Aarde, R.J. and Pimm, S.L. (1998) Managing the elephants of Kruger National Park. *Animal Conservation* 1, 77–83.

Whyte, I.J., van Aarde, R.J. and Pimm, S.L. (2003) Kruger's elephant population: Its size and consequences for ecosystem heterogeneity. In J.T. du Toit, K.H. Roger and H.C. Biggs (eds) *The Kruger Experience: Ecology and Management of Savanna Heterogeneity* (pp. 332–348). Washington, DC: Island Press.

Wiegmann, S.M. and Waller, D.M. (2006) Fifty years of change in northern upland forest understoreys: Identity and traits of 'winner' and 'loser' plant species. *Biological Conservation* 129, 109–123.

Wiener, C., Needham, M. and Wilkinson, P. (2009) Hawaii's real life marine park: Interpretation and impacts of commercial marine tourism in the Hawaiian Islands. *Current Issues in Tourism* 12, 489–504.

Wight, P. (1988) *Tourism in Alberta*. (Discussion paper prepared for the Alberta Conservation Strategy Project.) Edmonton: Environmental Council of Alberta.

Wight, P. (1994) Environmentally responsible marketing of tourism. In E. Cater and G. Lowman (eds) *Ecotourism: A Sustainable Option?* (pp. 39–55). Chichester: Wiley.

Wilkinson, P.F. (1997) *Tourism Policy and Planning: Case Studies from the Commonwealth Caribbean* (Tourism Dynamics Series). New York: Cognizant Communication Corporation.

Wilkinson, P.F. (2011) The panel on the ecological integrity of Canada's National Parks: Ten years later. *Journal of Tourism and Leisure Studies* 17 (2), 335–360.

Williams, D.R., Roggenbuck, J.W., Patterson, M.E. and Watson, A.E. (1992) The variability of user-based social impact standards for wilderness management. *Forest Science* 38 (4), 738–756.

Williams, D.R., Russ, G. and Doherty, P.J. (1986) Reef fish: large scale distribution and recruitment. *Oceanus* 29 (2), 76–82.

Williams, D.R., Champ, J., Lundy, C. and Cole, D. (2010) Using visitor generated internet content as a recreation monitoring tool. In M. Goossen, B. Elands and R. van Marwijk (eds) *Recreation, Tourism and Nature in a Changing World. Proceedings of the Fifth International Conference on Monitoring and Management of Visitor Flows in Recreational and Protected Areas* (pp. 128–129). Wageningen: Wageningen University.

Williams, J.C., ReVelle, C.S. and Levin, S.A. (2005) Spatial attributes and reserve design models: A review. *Environmental Modeling and Assessment* 10, 163–181.

Williams, P.B. and Marion, J.L. (1992) Trail inventory and assessment approaches applied to trail system planning at Delaware Water Gap National Recreation Area. In G.A. Stoep (ed.) *Proceedings of the 1992 Northeastern Recreation Research Symposium, April 5–7, 1992, Saratoga Springs, NY* (pp. 80–83) (General Technical Report NE-176). USDA Forest Service, Northeastern Forest Experiment Station.

Wills, R. (1992) Ecological impact of *Phytophthora cinnamomi* in the Stirling Range National Park. *Australian Journal of Ecology* 17, 145–159.

Wills, R. and Kinnear, J. (1993) Threats to the Stirling Range. In C. Thomson, G. Hall and G. Friend (eds) *Mountains of Mystery: A Natural History of the Stirling Range* (pp. 135–141). Perth: Department of Conservation and Land Management.

Wills, R. and Robinson, C.J. (1994) Threats to the flora based industries in Western Australia from plant disease. *Journal of the Royal Society of Western Australia* 77, 159–162.

Wilson, B.A., Newell, G., Laidlaw, W.S. and Freind, G. (1994) Impact of plant disease on animal communities. *Journal of the Royal Society of Western Australia* 77, 139–143.

Wilson, E.O. (1984) *Biophilia*. Cambridge, MA: Harvard University Press.

Wilson, J.P. and Seney, J.P. (1994) Erosional impact of hikers, horses, motorcycles and off-road bicycles on mountain trails in Montana. *Mountain Research and Development* 14 (1), 77–88.

Wilson, R. (1999) Possums in the spotlight. *Nature Australia* (autumn), 35–41.

Wilson, R. and Turton, S. (2011) The impact of climate change on reef-based tourism in Cairns, Australia: Adaptation and response strategies for a highly vulnerable destination. In A. Jones and M. Philips (eds) *Disappearing Destinations: Climate Change and Future Challenges for Coastal Tourism* (pp. 233–253). Cambridge, MA: CABI.

Wimpey, J.F. and Marion, J.L. (2010) The influence of use, environmental and managerial factors on the width of recreational trails. *Journal of Environmental Management* 91, 2028–2037.

Wimpey, J.F. and Marion, J.L. (2011) A spatial exploration of informal trail networks within Great Falls Park, VA. *Journal of Environmental Management* 92, 1012–1022.

Wober, K.W. (2002) *Benchmarking in Tourism and Hospitality Industries: The Selection of Benchmarking Partners*. Wallingford: CABI Publishing.

Wolcott, T.G. and Wolcott, D.L. (1984) Impact of off-road vehicles on macro-invertebrates of a mid-Atlantic beach. *Biological Conservation* 29 (3), 217–240.

Woldendorp, G. and Keenan, R.J. (2005) Coarse woody debris in Australian forest ecosystems: A review. *Austral Ecology* 30, 834–843.

Woodland, D.J. and Hooper, N.A. (1977) The effect of human trampling on coral reefs. *Biological Conservation* 11, 1–4.

WTO (World Tourism Organization) (1995) *Collection of Tourism Expenditure Statistics* (Technical Manual No. 2). Madrid: World Tourism Organization.

WTO (1998a) Protection: From the Amazon to Antarctica. *World Tourism Organization News* July–August, 10–11.

WTO (1998b) Ecotourism now one-fifth of market. *World Tourism Organization News* 1, 6.

WTO (1999) Tourism sector takes steps to ensure future growth. *World Tourism Organization News* 4, 5.

WWF (World Wide Fund for Nature) (2012) 'Choose Wisely' campaign in Starwood Abu Dhabi restaurants. http://wwf.panda.org/who_we_are/wwf_offices/united_arab_emirates/¿uNewsID =198771. Accessed 1 March 2012.

Yalden, P.E. and Yalden, D.W. (1990) Recreational disturbance of breeding golden plovers (*Pluvialis apricarius*). *Biological Conservation* 51, 243–262.

Yardstick Board (2010) *Yardstick ParkCheck Parks Survey Report 2009/2010* (Report prepared by Dr Virgil Troy for NZRA and Yardstick ParkCheck Technical Group). Wellington: New Zealand Recreation Association.

Yeatman, A. (1987) The concept of public management and the Australian state in the 1980s. *Australian Journal of Public Administration* 46 (4), 339–353.

Yuan, M. and Fredman, P. (2008) A call for a broad spatial understanding of outdoor recreation use. In A. Raschi and S. Trampetti (eds) *Management for Protection and Sustainable Development. Proceedings of the Fourth International Conference on Monitoring and Management of Visitor Flows in Recreational and Protected Areas* (pp. 169–173). Montecatini Terme, Italy. October 14–19, 2008. Firenze: Istituto de Biometeorologia (Ibimet).

Zabinski, C.A. and Gannon, J.E. (1997) Effects of recreational impacts on soil microbial communities. *Environmental Management* 21 (2), 233–238.

Zabkar, V., Brencic, M. and Dmitrovic, T. (2010) Modelling perceived quality, visitor satisfaction and behavioural intentions at the destination level. *Tourism Management* 31 (4), 537–546.

Zakai, D. and Chadwick-Furman, N.E. (2002) Impacts of intensive recreational diving on reef corals at Eilat, northern Red Sea. *Biological Conservation* 105, 179–187.

Index

In this index *fig* denotes a figure and *t* a table.